Agriculture under Climate Change
Threats, Strategies and Policies

Agriculture under Climate Change
Threats, Strategies and Policies

EDITORS

V.V. Belavadi
Department of Entomology,
University of Agricultural Sciences Bangalore,
GKVK Campus, Bengaluru, Karnataka, India

N. Nataraja Karaba
Department of Crop Physiology,
University of Agricultural Sciences Bangalore,
GKVK Campus, Bengaluru, Karnataka, India

N.R. Gangadharappa
Department of Agricultural Extension and Directorate of Research,
University of Agricultural Sciences Bangalore,
GKVK Campus, Bengaluru, Karnataka, India

ALLIED PUBLISHERS PVT. LTD.

New Delhi • Mumbai • Kolkata • Lucknow • Chennai
Nagpur • Bangalore • Hyderabad • Ahmedabad

ALLIED PUBLISHERS PRIVATE LIMITED

1/13-14 Asaf Ali Road, **New Delhi**–110002
Ph.: 011-23239001 • E-mail: delhi.books@alliedpublishers.com

87/4, Chander Nagar, Alambagh, **Lucknow**–226005
Ph.: 0522-4012850 • E-mail: appltdlko9@gmail.com
17 Chittaranjan Avenue, **Kolkata**–700072
Ph.: 033-22129618 • E-mail: cal.books@alliedpublishers.com

15 J.N. Heredia Marg, Ballard Estate, **Mumbai**–400001
Ph.: 022-42126969 • E-mail: mumbai.books@alliedpublishers.com

60 Shiv Sunder Apartments (Ground Floor), Central Bazar Road,
Bajaj Nagar, **Nagpur**–440010
Ph.: 0712-2234210 • E-mail: ngp.books@alliedpublishers.com

F-1 Sun House (First Floor), C.G. Road, Navrangpura,
Ellisbridge P.O., **Ahmedabad**–380006
Ph.: 079-26465916 • E-mail: ahmbd.books@alliedpublishers.com

751 Anna Salai, **Chennai**–600002
Ph.: 044-28523938 • E-mail: chennai.books@alliedpublishers.com

Hebbar Sreevaishnava Sabha, Sudarshan Complex-2
No. 24/1, 2nd Floor, Seshadri Road, **Bangalore**–560009
Ph.: 080-22262081 • E-Mail: bngl.books@alliedpublishers.com

3-2-844/6 & 7 Kachiguda Station Road, **Hyderabad**–500027
Ph.: 040-24619079 • E-mail: hyd.books@alliedpublishers.com

Website: www.alliedpublishers.com

Agriculture under Climate Change: Threats, Strategies and Policies

Editors: Belavadi, V.V., Nataraja, K.N. and Gangadharappa, N.R.

Cover design: Prasanna Kolte; *Cover photo:* R. Ganesan

The papers published in this book are those presented at the XIII Agricultural Science Congress jointly organized by the National Academy of Agricultural Sciences, New Delhi and the University of Agricultural Sciences, Bangalore during February 21–24, 2017.

NAAS

University of Agricultural Sciences
UAS, Bangalore

ISBN: 978-93-85926-37-2

Disclaimer: The data presented and views expressed in this volume by the authors are their own and these do not reflect those of the editors/publisher.

Published by Sunil Sachdev and printed by Ravi Sachdev at Allied Publishers Pvt. Ltd., (Printing Division), A-104 Mayapuri Phase II, New Delhi-110064

Foreword

Agriculture sector contributes significantly to the Gross National Product (GNP) of India. Nearly 70 per cent of the country's population still depends primarily on agriculture and naturally it plays a crucial role in the country's development. As the demand for food increases, agriculture will match other sectors as the engine of growth and find its place in national economy. However, the fact that farming will be significantly affected by the changing climate is beyond any reasonable doubt. Increase in temperature and changes in rainfall pattern would affect the diversity and intensity of crops cultivated in any given area. Several studies carried out across the country have attempted to assess the impact of climate change on agriculture. While these studies give us a broad picture of vulnerability to climate change, the task of predicting the future course of action for agriculture in a changing world are compounded by the fundamental complexity of natural agricultural systems, and socio-economic systems governing the food supply and demand.

The vulnerability of the agricultural sector to both climate change and variability is well established in the literature. Research has also shown that impacts on agricultural productivity are expected to be particularly harmful. The vulnerability is also especially likely to be acute in light of technological, resource, and institutional constraints. Although estimates suggest that global food production is likely to be robust, experts predict tropical regions will see both a reduction in agricultural yields and a rise in poverty levels as livelihood opportunities for many engaged in the agricultural sector become increasingly susceptible to expected climate pressures. The impacts are expected to be particularly severe at the local spatial scale due to extreme weather events.

The XIII Agricultural Science Congress with the theme of Climate-Smart Agriculture was held between 21st and 24th February 2017, at University of Agricultural Sciences Bangalore. Over 75 eminent scientists working in the area were invited to speak in their specific areas of expertise. The extended abstracts of these invited lectures, highlighting the salient points, both descriptive and prescriptive, have been brought out in this publication.

We hope the book will serve as a beacon to experts sailing in the vast and un-charted territory of global climate change and its impacts on agriculture and guide efforts at adaptation to, and mitigation of, climate change.

S. Ayyappan
Former DG, ICAR &
Former President NAAS

H. Shivanna
Vice-Chancellor
University of Agricultural Sciences,
Bangalore

Preface

The spectre of global climate change and variability is a major issue that has raised concerns among the farmers, general public, and policy makers alike. It is predicted that global temperature is increasing at an alarming rate, and impacts of global warming of 1.5°C above pre-industrial levels are being continuously assessed. The impacts of climate change would be more pronounced in developing countries and might have direct effects on poverty, food, and nutritional security. While the challenge of feeding ten billion by 2020 itself was considered a tough challenge, the rapidly changing climate has turned it into a herculean prospect. During the 20th century, the population saw its greatest increase, from about 1.6 billion (in 1900) to over 6 billion (in 2000). According to FAO, if the current trends continue, by 2050, the agricultural production will have to increase by 60 per cent to meet the expected demands for food and feed, under a favorable climate. There is an urgent need to support agriculture sector as the crop and animal production would be under pressure imposed by the changing climate and associated problems even to sustain current levels of production. While looking for opportunities to increase agricultural production, we need to look for options to reduce greenhouse gases (GHGs) emissions from agriculture sector by adopting climate-smart agriculture.

In recent years efforts are being made to highlight the importance of agriculture sector prominently in the climate-change discussions. This book is a compilation of lead papers on diverse subjects related to climate smart agriculture contributed by the experts. The book provides an insight into a range of issues including potential impacts, strategies to mitigate climate change, genetic enhancement of crops for climate resilience, water resource management, farmers' innovations to adapt to climate change, and agricultural education and policy to meet challenges raised by climate-change. The issues and ideas discussed in this book would help address climate change-related challenges that the agricultural sector is facing or likely to face. The ideas discussed here also explore options for conservation of energy and reducing GHG emissions to make farming climate-smart. We hope the information compiled and presented in this volume would aid policy makers to plan for strategies for sustainable agricultural production under changing climate and management of natural resources which is crucial for mitigation of GHG emissions, and also its adaptation in the agricultural sector in a climate-friendly manner.

We have taken efforts to compile the information presented in this book. However, it would not have been possible without the kind support and help of many individuals. We would like to extend our sincere thanks to all of them. We would like to express our special gratitude to Prof. R. Uma Shaanker, and Prof. K. Chandrashekara of UASB, who are instrumental in bringing this book, and we thank them for their valuable suggestions and guidance. We also thank Dr. Brahma Prakash and Dr. Ramakrishna Param for helping us with the editing. We are highly indebted to Prof. H. Shivanna, Vice Chancellor, University of Agricultural Sciences, Bangalore (UASB) and Dr. S. Ayyappan, Former Secretary & Director General, Indian Council of Agricultural Research (ICAR) and Ex-President, National Academy of Agricultural Sciences (NAAS), New Delhi for their support in compiling this volume.

Editors

Contents

SECTION-II: **Adaptations to Climate Change**

SECTION-III: **Mitigating Climate Change**

SECTION-I

Climate Change and Climate Variability

Combating Effect of Climate Change and Climatic Variability on Indian Agriculture through Smart Weather Forecasting and ICT Application

N. Chattopadhyay

Agricultural Meteorology Division, India Meteorological Department, Pune 411 005, India
E-mail: agrimet@imd.gov.in

ABSTRACT: *Weather forecasting at different spatial and temporal scales would be a significant tool for adaption in agriculture under future climate change scenario. In the future, we would expect larger, longer more intense extreme events than those of the last few years. Weather and climate events continue to exact a toll on society despite the tremendous advances and investment in science, especially on operational forecasting, over the past century. Weather-related hazards including early/late onset of rainy seasons and chronic events such as drought and extended periods of extreme cold or heat, trigger and account for a great proportion of disaster losses. Improvements in availability of weather data through satellite data collection and the science of weather forecasting using increasingly sophisticated models will provide an opportunity to make a quantum leap in improving both the access and quality of information to farmers in almost any part of the world. The challenge is to develop cost effective models that are able to deliver the information in a timely and understandable form. Climatic conditions and seasonal forecasts can help the farmers in planning for the upcoming season to maximize productivity based on expected weather patterns. Seasonal forecast can also be used to decide which crops to grow, varieties for planting, purchase seed and inputs and prepare their land accordingly. Shorter, real-time meteorological information of less than ten days and daily forecasts further help in determining timing of various activities such as sowing, weeding, spraying and harvesting. Real-time daily or 2–3 day forecasts can help farmers make practical decisions that can save them time and money or protect them from weather-related damage. Through Information Communication Technology (ICT), it is possible to transmit simple weather forecasts. Warnings or alerts can also be given in case of floods, storms and lightning. ICT has the potential to become a strategic enabler of climate and weather information systems by not only providing a platform to scale the dissemination of information to the farmers at unprecedented levels but also do so at the level of localization and temporal specificity that is an important element of effective and actionable weather* *information. In the present paper, state-of-art technologies of weather forecasting, including climate projections that help reduce the negative impacts of climate change and climatic variability are elaborated.*

Keywords: Climate Change, Climatic Variability, Adaptation, Weather Forecasting, ICT.

According to the IPCC (2007), natural systems around the world are being affected by regional climate changes, particularly by temperature increases. Increase in temperature is likely to be the result of anthropogenic emissions of greenhouse gases. In general, the impacts of climate change are expected to be broadly negative, including reduced water availability, damage to crops and increased potential for diseases, especially those transmitted by insect vectors. In the present paper, a projection of climate change in India and how the country is preparing to face the challenges by generation of improved forecasting methods and communicating to the farmers, is provided.

Using a number of climate models, different scenarios have been generated for the future climate change in India. It has been projected that average surface temperature will increase by 2–4°C during 2050s, with marginal changes in monsoon rain (JJAS) and large changes of rainfall during non-monsoon months. The number of rainy days is set to decrease by more than 15 and the intensity of rains is likely to increase by 1–4 mm/day. Increase in frequency and intensity of cyclonic storms is projected (Figure 1). The hydrological cycle is predicted to become more intense, with higher annual average rainfall as well as increased drought (Bhattacharya, 2006).

According to Lal (2001), annual mean area-averaged surface warming over the Indian subcontinent will range between 3.5 and 5.6°C by 2080. These projections show more warming in winter season over summer monsoon. The spatial distribution of surface warming suggests a mean annual rise in surface temperatures in north India by 3°C or more by 2050. In case of rainfall, a marginal increase of 7 to 10 per cent in annual rainfall is projected over the subcontinent by 2080. However, the study suggests a fall in rainfall by 5 to 25% in winter while it would be 10 to 15% increase in summer monsoon rainfall over the country.

Although the effects of climate change from anthropogenic activities, forcing on the use of water resources in the world, remain difficult to project, the anticipated climate change combined with other drivers of change, is likely to intensify current agricultural water management challenges in India. Higher temperatures and more frequent drought are expected to reduce water availability, hydropower potential, and in general, crop productivity. The effects of population growth and increasing water demand, which are often but not always coupled, are likely to be more significant sources of water stress than climate change when considering changes to mean precipitation and runoff. Increasing temperatures in all regions

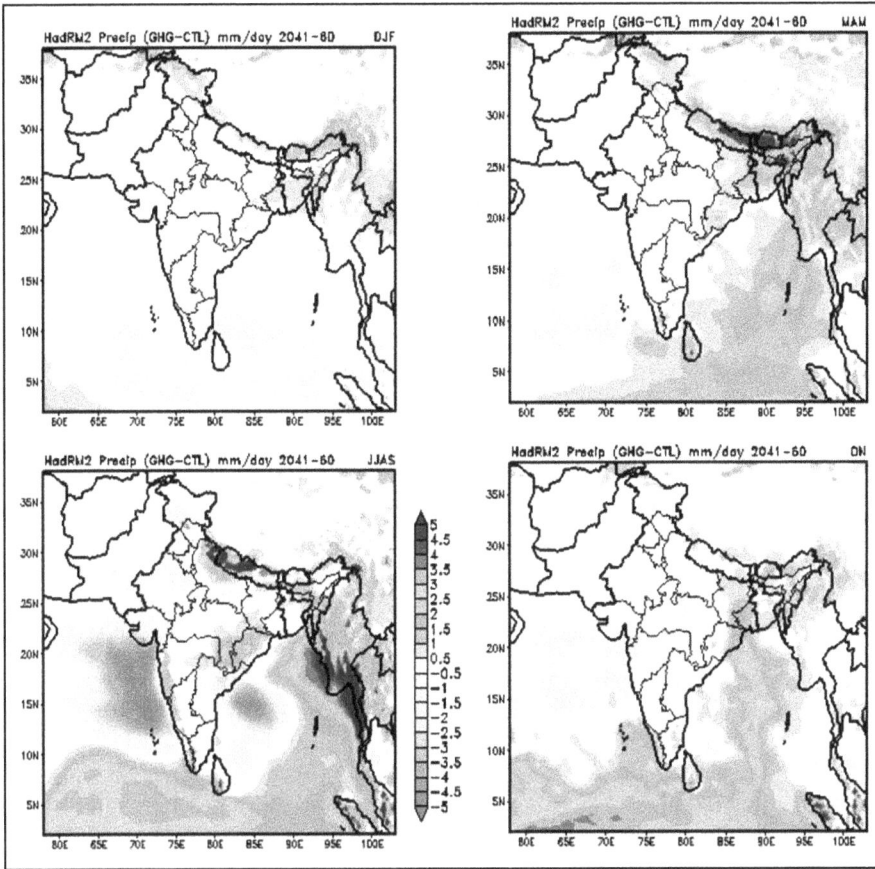

Fig. 1: Rainfall Projections at Different Seasons
(*Source:* Bhattacharya, 2006)

are expected to increase evaporative demand, which would tend to increase the amount of water required to achieve a given level of plant production if crop phenology and management are to be held constant.

Under the changing climate and climatic variability, accurate weather forecasts not only helps the farmers protect themselves against natural factors but they can also benefit significantly as long as they are aware of the actions they can take to the leverage of good weather patterns. It is no exaggeration to describe the advances made over the past half a century as revolutionary. Operational forecasting today uses guidance from a wide range of models. The forecasts of monsoon in intra-seasonal (up to three weeks and monthly) and seasonal (JJAS; seasonal average) scales are very crucial for agriculture purpose. The agriculture sectors, like farmers, are very much benefited if accurate outlook of monsoon conditions is provided to them in the extended range. The forecast of active/break cycle of monsoon, commonly known as the Extended Range Forecasts (ERF) or forecast of monsoon up to around three weeks is of great importance for agricultural planning (sowing, harvesting etc.). This can enable tactical adjustments to the strategic decisions that are made based on the longer-lead seasonal forecasts, and also will help in timely review of the ongoing monsoon conditions for providing outlooks to farmers. Forecasts of precipitation on this intermediate time-scale are critical for the optimization of planting and harvesting.

In order to prepare good forecast at different temporal and spatial scale, IMD is having different kinds of network of observatories in India to monitor and assess the weather which are Conventional Observational Network, Automatic Weather Stations (AWS), Buoy/Ship Observations, Cyclone Detection Radars, Doppler Weather Radars and Satellites observations (Figure 2). Satellite and radar observations are very crucial for monitoring and assessment of hazards, especially Himalayan region and North Indian Ocean.

Accurate weather forecasts not only help farmers protect themselves against natural factors but they can also benefit significantly as long as they are aware of the actions they can take to leverage of good weather patterns. Weather information is also an input for developing risk mitigation tools such as index insurance products for farmers. With limited external information, farmers rely on historical weather patterns for farming but the growing unpredictability in weather systems due to climate change has increased the risk for farmers.

Short-term weather patterns can also be significantly different within small geographic areas, especially in tropical zones. Providing granular information for these micro-climate systems is an important element of effective weather information. Multiple stakeholders in the agriculture ecosystem use weather information. Climatic conditions and seasonal forecasts help farmers in planning for the upcoming season to maximize productivity based on expected weather patterns.

Shorter real-time meteorological information of less than ten days and daily forecasts further help determine timing of various activities such as sowing, weeding, spraying and harvesting. Weather information can be especially impactful if combined with specific advice on the actions that need to be taken by farmers to address weather patterns. Real-time daily or 2–3 day forecasts can also help farmers make practical decisions that can save time and money or protect them from weather-related damage. Some of the most common uses of ICT for disseminating agro-meteorological information include:

1. Transmitting simple weather forecasts including seasonal, ten-day, or short-term forecasts.
2. Reminders or tips that are related to the actions that farmers can take in response to expected weather conditions (e.g. sunshine tomorrow, spray fertilizer or pesticide for a given crop).
3. Warnings or alerts for disasters or extreme weather events.

Extreme weather events such as storms and lightning are particularly prevalent in tropical climates and especially so in coastal areas where low-lying land is susceptible to frequent flooding due to inclement weather. Flood warnings are also important for farmers situated along major river systems.

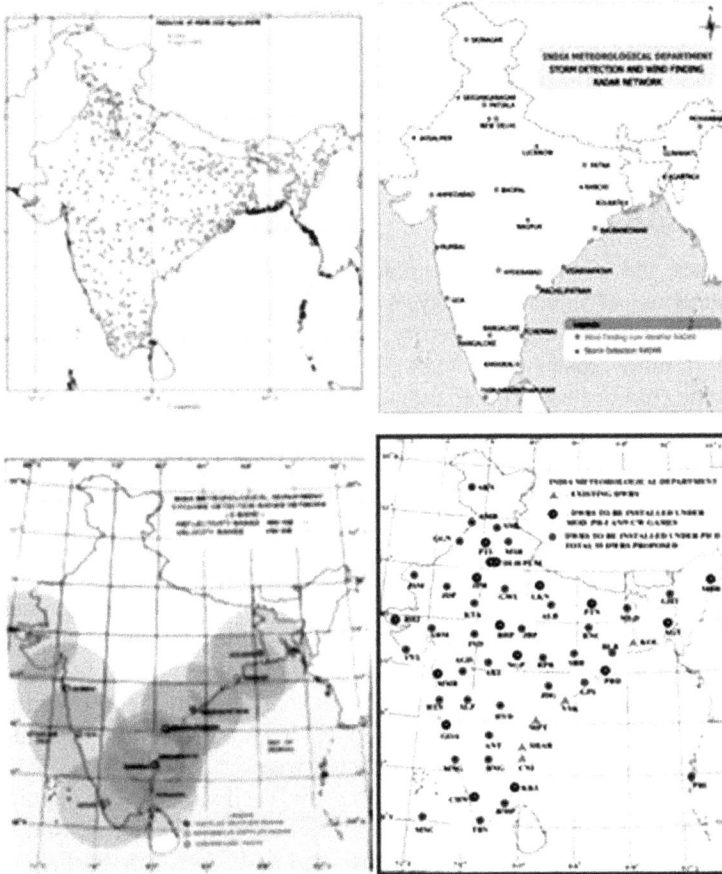

Fig. 2: Weather Observational Network in India
(*Source:* IMD)

ICT-based data collection also provides a mechanism to understand localized needs. Text surveys with simple yes/no responses or longer surveys completed through an intermediary on a smartphone platform can be used to customize information to be relevant for specific micro-climate areas and the challenges farmers are facing based on their crops or farming practices. Along with this ICT can play a great role in climate change adaptation. Linking of successful ICT pilot projects can be done with farmer field schools, farmer awareness programmes, farmers' clubs. This could also include those related to climate and building capacity on a wider scale to the farming communities. Such approaches enhance the understanding of farmers' needs and improve the existing state of dissemination of agromet advisory services. There is a need for strengthening of ICTs in the country as well as in the world in terms of infrastructure and capacity building.

In India, IMD in collaboration with 127 Agromet Field Units and Krishi Vigyan Kendras has taken initiatives to disseminate agromet advisories at district and block level, presently disseminating agromet advisories to 11.46 million farmers through SMS and IVRS system under Public Private Partnership mode like Reuter Market Light, NOKIA-HCL, Reliance Foundation, Mahindra Samriddhi, NABARD, IFFCO, Kisan Sanchar Limited and also through Kisan Portal launched by the Ministry of Agriculture.

SUMMARY

Weather and climate events continue to exact a toll on society, despite the tremendous advances and investment in prediction science and operational forecasting over the past century. Weather-related hazards including early/late onset of rainy seasons and chronic events such as drought and extended periods of extreme cold or heat, trigger and account for a great proportion of disaster losses. Improvements in availability of weather data through satellite data collection and the science of weather forecasting using increasingly sophisticated models will provide an opportunity to make a quantum leap in improving both the access and quality of information to farmers in almost any part of the world.

The challenge is to develop cost effective models that are able to deliver the information in a timely and understandable form at the level of granularity that is needed to make it actionable. A multi-stakeholder approach including public— private partnerships and civil society to coordinate efforts in facilitating efficient weather information systems may help support effectiveness. Climatic conditions and seasonal forecasts can help the farmers in planning for the upcoming season to maximize productivity based on expected weather patterns. Seasonal forecast can also be used to decide which crops to grow, seed varieties for planting, purchase seed and inputs and prepare their land accordingly. Shorter real-time meteorological information of less than ten days and daily forecasts further help in determining timing of various activities such as sowing, weeding, spraying and harvesting.

ICT has the potential to become a strategic enabler of climate and weather information systems by not only providing a platform to scale the dissemination of information to the farmers at unprecedented levels but also do so at the level of localization and temporal specificity that is an important element of effective and actionable weather information.

REFERENCES

Bhattacharya, S. (2006). Climate Change and India, *Proc. International Workshop on Future International Climate Policy*, University of Sao Paulo, Brazil.

Chattopadhyay, N. and Rathore, L.S. (2013). Extreme Events: Weather service for Indian Agriculture. *Geography and You*, 13(79): 12–16.

IPCC (2007). Impacts, Adaptation and Vulnerability. Fourth Assessment Report, Working Group II.

Lal, M. (2001). Future climate change: Implications for Indian summer monsoon and its variability. *Current Science*, 81(9): 1205.

Regional Climate Studies with Dynamical Downscaling

Sushil K. Dash

Centre for Atmospheric Sciences, Indian Institute of Technology Delhi,
Hauz Khas, New Delhi 110 016, India
E-mail: skdash@cas.iitd.ernet.in

ABSTRACT: Although climate change is a global phenomenon, its impacts are felt at the regional and local levels. So far as the impacts of climate changes on the society are concerned, the crucial issues are water resources, agriculture and human health. The related important climatic parameters are the surface temperature, relative humidity and rainfall. It is important to have accurate information on the spatio-temporal distribution of these climatic parameters and their future projections in order to be able to estimate the impacts of climate changes on agriculture and human health at any place. This information also help in appropriate policy formulation and adaptation measures. Today, it is possible to get the required climatic information and their future projections from several global and regional climate models. All the models have their biases. Further, the climate projections have a lot of uncertainties. Nevertheless, both statistical and dynamical downscaling techniques can be judiciously used to get spatio-temporal distribution of relevant climatic parameters at any place which will be eventually useful in the context of global change. This paper demonstrates the usefulness of a regional climate model in getting some robust climatic information at four cities in India.

Keywords: Global and Regional Climate Models, RegCM, CORDEX Simulations, Indian Summer Monsoon and Extreme Temperature and Rainfall.

India has unique geographical location and the country spreads over a large area. It experiences six different types of climates. The Southwest monsoon (June to September) accompanied by various types of cyclonic disturbances and the Northeast monsoon (October to December) along with Western Disturbances are the most important systems which dominate the weathers in this part of the world. The summer monsoon seasonal rainfall constitutes about 80% of the annual rainfall over India and it has a large temporal as well as spatial variability. The surface air temperature has also a large temporal and spatial variability. For suitable scientific analysis, the whole of India is usually divided into six homogeneous rainfall zones and seven homogeneous temperature zones (Figure 1).

Regional climate changes can be best examined by analysing the climatic parameters in different homogeneous zones. In the next step, one can go for investigating city level climatic parameters. Climate change at any place is felt by the people when weather extremes manifest at that place. Rainfall and temperature related extreme weather events include heavy precipitation, flash floods, droughts, heat and cold wave conditions. These extreme weather events have been responsible for heavy loss in terms of infrastructure, humans and other living beings. In the context of climate change, in addition to these extreme cases, it is important to examine all the weather events above their respective threshold levels in terms of frequencies of occurrences. Instead of focusing on the most severe extremes, it is always prudent to categorise the extremes of different intensities. A heat wave of less intensity but of longer duration can be equally or more dangerous than a severe heat wave of less duration. From this point of view, the most suitable way of defining and analysing temperature and rainfall extremes is following the method devised by the expert team on climate change detection and indices (ETCCDI). Different global and regional studies have used ETCCDI definitions for maximum and minimum temperatures and precipitation analysis (Klein Tank *et al.*, 2003, 2006). ETCCDI modified some guidelines that are used to examine the changes in climatic extremes over a period of time and is explained by Klein Tank *et al.* (2009).

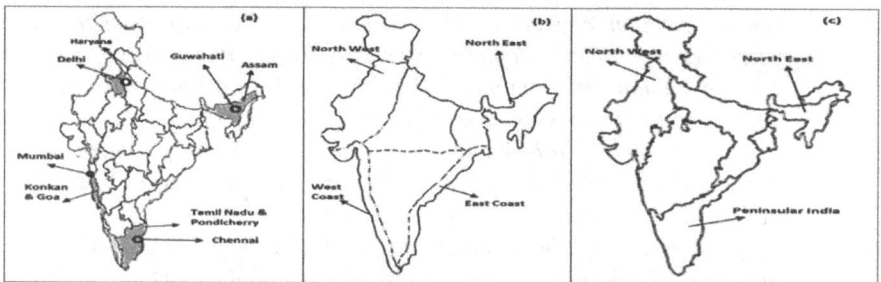

Fig. 1: (a) Location of Cities with Corresponding Meteorological Subdivisions (b) Temperature Homogeneous Zones, and (c) Rainfall Homogeneous Zones

ANALYSIS OF OBSERVED TEMPERATURE AND RAINFALL

It is well known that climate science has grown based on the analysis of archived weather observations and development and use of numerical models. India Meteorological Department (IMD) has archived more than 100 years of valuable meteorological data from several observatories in India. In the recent past, observed rainfall (Rajeevan *et al.*, 2006) and temperature (Srivastava *et al.*, 2008) data have been analysed and gridded data sets have been generated which are useful for climate studies and model validation over India. Studies by Dash and Hunt (2007), Dash and Mamgain (2011) and Dash *et al.* (2007, 2009, 2011) based on observed Indian data reveal interesting aspects of spatio-temporal variations of temperature and rainfall and their extremes. Results of these studies show that the atmospheric surface temperature has enhanced in all the homogeneous regions of India with a maximum value of about 1°C during winter and post-monsoon months. There is a significant seasonal asymmetry in the temperature rise. Also extreme temperature events of different types have enhanced over all the regions. It is found that the total precipitation during the summer monsoon months of June to September does not show any statistically significant trend. However, the numbers of short spell high intensity rain events and dry spells have increased in the last half century. Long spell rain events, on the other hand, show decreasing trend. The decrease in the number of long spell rain events associated with similar tendencies in the number of monsoon depressions, the mean monsoon wind and its shears over India suggests that the Indian summer monsoon circulation might be weakening.

DYNAMICAL DOWNSCALING USING RegCM

Climate change study in India needs thrust on its regional aspects. The mathematical models are the best tools for such study. In order to conduct sensitivity studies with regional models one needs the initial and boundary conditions from global model integrations. Since global model outputs are available from several well recognized international modeling communities, it may be worth using those model outputs to run regional models over India at higher resolutions and thus dynamically downscale the global model outputs. These regional model outputs in turn may be used to generate some important meteorological parameters such as temperature, rainfall and humidity at required resolutions by statistical downscaling.

In this study, some important climate change signals and their future projections have been examined over four Indian cities using the Regional Climate Model (RegCM) of the Abdus Salam International Centre for Theoretical Physics (ICTP). These simulations were done under the coordinated regional downscaling experiments (CORDEX) programme over the South Asia domain. Here, the state-of-the-art version 4 of RegCM has been integrated from 1970 to 2099 at 50 km horizontal resolution driven by the global model GFDL-ESM2M. The simulated mean summer monsoon circulation and associated rainfall by RegCM4 are validated against the observed values in the reference period 1975 to 2005 based on GPCP and IMD data sets. Regional model results are also compared with those of the global model GFDL which forces RegCM4. Future projections are categorized as near-future (2010–2039), mid-future (2040–2069) and far-future (2070–2099). Comparison of projected seasonal (June–Sept.) mean rainfall from the different time slices indicates gradual increase in the intensity of changes over some of the regions under both the scenarios RCP 4.5 and 8.5. RegCM4 projected rainfall over most of the Indian land mass and equatorial and northern India Ocean decreases and on the contrary it increases over the Arabian Sea, northern Bay of Bengal and the Himalayas. Results (Dash *et al.*, 2014) show that the monsoon circulation may become weaker in the future, associated with decrease in rainfall over Indian land points. The RegCM4 projected decrease in JJAS rainfall under RCP 8.5 scenario over the central, eastern and peninsular

India by the end of the century is in the range of 25–40% of their mean reference period values significant at 5% level. This study also projects the changes in temperature and rainfall extremes based on RegCM CORDEX simulation.

The study is based on the changes in temperature and rainfall extremes at Delhi, Mumbai, Chennai and Guwahati and in the respective meteorological subdivisions and homogeneous zones (Figure 1). Climate change signals are categorized as robust when the characteristics of changes remain the same in a city as well as in the respective meteorological subdivision and homogeneous zone. Results project robust signals in the increase in maximum and minimum temperatures at almost all cities under consideration. There is also robust signal of decrease in the occurrence of cold nights as in Mumbai (Table 1).

Table 1: Summary of temperature extremes at Mumbai, in Konkan and Goa and West Coast during 1975–2005 based on RegCM4.3 simulated fields. Increasing (decreasing) trends significant at 90%, 95% and 99% confidence levels are marked by the symbols △, ▲ and (▽,▼ and ⇩) respectively and +, (–) indicates increasing (decreasing) trend without any statistical significance.

Categories of Temperature Extremes		Mumbai	Meteorological Subdivision Konkan & Goa	Temperature Homogeneous Zone West Coast
Warm Days	TX90p	+	+	+
	TX95p	+	+	▲
	TX99p	+	+	+
Cold Days	TX10p	–	⇩	⇩
	TX05p	–	⇩	⇩
	TX01p	–	▽	▼
Warm Nights	TN90p	+	+	▲
	TN95p	+	+	△
	TN99p	–	–	+
Cold Nights	TN10p	▼	⇩	⇩
	TN05p	⇩	⇩	⇩
	TN01p	▽	▽	–

SUMMARY

This study is based on the analysis of IMD observed temperature and precipitation over India and the outputs from the simulation of RegCM over the South Asia CORDEX domain. Interesting results concerning the characteristics of temperature and rainfall changes over India are brought out here based on the papers published by the author along with his coauthors in the last decade. Results show differences in the characteristics of temperature and rainfall changes over different regions of India. However, the most important results over India at large include increase in the frequencies of occurrence of heavy rainfall incidents of shorter durations and warmer nights. This study further indicates that it is possible to use downscaling techniques so as to generate robust climate change signals at different places in India which can be used for the benefit of the people.

REFERENCES

Dash, S.K. and Hunt, J.C.R. (2007). Variability of climate change in India, *Current Science*, 93, 782–788.

Dash, S.K. and Mamgain, Ashu (2011). Changes in the Frequency of Different Categories of Temperature Extremes in India, *J. Appl. Meteor. Climatology*, 50, 1842–1858. doi: http://dx.doi.org/10.1175/2011JAMC2687.1.

Dash, S.K., Jenamani, R.K., Kalsi, S.R. and Panda, S.K. (2007). Some evidence of climate change in twentieth-century India, *Climatic Change*, 85, 299–321.

Dash, S.K., Kulkarni, M.A., Mohanty, U.C. and Prasad, K. (2009). Changes in the characteristics of rain events in India, *J. Geophys. Res.*, 114, D10109, doi:10.1029/2008JD010572.

Dash, S.K., Nair, A.A., Kulkarni, M.A. and Mohanty, U.C. (2011). Changes in the long and short spells of different rain intensities in India, *Theoretical and Applied Climatology*, DOI: 10.1007/s00704-011-0416-x.

Dash, S.K., Mishra, S.K., Pattnayak, K.C., Mamgain, Ashu, Mariotti, L., Coppola, E., Giorgi, F. and Giuliani, G. (2014). Projected Seasonal Mean Summer Monsoon over India and Adjoining Regions for the 21st Century, *Theoretical and Applied Climatology*, DOI 10.1007/s00704-014-1310-0.

Klein Tank, A.M.G. and Können, G.P. (2003). Trends in indices of daily temperature and precipitation extremes in Europe, 1946–99, *J. Clim.*, 16, 3665–3680.

Klein Tank, A.M.G., *et al.* (2006). Changes in daily temperature and precipitation extremes in central and south Asia. *J. Geophys. Res.*, 111, D16105, doi:10.1029/2005JD006316.

Klein Tank, A.M.G., Zwiers, F.W. and Zhang, X. (2009). Guidelines on Analysis of extremes in a changing climate in support of informed decisions for adaptation. Climate Data and Monitoring WCDMP-No. 72, WMO-TD No. 1500, 56pp. Available from: http://www.clivar.org/organization/etccdi/etccdi.php

Rajeevan, M., Bhate, J., Kale, J.D. and Lal, B. (2006). High resolution daily gridded rainfall data for the Indian region: Analysis of break and active monsoon spells, *Current Science*, 91(3), 296–306.

Srivastava, A.K., Rajeevan, M. and Kshirsagar, S.R. (2008). Development of a high resolution daily gridded temperature data set (1969–2005) for the Indian region. NCC research report No. 8/2008. *National Climate Centre, India Meteorological Department*. Available from: http://www.imdpune.gov.in

Climate-Smart Agriculture: The Need of the Hour for Pulse Production

Sulochana Gadgil

Centre for Atmospheric and Oceanic Sciences, Indian Institute of Science, Bengaluru 560 012, India
E-mail: sulugadgil@gmail.com

ABSTRACT: *With the first quantitative assessment of the impact of the Indian summer monsoon rainfall on the Indian food-grain production and the GDP, Gadgil and Gadgil showed that, while the adverse impact of droughts has remained large since the 1950s, the magnitude of the favourable impact of a normal or a good monsoon is smaller. Furthermore, they showed that this asymmetry has increased markedly after the 1980s. In this study, I suggest possible reasons for this asymmetric response to rainfall variability and address the problem of identifying strategies which are appropriate for the entire spectrum of variability of the rainfall from severe droughts to excess monsoon seasons.*

Keywords: Monsoon, Rainfed Agriculture, Pulse Production.

Variability of climate and particularly rainfall is one of the most dominant factors in determining the production of rainfed agriculture in the tropics. The inter-annual variation of rainfall in the growing season is known to have a large impact on the production of rainfed crops such as pulses. For example, the drought during 2015, led to a massive decrease of 52% (compared to the normal) in the *kharif* season production of pulses in Maharashtra, with 42% decrease in the production of pigeonpea (Govt. of Maharashtra, 2015). One of the major impacts for consumers was the high price of pulses and particularly of pigeon-pea, which reached an all-time high of ₹ 200 per kilo. This was a culmination of the price rise of pulses experienced for several years, which is considered to be a reflection of the demand–supply gap continuously increasing due to the growth rate of pulse production being much smaller than that of the population (Reddy, 2009). In fact, whereas a phenomenal increase in the Indian food-grain production was achieved during the green revolution associated with a rapid increase in yields due to the adoption of new dwarf, high yielding and fertilizer-responsive varieties (of rice and wheat, in particular), and a substantial increase in irrigation, fertilizer and pesticide application, there has been hardly any increase in the production of rainfed crops such as pulses in the country since the fifties (Abrol 1996, Gadgil *et al.*, 1999). The sharp contrast between the observed variation of the yields of Tur and wheat since the fifties is clearly brought out in Figure 1. India is the largest producer and consumer of pulses in the world, yet the yield is almost the lowest in the world and less than half of that of Canada, US and Russia (Singh 2009). Climate-smart agriculture clearly involves identification and adoption of strategies which can lead to enhancement in the yield of crops such as pigeon-pea in the face of climate variability of the type experienced. The first step in deriving such strategies is an assessment of the nature of the impacts of the experienced variability of the climate, which in the case of the Indian region implies assessing the impact of monsoon variability.

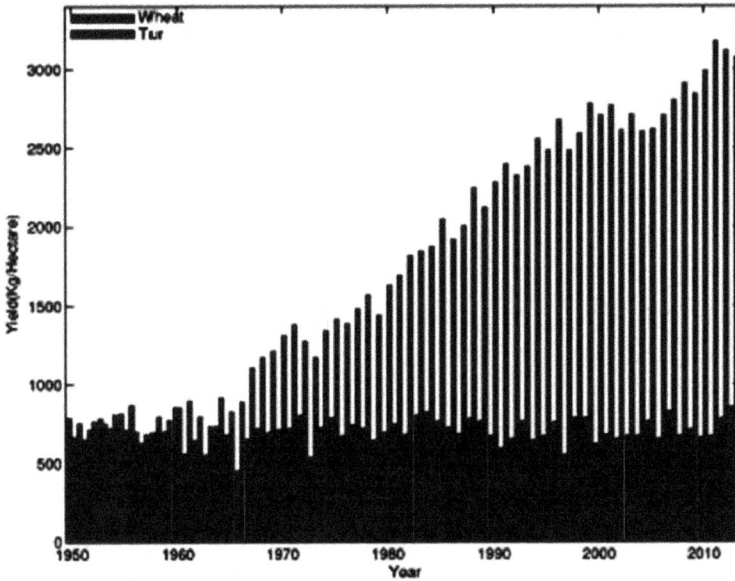

Fig. 1: Variation during 1950–2014 of All-India Yields of Wheat and Tur (pigeonpea) based on Data from Agricultural Statistics at a Glance (2014)

The first quantitative assessment of the impact of the Indian summer monsoon rainfall (ISMR) on the Indian food-grain production (FGP) and the GDP (Gadgil and Gadgil, 2006) revealed that the adverse impact of droughts on FGP as well as GDP has remained large (about 2 to 5% for GDP) throughout the period of the analysis, i.e. 1950–2003. Such a large impact on GDP was not expected because, with rapid development since independence from the colonial rule, the contribution of agriculture to the GDP declined from around 50% in 1950 to less than 20% in 2003, and it was believed that the economy would have become drought proof. However, as the livelihood of the majority of people is still linked to agriculture, in years of low agricultural production, there is an adverse effect on the purchasing power and hence the GDP. The total food grain production still depends on the rainfall because, although the area under rainfed cultivation has decreased over the years from around 82% in 1950, it is still slightly over 50%. Thus over 57% of the year-to-year variation in food grain production is associated with the year-to-year variation in ISMR. An important result of the Gadgil and

Gadgil (2006) study is the demonstration of an asymmetry in the response of FGP and GDP to the inter annual variation of ISMR, with the magnitude of adverse impact of a deficit monsoon being larger than that of the favourable impact of a good monsoon with comparable magnitude of the monsoon rainfall anomaly. Furthermore, they showed that this asymmetry has increased markedly between the earlier three decades 1950–80 and the post-1980 period (Figure 2). For example, while the expected adverse impact of a 15% deficit in monsoon rainfall is a decrease in FGP of around 10% during both the periods, the expected impact of the monsoon rainfall being 15% above the average, is a gain of about 6% in 1950–80 whereas it is less than 1% in the post-1980 period. Obviously, this has serious implications for food security.

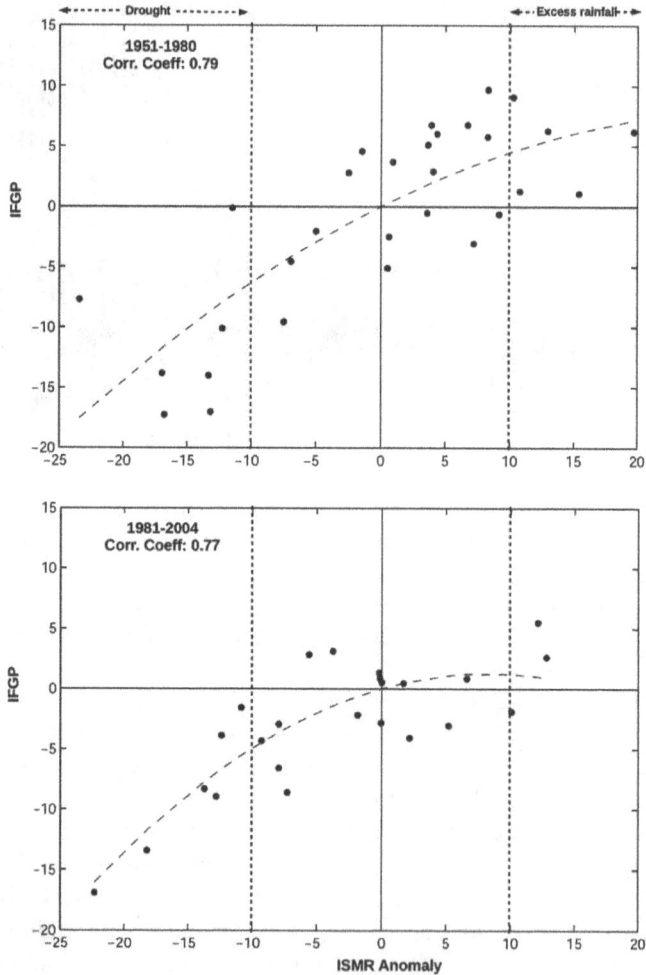

Fig. 2: Impact on FGP versus the Anomaly of the Indian Summer Monsoon Rainfall (ISMR) for the Period 1951–1980 (top) and 1981–2004 (bottom) (after Gadgil and Gadgil, 2006)

Thus, while we suffer large losses in FGP (and GDP) in droughts, we are not able to reap the benefits of a good monsoon and get enhancement of comparable magnitude in FGP (or GDP), particularly in the post-1980 period. We note that on the all-India scale, droughts occur in only about 17% of years and even for smaller regions the frequency does not exceed 25%. Thus the strategies adopted for varieties and management practices for rainfed regions are not appropriate for the majority of the years i.e. years in which the monsoon is normal or above average. Here I suggest possible reasons for this asymmetric response to rainfall variability and address the problem of identifying strategies which are appropriate for the entire spectrum of variability of the rainfall from severe droughts to excess monsoon seasons.

WHY ARE WE NOT ABLE TO REAP THE BENEFITS OF GOOD MONSOONS?

Gadgil and Gadgil (2006) suggested that 'an asymmetry in response is not surprising in the light of *Liebig's law of the minimum,* which says that the yield of a crop is determined by the scarcest resource (the so-called *limiting resource*). During a drought, one expects water to be the limiting resource, but this need not be the case when the rainfall is normal or surplus. However, one can draw a significant conclusion from the observation that the impact of surplus rainfall has diminished with time. This suggests that while in the earlier era water was the primary limiting resource, in recent times other factors determine the yield in years of normal or surplus rainfall. Identifying these factors can play a crucial role in increasing yields.'

Since the 70s there have been major changes in rainfed agriculture, in the cropping patterns as well as varieties with the new fertilizer responsive high yielding varieties often replacing the traditional ones. Also, the traditional complex cropping has been replaced by monocropping over large tracts of land. This has led to a large number of pests and diseases becoming endemic. Furthermore, intensive farming has led to a decrease in the fertility of the soil. Hence chemical inputs as fertilizers and pesticides, whose prices have been continually increasing, have become more and more essential to realize a good yield.

Insight into the poor response of rainfed agriculture to increasing rainfall can be obtained from a series of studies at the International Crops Research Institute for the Semi-Arid Tropics beginning with that of Sivakumar *et al.* (1983). They showed that while the observed yields of several rainfed crops on the farmers' fields increase rather slowly with increasing rainy season rainfall, the yields at agricultural stations (of the same crops, also under rainfed conditions) increase more rapidly with increasing rainfall (Figure 3). It is seen that the yields at the agricultural stations are comparable to the farmers' yields when the rainfall is low, hence the yield gap, i.e. the gap between the yields achievable with existing technology at agricultural stations and the yield of the farmers' fields, is small. The rapid increase of the yield at the agricultural stations with increasing rainfall leads to a large yield gaps for years with normal or good rainfall. Bhatia *et al.'s* (2006) comprehensive study of the gaps between the yields on the farmers' fields and potential yields from model simulations or the yields achieved at agricultural stations for soybean, groundnut, chickpea and pigeonpea, shows that the situation is similar to that shown in Figure 3. When the rainfall in the growing season is good, the potential yield simulated with models and the yield achieved at agricultural stations is much higher than the district average yield.

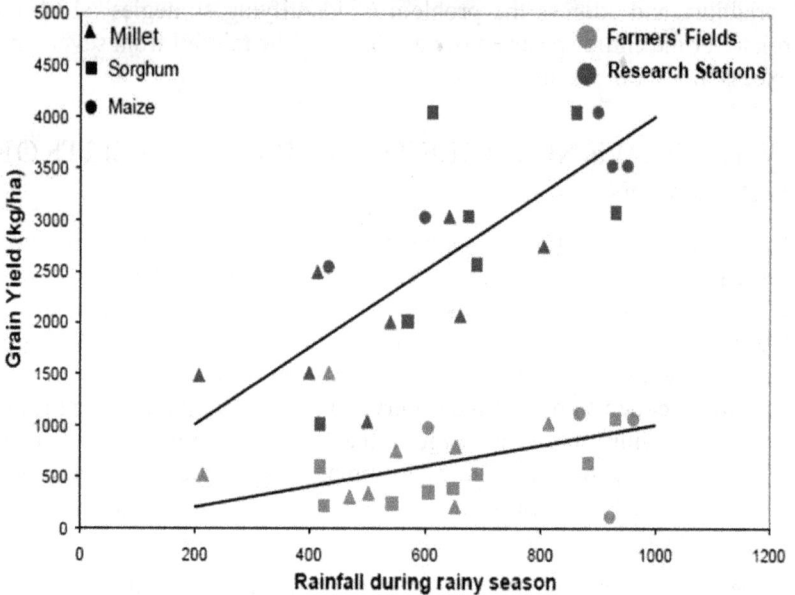

Fig. 3: Relationship between Rainfall during the Rainy Season and Yield of Maize, Sorghum and Millet at 15 Dryland Locations in India (after Shivakumar *et al.*, 1983)

The major management differences between the agricultural stations and farmers' fields are fertilizer and pesticide applications. While at agricultural stations, the applications are as per the recommendations, almost none of the farmers involved in rainfed cultivation invests at the recommended level in fertilizers and pesticides. Bhatia *et al.* (2006) also suggest that optimum use of nutrients and improved management practices are the main factors responsible for higher yields in simulation (and also at experimental stations) relative to the yields on the farmers' fields, during good rainfall years.

If it is indeed possible to get higher yields in good rainfall years with existing technology, the question arises, why is the technology not adopted by the farmers in rainfed cultivation? After all, the green revolution was achieved by farmers implementing the recommendations of the agricultural scientists about cultivation of high-yielding fertilizer-responsive varieties, application of fertilizers, pesticides, etc. It should be noted that the major difference between rainfed cultivation and cultivation over irrigated areas is that while for irrigated areas (over which substantial enhancement of production occurred during the green revolution) the yields are assured, there are large fluctuations in the rainfed crop yields from year to year (in response to the inter annual variation of rainfall) which leads to the farmers facing special resource constraints. Thus, for each recommendation for rainfed cultivation involving additional investment by farmers, (e.g. regarding the application of fertilizers and pesticides), before a decision in favour of the recommended practice can be made, the farmers have to assess the expected benefit in terms of the enhancement of yields/profits in the face of rainfall variability of the specific region, vis-à-vis the additional cost involved in implementing the recommendation. In rainfed cultivation, applications of fertilizers and pesticides enhance yields substantially (and hence are cost-effective) only when the rainfall is normal or good. Bhatia *et al.* (2006) also point out that, the positive impact of optimum use of nutrients is maximum when enough soil moisture is available in the soil; and that under sub-optimal soil moisture conditions due to low levels of rainfall in a given environment, the impact is reduced greatly. Although generally the farmers have the know-how, they do not invest in such applications in the absence of reliable information/prediction about the rainfall being normal or above normal, because it is not economically viable in years of poor rainfall (Gadgil *et al.,* 2002). So, the farmers are adopting a strategy which may be considered appropriate only for droughts and is clearly not appropriate for a majority of the years. Hence the farmers do not get an enhancement in yields commensurate with the rainfall, in good rainfall years.

WAY FORWARD

Consider first what could be done to enhance production of rainfed crops such as pulses, so as to meet the ever-increasing demand of protein, with efforts at adopting strategies which have been shown to enhance yields at agricultural stations in a majority of the years. Clearly, to get enhanced yields during normal or good rainfall years for the varieties cultivated, cost-effective strategies of application of fertilizers and pesticides for each specific region need to be worked out. For this, an important piece of information is the chance of occurrence of low rainfall, i.e. rainfall below the threshold above which fertilizers can improve the yield, for the specific crop/variety and specific region. With the meteorological data sets for long periods available in the country, it is possible to determine the probability of the growing season rainfall being in different ranges and, in particular, being above such a threshold. Depending on the costs involved, it may turn out to be profitable to invest in an appropriate fraction of the recommended level in fertilizers or on an appropriate fraction of the cultivated area every year so that substantial enhancement of yields in a majority of years will offset the lack of benefit in the few years in which the rainfall is below the threshold. I hope that in not too distant a future, appropriate implementable strategies, tailored to the entire spectrum of rainfall variability, leading to maximum gains in normal and good monsoon seasons, while minimizing the adverse impact of droughts, can be identified and adopted to meet the challenge of achieving a substantial enhancement of pulse production.

On the longer term, for sustained high yields it is essential to learn from our experience over the last few decades over irrigated and rainfed areas. The green revolution strategies adopted for wheat and rice, which substantially enhanced yields over irrigated regions, have led to a depleting the fertility of the soil and hence the growth rate of the rice and wheat yields has markedly decreased in Punjab and Haryana by 1985–94 (Abrol et al., 2002). Over the rainfed areas also, we noted that the soil fertility and the water retention capacity of the soils have decreased and in the modern era, fertilizer application has become necessary for realizing good yields during good monsoon years. Unless this trend of depletion of soil fertility is arrested, larger and larger quantities of fertilizers would be required to get the same level of yields as has been the experience over a large part of the irrigated regions.

Bhatia et al. (2006) suggest that an integrated approach including improved input use and adoption of improved technology, adoption of proven technologies such as effective watershed management, switching to planting on effective land configurations and water conserving cultural methods is needed. It is important to note from the point of view of food and nutrition security enhancement of production of pulses is essential, for farmers to adopt any strategy, it must lead to profit. It is being increasingly realized that in rainfed conditions, it is important to consider the optimum combination of trees, shrubs, fodder and crops, with appropriate soil and water management practices, which can maximize profit for specific agroclimatic regions. Such strategies have been identified and successfully implemented by Sheshagiri Rao in Karnataka. He finds that the soil depth has increased by 50% in six years, moisture retention has also markedly increased and no pesticides are required (Rao, 2014). It has thus become possible to make reasonable profits in a semi-arid region over which groundnut cultivation used to result in loss in most years. The challenge before us is to identify and adopt such strategies for sustained enhancement of rainfed agricultural production.

REFERENCES

Abrol, I.P. (1996). India's agriculture scenario. In *Climate Variability and Agriculture* (eds Abrol, Y.P., Gadgil, S. and Pant, G.B.), Narosa Publishing House, London, pp. 19–25.

Abrol, Y.P., Sangwan, Satpal, Dadhwal, V.K. and Tiwari, M.K. (2002). Land use/land cover in Indo-Gangetic Plains–history of changes, present concerns and future approaches, in "Land use: Historical Perspectives, Focus on Indo-Gangetic Plains" ed. Abrol *et al.* Allied publishers.

Bhatia, V.S., Singh, P., Wani, S.P., Kesava Rao, A.V.R. and Bhatia, K.S. (2006). "Yield gap analysis of soy been, groundnut, pigeonpea and chikpea in India using simulation modeling." *Global theme on agro-ecosystems report* #31, ICRISAT, Patancheru, India.

Gadgil, Sulochana and Gadgil, Siddhartha (2006). The Indian Monsoon, GDP and Agriculture. *Eco. Poli. Weekly,* XLI(47), 487–489.

Gadgil, Sulochana, Abrol, Y.P. and Seshagiri Rao, P.R. (1999). On growth and fluctuation of Indian foodgrain production. *Curr. Sci.*, 76(4), 548–556.

Gadgil, Sulochana, Seshagiri Rao, P.R. and Narahari Rao, K. (2002). Use of climate information for farm-level decision making: Rainfed groundnut in Southern India. *Agric. Syst.*, 74, 431–457.

Govt. of Maharashtra (2015). "Memoires on Drought," November 2015.

Rao, P.R.S. (2014). Personal communication.

Reddy, A.A. (2009). Pulses production technology: Status and way forward review of agriculture. *Econ. Polit. Weekly*, XLIV (52), 73–80.

Singh, R.P. (2009). Status paper on pulses. Government of India, Ministry of Agriculture, Directorate of Pulses Development, Bhopal.

Sivakumar, M.V.K., Singh, P. and Williams, J.H. (1983). "Agroclimatic aspects in planning for improved productivity of Alfisols." In *Alfisols in the Semi-Arid Tropics: A Consultants Workshop*, 1–3, December 1983, ICRISAT Centre, India, pp. 15–30.

Climate Change and Indian Agriculture: Implications for Food Security

B. Venkateswarlu

Vasantrao Naik Marathwada Krishi Vidyapeeth, Parbhani 431 402, India
E-mail: vcmau@rediffmail.com

ABSTRACT: *Food security means when all people all the time have physical and economic access to sufficient, safe and nutritious food to meet their dietary needs. The medium term (2010–2039) impacts of climate change on food production are in the range of 4.5–9 per cent while long term (2070–2099) impacts could reduce yields by 25 per cent or more. Since agriculture makes up roughly 18 per cent of India's GDP, a 4.5–9 per cent negative impact on production implies a cost of climate change to be roughly 1.6 per cent of GDP per year. Technology and policy options to mitigate the impacts of climate change are discussed in this paper.*

Keywords: Climate Change, Global Warming, Adaptation, Mitigation, Indian Agriculture.

Climate change impacts on agriculture are being witnessed all over the world, but countries like India are more vulnerable in view of the high population depending on agriculture, excessive pressure on natural resources and poor coping capabilities. The projected impacts are likely to further aggravate yield fluctuations of many crops impacting food security. For every 1° increase in temperature throughout the growing season, the production of wheat in the country may reduce by 4–5 million tones according to the study conducted by IARI, New Delhi. Similarly, rice yields may decline by 6% for every 1°C rise in temperature. Bapujirao *et al.* (2014) indicated that rise in minimum temperature is happening during *kharif* season at a rate of 0.19°C/10 years and this rise was found to have a negative impact on paddy yields over 50.3% of cultivated area. Warming during *rabi* season has serious implications for production of crops like wheat, mustard and chickpea in the Indo-Gangetic plains. It is to be noted that the contribution of *rabi* season to total food grain production is increasing in recent years. Negative impacts on mustard, groundnut, potato, maize and sorghum are also reported due to warming. Horticulture crops are particularly vulnerable to temperature, unseasonal rainfall, hailstorms and pest and diseases incidence caused by climate variability. Besides direct effects on crops, climate change is likely to impact natural resources like soil and water. Increased rainfall intensity in some regions would cause more soil erosion leading to land degradation. Increased temperatures will also increase crop water requirement. A study by CRIDA on major crop growing districts in the country for four crops, viz. groundnut, mustard, wheat and maize indicated a 3% increase in crop water requirement by 2020 and 7% by 2050 across all the crops/locations. Irrigation requirement in arid and semi-arid regions is estimated to increase by 10% by every 1°C rise in temperature.

ECONOMIC IMPACT

Though the medium term (2010–2039) impacts on food production are in the range of 4.5 to 9 per cent, long term (2070–2099) impacts could reduce yields by 25 per cent or more. Since agriculture is about 15 per cent of India's GDP, a 4.5–9 per cent negative impact on production implies a cost of climate change to be roughly up to 1.6 per cent of GDP per year. Since maintaining agricultural productivity is critical for the well-being of the poor, climate change losses could hit the poor most. In the absence of rapid and full adaptation, the consequences of long-run climate change could be even more severe on livelihood security of the poor (Guiteras *et al.,* 2007). Rainfed agriculture in particular, is likely to be more vulnerable. Reduction in number of rainy days will result in longer dry spells—affecting rainfed crops in semiarid and arid regions. Vulnerability mapping of Indian agriculture to climate change by CRIDA indicated that districts with very high and high vulnerability to climate change are in Rajasthan, Gujarat, Maharashtra, Karnataka and MP (Figure 1). These areas are also witnessing high ground water exploitation and facing water quality issues.

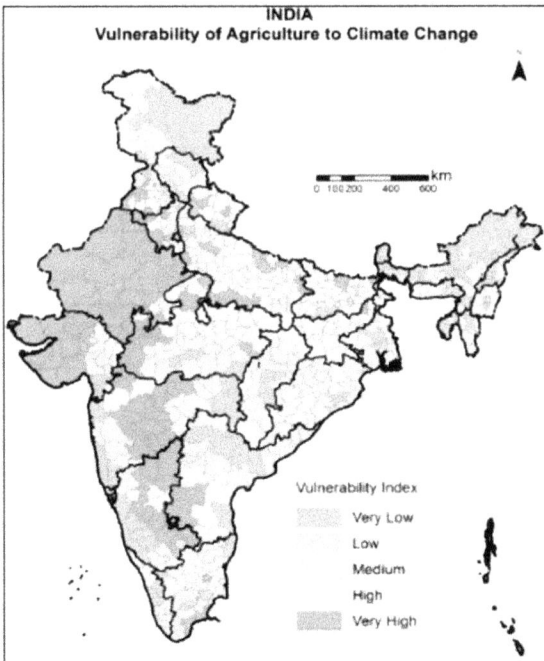

Fig. 1: Vulnerability Atlas of Indian Agriculture to Climate Change, 2021–2050
(*Source:* Rama Rao *et al.*, 2013)

IMPLICATIONS FOR FOOD SECURITY

Food security means when all people all the time have physical and economic access to sufficient, safe and nutritious food to meet their dietary needs. Food security has three components, i.e. availability, access and absorption. Several modelling studies were carried out on climate change impacts on food production, but mostly on wheat and rice and to some extent on pulses and coarse cereals. Short-term impacts on many food crops in the range of 6–10 per cent can be offset by adaptation measures based on available technologies like advancing planting dates, heat-tolerant varieties, efficient water management, conservation agriculture and protected cultivation, etc., but addressing long-term impacts requires investments on strategic research and extension and helping farmers in risk management through innovative insurance products. However, there are very few studies on how climate change impacts the access to food, particularly in India. The fourth assessment report of IPCC puts the number of people in the world who are likely to suffer from hunger by 2080 at 200–600 million depending on the scenario used. No such specific data is available for India. Undoubtedly, future climate change and variability could significantly influence the economic access to food in India both in rural and urban areas. Droughts, floods and extreme weather events not only reduce crop yields but also affect the incomes and livelihoods of farmers and farm labourers due to loss of wages and increase in prices. The nutritional quality of foods like protein and mineral content could also be influenced by climate change impacts like elevated carbon dioxide and depletion of carbon and micro nutrients in the soils. It is well known that deficiency of micro nutrients like iron and zinc causes hidden hunger. Rattan Lal (2013) described the complex relationship between climate change and food security in the flow chart (Figure 2).

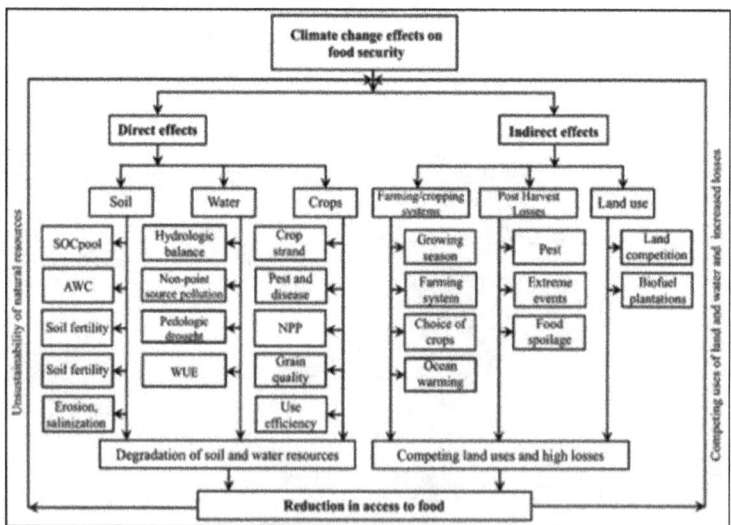

(*Source:* Rattan Lal, 2013), SOC: Soil Organic Carbon, AWC: Plant Available Water Capacity, WUE: Water Use Efficiency, NPP: Net Primary Production

ADAPTATION AND MITIGATION

A comprehensive strategy of utilization of existing knowledge, strengthening R&D in key areas and evolving a policy framework that builds on risk management and providing incentives to sustainable use of natural resources will be required for successful adaptation by farm sector to climate change. The goal of this strategy is to minimize the risks associated with farming and enable farmers to cope with these risks.

The main adaptation strategies include development of new genotypes; search for genes for stress tolerance across plant and animal kingdom; intensifying research efforts on marker assisted selection and transgenics development for biotic and abiotic stress management; development of heat and drought tolerant genotypes; development of new land use systems; evolving new agronomy for climate change scenarios; explore opportunities for restoration of soil health; use multipurpose adapted livestock species and breeds; development of spatially differentiated contingency plans for weather-related risks, supply management through market and non-market interventions in the event of adverse supply changes; research on short, medium and long range weather forecasts for reducing production risks; development of pest and disease forewarning systems covering a range of parameters for contingency planning. Documentation and utilization of indigenous knowledge also forms an important strategy towards climate change adaptation.

While adaptation measures are important, we must also focus simultaneously on mitigation measures so that we contribute to a reduction in the pace of global climate change (Venkateswarlu and Arun Shanker, 2009). The important mitigation options include efficient water and nutrient management options to enhance use efficiency; evaluate carbon sequestration potential of different land use systems including opportunities offered by conservation agriculture and agroforestry; identify cost effective methane emission reduction practices in ruminants and in rice paddy. However, we need to assess the socio-economic implications of proposed mitigation options before developing a policy framework.

Globally, weather insurance plays an important role in mitigating climatic risks. In several developed countries this strategy has worked successfully as these countries have excellent long term weather data, farmers have large holdings and have a business approach for farming. In India, the small holders are generally more prone to risks but they are averse to buy insurance policies. The crop insurance scheme has made some progress but it is a long way to go. Considering the climate trends being witnessed in recent years all over the country, weather-based insurance is an important alternative for mitigating risks in agriculture for Indian farmers. Research institutes and insurance companies should develop crop-wise data on the weather sensitivity so that appropriate policies can be designed which are friendly to farmers, and at the same time, keep the insurance companies viable. The Government also should share the premium burden. Instead of spending huge amounts of money on rehabilitation after the disaster, it is prudent to spend on premium subsidy.

Building state of the art infrastructure for research and training of scientists in frontier areas and tools, increasing climate change literacy to different levels of stakeholders, mainly farmers; enhancement of national capacity on decision support systems, developing best weather insurance products for vulnerable areas and farmers and carbon trading in agriculture; and international collaboration are some other key areas through which we can tackle challenges of climate change.

TECHNOLOGY AND POLICY OPTIONS

No single strategy is adequate to address the complex issue of climate change and its nexus with food security. The National Academy of Agricultural Sciences has brainstormed the issue in 2013 and brought out key recommendations (NAAS, 2013). Following are few technology and policy directions which need immediate attention.

1. ***Prioritization and Focus on Vulnerable Hotspots:*** Investments on climate change adaptation have to be focused in scientifically identified vulnerable regions. Unlike other development schemes, climate change adaptation programmes need to be tailored to the needs of the vulnerable regions and communities. To enable this, a robust micro-level collection of weather data on a continuous basis is a pre-requisite. The Government of India should strengthen the weather observatory network in the country and also initiate documenting extreme weather events and their economic impacts. This is essential for both short and long term planning of climate resilient agriculture. Data availability is a major constraint in India, a national data sharing policy is urgently needed for easy access of weather, crop and market data for all user groups.

2. ***Investments on Technology Generation and Dissemination:*** Climate change and variability manifest in terms of increased incidence of abiotic stresses like droughts, floods, cyclones, cold and heat-waves etc. Research on development of multiple abiotic stress tolerant crop varieties needs to be strengthened with adequate financial support and state-of-the-art research infrastructure. The government of India made a good beginning by launching National Innovations on Climate Resilient Agriculture (NICRA), but this needs to be strengthened and expanded with focus on capacity building of young scientists. The best output can be achieved if public–private partnership is ensured in development and dissemination of stress tolerant crops and varieties.

3. ***Prudent Use of Natural Resources:*** Prudent management of natural resources like land and water are key to achieving resilience in agriculture against climate change. Continued emphasis on on-station and on-farm research is critical in land use planning, environmental services and

efficient use of surface and ground water. The best-bet practices in natural resource management can only be promoted widely with a matching policy by the Government. This is largely a public sector domain and effective inter-institutional and departmental collaboration will produce the best results.

4. *Leveraging the Ongoing Schemes and Missions:* Several on-going schemes of the Government of India under the Ministries of Agriculture, Rural Development, Environment and Forests, Earth Sciences and Water Resources have strong components which contribute to climate resilient agriculture. What is important is to prioritize region-specific interventions which can minimize climatic risks at farm level and achieve economy of scale in implementation and create impact through convergence of resources from these schemes. Clear-cut convergence guidelines at state, district, block and village level are to be prepared for achieving this goal.

5. *Livelihood Diversification and Risk Transfer:* Despite the best efforts in climate change adaptation, extreme climatic events pose immense risks to small and marginal farmers, which leads to total loss of income at times. Promotion of non-farm source of income and enterprises through credit flow and skill development is the only way to minimize such negative impacts. The livelihood diversification efforts should be linked to the national livelihood and skill development missions. Weather-based crop insurance needs to be promoted as a risk transfer mechanism widely with built-in incentives and awareness creation among farmers.

6. *Financing Adaptation Costs:* It is essential to pool financial resources under different ministries for promoting technology and institutional interventions towards climate resilient agriculture through National Mission on Sustainable Agriculture (NMSA). Additional funds from private sector and international organizations can be ploughed in to upscale successful adaptation strategies. The state governments should be persuaded to commit adequate resources for financing climate change adaptation as the issue has gained significant attention in most of the states.

7. *Capacity Building:* There is an urgent need to build the capacity of different stakeholders on climate resilient agriculture; structured training programmes need to be designed and implemented at all levels with a time frame. Climate resilient agriculture should become an important component of all farmers' training programmes, organized by primary extension service providers. There is a need to utilize the indigenous knowledge of farmers in coping with climate variability.

8. *Mainstreaming CC Adaptation in Planning:* The key principles of climate change adaptation have to be mainstreamed into the planning system at the state and central level. Each policy initiative needs to be analysed in terms of its implication in promoting or negatively impacting climate resilient agriculture. Investments in agriculture, rural infrastructure and water resources

have to be tailored to the needs of the vulnerable regions and communities in all such plans.

REFERENCES

Bapuji, Rao B., Chowdary, P. Santhibhushan, Sandeep, V.M., Rao, V.U.M. and Venkateswarlu, B. (2014). Rising minimum temperature trend over India in recent decades: Implications for agricultural production. Global and Planetary Change, 117 (2014): 1–8.

Guiteras, R. (2007). The Impact of Climate Change on Indian Agriculture. Department of Economics, Massachusetts Institute of Technology, December 2007 Working Paper.

IPCC, WGII (2007). In: Parry, M.L., Canziani, O.F., Palutik of, J.P., van der Linden, P.J., Hanson, C.E. (Eds.), Climate Change 2007: Impacts, Adaptation and Vulnerability. Contribution of Working Group II to the Fourth Assessment Report of the Intergovernmental Panel on Climate Change. Cambridge University Press, Cambridge, UK and New York, NY, USA, pp. 976.

NAAS (2013). Climate resilient agriculture in India. NAAS policy paper 65, p. 20.

Rama Rao, C.A., Raju, B.M.K., Subba Rao, A.V.M., Rao, K.V., Rao, V.U.M., Kausalya, Ramachandran, Venkateswarlu, B. and Sikka, A.K. (2013). Atlas on Vulnerability of Indian Agriculture to Climate Change. Central Research Institute for Dryland Agriculture, Hyderabad, p. 116.

Rattan, Lal (2013). Food security in a changing climate. *Ecohydrology and hydrobiology*, 13, 8–21.

Venkateswarlu, B. and Shanker, A.K. (2009). Climate change and agriculture: Adaptation and mitigation strategies. *Indian Journal of Agronomy*, 54(2), 226.

Climate Change and Variability and their Impacts on Different Crops in Gujarat

Vyas Pandey*, A.K. Misra and S.B. Yadav

Department of Agricultural Meteorology, B.A. College of Agriculture,
Anand Agricultural University, Anand 388 110, India
*E-mail: pandey04@yahoo.com

ABSTRACT: *The long term (> 30 years) climatic data (temperature and rainfall) of different stations of Gujarat have been used to ascertain the variability and trend of rainfall and temperature in Gujarat. The PRECIS model outputs for climate projection for 2071 to 2100 under A2 scenario have been generated for different regions of Gujarat. The impact of climate variability on crop was studied using the validated crop simulation models viz. DSSAT and InfoCrop for the growth and yield of different crops (wheat, rice, maize, pearl millet, groundnut etc.) of Gujarat. The results show that there is large variability in trends of temperature and rainfall on monthly, seasonal and annual basis across the locations in Gujarat. Some trends are significant and others are non-significant. The PRECIS model projection shows 15 to 100% increase in rainfall in different districts by 2100 while maximum and minimum temperature would increase by 2 to 5°C during these periods. The impact of these changes on different crops suggested negative effect mainly due to increase in temperature.*

Keywords: Trend, PRECIS, DSSAT, InfoCrop, Impact.

Natural fluctuations or variability is an established nature of the climate and the way it fluctuates above or below a long-term average value is termed as climatic variability. This variability could be the result of natural processes within the climate system or from variations in natural or anthropogenic external forces. On the other hand, climate change refers to a change in climate for a time series either due to natural variability or anthropogenic forces and it is imperative to have a clear distinction between climate variability and change. A changing climate leads to changes in the frequency, intensity, spatial extent, duration, and timing of extreme weather and climate events, and can result in unprecedented extreme weather and climate events (IPCC, 2014). Climate change is now a reality. However, the natural variability within the system is causing more vulnerability to agricultural system than climate change itself.

There is large variability in climate of Gujarat ranging from arid, semi-arid to sub humid due to large variability in rainfall which varies from 250 mm in Kutch district in the North West to more than 1500 mm in South Gujarat (Pandey and Patel, 2011). Lunagaria *et al.* (2012 and 2015) analysed long period climatic parameters to find the trends and variability and also of extreme weather events. They found that a large variability in climatic trends for various climatic parameters/extremes over the location in Gujarat. Kumar *et al.*, (2014) has computed the decadal seasonality index for 20 major districts of Gujarat from 1932 to 2011 to observe variability in rainfall and change in spatial and temporal pattern of seasonality index.

Impacts of climate variability have been studied by sensitivity analysis of validated simulation models viz WOFOST for wheat (Mishra *et al.,* 2015), DSSAT for groundnut (Parmar *et al.,* 2013) and InfoCrop for maize (Chaudhary *et al.,* 2015). The present paper describes the work done on trend analysis for climate change, its projection for 2071–2100 and impact on different crops in different districts of Gujarat.

The historical climatic data for maximum and minimum temperature and rainfall for different meteorological stations viz. Anand, Vadodara, Junagadh, Bhavanagar, Bhuj, Rajkot and Kesod were used for climate change and trend analysis study. The climate change projection under A2 scenario was obtained from PRECIS downscaled model output provided by IITM Pune in a grid size of 0.4 degree. The climate projection under A2 scenario for period 2071–2100 were considered for climate change study. The well calibrated and validated InfoCrop and DSSAT v 4.5 were used to study the crop response with the weather data generated using first approach i.e., day to day sum of actual weather data of 1961–90 and changes calculated using PRECIS baseline and A2 scenario projection data.

The results of trend analysis as presented in Table 1 show that there is positive trends in rainfall and temperature in different seasons though non-significant except rainfall is significantly decreasing during winter season at Rajkot. Bhavanagar and Vadodara were found to have non-significant trend for all the parameters and in all the seasons. Maximum temperature was found to be increasing significantly during winter, monsoon and post-monsoon seasons at Bhuj station and during post monsoon and annual basis at Anand, while at Rajkot it showed significantly negative trend during summer season. The minimum temperature was found to increase significantly in different seasons at Keshod, Junagadh, Rajkot and Anand while it showed decreasing trend at Bhuj. Except Rajkot none of the station showed any significant trend for rainfall. This clearly suggests that during 1961–90 period there was not uniform climatic trend or variation in any of the climatic parameters which was also stated by (Lunagaria *et al.*, 2012).

Table 1: Trend Analysis of Selected Stations of Gujarat (1961–90)

Weather parameters	Period/Season	Bhavnagar	Bhuj	Kesod	Rajkot	Junagadh	Vadodara	Anand
Maximum temperature	Winter	-0.006	0.031*	-0.002	-0.012	-0.002	0.036	0.030
	Summer	0.004	0.033	-0.012	-0.0001**	-0.012	0.033	0.017
	Monsoon	0.001	0.044*	0.022	0.015	0.022	0.018	0.016
	Post-monsoon	-0.003	0.052*	0.001	0.012	0.001	0.039	0.039**
	Annual	-0.001	0.039	0.021	0.005	0.021	0.031	0.027**
Minimum temperature	Winter	0.007	-0.032	0.009	0.082	0.009	0.021	0.017*
	Summer	0.018	-0.020*	0.007	0.045*	0.007	0.040	0.027**
	Monsoon	0.016	0.014	0.021*	0.037	0.021*	0.017	0.017**
	Post-monsoon	-0.012	-0.019	0.021	0.092*	0.021	0.026	0.025*
	Annual	0.007	-0.016	-0.467	0.064	-0.467	0.022	0.024**
Rainfall	Winter	0.075	-0.032	0.098	-3.28*	0.098	0.000	0.000
	Summer	-0.180	-0.045	-0.019	-0.158	-0.019	0.00	0.000
	Monsoon	-0.296	-1.508	-1.414	0.548	-1.414	10.98	16.51
	Post-monsoon	2.344	-0.216	-0.237	0.443	-0.237	0.000	0.000
	Annual	0.362	-0.449	0.002	-0.613	0.002	10.12	13.25

*: Significant at 5 % level, **: Significant at 1 % level

(*Source:* Patel *et al.*, 2015).

The mean of thirty years (2071–2100) projected climate data are compared with the base line (1961–90) periods in Table 2. It has been observed that highest change in projected rainfall (101%) is recorded at Bhuj whereas the lowest change in projected rainfall (15%) is at Vadodara. In general locations falling under Saurashtra region are expected to receive higher rainfall (> 50%) as compared to locations of middle Gujarat region (15–37%). The maximum temperature has been projected to increase between 2.8 to 7.2°C at different locations, the maximum increase being at Anand while lowest increase being at Bhuj. Similarly, the minimum temperature has also been projected to increase between 3.8 to 5.2°C at different locations.

Table 2: Per cent Variation of Climate at Various Locations in Relation to Base Line

Station	Rainfall (mm)			Maximum temperature (°C)			Minimum temperature (°C)		
	Base line (1960-90)	Projected (2071-2100)	Percent change	Base line (1960-90)	Projected (2071-2100)	Difference	Base line (1960-90)	Projected (2071-2100)	Difference
Anand	919	1259	37	29.8	37.5	7.7	19.1	24.3	5.2
Vadodara	980	1127	15	29.7	33.3	3.6	19.3	23.5	4.2
Junagadh	836	1513	81	33.3	37.3	4.0	19.9	24.1	4.2
Bhavanagar	627	953	52	33.9	38.0	4.1	21.3	25.1	3.8
Bhuj	389	782	101	34.7	37.5	2.8	19.4	23.3	3.9
Rajkot	660	1181	79	33.7	37.7	4.0	20.1	24.5	4.4
Kesod	836	1456	74	33.3	37.0	3.7	19.9	24.0	4.1

The impact of projected climate on various crops viz. wheat, maize, paddy, pearl millet and groundnut were studied using DSSAT and InfoCrop models and projected climatic data of Anand and Junagadh. Junagadh was selected for groundnut crops while for other crops Anand was selected. The models were run for both the periods i.e., base line (1961–90) and projected (2071–2100). The mean of thirty years were worked out and are compared. The differences in yield are presented in Figure 1. It may be seen that the highest yield reduction (–61%) was noted under wheat crop followed by kharif maize (47%), kharif paddy (32%) groundnut (24%) and kharif pearl millet (14%), rabi maize (10%) and summer pearl millet (8%). The reductions in yield are mainly attributed to increase in temperature during projected period.

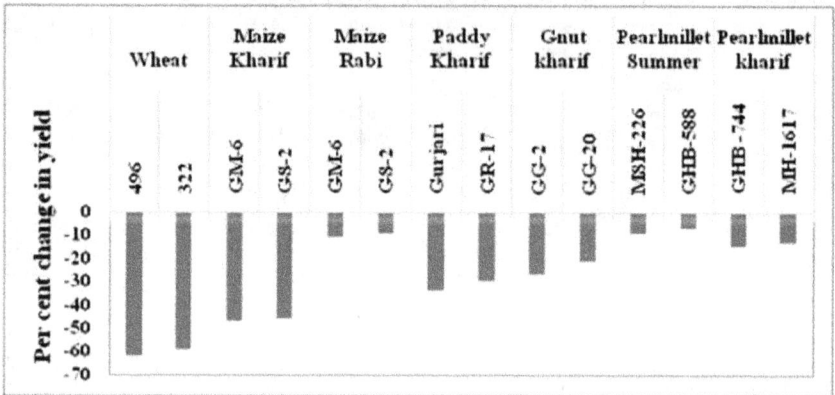

Fig. 1: Percent Change in Yield of Various Crops of Gujarat under Projected Climate

CONCLUSION

Climatic trend at selected stations of Gujarat indicated that the annual rainfall during projected period would be in the range of 15–101 per cent higher than the base line rainfall (1961–90), maximum temperature would increase by 2.8 to 7.7°C, while minimum temperature would increase by 3.8 to 5.2°C in different parts of Gujarat. The projected climate change will adversely affect the yields of different crops. The maximum yield reduction is projected in wheat and lowest in pearl millet. Maize during kharif season would be more affected than the rabi season. Similarly pearl millet in summer season will be affected less than kharif pearl millet.

REFERENCES

Chaudhary, D., Patel, H.R. and Pandey, V. (2015). Evaluation of adaptation strategies under A2 climate change scenario using InfoCrop model for *kharif* maize in middle Gujarat region. *Journal of Agrometeorology*, 17(1), pp. 98–101.

IPCC (2014). Climate change, AR5, Synthesis report.

Kumar, V., Samtani, B.K., Yadav, S.M. and Naresh, K. (2014). Decadal Comparison of Rainfall Seasonality Index in Gujarat. In G.C. Mishra, ed. *Global Sustainability Transitions: Impacts and Innovations*, Excellent Publishing House, New Delhi, pp. 238–245.

Lunagaria, M.M., Dabhi, H.P. and Pandey, V. (2015). Trends in the temperature and rainfall extremes during recent past in Gujarat. *Journal of Agrometeorology*, 17(1), pp. 118–123.

Lunagaria, M.M., Pandey, V. and Patel, H.R. (2012). Climatic trends in Gujarat and its likely impact on different crops. *Journal of Agrometeorology*, 14(1), pp. 41–44.

Lunagaria, M.M., Patel, H.R., Shah, A.V., Yadav, S.B., Karande, B.I. and Pandey, V. (2011). Validation of PRECIS baseline (1961–1990) simulation for middle Gujarat agroclimatic zone. *Journal of Agrometeorology*, 13 (2), pp. 92–96.

Mishra, S.K., Shekh, A.M., Pandey, V., Yadav, S.B. and Patel, H.R. (2015). Sensitivity analysis of four wheat cultivars to varying photoperiod and temperature at different phenological stages using WOFOST model. *Journal of Agrometeorology*, 17(1), pp. 74–79.

Pandey, V. and Patel, H.R. (2011). Climate Change and its Impact on Wheat and Maize Yield in Gujarat. *In.* Challenges and Opportunities in Agrometeorology (eds.) S.D. Attri, L.S. Rathore, M.V.K. Sivakumar; S.K. Dash, Springer, pp. 321–334.

Parmar, P.K., Patel, H.R., Yadav, S.B. and Pandey, V. (2013). Calibration and validation of DSSAT model for *kharif* groundnut in north-Saurashtra agroclimatic zone of Gujarat. *Journal of Agrometeorology*, 15(1), pp. 62–65.

Patel, H.R., Lunagaria, M.M., Karande, B.I., Yadav, S.B., Shah, A.V., Sood, V.K. and Pandey, V. (2015). Climate change and its impact on major crops in Gujarat. *Journal of Agrometeorology*, 17(2), pp. 190–193.

Climate Resilient Livestock Farming

A.K. Srivastava* and Sohanvir Singh

ICAR-National Dairy Research Institute Deemed University,
Karnal 132 001, India
E-mail: dir.ndri@gmail.com; dir@ndri.res.in

ABSTRACT: *Livestock is a critical source of livelihood for the poor farmers. Livestock population is threatened by the present scenario of climate change. Climate change is the planet's biggest threat affecting livestock welfare and production. Most of the livestock in India is owned by small farmers, marginal farmers, and landless labourers. Livestock owners of such animals with poor resource are more vulnerable to climate change. There is a requirement for adaptation pathways to withstand the effects of climate change (short and long term) as well as effective implementation of policies that promote sustainable development. Genetic, management and environmental modifications may be helpful to maintain sustainable development of livestock. Vulnerable livestock population may require genetic modification to meet the challenges of future climate change. Faster genetic gains can be achieved with new technologies, including genomic selection and advanced reproductive technologies. However, in order to realise the full benefits of the tested solutions, livestock farmers require financial, institutional and intellectual supports from stakeholders to remain climate-resilient.*

Keywords: Livestock, Resilience, Climate Change.

Livestock plays a major role in the agriculture sector of developing countries, contributing 40% to the agricultural GDP. India has held a huge population of livestock in the world. It has 56.7% of world's buffaloes, 12.5% cattle, 20.4% small ruminants, 2.4% camel, 1.4% equine, 1.5% pigs and 3.1% poultry. In 2010–11 livestock generated outputs worth ₹ 2075 billion (at 2004–05 prices) which comprised 4% of the GDP and 26% of the agricultural GDP. The total output worth was higher than the value of food grains (12th Five Year Plan— 2012–17). It provides livelihood to many of the world's poorest people in rural areas providing both food and income. It provides food, income, employment and many other contributions to rural development such as draught power, means of transport, organic manure for crop production and domestic fuel, hides and skin. They contributed about 16% of income, more so in states like Gujarat (24.4%), Haryana (24.2%), Punjab (20.2%) and Bihar (18.7%). Due to economic growth and urbanization, global demand for foods of animal origin is growing and it is apparent that the livestock sector meet this demand of the growing population (FAO, 2009). Although the prospect of animal productivity and its share to food security and national GDP is on the upward swing, livestock is adversely affected by the prevailing scenario of heat stress, land and water scarcity and the ensuing climate change.

Global warming and climate change have become the major threats to the sustainability of livestock production systems (Shelton, 2000). In tropical and sub-tropical regions, high ambient temperature is the major constraint on animal production (Marai, *et al.,* 2007; Nardone *et al.,* 2010). High environmental temperature exerts a negative influence on the performance of livestock population (Liu *et al.,* 2011; Singh and Upadhyay, 2009). By 2100, the temperature will be about 1.4–5.8°C over 1990 levels (IPCC, 2007). Nearly 20 to 30 per cent of plant and animal species are expected to be at risk of extinction due to this increase of 1.5°C to 2.5°C (FAO, 2007). The impacts of climate change are visible all over the world, but India is categorized among the most vulnerable areas, as rural economy is primarily dependant on crop-livestock production systems, and almost 70 per cent of livestock in India is owned by small farmers, marginal farmers and landless labourers. Animals of such livestock owners with poor resource are more vulnerable to climate change, and thus are at greater risk. India is currently losing nearly two per cent of the total milk production, amounting to a whopping over ₹ 2,661 crore due to rise in heat stress among cattle and buffaloes because of the global warming (Upadhyay *et al.,* 2007). Majority of the areas in India show higher Temperature Humidity Index (THI = 75) or more and 85% of places in India experience moderate to high heat stress during April, May and June (NDRI Vision 2030, 2011).

There is an urgent need to increase the adaptive capacity of the livestock to heat stress. However, adaptation to climate change is unlikely to be achieved with a single strategy (Hoffmann, 2010). Therefore, genetic, management and environmental modifications will be helpful in building livestock resilience to climate change among the vulnerable populations. Genetic modifications may be done by increasing the gene flow and introduction of breeds more adapted to the environment (Hoffmann, 2008). Zebu cattle are more heat tolerant than European cattle (Lacetera *et al.,* 2006; Collier *et al.,* 2008; Singh and Upadhyay, 2009). Preparedness for such transformations will require a significant research commitment. Many studies confirm that animal health and welfare are integral to environmental sustainability. Intensive high-input, high-output systems that appear highly efficient at first glance are in fact energy and resource hungry. Selection of animals for high yield is often directly associated with poor welfare which in itself can significantly contribute to increasing carbon emissions that may further increase the atmospheric temperature which is already a threat to the livestock. Pasture-based systems of livestock rearing can also reduce GHG emissions through grassland's capacity for carbon storage (sequestration). Land

and vegetation has the capacity to store carbon at different concentrations. Amount of metabolic heat production is affected by quality as well as particle size of feeds and fodder. Nutritional modifications can be used to reduce the internal heat load on animal. Highly digestible feeds are recommended because poor quality roughage generates a lot more heat than highly digestible rations. If feeding is done during mid-day, feeding under shade can be suggested to minimize exposure of cattle to heat stress (ATPS, 2013). Simple feed technologies like incorporation of good quality green fodder, increment of nutrient density by replacing poor quality roughage with concentrate, feeding properly chaffed dry fodder and hydration of dry straws during hot dry period reduce the internal heat load on animal body. Some of the feed additives like antioxidants and minerals can also be supplemented to minimize the impact of climate.

Animal shelters should be designed to reduce heat load from the external environment in tropical and sub-tropical climate. Design, orientation and height of the shelters, choice of roofing material, open space ventilation and provision of adequate space per animal are some of the important aspects for cooler microenvironment of the animal. During the period of high temperatures water can be used to bring down the micro-environmental temperature within the animal shelters. Use of air cooling systems is also an efficient method. Efficient and affordable adaptation practices for rural poor who are not able to buy expensive adaptation technologies include shading, sprinkling and ventilation to reduce heat stress from elevated temperature. The animals in arid zone are usually reared under extensive system of farming, where the animals go for grazing in the fields during day which are exposed to peak heat. Provision of community shelters in these areas will give a space to the animals to take rest during peak hot hours.

CONCLUSION

In the future, climate change is expected to be a major force testing resilience of global food production systems. Climate change will have far-reaching consequences for dairy and meat production, especially in vulnerable parts of the world where it is vital for nutrition and livelihoods. Animal well-being issue merits consideration in management and breeding programmes. However, climate change is a multi-faceted challenge that can be mitigated utilizing a variety of available tools and resources as well as a holistic approach. Ensuring good animal welfare will be paramount to address these challenges. Breeds suited to the local environment are often more robust and resilient than industrially farmed breeds. Vulnerable livestock population may require genetic modification to meet the challenges of future climate change. Different selection strategies will result in different adaptation rates to new production conditions. Faster genetic gains for such adaptive traits can be achieved with new technologies, including genomic selection and advanced reproductive technologies. But, in order to realise the full benefits of the tested solutions, livestock farmers require financial, institutional and intellectual support from stakeholders to remain climate-resilient.

REFERENCES

12th Five Year Plan (2012–17). Report of the working group on animal husbandry and dairying, Planning Commission, Government of India, New Delhi.

African Technology Policy Studies Network, ATPS (2013). Vulnerability and adaptation strategies to climate variability and change of the Bos-taurus dairy genotypes under diverse production environments in Kenya, ATPS research paper no. 23.

Collier, R.J., Collier, J.L., Rhoads, R.P. and Baumgard, L.H. (2008). Genes involved in the bovine heat stress response. *J. Dairy Sci.,* 91, 445–54.

Food and Agriculture Organization (FAO), (2007). The state of the world's animal genetic resources for food and agriculture, Ed. B. Rischkowsky and D. Pilling. Rome, Italy.

Food and Agriculture Organization (FAO), (2009). The state of food and agriculture, Rome, Italy http://www.fao.org/docrep/012/i0680e/i0680e.pdf

Hoffmann, I. (2008). Livestock genetic diversity and climate change adaptation. In Livestock and Global Climate Change. P. Rowlinson, M. Steele and Y.A. Nefzaoui (Eds.). pp. 76– 80. *BSAS Proceedings,* Cambridge University Press.

Hoffmann, I. (2010). Climate change and the characterization, breeding and conservation of animal genetic resources. *Anim. Genet.,* 41(1): 32–46.

IPCC (2007). Inter Governmental Panel on Climate Change.

Lacetera, N., Bernabucci, U., Scalia, D., Basirico, L., Morera, P. and Nardone, A. (2006). Heat stress elicits different responses in peripheral blood mononuclear cells from Brown Swiss and Holstein cows. *J. Dairy Sci.,* 89: 4606–4612.

Liu, Y., Li, D., Li, H., Zhou, X. and Wang, G.A. (2011). Novel SNP of the ATP1A1 gene is associated with heat tolerance traits in dairy cows. *Mol. Bio. Rep.,* 38: 83–88.

Marai, I.F.M., E.L.-darawany, A.A. and Fadiel, A. (2007). Physiological traits as affected by heat stress in sheep. A review: *Small Ruminant Res.,* 71: 1–12.

Nardone, A., Ronchi, B., Lacetera, N., Ranieri, M.S. and Bernabucci, U. (2010). Effects of climate changes on animal production and sustainability of livestock systems. *Livest. Sci.,* 130: 57–69.

NDRI Vision (2030). published by ICAR-NDRI, Karnal.

Shelton, M. (2000). Reproductive performance of sheep exposed to hot environments. In, *Sheep Production in Hot and Arid Zones.* Malik R.C., Razzaque M.A., A.l.-Nasser A.Y. (eds). *The Kuwait Institute for Scientific Research,* pp. 155–162.

Singh, S.V. and Upadhyay, R.C. (2009). Thermal stress on physiological functions, thermal balance and milk production in Karan Fries and Sahiwal cows. *Indian vet. J.,* 86: 141–144.

Upadhyay, R.C., Singh, S.V., Gupta, A.K. and Ashutosh, S.K. (2007). Impact of climate change on milk production of Murrah buffaloes. *Ital. J. Anim. Sci.,* 6: 1329–1232.

Climate and Climate Variability: Impact on Fisheries in Food and Nutritional Security

J.K. Jena[1] and Grinson George[2]

[1]Fisheries Science Division, Indian Council of Agricultural Research,
 Krishi Anusandhan Bhawan-II, New Delhi 110 001, India
[2]ICAR-Central Marine Fisheries Research Institute, Kerala 682 018, India
[1]E-mail: jkjena2@rediffmail.com; jkjena2@gmail.com

ABSTRACT: *With the total fish production of 10.8 million tones today, its demand in India has been increasing consistently with the increased income of the consumers and their growing health consciousness. It is estimated that to meet the domestic demand the country requires additional 10 million tons of fish by 2050. To maintain the growth pace of the sector, the fisheries and aquaculture, therefore, need to tackle the challenges of resource competition, water scarcity, climate change, growing disease problems and several other technical, environmental and social issues. Climate change impacts such as variations in rainfall and temperature, drought, extreme weather events and changes in oceanographic features, affecting the fisheries resources, farming conditions, fishery distribution and habitat changes, have been found to be quite significant and alarming. Therefore, identification of the vulnerabilities associated with climate change events on fishery resources and coming up with resilience measures as well as adaptation strategies are urgent necessities. An attempt has been made to collate and synthesize information on future climate change impacts and challenges posed on the fisheries and aquaculture sector and explain various mechanisms for addressing these challenges. Immediate necessity of developing an action plan to mitigate the impact of such climate change scenario in fisheries and aquaculture has been emphasized. Further, the paper also envisioned the use of Representative Concentration Pathway (RCP) scenarios (in 2030, 2050 and 2080) developed by the IPCC in concert with predictive modelling techniques to gauge the intensity of future changes likely to occur in the fisheries sector.*

Keywords: Climate Change, Aquaculture, Fisheries, Marine, Freshwater.

Climate change is one of the most important global environmental challenges having significantly higher implications on agricultural sectors, including natural ecosystems, fisheries and aquaculture. The changes include increase in air and water temperature, regional monsoon variation, non-seasonal rains, recurrent droughts and rise in extreme weather incidences in coastal regions and recession of Himalayan glacier. While changing climate poses a challenge to humanity as a whole, the available evidence suggests that the developing countries in particular are more vulnerable. Although it is difficult to predict the climate outcomes precisely, the probability towards greater impacts of climate challenge is clear. The projected increase of global population from the present level of 7.4 billion to 9.0 billion by the year 2050, together with increasing awareness of fish as a health-food, it is certain that the demand for fish and fisheries products would increase substantially. Further, India is emerging as the most populous country in the world; the estimated demand of fish for the country by 2050 will be over 20.0 million tonnes. Therefore, development of research strategies with regard to climate resilient aquaculture in inland or open sea, and other relevant fisheries management approaches is of high significance today.

POTENTIAL CLIMATE CHANGE SCENARIO AND ITS IMPACT IN FISHERIES SECTOR

Seasonal and annual extremities in precipitation, extremities in temperatures, late arrival and early withdrawal of monsoon and other Extreme Weather Events (EWEs) such as El-Nino, La-Nina, Tsunamis, cyclones, floods and drought are the adverse weather scenarios the fish farmers/fishermen are being confronted with. Climatic scenarios generated by computer models further show that India could experience warmer and wetter conditions as a result of climate change, including an increase in the frequency and intensity of heavy rains and EWEs. The stress being imposed due to climate change impacts is leading to uncertainty of fish catch in open-water systems, and growth retardation and disease out-breaks in aquaculture systems, affecting livelihoods of fish farmers.

Ocean-atmospheric climate models predict changes in the ocean circulation and further such changes will stimulate phytoplankton biomass production in the nutrient-depleted areas in the open ocean. The challenge is to study the magni-tude and variability of primary productivity, and resultant secondary and tertiary productivity. The distribution, abundance and phenology of fish stocks are also expected to change and there would be a novel mix of organisms, impacting the structure and functions of ecosystems. Studies have shown with clear indications of extension of spread of small pelagic fishes such as Indian oil sardine and mackerel in Indian waters. 'Pelagification', the terminology coined by ICAR-CMFRI, is probably going to be more pronounced, due to the surplus production that is happening in pelagic realms with a definite reduction in the demersal resources.

Small pelagic fishes (sardines, anchovies and mackerels) have shorter life span and are the best indicators of climate change, as the pelagic coastal habitat is more influenced by ocean-atmosphere variability. The distribution of sardines and mackerels which were restricted to the Malabar upwelling system along the southwest coast of India, have shown a clear shift since 1989. Distinct decline of population and sudden recoveries, and a strong inverse relationship with extreme events are witnessed in the dominant pelagic species in India (Manjusha *et al.*, 2010).

Studies conducted by ICAR Fisheries Research Institutes in the maritime states of the country under the National Innovations on Climate Resilient Agriculture (NICRA) project, being implemented by ICAR, indicated that seasonal weather variations cause 20–40% loss and EWEs cause 50–100% loss in coastal aquaculture of shrimps. It may be stated that the coastal aquaculture in India is centered on shrimp farming and largely carried out by small-scale farmers (CIBA, 1996). While it has been projected that bulk of the fish production of the country in coming decades to be largely met from aquaculture, the likely impacts of climate change will have great bearing on production and productivity of the sector. Therefore, it would necessitate greater research thrust on mitigating the challenges of associated biotic and abiotic stressors.

POSSIBLE IMPACT OF VARIATIONS IN RAINFALL

Variations in intensity of annual rainfall and resultant water availability are likely to have impact on intensity and duration of fish farming, and could lead to conflict with other agricultural, industrial and domestic users in water-scarce areas. The seasonally reversing monsoons over the Indian sub-continent also play a big role in capture fisheries. Weaker rainfalls indicate weak upwelling, which are one of the major drivers of marine fish productivity in Indian exclusive economic zone (EEZ). Further, the changes in suspended sediment and nutrient loads resulting from altered rainfall patterns will affect aquaculture in marine environment, brackish/freshwater ponds and reservoirs. The lack of rains during summer months may lead to an increase in the salinity of the creek beyond the tolerable limits of the cultured shrimps. Further, high rainfall would result in rapid drop in salinity to levels that may be lethal for the cultured species, as observed in *Penaeus japonicus,* causing mass mortality of the farm crop (Preston *et al.*, 2001).

POSSIBLE IMPACT OF TEMPERATURE RISE

Sea surface temperature has risen significantly in India since over the last 40 years, i.e. by 0.602°C along north-east India, by 0.597°C along north-west India, by 0.690°C along south-east India and by 0.819°C along south-west India (CMFRI-NICRA, 2016). Increase in mean air temperature will not necessarily equate to increases seen in temperature of inland and sea waters. In fact, they have contrasting effects on capture and culture conditions.

Issues in Marine Environment

Warming of water and sea level rise are the two universal features, which impact the marine fishery. The challenges imminent are changing spawning behaviour in fishes (Vivekanandan and Rajagopalan, 2009b), inter-annual variability in fish abundance (Sathianandan *et al.*, 2011), changes of community structure in marine biodiversity (Krishnan *et al.*, 2013), bleaching of corals (Krishnan *et al.*, 2010) and productivity variations in coastal waters (George, 2014). Changes in sea surface temperature due to global warming could result in changes in the seasonal distribution of certain trans-boundary species such as tunas, and ultimately result in disruption of their harvest, which is usually based on indigenous knowledge. Warmer water temperatures can result in massive coral bleaching that results in the expulsion of the symbiotic zooxanthellae from the tissues of coral. Between 1979 and 1990, sixty major episodes of coral bleaching were recorded, and in 2016 the longest coral bleaching event on record was observed. Different species of coral have differing tolerances to bleaching events, with some species being able to withstand the thermal shocks that would greatly affect more fragile corals (Lix *et al.*, 2016). This could result in the reduction in the species richness of corals in certain global warming hotspots.

Issues in Inland Environment

It is evident that inland waters are warming, which is leading to visible changes in the distribution and breeding pattern of the fish (Vass *et al.*, 2009). From the analysis of 30 years' time series data, an increase in annual mean minimum water temperature is observed in the upper cold-water stretch of the Ganga (Haridwar) by 1.5°C and in the aquaculture farms in the lower stretches in the Gangetic plains by 0.2–1.6°C (Vass *et al.*, 2009). The study on the distribution and abundance pattern of fishes has shown that a number of fish species which were never reported in the upper stretch of the river and were mostly confined to the lower and middle stretches have been recorded from the upper cold-water region now. In order to assess the vulnerability of inland fisheries to climate variations, Das *et al.* (2014) have prepared a framework in 13 districts of West Bengal, the data of which showed different spatial combinations of climate exposure, sensitivity and adaptive capacity among the districts.

Issues in Culture Environment

A change in temperature of only a few degrees might mean the difference between a successful aquaculture venture and an unsuccessful one. The production efficiency of tropical and sub-tropical species of farmed shrimps, such as *Penaeus monodon* and *P. merguiensis*, can be increased by a rise in water temperature. Rising temperatures may not only enhance growth rates at existing sites, but also extend the area suitable for farming these species. Increased water temperatures and other associated physical changes, such as shifts in dissolved oxygen levels, have been linked to increases in the intensity and frequency of disease outbreaks (Goggin and Lester, 1995; Harvell *et al.*, 2002; Vilchis *et al.*, 2005). In 2002, approximately 40–45% of brackish water and 60% of fresh water area were found to be affected with the severe drought in Krishna District, Andhra Pradesh. Changes in temperature can also have positive impacts for aquaculture. Elevated temperatures of coastal waters also could lead to beneficial impacts with respect to growth rate and feed conversion efficiency (Lehtonen, 1996), and increased production.

Das and Saha (2008) studied the effect of temperature on the reproductive integrity on mature female *Cyprinus carpio*. They showed that the fish subjected to enhanced temperature of 34°C resulted in a decrease in the Gonado-Somatic Index and accumulation of liver and ovarian cholesterol. While increase in the temperature in the sub-tropical region may lead to early maturity of tropical species, it would have adverse impact on the breeding performance of cold-water species in temperate region and warm-water species in tropical region.

IMPACT OF EXTREME WEATHER EVENTS

The east coast of India is subject to frequent cyclonic storms and occasional tidal waves, which cause loss of aquaculture stock and damage to aquaculture facilities (Muralidhar *et al.*, 2006; 2009). The farms were inundated almost one meter above bund level and the damage included erosion of bunds, heavy siltation, damage to electrical installations, sluices, shutters and screens, communications failure and loss of stored feed. Changes in pond water quality, introduction of disease into aquaculture facilities along with the flooded water resulted in yield reduction and crop losses. The threat of sea-level rise (SLR) makes the coastal resources, infrastructure and population highly vulnerable. Shift of aquaculture to brackish water species due to reduced freshwater availability is a possibility. Increased saline-affected areas might help in expansion of brackish water culture of high-value species such as shrimps and mud crab. Survey by ICAR-Central Island Agricultural Research Institute (Dam Roy and Grinson George, 2009) and ICAR-CIBA revealed that after the tsunami of 2004, in the Andaman and Nicobar Islands, about 830 ha of seawater inundated areas become suitable for brackish-water aquaculture (Pillai and Muralidhar, 2006).

CHANGES IN OCEANOGRAPHIC VARIABLES

Fishery resources have a strikingly important place of prominence in the bio-diversity map of the earth. The socio-economics and nutritional sustenance in large extent of our coastal region is centered on fish as a nutritional requirement and foreign exchange earner. The spatio-temporal fluctuations of the phytoplankton richness can be remotely sensed for predicting resource richness in aquatic systems. Taking cue from established models, the potential fish availability in our EEZ could be predicted from the ocean colour data (satellite chlorophyll) after a validation of the prediction with the estimated fish catch. ICAR-CMFRI has come up with a flagship programme named Chlorophyll based Remote Sensing assisted Indian Fisheries Forecasting System (ChloRIFFS) which is operationalizing the primary productivity to fish production model. It is envisaged that such models would help in forecasting the fisheries resource potential regions, even in the context of climate-mediated changes in hydro-biological characteristics.

Coastal red tides could affect operations and production of shrimp farming. Variations in wind velocity and currents could have impacts on decreased flushing time and dilution rate in the source water bodies resulting in alternations in water exchanges and waste accumulation affecting the carrying capacity.

VULNERABILITY

The various vulnerability issues related to climate change as compiled from the NICRA based studies are included in the following table:

Vulnerability in fisheries due to Climate Change	*Possible measures for resilience*	*Indicators of measurement of resilience*
Highly vulnerable fish stocks	Regulation of fishing harvesting units, mesh size, spatio-temporal closure/ habitat restoration	1. Increase in Catch Per Unit Effort (CPUE) 2. Increase in mean length in the catch 3. Increase in fecundity 4. Increase in size at maturity 5. Reduction in fleet size
Reduction in fecundity/ size at maturity in wild stocks	Implementation of MLS to increase mean size in the catch	1. Increase in size at maturity 2. Increase in fecundity 3. Implementation of MLS regulations
Extension of distributional boundaries of small pelagics due to increase in SST	Better exploitation and utilisation of small pelagics in all the maritime zones	1. Increase in the landings of pelagic extended species 2. Increase in CPUE of small pelagics
Increased carbon footprint of mechanised fishing operations	Use of potential fishing zones (PFZs) to reduce scouting time, Use of wind/solar energy in fishing vessels (Green fishing)	1. Availability of PFZ advisory for the region 2. Number of vessels utilise PFZ advisories 3. Number of vessels use low energy alternatives for fishing
Reduced grow-out period in culture ponds	Water retaining structures to tide over extreme droughts	1. Increased water retention 2. Prolonged culture period

ADAPTATION STRATEGIES

It is time to use representative concentration pathway (RCP) scenarios (for 2030, 2050 and 2080) developed by the IPCC in concert with predictive modelling techniques to gauge the intensity of future changes likely to occur in the fisheries sector. The implications of climate change are far reaching and hence there is a need to develop and implement management actions to increase the resilience of freshwater, brackish water and marine systems. Necessary action plan has been chalked out by the fisheries research Institutes of the country (ICAR-CMFRI, 2015) and comprises of the following key elements:

- *Research thrust*: Address critical knowledge gaps about climate change impacts; identify thresholds, improve monitoring, and evaluate strategies; and translate information into active management responses.
- *A resilient aquatic ecosystem*: Minimize impacts through local management actions; adapt existing management to incorporate climate change considerations; and maximize resilience by protecting vulnerable ecosystems and species.
- *Adaptation of industries and communities*: Identify risks and resilience of fisheries, industries and communities; maximize resilience by planning regulations, policies and guidelines and assist in adaptation responses.
- *Reduced climate footprints*: Increase knowledge and involvement of stakeholders; and work with organizations and individuals to reduce their climate footprint.

The complex nature of physical, ecological and social systems challenges our capacity to accurately predict changes and consequently develop adaptation strategies based on realistic forecasts. Quantification of feedbacks between the biophysical environment (climate and limnology/oceanography and species biodiversity and abundance) and socio-economic environment (fishermen communities, market drivers and policy and governance arrangements) provides a means to advance our understanding of ecosystem health, resilience and productivity. Thus, strategically designed monitoring programmes at spatial and temporal scales that will capture processes driving these systems and links between the biophysical and socio-economic arenas are integral to advance our modelling capabilities and to generate innovative assessment tools.

In spite of carbon sequestering ecosystems in India, the opportunity for using blue carbon has not been adequately realized. It is, therefore, important to seriously examine the role and potential of blue carbon at national level. This would lead to financial incentives through carbon trading to protect and sustainably manage all blue carbon ecosystems as part of wider climate change adaptation and mitigation strategies with a core focus on local communities. There is a growing awareness among stakeholders regarding 'blue carbon' stored in our mangroves, seagrass meadows and marshy coastal wetlands. The day when carbon trading starts in the marine fisheries sector is not far away and when it happens it is to be ensured that local fishing communities harvest the benefits.

Individual farmers face risks associated with climate variability and climate change. Their livelihoods are exposed to climate risks and associated impacts. Risk identification and assessment are the two important steps that form the basis for successful implementation of adaptation practices. Knowledge on the nature of risks, their geographic coverage and their potential future behaviour is fundamental for designing a viable adaptation practice to reduce the impact of climate change in the aquaculture sector. It is important to plan the culture as per the changes in weather parameters in different seasons. In order to understand this concept, the seasonal changes in every month including EWEs are to be matched with month-wise farming activities. In areas more prone to higher frequency and/or intensity of storms, remedial measures should be incorporated during the site selection, designing and construction of farms. A strong focus on building general adaptive capacity can help the poor aquaculture communities to cope with new challenges.

CONCLUSION

Increasing human population, growing purchasing power and increasing health consciousness of the consumers would result in greater demand for fish protein in days to come. Therefore, the challenges those may arise in the aquatic system due to the climate change impact, both in aquaculture and in open-water systems, need to be addressed with great understanding. Changes in phenology, trophodynamics of fish, spawning season shift, maturity, mean length and distributional shift of fish species were identified to have implications on stock availability, catch, live-lihoods and national economy. Scientific interventions need to be made at national level to identify the impacts and to develop adaptation and resilient strategies on a continuous basis to support fisheries and aquatic ecosystems. The fishermen communities and other stakeholders associated with the sector need to be empowered to cope with climate change through regular training, workshops, awareness programmes, etc. Technologies ought to be developed to adapt with climate change and demonstrated directly to fishermen communities. Existing management plans for the fisheries and aquaculture sectors, coastal zones and watersheds need to be reviewed and, if needed, further developed to ensure that they cover potential climate change impacts, mitigations and adaptation responses.

ACKNOWLEDGEMENTS

The research activities on impact of climate change in fisheries sector taken up by various fisheries research institutes in the country is acknowledged. Special thanks are due to the team members of NICRA for the scientific inputs.

REFERENCES

CIBA (1996). Comprehensive Survey of Impact Assessment of Shrimp Farms in Nellore District. ICAR-Central Institute of Brackish Water Aquaculture, Chennai.

CMFRI-NICRA Annual Report, 2015–2016 (2016). Marine Fisheries, Report of work done at *CMFRI* submitted to CRIDA, p. 27.

Dam Roy, S. and George, Grinson (2009). Prospect and scope of extensive shrimp farming in tsunami submerged areas. *J. Indian Soc. Coastal Agric. Res.*, 27(2): 54–57.

Das, M.K. and Saha, P.K. (2008). Impact and adaptation of inland fisheries to climate change in India. Bull. No. 151, CIFRI, Kolkata, India.

Das, M.K., Srivastava, P.K., Rej, A., Mandal, M.L. and Sharma, A.P. (2014). A framework for assessing vulnerability of inland fisheries to impacts of climate variability in India. *Mitig. Adap. Strateg. Glob. Change*, DOI: 10.1007/s11027-014-9599-7.

George, Grinson (2014). Numerical modelling and satellite remote sensing as tools for research and management of marine fishery resources. In: W. Finkl and C. Makowski (Eds), Remote Sensing and Modelling: Advances in Coastal and Marine Resources. Chapter-18, *Springer International Publishing*, Switzerland, pp. 431–452.

Goggin, C.L. and Lester, R.J.G. (1995). *Perkinsus*, a protistan parasite of abalone in Australia—Review. *Marine Freshwater Res.*, 46: 639–646.

Harvell, C.D., Mitchell, C.E. Ward, J.R. Altizer, S. Dobson A.P., Ostfeld and R.S. Samuel, M.D. (2002). Climate warming and disease risks for terrestrial and marine biota. *Science*, 296: 2158–2162.

ICAR-CMFRI (2015). Vision 2050. Central Marine Fisheries Research Institute, Kochi, pp. 47.

Krishnan, P., Dam-Roy, S., George, G., Anand, A., Murugesan, S., Kaliyamurthy, M. and Soundararajan, R. Vikas (2010). Elevated Sea Surface Temperature (SST) induces mass bleaching of corals in Andaman. *Curr Sci.*, 100(1): 1800–1804.

Krishnan, P., George, G., Immanuel, T., Bitopan-Malakar, B. and Anand, A. (2013). Studies on the recovery of bleached corals in Andaman fishes as indicators of reef health. In: K. Venktaraman, C. Sivaperuman and C. Raghunathan (Eds.), *Ecology and Conservation of Tropical Marine Faunal Communities.* Chapter 25. Springer Verlag Berlin, pp. 395–408.

Lehtonen, H. (1996). Potential effects of global warming on northern European freshwater fish and fisheries. *Fish. Manag. Ecol.*, 3: 59–71.

Lix, J.K., Venkatesan, R., Grinson, George, Rao, R.R., Jineesh, V.K., Arul, M.M., Muthiah, Vengatesan, G., Ramasundaram, S., Sundar, R. and Atmanand, M.A. (2016). Differential bleaching of corals based on *El Nino* type and intensity in the Andaman Sea, southeast Bay of Bengal. *Environ. Monitor. Assessm.*, 188(175): 1–13.

Manjusha, U., Ambrose, T.V., Remya, R., Paul, S., Jayasankar, J. and Vivekanandan, E. (2010). Seasonal and inter-annual changes inoceanographic features and their impact on small pelagic catches off Kerala. In: *International Symposium on Remote sensing and Fisheries, SAFARI*, Cochin, Book of Abstracts, 42, p 83.

Muralidhar, M., Gupta, B.P., Jayanthi, M., Krishnan, M. and Ponniah, A.G. (2009). Impact of extreme climatic events on brackish water aquaculture. In: Marine Ecosystems Challenges and Opportunities, Book of Abstracts, *Marine Biological Association of India*, February 9–12, 2009, Cochin, pp. 249–250.

Muralidhar, M., Gupta, B.P., Saraswathy, R., Abraham, Abey Varampath and Nagavel, A. (2006). Has tsunami affected the quality of seawater and backwaters? Case studies in and around Chennai. *J. Indian Soc. Coastal Agric. Res.*, 24(2): 386–390.

Pillai, S.M. and Muralidhar, M. (2006). Survey and demarcation of seawater inundated areas for eco-friendly aquaculture in Andaman and Nicobar Islands. Report submitted to Andaman and Nicobar Administration, Port Blair. *Central Institute of Brackish Water Aquaculture*, Chennai, 69 pp.

Ponniah, A.G. and Muralidhar, M. (2009). Research requirements to understand impact of climate change on brackish water aquaculture and develop adaptive measures. In: *Proceedings of 96ᵗʰ Session of the Indian Science Congress, Part II: Session of Animal Veterinary and Fisheries Sciences,* January 3–7, 2009, pp. 46–47.

Preston, N.P., Jackson, C.J., Thompson, P., Austin, M., Burford, M.A. and Rothlisberg, P.C. (2001). Prawn farm effluent: Composition, origin and treatment. *Fisheries Research and Development Corporation, Australia, Final Report,* Project No. 95/162.

Sathianandan, T.V., Jayasankar, J., Kuriakose, Somy, Mini, K.G. and Mathew, W.T. (2011). Indian marine fishery resources: Optimistic present, challenging future. *Indian J. Fish.,* 58(4): 1–15.

Vass, K.K., Das, M.K., Srivastava, P.K. and Dey, S. (2009). Assessing the impact of climate change on inland fisheries in river Ganga and its plains in India. *Aqua. Eco. Health Manag,* 12(2): 138–151.

Vilchis, L.I., Tegner, M.J., Moore, J.D., Friedman, J.D., Friedman, C.S., Riser, K.L., Robbins, T.T. and Dayton, P.K. (2005). Ocean warming effects on growth reproduction and survivorship of southern California abalone. *Ecolog. Applic.,* 15: 469–480.

Vivekanandan, E. and Rajagopalan, M. (2009). Impact of rise in sea water temperature on the spawning of threadfin breams. In: P.K. Aggarwal (Ed), Global Climate Change and Indian Agriculture, *ICAR,* New Delhi, India, pp. 93–96.

Climate Change and its Impacts on Food and Nutritional Security in India

S. Naresh Kumar

Centre for Environment Science and Climate Resilient Agriculture
ICAR-Indian Agricultural Research Institute, New Delhi 110 012, India
E-mail: nareshkumar.soora@gmail.com; nareshkumar@iari.res.in

ABSTRACT: *Climate change impacts various sectors including agriculture. Indian agriculture is projected to be affected by the climate change impacts and adaptation is a must for developing resilient agricultural systems. Adaptation costs and gains may vary depending on the nature of the agricultural system and the size and character of farm. Climate smart agriculture is a step towards making Indian agriculture knowledge based. Increasing food basket and changing food habits are some steps that are required to achieve food and nutrition security in future climates in India.*

Keywords: Climate Scenario, Season, Adaptation, Crops.

The global mean temperatures during the 1951–2010 period increased by 0.6 to 0.7°C, out of which natural variability contributed to ±0.1°C change in temperature, while 0.6°C increase is due to anthropogenic activities (AR 5-WG-I, IPCC, 2013). In Indian region, the warming has been at a rate of 0.17°C/10 years (for maximum temperatures) and 0.29°C/10 years (for minimum temperature) since 1970s (INCCA, 2010). During the 1871–2016 period, India has faced 28 deficit and 20 excess monsoon years. Of these, as many as 16 deficient monsoon years and 6 excess monsoon years fell in post-1960 period. All these have been affecting the Indian agriculture posing a serious threat to food and nutritional security. Areas encompassed by climatic stresses and magnitude of losses have been increasing recently. For Indian region, the AR5-WGII (IPCC, 2014) report projects an increase in frequency of extreme temperatures, rainfall, heat wave, flood and drought events and skewed monsoon years. Further, it projects an increased risk of drought-related water and food shortage if agriculture is not adapted to changing climates.

The World Food Summit in 1996 defined food security as a situation in which "all people, at all times, have physical and economic access to sufficient, safe and nutritious food to meet their dietary needs and food preferences for an active and healthy life". This definition on food security focuses on the linkage between food, nutrition and health along with availability of food, accessibility (economically and physically), utilization (the way it is used and assimilated by the human body) and stability of these three dimensions (FAO, 2016). Out of over 20,000 species of edible plants in the world, fewer than 20 species provide 90 per cent of our food. Protein and energy malnutrition, micronutrient deficiencies, vitamin A deficiency, iron deficiency anemia, iodine deficiency disorders and vitamin B-complex deficiencies are major nutritional problems encountered frequently among the rural poor and even urban slum communities in India (NIN, 2011). India has attained self-sufficiency in food grains but still has an alarming gap in per capita intake of calories, ranging from 350 to 1100 kcal/day, from the recommended daily allowance depending on age and sex (Ramachandran, 2012).

Availability of food primarily, though not entirely, depends on its production. This paper presents the i) seasonal climate scenarios (since majority of agricultural activities are seasonal: monsoon-*kharif*; winter-*rabi*; summer-*zaid*), ii) impact of climate change on major crops, adaptation gains, iii) a case study on adaptation costs and gains across farm size and iv) options and challenges for food and nutritional security in changing climates. Such information becomes important for knowledge-based adaptation decisions to make Indian agriculture resilient to climatic stresses and achieve food and nutritional security.

Climate scenario data sets from over 30 Global Climate Models (GCMs) were analysed for seasonal climate change projections for India. Since the GCM outputs had significant cold bias for northern regions of India and hot bias for southern parts, the bias correction was carried out. Thereafter, the probabilistic ensemble scenarios were developed and analysed for seasonal projections.

Impact of climate change on various crops and the adaptation gains were assessed using the InfoCrop model and reported from the published work. Similarly, impacts on various other commodities and food production components, adaptation gains and costs at house hold level are reported from the published work.

PROJECTING SEASONAL CLIMATE CHANGE FOR INDIA

Climate change projections for India indicated that during *kharif* minimum temperatures to increase in the range of 0.946–4.067°C (2020 to 2080) in different RCPs over baseline (1976–2005) temperatures, while maximum temperatures were projected to increase in the range of 0.741–3.533°C (2020 to 2080). Similarly, during *rabi*, projected rise in minimum temperatures was found to be in the range of 1.096–4.652°C (2020 to 2080); while the projected increase in maximum temperatures was in the range of 0.882–4.01°C. Rise in temperatures was projected to be more for *rabi* than for *kharif* season; minimum temperatures to rise more than rise in maximum temperatures; rise in temperatures to be more in northern parts of India than in southern parts. Rainfall during *kharif* is projected to increase in the range of 2.3–3.3% (2020), 4.9–10.1% (2050); while *rabi* rainfall was projected to increase in the range of 12% (2020), 12–17% (2050). Variability in minimum and maximum temperatures as well as that of rainfall is expected to increase significantly. The CO_2 concentrations were projected to increase in the range of 419–432 µmol (2020), 441–572 µmol (2050) and 429–799 µmol (2080) in different RCPs.

PROJECTED IMPACTS ON FOOD PRODUCTION (CROPS, LIVESTOCK AND FISH)

Climate change impact projections were worked out for major crops such as rice, wheat, maize, sorghum, mustard, soybean, potato, cotton and coconut. Summary of the projections for these crops is given below:

- Climate change is projected to reduce irrigated rice yield by ~4% in 2020 (2010–2039) and rainfed rice yield by 6% (Figure 1a and b). Rainfed rice yields in India to reduce by ~6% in the 2020 and < 2.5% in 2050 (2041–2070) and 2080 (2070–2099) scenarios. With adaptation, however, irrigated rice yield is projected to increase by about 17% and rainfed rice yield by about 20% (Naresh Kumar *et al.*, 2013).

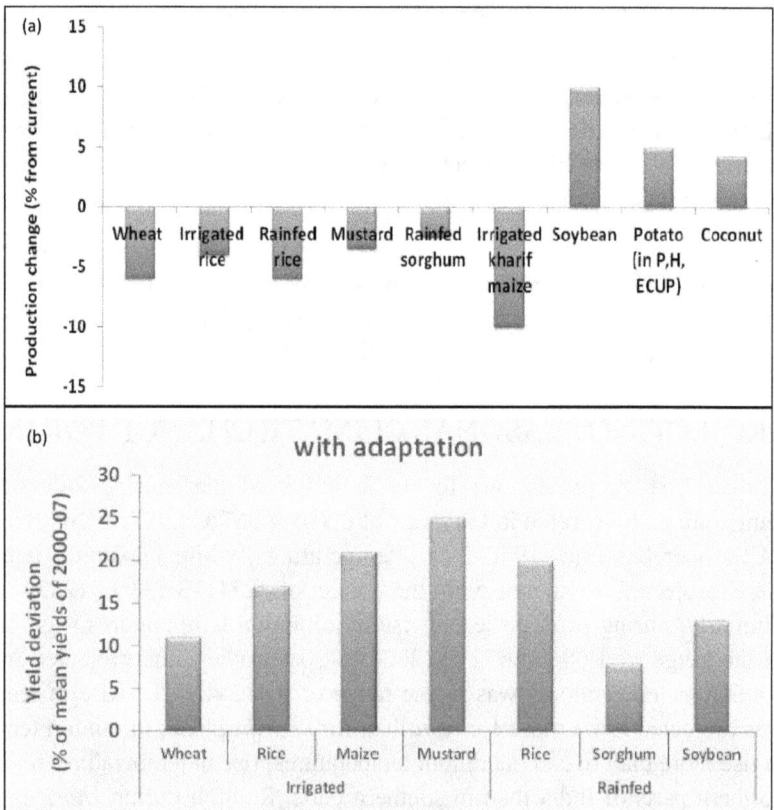

Fig. 1: Impact of Climate Change on Major Crops in India without (a) and with Adaptation (b) in 2020 Scenario

- Wheat yield in India would reduce by 6 to 23% by 2050 scenario, if no adaptation is followed. Yield would reduce in areas with mean seasonal maximum and minimum temperatures more than 27 and 13°C, respectively. Adjusting the time of sowing, suitable variety and input (fertilizer and irrigation) management may be a practical low-cost adaptation strategy to increase the yield (by > 10%) in future climates (Naresh Kumar *et al.*, 2014a, b).
- Maize yield in kharif season is projected to decrease by 18% but adaptation can increase the yield up to 21% in 2020 scenario (Naresh Kumar *et al.*, 2012).
- Rainfed sorghum yield is projected to reduce by 2.5% in 2020 (2010–2039). Adaptation, however, can increase the productivity by 8% in 2020 (Naresh Kumar *et al.*, 2012).
- Mustard yield is projected to reduce by <"2% in 2020 (2010–2039). Regions with mean seasonal temperature regimes above 25/10°C to lose due to temperature rise. As climatically suitable period for mustard cultivation may reduce in future, short-duration (< 130 days) cultivars with 63% pod filling period will become more adaptable (Naresh Kumar *et al.*, 2015).
- Increase in soybean yield in the range of 8–13% under different future climate scenarios (2030 and 2080) is projected. In case of rainfed groundnut, except in the climate scenario of A1B 2080 (–5%), in rest of the scenarios yield is projected to increase by 4–7% (Bhatia *et al.*, 2012; Naresh Kumar *et al.*, 2012).
- The potato crop duration in the Indo-Gangetic Plains is projected to decrease and yield to reduce by ~2.5, ~6 and ~11% in 2020 (2010–2039), 2050 (2040–2069) and 2080 time periods, respectively. Change in planting time could be the most important adaptation option which may lead to yield gain by ~6% in 2020 (Naresh Kumar *et al.*, 2016).
- Cotton productivity in northern India may marginally decline due to climate change while in central and southern India, productivity may increase. However, at the national level, cotton productivity may not be affected (Hebbar *et al.*, 2013).
- Climate change is projected to increase coconut productivity in western coastal region, Kerala, parts of Tamil Nadu, Karnataka and Maharashtra (provided current level of water and management is made available in future climates as well) and also in North-Eastern states, islands of Andaman and Nicobar and Lakshadweep while negative impacts are projected for Andhra Pradesh, Odisha, West Bengal, Gujarat and parts of Karnataka and Tamil Nadu. On all India basis, even with current management, climate change is projected to increase coconut productivity by 4.3% in the A1B 2030, 1.9% in A1B 2080, 6.8% in A2 2080 and 5.7% in B2 2080 scenarios. Adaptation can increase the productivity by ~33% in 2030, and by 25–32% in 2080 climate scenarios. Productivity in India can be improved by 20% to almost double if all plantations in India are provided with location-specific agronomic and genotype intervention in current climates (Naresh Kumar and Aggarwal, 2014).

- Apple productivity is likely to be affected and its cultivation is projected to shift to higher latitudes to 2500 m amsl from 1250 m amsl in Himachal Pradesh (Bhagat *et al.,* 2009).
- Climatic stresses such as heavy rainfall events damage horticultural crops. Flooding for 24 h affects tomato with flowering period being sensitive. Similarly, onion bulb initiation stage is sensitive to flooding causing a 27 and 48% reduction in bulb size and yield, respectively (Rao *et al.,* 2009).
- Climate change is projected to affect the quality in terms of reduced concentration of grain protein (under low fertilizer input conditions), and some minerals like zinc and iron due to elevated CO_2 (Porter *et al.,* 2014).
- Elevated CO_2 caused reduction in the concentration of protein, secondary metabolites while rise in temperature enhanced their concentration in pulse, several vegetable and fruit crops.
- Global warming is likely to lead to a loss of 1.6 MT of milk production by 2020 and 15 MT by 2050 if no adaptation is followed. The losses may be highest in UP followed by Tamil Nadu, Rajasthan and West Bengal. Increased number of heat stress days and probable decline in availability of water may further impact animal productivity (Upadhyay *et al.,* 2013).
- Rise in temperature caused latitudinal extension in abundance of oil sardine along the Indian coast. Marine fish availability has extended to deep waters and the spawning activity of *Nemipterus* spp. reduced in summer months and shifted towards cooler months (Vivekanandan *et al.,* 2013).
- Breeding season of Indian major carps extended from 110–120 days (pre 1980–85) to 160–170 days, making it possible to breed them twice in a year at an interval ranging from 30–60 days (Das *et al.,* 2013).
- Rise in temperature causes reduction in egg and meat production of poultry birds. The critical body temperature at which the poultry birds succumb to death would be 45°C which was observed at the shed temperature of 42°C (Reddy *et al.*, 2013).
- Climate change is projected to increase soil erosion, affect water availability and quality (IPCC, 2014).
- Enhanced frequency and duration of extreme weather events such as flood, drought, cyclone, cold and heat wave may adversely affect agricultural productivity.
- However, rise in temperature may decrease cold waves and frost events leading to reduced damage to frost-sensitive crops such as chickpea, mustard, potato and other vegetables.

Indian population is projected to stabilize around 1.6 billion by 2060 with corresponding increase in demand for quantity and quality food. Annual demand for food grains is estimated to be about 333 MT by 2050. All these projections indicate that food and nutritional security needs to be addressed based on scientifically derived adaptation strategies so as to meet the projected increase in demand for food. While doing so, it is also important to reduce emission of greenhouse gases and sequester them in agricultural soils, agro-forestry systems, perennial horticultural systems and plantation crops.

ADAPTATION OPTIONS, GAINS AND COSTS: EXPERIENCES FROM HOUSEHOLD ANALYSIS

Indian farmers have been adapting to climatic risks. A detailed household level analysis in a village indicated that the cost of adaptation and gains varies with type of climatic risk, type of agriculture and farm size. Strategies such as improved varieties, crop diversification, crop water and livestock management, value additions, etc. have led to farm resilience to climatic risks. Small and marginal farm (< 4 acre) families cannot support themselves with agricultural income alone, however with adaptation a self-sustaining agriculture could be achieved. Additional cost is not always required for adaptation (particularly in mid and large farm holdings) and rationalizing agricultural expenditure through scientific crop management is essential for adapting to climatic risks (Naresh Kumar *et al.,* 2016).

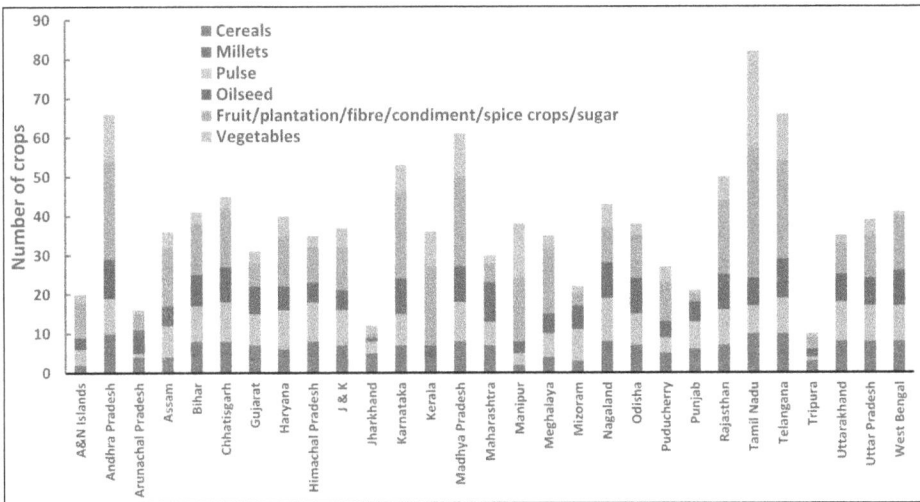

Fig. 2: Crop Diversity in Different States of India

OPTIONS AND CHALLENGES FOR FOOD AND NUTRITIONAL SECURITY IN CHANGING CLIMATES

- Apart from climate change, rapidly changing food preferences and habits have been challenging agriculture to produce quality food across the seasons and around the year.
- Increasing demand for animal products and animal meat causes additional pressure on agro-ecosystem resources.
- Preference for proteins, carbohydrates, specific fatty acids, vitamins, minerals and other secondary metabolites, fiber, etc., has been in increasing trend than for 'food' – a complete diet. There is a need to learn lessons from societies

which followed such unhealthy food habits and prefer 'food' intake rather than specific nutrients.

- The food basket, which was highly diversified earlier, has now narrowed down to only a few crops. A look at the number of crops with substantial area in each state indicates that states like Tamil Nadu, Telangana and Andhra Pradesh to have more diversified cultivation than other states with least being in Punjab, Jharkhand and A&N islands.
- Growing perennial fruit trees at farm house holds (as in A&N islands, Kerala, NE states), in kitchen garden, along farm borders and roads can provide easy access to fruits. This also will help in carbon sequestration and achieving environmental balance.
- Expanding food basket and exploiting other edible but neglected or under-utilized species.
- 'Input–output service' providers for crop-diversification zones, incentiviz-ation, expansion of minimum support price, safety net backup, etc. can encourage the farmers towards diversification.

REFERENCES

Aditi, Srivastava, Naresh Kumar, S. and Aggarwal, P.K. (2010). Assessment on vulnerability of sorghum to climate change in India. *Agric. Ecosyst. Environ.* Doi:10.1016/j.agee.2010.04.012; 138:160–169.

Bhagat, R.M., Rana, R.S. and Kalia, V. (2009). Weather change related shift in apple belt in Himachal Pradesh. In. *In Climate change and Indian Agriculture* (Editor: P.K. Aggarwal) pp. 48–53.

Byjesh, K., Naresh Kumar, S. and Aggarwal, P.K. (2010). Simulating impacts, potential adaptation and vulnerability of maize to climate change in India. *Miti. Adap. Strat. for Global Change.* 15: 413–431.

Das *et al.* (2013). Effect of climate change on fresh water fisheries In *Network Project on Climate Change* Final Report (Editor: Naresh Kumar ed).

FAO (2016). Climate change and food security: Risks and responses, p. 98.

Hebbar, K.B., Venugopalan, M.V., Prakash, A.H. and Aggarwal, P.K. (2013). *Climate Change*, 118(3), 701–713.

INCCA (2010). Climate change and India: A 4×4 assessment—A sartorial and regional analysis for 2030s: Chapter on agriculture, MoEFpub. pp. 67–88.

IPCC (2013). Assessment Report 5 WG I report—Climate Change: The Physical basis.

IPCC (2014). Assessment Report 5 WG II report on Climate Change: Impacts, Vulnerability and Adaptation.

Naresh Kumar, S., Anuja, Rashid, Md., Bandyopadhyay, S.K., Padaria, Rabindra and Khanna, Manoj (2016). Adaptation of farming community to climatic risk: Does adaptation cost for sustaining agricultural profitability? *Curr. Sci.* 110(10): 1216–1224.

Naresh, Kumar S., Govindakrishnan, P.M., Swarooparani, D.N., Nitin, Ch., Surabhi, J. and Aggarwal, P.K. (2015). Assessment of impact of climate change on potato and potential adaptation gains in the Indo-Gangetic Plains of India. *International Journal of Plant Production* 9(1), 151–170.

Naresh Kumar Soora, Pramod Kumar Aggarwal, Kumar Uttam, Jain Surabhi, Swaroopa Rani, D.N., Nitin Chauhan and Rani Saxena (2014a). Vulnerability of Indian mustard (*Brassica juncea* (L.) Czernj. Cosson) to climate variability and future adaptation strategies. *Miti. Adap. Strat. for Global Change.* 10.1007/s11027-014-9606-Z.

Naresh Kumar S., Aggarwal, P.K., Swarooparani, D.N., Rani Saxena, Nitin Chauhan and Surabhi Jain (2014b). Vulnerability of wheat production to climate change in India. *Climate Research.* doi: 10.3354/cr01212, Vol. 59: 173–187, 2014.

Naresh Kumar, S., Aggarwal, P.K., Rani Saxena, Swaroopa Rani, Surabhi Jain and Nitin Chauhan (2013). An assessment of regional vulnerability of rice to climate change in India. *Clim. Change.* DOI 10.1007/s10584-013-0698-3.118 issue 3–4 June 2013. pp. 683–699.

Naresh Kumar, S. and Aggarwal, P.K. (2013). Climate change and coconut plantations in India: Impacts and potential adaptation gains. *Agril. Syst.* http://dx.doi.org/10.1016/j.agsy.2013.01.001.

Naresh Kumar, S., Aggarwal, P.K., Swaroopa Rani, Surabhi Jain, Rani Saxena and Nitin Chauhan (2011). Impact of climate change on crop productivity in Western Ghats, coastal and northeastern regions of India. *Current Sci.* 101(3): 33–42.

Naresh Kumar, S., Singh, A.K., Aggarwal., P.K., Rao, V.U.M. and Venkateswarlu, B. (2012). Climate change and Indian Agriculture: Salient achievements from ICAR network project. IARI Pub., 32 p.

NIN (2011). Dietary Guidelines for Indian—A Manual, p. 127.

Porter, J.R., Xie, L., Challinor, A.J., Cochrane, K., Howden, S.M., Iqbal, M.M., Lobell, D.B. and Travasso, M.I. (2014). Food security and food production systems. In: *Climate change 2014: impacts, adaptation, and vulnerability. Part A: global and sectorial aspects*, pp. 485–533.

Ramachandran (2012). Food & nutrition security: Challenges in the new millennium. *Indian J Med Res* 138, September 2013, pp. 373–382.

Rao, N.K.S., Laxman, R.H. and Bhatt, R.M. (2009). Impact of elevated CO_2 and temperature on growth and yield of onion and tomato, In: *Climate change and Indian Agriculture* (P.K. Aggarwal ed.) pp. 35–37.

Reddy *et al.,* (2013). Effect of climate change on poultry, In: Network Project on Climate Change Final Report (Naresh Kumar ed.).

Upadhyay *et al.* (2013). Effect of climate change on livestock, In: Network Project on Climate Change Final Report (Naresh Kumar ed.).

Vivekanandan *et al.* (2013). Effect of climate change on marine fisheries, In: Network Project on Climate Change Final Report (Naresh Kumar ed.).

Climate Change and its Impact on Pest Control and Pesticide use in Ensuring Food Security

Padmaja R. Jonnalagadda

Food & Drug Toxicology Research Centre, National Institute of Nutrition, Indian Council of Medical Research, Tarnaka, Hyderabad 500 007, India
E-mail: jprambabu@yahoo.com

ABSTRACT: *Excessive deforestation, urbanization and industrialization enhanced anthropogenic activity in the CO_2 emission in the atmosphere. These changes in CO_2 emission lead to increase in the temperature and thereby causing global warming. Further, these influence the outbreaks of pests/pest-related diseases in agricultural crops, efficiency of the natural enemies and ultimately impacting agricultural production and food security. Additionally, the impact due to climate change on the fisheries, livestock and biological systems and humans is also no less than as compared to others. Therefore, attention is required to combat the climate change and to mitigate the associated risks to ensure food security.*

Keywords: Global Warming, Natural Enemies, Agricultural Production, Pests, Biological Systems.

Deforestation, urbanization and industrialization have led to an increasing trend in the concentration of CO_2 levels. As a result, adverse effects on the weather which further intensified hot/cold waves, thus delayed onset/insufficient rainfall/excessive monsoon etc. All these adverse effects have resulted in frequent droughts/floods, etc. Further, on many biological systems and medical systems and ultimately on humans. Changes in climate also have impact on global warming. Global warming and climate changes are independent factors and considered as man made activities. As per the reports on climate change, there is an increase in the temperature by 0.85°C (0.65 to 1.06°C) during the period 1880–2012 and CO_2 concentration from 280 ppm to 401 ppm in 2015 (War *et al.*, 2016).

Since recent past, one of the defining factors for food security is climate. With the increase in the demand for energy, food and due to the intense agricultural practices, there is also a change in the pattern of plant diseases due to newer pests. One of the major challenges ahead in the coming centuries is minimizing the risk to food security due to changes in the climate.

Agricultural crops are posed to risk by about 10,000 insect pests causing an annual global loss of 13.6%, of which India accounts for 23.3% loss. The global climate change on pest and disease epidemics is crucial in assessing the future spread and development and ultimately affects the global crop production (War *et al.,* 2016).

EFFECT OF CO_2

There has been an increase in the atmospheric concentration of CO_2 by about 35% over the past 200 years (Sharma and Prabhakar, 2014). Sudden emergence of insect pests like *Spodoptera* species due to higher levels of CO_2 was noticed and some such examples are wooly aphids in Maharashtra, West Bengal, and in central and south India too, due to climate change. Sometimes, the climate change may protect the crops and however, many times it may be destructive and damage the crops. The information on the long term effects of change in CO_2 and temperature either in single or in combination is limited. Review of the literature on elevated levels of CO_2 indicated that some of the insects may either be unaffected or decreased and do not increase, while others like sap sucker density may increase (Reddy, 2013). One of the interesting observations made by the earlier workers, that the *Helicoverpa armigera* on spring wheat was developed under elevated levels of CO_2 and also lived longer, but showed significant reduction in number of eggs laid. However, another study revealed that the damage by the cotton bollworm, was more due to elevated CO_2 to the cotton crop (Reddy, 2013). The increased rate of biomass due to elevated CO_2 declines the decomposition rate of litter paving the way for prolonged survival period for pathogens, thereby making them more conducive for the availability of inoculum to cause subsequent infections in the plant. Increased CO_2 may also increase the efficiency of plants using more water, which suits many of the herbivorous insects, due to the assimilation of nutrients (Fand *et al.,* 2012).

EFFECT OF TEMPERATURE AND RELATIVE HUMIDITY

Increasing temperature may act on the physiology of the plant leading to stress especially during its time of growth. This growth again depends on the emergence of new breeds, prolonged time taken for the plant growth, increased use of pesticides with the increase in the host susceptibility, host resistance and changing time in sowing the seeds and changing pattern of the crops. Excessive increase in the temperature and direct exposure to sunlight may pave way to newer pests and their niches. Further, as the insects do not have any mechanism of regulation in temperature, their development depends on the external variation in temperature. Delayed effects such as sowing/ploughing with varying temperature and relative humidity due to climate change may influence the occurrence of insect pests causing intensive damage due to *Helicoverpa* and *Spodoptera* species in pigeon pea plants. The crops undergo tremendous stress, due to the reduction in the pesticide activity, may necessitate for the use of different/newer molecules of pesticides (War *et al.,* 2016). Humidity due to climate change affects many of the pest management programe like natural plant products, efficacy of synthetic chemicals, etc. In a topical country like India, the elevated temperature will lead to the rapid growth of nematode population due to the warm temperature in the soil. However, disease development in the plants due to higher temperature varies with different agro ecological zones. Some of the wheat plants showed more susceptibility to rust diseases (Petzoldt and Seamman, 2010).

IMPACT ON INSECT PESTS OF AGRICULTURE/ HUMAN IMPORTANCE

In the animal kingdom of invertebrates, Phylum Insecta is the largest group. They are also considered as good indicators of environmental changes, as well as pollinators for many crops. Climate change will have a direct effect on the biology of insects and indirect effect on the hosts (Sharma and Prabhakar, 2014). Some of them are rapid reproduction rate and virulence in the pest population, behavioural changes in the pests, increasing rate of biodiversity loss, and biotic stress etc. lead to increasing burden on the agricultural crops. The biodiversity loss will not only leave an adverse effects on the biological and ecosystems, but also escalates disease outbreaks of insect pests leading to destruction of the agricultural crops/yields. The global climate change on the dynamics of the pests in tropics depends on the duration of both wet and dry seasons (War *et al.,* 2016). Change in climate had drastic impact on changes in the phenology of *lepidopteran* insects and their arrival times, the nutritional resources of the hosts, predators and also the parasitoids. Further, it is necessary to understand response and relationship between the pests and natural enemies due to climate change, as, this, changes due to global warming. There seems to be a decrease in the rate of predation and parasitism of these herbivores by natural enemies, as the temperature and CO_2 influence the changing lifestyle of herbivores. Further, increase in the foliage pattern of plants, increase in the number of foliage pathogens (rusts, mildews, leaf spots and blights), the detection of pests by natural enemies was also a limiting factor, and play an important role in contributing to human economy. Change in the climate not only influence the effective functioning of the natural enemies in reducing the pests, but also the susceptibility of plants to such pests/ herbivores through increased CO_2 levels in reducing some of the aphid parasitoids viz., *Lysiphlebia japonica,* but some other like *Aphidiuspicipes* showed increased rate. There are reported evidences of higher rate of nematodes like *Radopholus* species causing damage to banana at an elevated temperature (Reddy, 2013). It also led to mismatch and synchrony with the host and parasitoid selection. Some studies have also found variation in the selection of the food from leaves to plants under elevated levels of CO_2 (Whittaker, 1999).

EFFECT ON PESTICIDE USE

Changing climate will also have an impact on the pesticides, their effective functioning on the targeted pests and on their residues on the agricultural crops and on the health of consumers. Excessive rainfall in monsoon due to their changing levels may damage the younger plants. There appears to be an impact on the absorption, uptake and decrease in the transportation of the pesticides in order to counter the target pests due to precipitation. This eventually decreased the yields. Some of the synthetic pyrethroids and organophosphorus pesticides, because of their high unstable nature, do degrade at a rapid rate due to elevated temperatures. As a result, the efficacy of their functioning in controlling the targeted pests may be reduced and thereby leading to the increased rate of their applications on the crops more than what is required. This is quite evident in the case of pests attacking the sweet corn crop. With the increased number of applications, there may also be an increase in the resistance of the pests to pesticides (Petzoldt and Seaman, 2010). Increase in the moisture will infect some of the vegetable root plants with late blight. Precipitation due to climate change may be reduced with the application of fungicide, however, the increased rainfall reduced the sustenance of the fungicide residues on the foliage of the plants, thereby leading to its repeated/frequent applications (Reddy, 2013).

EFFECT ON AGRICULTURAL YIELD

There seems to be a significant impact on crop area and crop production with respect to the change in climate (FAO, 2011). As per the estimates, it is expected to increase in the harvested irrigated area by more than double in another two decades in the developing countries resulting in increased demand for water. Further, due to degradation of natural enemies, decaying pattern of pest population, attack due to sudden emergence/range/newer pests, interaction between pests and natural enemies and frequent pest-related disease outbreaks at every

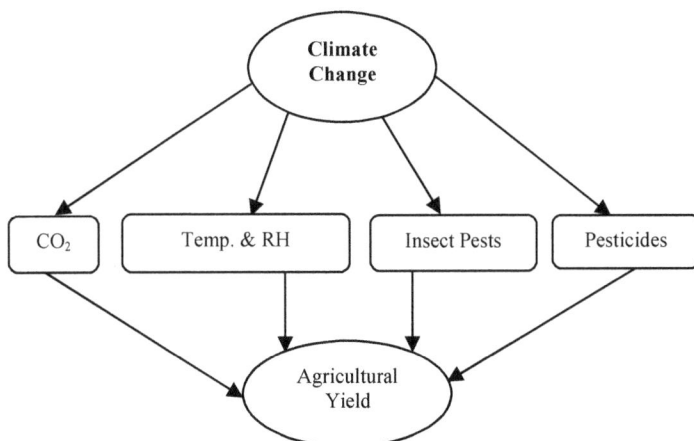

Fig. 1: Impact Due to Climate Change on Agricultural Yield

stage, impact the crop production. As per the millennium ecosystem assessment report (2005), about one-third of the species has become extinct and degradation of 60% of earth ecosystem over the past half century. Of the 99% of all the species which have become extinct, the Asia's risk of total biodiversity is about 50% due to climate change (Fand *et al.,* 2012). A wide variety of rare, exotic and endemic species of butterflies in the world appears in Western Ghats of India, but such species are now at risk due to climate change.

STRATEGIES

- Need to create awareness among the farmers/farm workers, stakeholders who are playing a key role in the food supply/production chain on the impact of climate change through extensive education.
- Adaptation of strategies to combat the incidences on emergence of newer pests/disease outbreaks through Krishi Vigyan Kendras.
- To provide the education material and guidelines to different types of risks associated through climate change.
- To ensure and enrich the farmer's adaptive capacity through integration mechanism such as making them involve in multi-disciplinary approach/ activities by way of communication between researchers and farmers.
- To focus the research more on the protection of the agricultural crops from the emergence of resistant varieties of insects belonging to different classes in order to reduce the plant infestation rates and to ensure the food security.

CONCLUSION

Though studies on the effect of climate change on the insect pests and food security (Birch *et al.,* 2011; Popp *et al.,* 2012; Delcour *et al.,* 2015; McLaughlin and Kinzelbach, 2015; Nadal *et al.,* 2015; Campbell *et al.,* 2016) have been reported earlier, more attention is required on the impact of climate change on livestock, fisheries production systems, emergence of newer virulent strains with resistance to the plant protection measures and finally to food security dimensions. It is more evident from the review of literature that, change in the climate has an impact on the pests, hosts and biodiversity, crop production and food security. Therefore, in order to combat the impact due to climate change, it is essential to focus on the research to address the adaptation mechanism through development of various intrinsic factors to counteract the pests attack. Also, necessary steps need to be strengthened at the production and consumer levels to deliver the quality foods as part of food safety and security. There needs to be a multidisciplinary approach to focus the research in understanding the host–pathogen interactions, making the natural enemies more effective, insect–plant resistance, corrective measures in the adaptation of pest management practices and proper protective measures in order to combat the climate change and to ensure the food security.

REFERENCES

Birch, E.A.N., Begg, G.S. and Squire, G.R. (2011). How agro-ecological research helps to address food security issues under new IPM and pesticide reduction policies for global crop production systems. *Journal of Experimental Botany*, 62(10), pp. 3251–3261. doi: 10.1093/jxb/err064.

Campbell, B.M., Vermeulen, S.J., Aggarwal, P.K., Corner-Dolloff, C., Girvetz, E., Loboguerrero, A.M., Ramirez-Villegas, J., Rosenstock, T., Sebastian, L., Thornton, P. and Wollenberg, E. (2016). Reducing risks to food security from climate change. *Global Food Security*, doi: 10.1016/j.gfs.2016.06.002.

Delcour, I., Spanoghe, P. and Uyttendaele, M. (2015). Literature review: Impact of climate change on pesticide use. *Food Research International*, 68, pp. 7–15. doi: 10.1016/j.foodres.2014.09.030.

Fand B., B., Kamble, L., A. and Mahesh, K. (2012). Will climate change pose serious threat to crop pest management: A critical review? *International Journal of Scientific and Research Publications*, 2(11).

Food and Agriculture Organization (FAO) (2011). Crop Prospects and Food Situation.

McLaughlin, D. and Kinzelbach, W. (2015). Food security and sustainable resource management. *Water Resources Research*, 51(7), pp. 4966–4985. doi: 10.1002/2015 wr017053.

Nadal, M., Marquès, M., Mari, M. and Domingo, J.L. (2015). Climate change and environmental concentrations of POPs: A review. *Environmental Research*, 143, pp. 177–185. doi: 10.1016/j.envres.2015.10.012.

Petzoldt, C. and Seaman, A. (2010). Climate change effects on insects and pathogens, New York State Agricultural Extension Station, Geneva, NY 14456.

Popp, J., Pető, K. and Nagy, J. (2012). Pesticide productivity and food security. A review. *Agronomy for Sustainable Development*, 33(1), pp. 243–255. doi: 10.1007/ s13593-012-0105-x.

Reddy, P.P. (2013). Impact of climate change on insect pests, pathogens and nematodes, *Pest management in horticultural ecosystems*, 19(2), pp. 225–233.

Sharma, C.H. and Prabhakar, S.C. (2014). Impact of climate change on Pest Management and Food Security, In, *Integrated Pest Management: Current Concepts and Ecological Perspectives*. Edited by Dharam P Abrol. Elseiver Inc.

War, A.R., Taggar, G.K., War, M.Y. and Hussain, B. (2016). Impact of climate change on insect pests, plant chemical ecology, tritrophic interactions and food production. *International Journal of Clinical and Biological Sciences*, 1(2), pp. 16–29.

Whittaker, B.J. (1999). Impact and responses at population level of herbivorous insects to elevated CO_2. *European. J. Entomol.*, 96, pp. 149–156.

Vulnerability of Farming Systems— Understanding Climate Impacts at Local Scale Encompassing Uncertainties

V. Geethalakshmi*, A.P. Ramaraj, R. Gowtham,
K. Bhuvaneswari, N. Manikandan, K. Senthilraja,
K. Vinoth Kumar and S. Panneerselvam

Tamil Nadu Agricultural University, Coimbatore 641 003, India
*E-mail: geetha@tnau.ac.in

ABSTRACT: *Climate change presents a challenge for researchers attempting to quantify its local impact due to the global scale of likely impacts and the diversity of agricultural systems. Simulation of the effects of climate change on agriculture usually employs a climate model coupled with a mechanism-based crop growth model. Furthermore, a full assessment of the climate change impacts on agricultural production should take comprehensively into account available adaptation strategies, and account for the uncertainties from many physical, biological, and social-economic processes. The work presented here builds upon, and largely supersedes, the study of climate change impact assessment at local scale considering uncertainties associated with it. The study presented here is a part of Agricultural model Inter-comparison and Improvement Project (AgMIP). Procedures followed here are developed by the AgMIP climate modeling team. The maximum and minimum temperature over Trichy is projected to increase while rainfall is projected to vary widely. The uncertainty in projections of climate models and its cascading effect on crop models were analyzed to understand their variability and to bring certainty to the projections through multi-model integrated approach.*

Keywords: Agriculture, Assessment, Uncertainty, Adaptation.

Climate change is one of the most pressing global problems of recent times and it has become a more complex issue than it was before. While climate change is a global problem, the impacts are felt by the human society mainly on regional to local scales (IPCC, 2007). Adverse effects of climate change are expected to affect in particular the poorest people because of their high vulnerability and low adaptive capacity. Furthermore, climate change impacts are manifold and multidimensional, and hence, developing and implementing adequate adaptation measures is complex (Barnett *et al.*, 2005). Thus, it is of great importance to understand and project the nature and magnitude of the changes in the climate. In order to provide a basis for adaptation and reduction of adverse impacts, climate change impact assessments at the regional to local level are needed.

Agriculture is inherently sensitive to climate conditions, and is one of the sectors most vulnerable to the risks and impacts of global climate change (Reilly 1995; Smith and Skinner 2002). The long-term challenge of avoiding a perpetual food crisis under conditions of global warming is serious. Adaptation is a key factor that will shape the future severity of climate change impacts on food production (Lobell *et al.*, 2008), and has recently received increasing attention.

Each of the last three decades has been successively warmer at the Earth's surface than any preceding decade since 1850. The period from 1983 to 2012 was likely the warmest 30-year period of the last 1400 years in the Northern Hemisphere. The globally averaged combined land and ocean surface temperature data show a warming of 0.85°C over the period 1880 to 2012 (IPCC, 2014). Natural hazards are purely natural, but climate change may exacerbate it. Anthropogenic interventions convert hazards into disasters (Mitra, 2007).

Anticipated projections based on the Coupled Model Intercomparison Project Phase 5 (CMIP5) ensemble, global mean surface temperature rise for the late 21st century relative to the 1986–2005 period is assessed. For the scenario RCP 2.6, it is likely between 0.3 and 1.7°C with a mean increase of 1°C. For RCP 4.5, an increase of 1.1–2.6°C with a mean increase of 1.8°C and an increase of 1.4 to 3.1°C with mean increase of 2.2°C for RCP 6.0 was projected. RCP 8.5 had the highest range with a range of 2.6 to 4.8°C and mean increase of 3.7°C.

Impact of climate change on agriculture will be one of the major deciding factors influencing the future food security of mankind on earth. Climate change is a global problem and India will also feel the heat, nearly 700 million rural people in India directly depend on climate sensitive agriculture sector (Mitra, 2007). In order to provide a basis for adaptation and reduction of adverse impacts, climate change impact assessments at the regional to local level are needed. There are uncertainties in each of the stages for assessing the impacts, vulnerability and adaptation options (Mearns *et al.*, 2001) and should be accounted.

Future climate projections were created by utilizing a "delta" approach, in which the mean monthly changes (from baseline) were applied to the daily baseline weather series as described by Villegas and Jarvis (2010). These scenarios of future projections were referred to as "mean change scenarios". Uncertainty in climate projections over the study location was analysed from all the climate models for RCP 4.5 and RCP 8.5. The daily output of the models was converted to decadal, seasonal *viz.,* southwest monsoon (June, July, August, September) and northeast monsoon (October, November, December).

The deviations from the base year (1980–2010) were calculated by obtaining the difference between the near, mid and end century with the base years. The deviations were calculated for each location using all the models and then the maximum, minimum and average deviations were computed. These ranges of maximum, minimum and average are given as uncertainty in climate projections. Climate and crop integrated assessment (IA) was carried out through interlinking the climate and crop models through scenario data generated. The future climate projections were injected into crop simulation models for simulating the crop yields to study the uncertainty cascaded in and to test the various adaptation strategies.

FUTURE PROJECTIONS FOR MID CENTURY THROUGH SELECTED GCMS

Maximum temperature of Trichy is projected to increase by all the scenarios and models studied with variation in their magnitude. For the scenario RCP 4.5, annual increase in maximum temperature is projected to vary between an increase of 0.8°C (inmcm4) to 1.9°C (CanESM2) by the end of mid century while 1.0°C (inmcm4) to 3.0°C (CanESM2) was projected by RCP 8.5. During SWM, an increase of 0.6°C (inmcm4 and MIROC 5) to 1.7°C (CanESM2) through RCP 4.5 by the end of mid century while 0.7°C (MIROC 5) to 2.5°C (CanESM2) was projected by RCP 8.5. During NEM, an increase of 0.5°C (inmcm4) to 1.9°C (CanESM2) through RCP 4.5 by the end of mid century while 0.9°C (inmcm4) to 2.9°C (CanESM2) was projected by RCP 8.5.

Minimum temperature of Trichy is projected to increase by all the scenarios and models studied with variation in their magnitude. For the scenario RCP 4.5, annual increase in minimum temperature is projected to vary between an increase of 0.5°C (inmcm4) to 1.8°C (CanESM2, IPSL-CM5A-MR and HadGEM2-AO) by the end of mid century while 1.3°C (inmcm4) to 2.6°C (IPSL-CM5A-MR) was projected by RCP 8.5. During SWM, an increase of 0.6°C (inmcm4) to 1.8°C (HadGEM2-AO) through RCP 4.5 by the end of mid century while 1.1°C (inmcm4) to 2.5°C (IPSL-CM5A-MR) was projected by RCP 8.5. During NEM, an increase of 0.5°C (inmcm4) to 2.0°C (IPSL-CM5A-MR) through RCP 4.5 by the end of mid century while 1.0°C (inmcm4) to 2.6°C (IPSL-CM5A-MR) was projected by RCP 8.5.

Rainfall of Trichy was projected through all the GCMs studied. Annual and seasonal rainfall deviation from observed period was worked out to understand the future changes. For RCP 4.5, annual rainfall of Trichy was projected to have variation ranging from a decrease of –2 per cent (CanESM2 and inmcm4) to an increase of +28 per cent (MIROC 5) by the end of mid century while –9 (CanESM2) to +28 (IPSL-CM5A-MR) per cent variation was projected through RCP 8.5. During SWM, a decrease of –1 (CanESM2) to 36 per cent (MIROC 5) through RCP 4.5 while about +1 (inmcm4) to +51 (MIROC 5) per cent variation was projected through RCP 8.5. During NEM, a decrease of –4 (inmcm4) to +38 per cent (IPSL-CM5A-MR) through RCP 4.5 while about –14 (CanESM2) to +52 (IPSL-CM5A-MR) per cent variation was projected through RCP 8.5.

INTEGRATED CLIMATE-CROP ASSESSMENT

Five representative models viz. MIROC 5 (O), inmcm4 (L), CanESM2 (D), IPSL-CM5A-MR (N) and HadGEM2-AO (Y) to signify cool/wet, cool/dry, hot/wet, hot/dry and Middle conditions respectively were selected from the 29 GCMs for impact assessment under mid century 8.5 scenario. Impact of maize yield under current and future climate is presented in Figure 1.

Both DSSAT and APSIM model simulations under RCP 8.5 climate conditions predicted the yield decrease of maize with varying magnitude for all the five models. Forcing CanESM2 (hot and wet) model with APSIM showed a deviation in maize productivity by (–)8 to (+)5 per cent under different farm conditions, while with DSSAT, the deviation was from (–)11 to (–)22 per cent. The inmcm4 model representing cool/dry condition indicates a deviation of (–)3 to (+)7 per cent for APSIM and (–)2 to (–)7 per cent for DSSAT model. The IPSL-CM5A-MR (hot and dry climatic condition) forcing showed (–) 2 to (–)8 per cent deviation in maize yield for APSIM model and (–)10 to (–)33 for DSSAT model. Forcing of the cool and wet climatic condition (MIROC 5) model showed (–)4 to (+)14.6 per cent deviation in maize yield for APSIM model and (–)2 to (–)7 per cent deviation for DSSAT model. The HadGEM2 model representing the middle quadrant indicated (–)6 to (+)9 per cent deviation in maize yield for APSIM model and for DSSAT (–)8 to (–)19 per cent deviation. These variations in yield were attributed to the climate projected by the models and their interaction with the crop models.

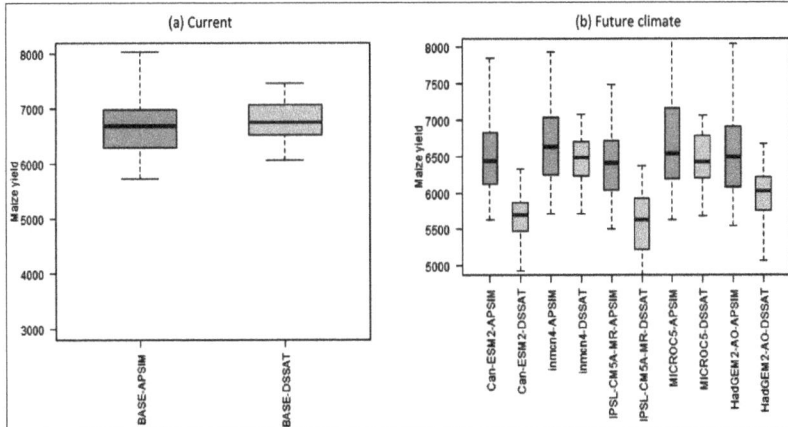

Fig. 1: Impact of Maize Yield

REFERENCES

Barnett, T.P., Adam, J.C. and Lettenmaier, D.P. (2005). Potential impacts of a warming climate on water availability in snow-dominated regions, Nature, 438, 303–309.

IPCC (2007). Summary for Policymakers. In: Climate Change 2007: The Physical Science Basis. Contribution of Working Group I to the Fourth Assessment Report of the Intergovernmental Panel on Climate Change [Solomon, S., D. Qin, M. Manning, Z. Chen, M. Marquis, K.B. Averyt, M. Tignor and H.L. Miller (eds.)]. Cambridge University Press, Cambridge, United Kingdom and New York, NY, USA.

Lobell, D.B., Burke, M.B., Tebaldi, C., Mastrandrea, M.D., Falcon, W.P. and Naylor, R.L., (2008). Prioritizing climate change adaptation needs for food security in 2030. *Science,* 319: 607–610.

Mearns, L.O., Hulme, M., Carter, T.R., Leemans, R., Lal, M. and Whetton, P. (2001). Climate Scenario Development in: IPCC WG1 TAR, 2001.

Mitra, S. (2007). Climate change will pose new challenges to disaster management. *Cur. Sci.,* 92(11): 1474.

Reilly, J. (1995). Climate change and global agriculture: Recent findings and issues. *American Journal of Agricultural Economics,* 77, 727–733.

Smith, B. and Skinner, M. (2002). Adaptation options in agriculture to climate change: A topology. *Mitigation and Adaptation Strategies for Global Climate Change,* 7, 85–114.

Villegas, J.R. and Jarvis, A. (2010). Downscaling Global Circulation Model Outputs: The Delta Method Decision and Policy Analysis Working Paper No. 1. International centre for tropical agriculture.

Inherent Vulnerability Assessment of Agriculture Communities in the Himalayas

P.K. Joshi

School of Environmental Sciences, Jawaharlal Nehru University,
New Delhi 110 067, India
E-mail: pkjoshi27@hotmail.com; pkjoshi@mail.jnu.ac.in

ABSTRACT: *The current practices of vulnerability assessment for agriculture communities in the Himalayas and the findings of village level assessment (n = 15,285) carried out for the entire state of Uttarakhand are presented here. First, a systematic review of peer-reviewed literature was done to gauge the current approaches and methods being used in vulnerability assessment. Following which, a holistic inherent vulnerability assessment of agriculture communities was done using data on 36 indicators, reflecting social (n = 22) and ecological (n = 14) dimensions of sensitivity and adaptive capacity. Analytical hierarchical process (AHP) was used for weighing the indicators, and finally aggregated to map spatial distribution of inherent vulnerability under five classes and identify the hotspots of inherent vulnerability. Overall the state observes high inherent vulnerability, with about 23.6% and 24.7% villages classified under very high and high vulnerability class respectively. The results highlight presence of very high biophysical vulnerability (0.82 ± 0.10) and high social vulnerability (0.65 ± 0.15). The spatial pattern of both biophysical and social vulnerability shows significant altitudinal gradient with most of the vulnerability hotspots villages located in middle altitudinal zone. Absence of dense forest, local institutions, negligible infrastructure facilities and high occupational dependence on agriculture were highly correlated with presence of high vulnerability. The findings would assist the policy interventions in prioritizing allocation of resources to enhance the capacities of agricultural communities in the identified hotspots. Additionally, the adaptation programmes in the region need to be more context-specific to accommodate the differential altitudinal vulnerability profiles.*

Keywords: Social Vulnerability, Biophysical Vulnerability, Village Level, AHP, Himalaya.

Agriculture forms the main source of livelihood and food security for the rural communities living in the most marginalized and remote mountain regions (Tiwari and Joshi, 2015). Agriculture in the Himalayan region, mainly subsistence-based, is heavily dependent on natural resources like forests and rainfall for its stability and long-term sustainability (Singh *et al.*, 2015). This high, often absolute, dependence on natural resources poses severe inherent limitations on agricultural productivity. In recent times, multiple stressors (variability in climate, environmental degradation, deforestation) acting along at different scales (global, regional and local) have adversely affected the stability of mountain agriculture systems. The declining stability and productivity of the mountain agriculture system are in turn eroding the local socio-cultural harmony (Chapagain and Gentle, 2015). Extensive loss in agro-biodiversity, agricultural production, livelihoods and changes in food consumption patterns have often been reported in the region. Therefore, to minimize the risks of these stressors, there is a need to holistically understand the present distribution of vulnerability of the agriculture communities, which depend on these social and ecologically coupled agriculture systems.

Vulnerability of a system determines its susceptibility to any harm and is often equated to concepts like marginalization, fragility and risk. The fifth Assessment Report (AR 5) of Intergovernmental Panel for Climate Change (IPCC) defines vulnerability as the 'propensity or predisposition to be adversely affected'. Broadly two approaches exist for vulnerability conceptualization, one that considers vulnerability to be a pre-existing state of a system (starting point or contexual) and the other that relates vulnerability to be an outcome (end point or outcome) of a hazard. Vulnerability analysis is often done using indicator-based approaches (O'Brien *et al.,* 2004; Rygel *et al.,* 2006; Preston *et al.,* 2011). The main steps of an indicator-based vulnerability assessment are selection of indicators using a defined vulnerability framework, data normalization, weighing, aggregation and mapping (Perston *et al.,* 2011). The advantage of using indicator-based approach is that the final composite index developed using this approach appropriately captures the multi-dimensional nature of vulnerability. Nonetheless, many studies have highlighted the challenges associated with weighing individual indicators as the weights of the indicator determine its relative importance in determining vulnerability (O'Brien *et al.,* 2004; Rygel *et al.,* 2006). Expert judgment-based approaches, which generally utilize multi criteria decision analysis (MCDA) such as analytical hierachical process AHP, can effectively address these weighing issues (Eakin and Tapia, 2008).

In this paper, first, we document the current knowledge on vulnerability of agriculture communities inhabiting the Himalayas through a systematic review of peer-reviewed literature. Further, we assess the social and the biophysical vulnerability of the agriculture communities in the state of Uttarakhand. Finally, we assess the distribution of inherent vulnerability and its components, i.e. sensitivity and adaptive capacity for the entire state.

Uttarakhand state is located between 28°43′24″ to 31°27′50″ N latitude and 77°34′27″ to 81°02′22″ E longitude covering a geographical area of 53484 km². According to 2011 census, there are 13 districts, 78 tehsils and 16826 villages in the state. The state supports a total population of 10.11 million, of which 70% is rural. Agriculture is the dominant livelihood sector despite the harsh geophysical conditions and ecological fragility of the region (Tiwari and Joshi, 2015). Agriculture is practiced on terraced agriculture fields (except in the plain lowland regions) and is mainly rain-fed, as only 45% of the net sown area is irrigated. Agriculture productivity and agro-diversity vary with the altitudinal gradient and are further influenced by ecological factors. Based on elevation gradient, three broad altitudinal zones are identified in the region namely lower (up to 1200 m), middle (1200–1700 m) and upper zones (above 1700 m).

Majority of the data representing demographic attributes was collected from village directory provided by the Census of India (2011). Data for few indicators was derived from remote sensing outputs such as elevation, aspect, slope, NDVI and NDWI. Data for financial capital was based on 2001 census due to the unavailability of any recent data.

A total of 15285 villages were selected (excluding un-inhabited and forest villages) for assessing vulnerability of agriculture communities. For indicator selection, initially a list of indicators was prepared based on the vulnerability conceptual framework. Final selection of indicators was based on experts' suggestions and availability of data (at village level).

Composite social (22 indicators, 8 for sensitivity and 14 for adaptive capacity) and biophysical (14 indicators, 9 for sensitivity and 5 for adaptive capacity) vulnerability index was calculated using the respective indicators. Data normalization was done using linear (minimum–maximum) scaling method, where all the normalized values lie between 0 and 1. As different indicators contribute differently to vulnerability, the study uses a hierarchical analysis design for calculating the weight of each indicator using AHP (Saaty, 1980). Based on the literature review and expert ratings, relative importance of each indicator was marked. A hierarchical weighted summation method was used to aggregate the indicators to compute the final vulnerability index (Figure 1). Two composite indices namely, sensitivity index and adaptive capacity index were also computed to understand their distribution across the villages in the study area. These indices were classified under five classes namely very low, low, moderate, high and very high based on quantile method of classification.

Spatial auto-correlation estimate, Local Indicator of Spatial Association (LISA), was used to identify the hotspots of inherent vulnerability. LISA identifies four association clusters namely, high–high (HH), high–low (HL), low–high (LH) and low–low (LL). The robustness of final vulnerability rank was assessed by consecutively excluding all the indicators under each sub-dimension using local sensitivity analysis. The mean change in the vulnerability ranks by exclusion of specific sub-dimension was plotted to identify the most influencing sub-dimension.

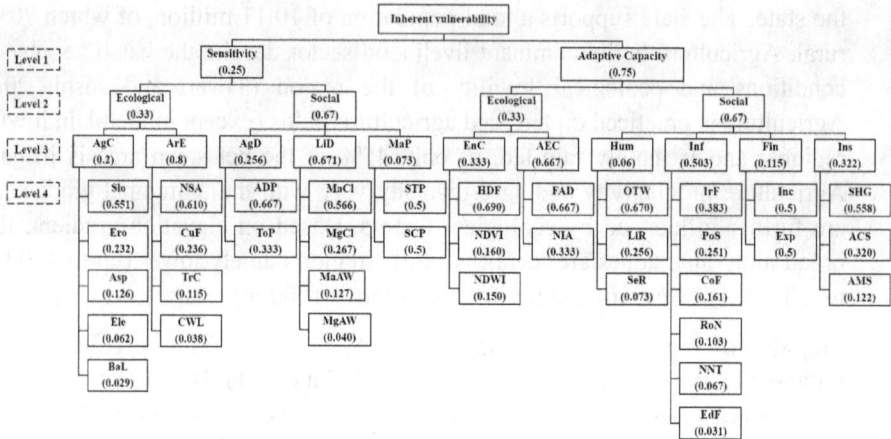

Fig. 1: The Four-Level Hierarchical Structure of Inherent Vulnerability Components (level 1), Dimensions (level 2), Sub-dimensions (level 3) and Selected Indicators (level 4)

CURRENT KNOWLEDGE AND GAPS

The review revealed that the highest number of vulnerability assessments focusing on agriculture communities has been in Nepal (~54%), followed by India (~34%). Single studies were available for Bhutan and the Himalayan parts of China (Tibet) and Pakistan. All assessments have been done after 2008 (60% between 2014 and 2016), indicating that the vulnerability assessment of the agriculture communities is rather a new topic of exploration in the Himalayan region. Assessments were most commonly done at local-level (62% studies). 50% of the papers have addressed vulnerability of agriculture communities to climate change and climate variability. Other stressors for which vulnerability was estimated are climate-induced hydrological changes (19%), droughts (4%), disasters (4%) and extreme events such as floods and landslides (4%). Socio-cultural stressors were addressed by only two of the studies mostly at local levels. Further, only one study addresses multiple stressors being experienced by the agricultural communities of the Himalayas.

BIOPHYSICAL AND SOCIAL VULNERABILITY

The mean of *biophysical vulnerability index* was high (0.824 ± 0.102). Such high mean value is reflexive of the critical state of the biophysical resources, which determine the productivity of agriculture systems. Biophysical vulnerability in lower zone is slightly higher, owing to greater proportion of agriculture fields and therefore more sensitivity to any external stress, as the assets that will be adversely affected, are higher. The biophysical adaptive capacity index is slightly higher in the upper zone due to the presence of dense forest, essential for the stability and productivity of mountain agriculture systems. The composite biophysical vulnerability index is significantly negatively correlated with the presence of high-density forest ($r > 0.7$, p-value = 0) and positively correlated with percent of net sown area ($r > 0.5$, p-value = 0).

The mean *social vulnerability index* was 0.651 (± 0.154). Social vulnerability score is highest in the middle altitudinal zones and lowest in the lower altitudinal zone. Most of the villages in the lower zone have low social sensitivity and high social adaptive capacity, therefore recording an overall low social vulnerability. Villages in the middle and upper altitudinal zones record higher social sensitivity, as occupational dependency on agriculture is high owing to limited livelihood options in these zones. Moreover, the values of social adaptive capacity in these zones are less, as compared to the lower altitudinal zone because of limited infrastructural capacity. Disparity exists in the availability of physical infrastructures like irrigation facilities, power supply for agriculture and road access. The social vulnerability index is significantly related with availability of Self-Help Groups (SHGs) ($r > 0.6$, $p = 0$), agricultural population density ($r > 0.4$, $p = 0$) and power supply for agriculture ($r > 0.4$, $p = 0$).

SENSITIVITY, ADAPTIVE CAPACITY AND INHERENT VULNERABILITY

The percentage distribution of villages based on the sensitivity, adaptive capacity and inherent vulnerability index is given in Table 1. The mean value of sensitivity index was 0.427 (± 0.13) and adaptive capacity index was 0.234 (± 0.18). The severity of vulnerability results from a combination of these high sensitivity and low adaptive capacity conditions. The mean value of the overall inherent vulnerability was 0.665 (± 0.15). Based on the results, Tehri Garhwal records the highest mean score of vulnerability followed by Garhwal, Rudraprayag, Almora and Bageshwar.

Table 1: Percentage Distribution of the Villages in Five Different Vulnerability Classess

	Very Low	Low	Moderate	High	Very High
Sensitivity Index	10.0	23.0	32.0	26.0	9.0
Adaptive Capacity Index	36.1	19.6	25.5	12.5	6.3
Inherent Vulnerability Index	7.4	18.3	26	24.7	23.6

The results of LISA analysis indicate that 23.19% of villages are classified in HH clusters, 16.03% of villages in LL cluster, 2.82% of villages in HL clusters and 4.97% villages in LH cluster type. The spatial distribution of HH cluster villages is mainly restricted to middle altitudinal zone. Disintegrating these hotspots based on their underlying social and ecological vulnerability dimensions provides an understating that these hotspots are characterized by very low social as well as ecological adaptive capacities. However, these hotspots differ in drivers of high sensitivity. Therefore, priority needs to be given to specific vulnerability-creating dimensions to effectively increase capacities of communities and reduce existing susceptibilities.

The results of local sensitivity analysis depict that elimination of adaptive capacity sub-dimensions leads to higher deviations in vulnerability rank. Among the six sub-dimensions of adaptive capacity, the exclusion of institutional capital leads to highest shift of ~16% in vulnerability class followed by the environmental capacity (14.6%) and agro-ecological capacity (10.5%). Out of the five sensitivity sub-dimensions, maximum shift in inherent vulnerability rank was induced by exclusion of area-exposed (6.5%) followed by livelihood dependency (4.5%). The results when seen in relation with the distribution of these sub-dimensions in the villages reveal that institutional capital, measured using the presence of SHGs, agriculture marketing and agriculture credit societies, is absent in 50% of the villages in the study region. Hence, its presence or absence strongly determines the rank of inherent vulnerability in the present study.

CONCLUSION

In summary, results from our review highlight that while formal vulnerability assessments have been attempted but their number remains very limited. There exists a clear gap in our understanding of vulnerability through integrated multiple-level assessment, which is utmost necessary, given the complexity of challenges that the agriculture communities face in the Himalayan mountains. Vulnerability assessments in the mountains need to acknowledge the complexity of relationship among vulnerability, poverty and environmental pressures. Therefore adopting a multi-stress approach is crucial. Given the data scarcity in the region, we need to consider synergies to maximize utility of various available datasets like census, participative and remote sensing data, so as to better inform adaptation planning.

Our study assessed the social, biophysical and inherent vulnerability of the agriculture communities using an indicator-based approach at village-level for the state of Uttarakhand. In order to appropriately weigh indicators according to their contribution in determining vulnerability of these communities, AHP method was used. The study recognizes that vulnerability of mountain agriculture communities arises due to inherent pre-conditions and is spread unevenly within communities. However, sharp contrast exists in their distribution across the three altitudinal zones. The study highlights that the overall score of biophysical and social vulnerability is high in the state. The study offers insights into the social and ecological determinants which contribute to high inherent vulnerability. The study also highlights the hotspots of inherent vulnerability where targeted policies for prioritizing adaptive management actions are immediately needed. Based on the findings of the study, it is recommended that there is a need to develop infrastructure facilities, promote establishment of local institutions like SHGs, create diversified livelihoods (conserving the present socio-cultural structures) and develop market linkages to foster adaptive capacity and reduce vulnerability of agriculture communities to the ongoing environmental, climatic, demographic and governance changes.

ACKNOWLEDGEMENT

Some of the text here overlaps with our earlier publications. Please refer *Applied Geography* (2016) 74:182–198 and *Journal of Mountain Science* (2016) 13: 2260–2271 for details.

REFERENCES

Chapagain, B. and Gentle, P. (2015). Withdrawing from Agrarian Livelihoods: Environmental Migration in Nepal. *Journal of Mountain Science,* 12:1–13.

Eakin, H. and Tapia, B., 2008. Insights into the composition of household vulnerability from multi criteria decision analysis. *Global Environmental Change,* 18:112–127.

Goverment of Uttarakahnd (GoU) (2014). State Action Plan on Climate Change "Transforming Crisis into Opportunity" Government of Uttarakhand.

Maikhuri, R.K. and Ramakrishnan, P.S., 1990. Ecological analysis of a cluster of villages emphasising land use of different tribes in Meghalaya in North-east India. *Agriculture, Ecosystems and Environment,* 31:17–37.

O'Brien, K., Leichenko, R., and Kelkar, U., Venema, H., Anandahl, G., Tompkins, H., Javed, A., Bhadwal, S., Barg, S., Nygaard, L. and West, J. (2004). Mapping vulnerability to multiple stressors: climate change and globalization in India. *Global Environmental Change,* 14(4): 303–313.

Preston, B.L., Yuen, E.J. and Westaway, R.M. (2011). Putting vulnerability to climate change on the map: a review of approaches, benefits, and risks. *Sustainability Science,* 6(2): 177–202.

Rygel, L., O'Sullivan, D. and Yarnal, B. (2006). A method for constructing a social vulnerability index: An application to hurricane storm surges in a developed country. *Mitigation and Adaptation Strategies for Global Change,* 11:741–764.

Saaty, T.L., 1980. The Analytic Hierarchy Process. McGraw-Hill, New York.

Singh, G.S., Rao, K.S. and Saxena, K.G. (2015). Energy and Economic Efficiency of the Mountain Farming System: A Case Study in the North-Western Himalaya. *Journal of Sustainable Agriculture,* 9: 2-3.

Tiwari, P.C. and Joshi, B. (2015). Local and regional institutions and environmental governance in Hindu Kush Himalaya. *Environmental Science and Policy,* 49:66–74.

Climate Change and Variability: Mapping Vulnerability of Agriculture using Geospatial Technologies

Vinay Kumar Sehgal*, Malti Rani Singh, Niveta Jain and Himanshu Pathak

ICAR-Indian Agricultural Research Institute, New Delhi 110 012, India
*E-mail: vk.sehgal@icar.gov.in; vksehgal@gmail.com

ABSTRACT: *Climate change and variability is of particular concern for India, since it has a direct impact on agriculture. It is expected to contribute to increase in temperature, rainfall variability and decrease in irrigation water supplies, greater frequency of extreme weather events and shifting seasons with serious effects on agriculture sector, forestry, food security, natural resources, economic activity, human health, and infrastructure which have deep influence on Indian national economy and livelihood (IPCC, 2014). Faced with these challenges, decision makers and planners need information to assist for preparedness, allocate resources effectively, and reduce impacts. To minimize the likely harm associated with climate change, people and society need an accurate assessment of the vulnerability of the ecosystem in which they inhabit. The key features of climate change vulnerability and adaptation are those related to variability and extremes, not specifically changed average conditions. The present study showed that geospatial technologies can be used to collate, reformat and standardize indicator level data at administrative level (here district) to generate vulnerability rating.*

Keywords: Climate Change Vulnerability, Composite Agricultural Vulnerability, Vulnerability Framework, Coping Abilities.

Researchers have analysed the concept and definition of vulnerability and concluded that vulnerability has three components: (i) the susceptibility of society (i.e. converse of adaptive capacity) which depend upon attributes of society (ii) exposure to hazard (e.g. water stress) and (iii) coping abilities (Kates, 1985; Chambers, 1989; Blaikie *et al.*, 1994; Bohle *et al.*, 1994; Downing and Bakker 2000). However, vulnerability assessments are commonly subjective because of the intricacy of the issue of vulnerability, and vary between regions and hazards. Many factors affect vulnerability but their inclusion in assessment depend on data availability and study context. IPCC has theorized the concept of vulnerability (V) as a function of exposure (E), sensitivity (S) and adaptive capacity (A).

Vulnerability assessment at different spatial scales has now become possible with the recent advances in geospatial technologies (i.e. remote sensing and GIS) integrating various spatial data (both short and long term), but there are still methodological problems in implementing model at larger scale. Besides, vulnerability maps convey a lot more to different stakeholders than tabulated categorical assessments. In view of the above, we present a case study to demonstrate a methodology to assess and map composite vulnerability of agriculture to climate variability and change for whole India at district level using geospatial technologies. It aimed at adopting a conceptual framework of vulnerability, generate spatial datasets of key factors contributing to vulnerability, estimate weights of factors and then generate vulnerability ranking maps of the districts. A novelty of this study was that it considered climatic, physical and socio-economic factors together to arrive at vulnerability rating.

Many approaches have been proposed for vulnerability assessment (Cutter *et al.*, 2003; Hayes *et al.*, 2004), however there is no universally applicable metrics for vulnerability or its components (Schroter *et al.*, 2005). Based on review of the literature and expert discussion, we enumerate the following broad sequential steps to be taken in any vulnerability assessment: 1) Set clear aim(s) of vulnerability assessment, 2) Decide on a framework of vulnerability assessment, 3) Decide on the workable definition of hazard, 4) Type of vulnerability to be addressed (environmental/social/economic/composite), 5) Decide spatial scale, 6) Decide temporal scale, 7) Select indicators, 8) Formulate indices, 9) Rating and weighting of indices, 10) Quantitative tools for combining indices to arrive at vulnerability rating, 11) Validation of assessment, 12) Communicating assessment, and 13) Adopt into decision support systems. In this study, we adopted the IPCC framework which considers vulnerability to climate change as a comprehensive multi-dimensional concept affected by a large number of indicators which can be defined as a function of exposure (E), sensitivity (S) and adaptive capacity (A) as shown in Figure 1.

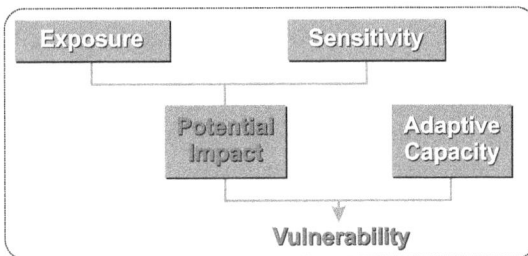

Fig. 1: Vulnerability Framework

Various indicators were identified for assessing each component of vulnerability and are shown in Figure 2. These indicators were generated for each district of the country using GIS. The exposure indicators were derived from monthly gridded temperature and rainfall layers and processed to derive trends for maximum and minimum temperature and while frequency of extreme high and low rainfall events was captured in terms of Standardized Precipitation Index (SPI) (McKee *et al.,* 1993), separately for *kharif* and *rabi* seasons. The study used gridded monthly precipitation, maximum and minimum temperature, time series data constructed by Climatic Research Unit (CRU TS 3.0) at a spatial resolution of 0.5×0.5 degree. The gridded temperature data were used to calculate the rate of change over the years by fitting the linear time trend and monthly rainfall data to calculate SPI, an index of rainfall deviation, over the 1951–2009 period. The Digital Soil Map of the World and Derived Soil Properties (Version 3.5) produced by FAO (1995) at a finer resolution of $5' \times 5'$ grid size by using the World Inventory of Soil Emissions (WISE) database was used. The digital maps of soil moisture storage capacity (mm) for 1 m profile depth and soil organic carbon content (kg m^{-2}) were extracted and their average value was calculated for each district in GIS. The study utilized the FAO Global Map of Irrigation Areas (GMIA) version 4.0.1 having a grid size of $5' \times 5'$ for calculating the irrigated area (Siebert *et al.,* 2005). The status of groundwater exploitation was taken from the maps produced by Central Ground Water Board. District-wise productivity of food grains and net sown area statistics were obtained from the Agricultural Statistics published by the Directorate of Economics and Statistics, Department of Agriculture and Cooperation, Ministry of Agriculture, for the period 2006 to 2009. Livestock density was compiled from the 18th Livestock Census (2007) published by the Department of Animal Husbandry and Dairying, Government of India. The district-wise statistics on human population density, number of villages electrified and the number of villages with paved roads were compiled from the Census of India (2011). The Human Development Index (HDI) was obtained from the Human Development reports of respective states, as produced by the UNDP. The annual NPK fertilizer consumption data was compiled from the reports of Fertilizer Association of India for the period 2006 to 2009.

A five-point ordered scale was used to rank from 'very-low' to 'very-high' for each indicator of exposure, sensitivity and adaptive capacity, according to their functional relationship with vulnerability, *i.e.* if the indicator was directly related to vulnerability; higher ranks were given for higher values and vice versa. For deriving weights of each indicator in their respective component of exposure, sensitivity and adaptive capacity, pair-wise comparison approach of Analytic Hierarchy Process (Saaty, 1980) was employed. The overall consistency ratio of 0.09 was achieved, suggesting that weights were generated randomly (Figure 2). Once the scores were standardized and weight established for factors, exposure, sensitivity, adaptive capacity and vulnerability maps were prepared varying from 1 to 5 ('very low', 'low', 'moderate', 'high' and 'very high') by taking weighted sum of the rank of all relevant indicators.

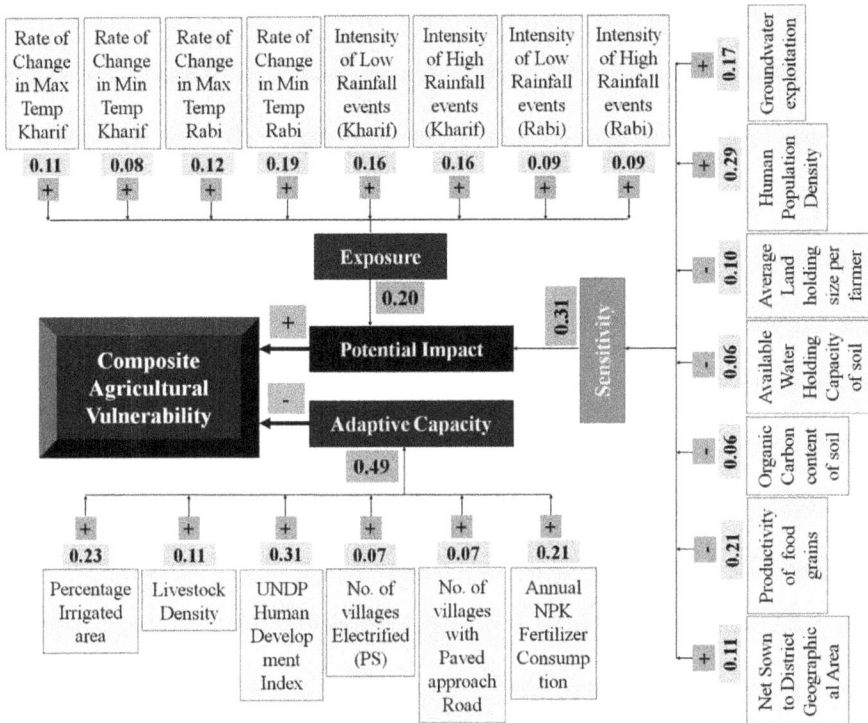

Fig. 2: Indicators, their Weightage and Nature of Relation (+ or −) used for Deriving Composite Agricultural Vulnerability

Using the district-wise data of indicators along with their weights maps of exposure, sensitivity and adaptive capacity were generated (not shown here). A 'very-high' to 'high' exposure was observed for most of Andhra Pradesh, Kerala, Tamil Nadu and southern Karnataka. High exposure was observed in J&K, Arunachal and Bundelkhand region of Uttar Pradesh and Madhya Pradesh. Punjab, Haryana and Uttarakhand districts showed low to very low exposure. The very high to high sensitivity was observed for most of the districts of Rajasthan and a few districts of southern Karnataka and northern Tamil Nadu. 'High' to 'moderate' sensitivity was also observed for most districts of Punjab and Haryana, parts of Bihar, MP and Maharashtra. The adaptive capacity was 'very-low' to 'low' in Rajasthan, MP, Chattisgarh, Odisha, Assam, Himachal Pradesh, Uttarakhand and parts of Maharashtra (Vidharbha region). 'Moderate' to 'high' adaptive capacity was found in Tamil Nadu, Kerala, Gujarat, Haryana and Uttar Pradesh. 'Very-high' adaptive capacity was found for most of the districts of Punjab and few in Haryana.

Combining the three component maps (E, S, and A) using weighted sum a vulnerability map of India was generated (Figure 3). A 'very-high' vulnerability was observed for most districts of Rajasthan and northern districts of Madhya Pradesh (including Bundelkhand), owing to very high sensitivity and low adaptive capacity of agriculture in these regions. Similar was the case for southern districts of Karnataka. High-vulnerability was obtained for most districts of central India comprising Maharashtra and Madhya Pradesh and also in most of Assam. Punjab, Haryana, Gujarat (Saurashtra), Uttarakhand, Kerala, southern Tamil Nadu, coastal Karnataka and Maharashtra, and parts of Western Uttar Pradesh showed very low to low vulnerability of agriculture to climate change.

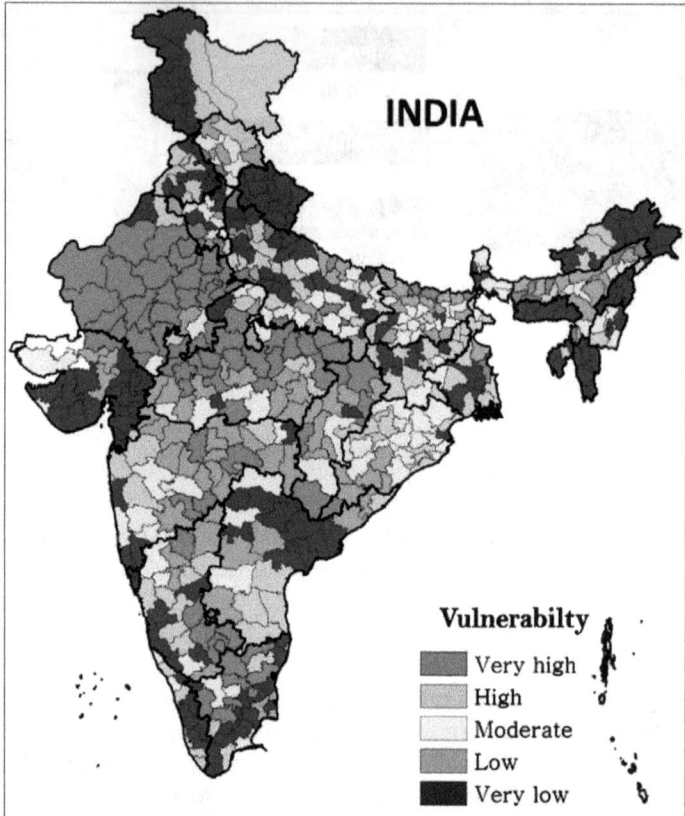

Fig. 3: Map of Composite Vulnerability of Agriculture
to Climate Change and Variability

Adaptation to climate change has the potential to substantially reduce many of the adverse impacts of climate change and enhance beneficial impacts. The key features of climate change vulnerability and adaptation are those related to variability and extremes, not specifically changed average conditions. Enhancement of adaptive capacity is a necessary condition for reducing vulnerability to climate-induced changes in availability of water resources, frequency and intensity of extreme events like floods, droughts, heat and cold waves, and the associated impacts on agriculture and other livelihood options. The study showed that geospatial technologies can be used to collate, reformat and standardize indicator level data at administrative level (here district) to generate vulnerability rating.

It identified the districts which should be prioritized for undertaking adaption measures keeping in view the underlying causes of vulnerability. Overall, this study on vulnerability is expected to lead to replication of proposed methodology to other agricultural areas of the world in the same or improved format so that better agricultural management plans could be attempted.

REFERENCES

Blaikie, P., Cannon, T., Davis, I. and Wisner, B. (1994). At risk: natural hazards, people's vulnerability, and disasters. Routledge publisher, London and New York.

Bohle, H.G., Downing, T.E. and Watts, M.J. (1994). Climate change and social vulnerability—Toward a sociology and geography of food insecurity. *Global Environment Change,* 4(1), 37–48.

Chambers, R. (1989). Vulnerability, Coping and Policy—Introduction. IDS Bulletin-Institute of Development Studies, 20(2), 1–7.

Cutter, S. L., Boruff, B.F., and Shirley, W.L. (2003). Social vulnerability to environmental hazards. *Social Science Quarterly*, 84(2), 242–261.

Downing, T.E., and Bakker, K. (2000). Drought discourse and vulnerability. Chapter 45, In: Wilhite, D.A., (Eds.), Drought: A Global Assessment, Natural Hazards and Disasters Series, Routledge Publishers, U.K.

Hayes, M.J., Wilhelmi, O.V. and Kautson, C.L. (2004). Reducing drought risk: Bridging theory and practice. *Natural Hazards Review,* 5(2), 106–113.

IPCC (2014). Climate Change 2014: Impacts, Adaptation, and Vulnerability. Part B: Regional Aspects. Contribution of Working Group II to the Fifth Assessment Report of the Intergovernmental Panel on Climate Change [Barros, V.R., C.B. Field, D.J. Dokken, M.D. Mastrandrea, K.J. Mach, T.E. Bilir, M. Chatterjee, K.L. Ebi, Y.O. Estrada, R.C. Genova, B. Girma, E.S. Kissel, A.N. Levy, S. MacCracken, P.R. Mastrandrea, and L.L. White (eds.)]. Cambridge University Press, Cambridge, United Kingdom and New York, NY, USA, 688 pp.

Kates, R.W. (1985). The interaction of climate and society, in: Kates, R.W., Ausubel, J.H., Berbarian, M., (Eds.), Climate Impacts Assessment, John Wiley, Chichester, pp. 3–36.

McKee T.B., Doesken N.J., and Kleist, J. (1993). The relationship of drought frequency and duration to time scales. In: Proc. 8th conference on applied climatology. Anaheim, California, pp. 179–184.

Saaty, T.L. (1980). The Analytic Hierarchy Process. New York, McGraw-Hill International. pp. 20–25.

Schroter, D., Polsky, C. and Patt, A.G. (2005). Assessing vulnerabilities to the effects of global change: An eight step approach. Mitigation and Adaptation Strategies for Global Change, 10, 573–596.

Siebert, S., Doll, P., Hoogeveen, J., Faures, J.M., Frenken, K. and Feick, S. (2005). Development and validation of the global map of irrigation areas. *Hydrology and Earth System Science,* 9, 535–547.

Satellite Remote Sensing for Climate Variability/Change Studies: Status and Scope

M.V.R. Sesha Sai* and Y.V.N. Krishnamurthy

National Remote Sensing Centre, Indian Space Research Organization, Hyderabad 500 037, India
*E-mail: seshasai_mvr@nrsc.gov.in

ABSTRACT: *Valuable spatio-temporal information on different climate science-related parameters is being generated exploring the satellite data. Nevertheless, satellite data often contain uncertainties caused by biases in sensors and retrieval algorithms for capturing robust long term trends of climate variables. Further, short duration series of satellite data and calibration/validation issues are some of the important constraints. Currently, the studies are concentrated to circumvent these issues by integration of observations from multi-level platforms and simulation modeling to supplement the information needs and address the gaps through international cooperation.*

Keywords: Spatio-Temporal Trends, Bhuvan, NICES, Path Finder Datasets, GIMMS NDVI.

Observational data and model simulations form the foundation for understanding the climate system. Satellite remote sensing allows for continuous monitoring on the global scale. It provides an independent source of observation to validate climate models and climate theories. It started with the Vanguard2 satellite, launched following the Sputnik. Since then, several satellites in both polar sun synchronous and the geo-stationary orbits have been launched to monitor land, atmosphere and oceans towards understanding the complex climate variability and response. Data from these satellites vary in spatial and temporal resolutions and often serve as complementary sets and augment the scientific analyses.

Observations from satellites enable deriving information on land, ocean and atmosphere at various spatial and temporal scales. Spatio-temporal changes of land cover have been studied by the research community in a variety of situations to understand the drivers that are responsible for such changes and use such information towards developing mitigation measures to cope with the adverse impacts, if any. Satellite data on land, water and atmosphere are integrated towards generating products for re-analysis of the data for studying a wide range of climate research towards better comprehension of the processes and their interactions. Some of the salient inferences drawn by the scientific community are briefly presented hereunder.

Among the characteristics of the terrestrial biosphere, vegetation and its phenology are the main characteristics. Phenology is highly sensitive to the climate change as the vegetation–climate feedbacks are mediated by phenology. While the trend of global average greenness is positive, the regional trends did exhibit considerable variations in direction and magnitude of change. Fortnightly global inventory modelling and mapping studies (GIMMS)–NDVI (normalized difference vegetation) series derived from a suite of NOAA AVHRR (advanced very high resolution radiometer) satellites for the period of 1981 to 2010 have been extensively used by the researchers to detect land surface phenology, change in land cover, drought frequency, NPP dynamics, greening and browning trends, etc. The processed data have been corrected for sensor degradation, and calibration of inter-sensor differences. The spatio-temporal trends in phenology and crop vigour, as indicated by the normalized difference vegetation index, over India are depicted in Figure 1.

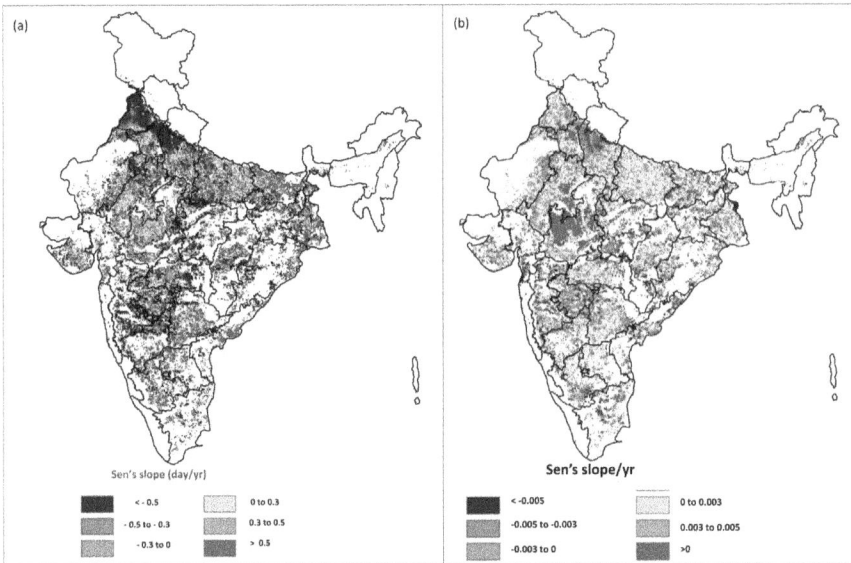

Fig. 1: Spatio-Temporal Trends (a) at Start of Growing Season; (b) of Crop Vigour

SSMI/DMSP (Special Sensor Microwave Imager/Defence Meteorological Science Programme) provided information on the variations in the spatial extent of the sea ice. During 1979–2010, the analysis revealed that the Antarctic Sea Ice extent increased by $1.5 \pm 0.4\%$ per decade. ERS-1 and 2, Envisat and Cryosat data provided valuable information on the mass losses of the Antarctic and Greenland ice sheets. Ocean surface topography observations from the TOPEX/Poseidon and other such missions indicated a global mean sea level rise of 3.2 ± 0.8 mm yr^{-1} between 1992 and 2006. The Global Climate Observation system has listed 26 out of 50 essential climate variables as significantly dependent on satellite observations. A train formation has been useful in collecting data for retrieval of several important climate parameters from a host of six sensors, four NASA missions (Aqua, Aura, CALIPSO and Cloudsat) and one each of JAXA (GCOM-W1) and of CNES mission (PARASOL), flying in close proximity to one another. This provided near-simultaneous observations of a wide variety of climate science-related parameters to aid the scientific community in advancing our knowledge of earth-system science and applying this knowledge for the benefit of society. The satellites are in a polar orbit, crossing the equator northbound at about 1:30 p.m. local time, within seconds to minutes of each other.

Extended Path Finder datasets of Atmosphere and Oceans (PATMOS-x) aims to derive atmospheric and surface climate records from the roughly 25 years of data from NOAA-AVHRR. PATMOS-x generates mapped and sampled results with a spatial resolution of 0.1° on a global longitude, latitude grid. This format avoids spatial or temporal averaging of data, thus maintaining the flexibility to conduct multi-dimensional analysis.

In India, over the past 3 decades Indian Remote Sensing Satellite are being used to generate information on natural resources, including infrastructure, and disaster management support. In addition, with the availability of global Earth Observation (EO) datasets and by participation in various global and national initiatives, geo-spatial information on various land, atmosphere and ocean parameters is being generated. The geo-portal of the Indian Space Research Organization, Bhuvan, showcases the Indian imaging capabilities in multi-sensor, multi-platform and multi-temporal domain. Bhuvan provides a range of services enabling visualization of various thematic data generated from the national missions and projects carried out by the National Remote Sensing Centre (NRSC). This Earth browser of Bhuvan gives a gateway to explore and discover virtual earth in 3D space with specific emphasis on Indian region. An information system, called the NICES (National Information System for Climate and Environmental Studies) using Indian Remote Sensing and geostationary satellites data and others is being developed. Additionally, the NICES will build specific observational networks, effectively coordinate among various departments and carry out climate and environmental impact assessment through formulation of a science plan.

Data from global satellite remote sensing products is being extensively used in combating climate change. Several international initiatives like Global Observing Systems Information Centre (GOSIC), Global Geodetic Observing System (GGOS), Global Earth Observing System of Systems (GEOSS) have been implemented to coordinate efforts to produce and disseminate high quality satellite climate records, that are of significance to predict, mitigate and adapt to climate variability and change as well as for better understanding of the carbon cycle. For example, soil moisture influences hydrological and agricultural processes, runoff generation, drought development and many other processes. It also impacts on the climate system through atmospheric feedbacks. Soil moisture, by being a source of water for evapotranspiration over the continents, and is involved in both the water and the energy cycles, has been recognised as an Essential Climate

Variable. Back scatter/brightness temperature from active and passive microwave sensors provide data for soil moisture estimation.

The European Space Agency has established the Climate Change Initiative (CCI) to create new climate data records for (currently) 13 essential climate variables (ECVs) and make these open and easily accessible to all. A climate modeling users' group provides a climate system perspective and a forum to bring the data and modeling communities together. Parameters in the CCI include clouds, aerosols, ozone, greenhouse gases, sea surface temperature, ocean colour, sea level, sea ice, land cover, fire, glaciers, soil moisture and ice sheets.

Increasing Atmospheric Carbon Dioxide and Temperature—Threats and Opportunities for Rainfed Agriculture

M. Vanaja*, S.K. Yadav, M. Maheswari and Ch. Srinivasarao

ICAR-Central Research Institute for Dryland Agriculture, Santoshnagar, Hyderabad 500 059, India

*E-mail: mvanaja@crida.in

ABSTRACT: *Field studies were carried out using OTCs, FACE and FATE facilities for a better understanding of rainfed crops response to enhanced CO_2, temperature and moisture deficit stress individually as well as in combination. Elevated CO_2 (eCO_2) resulted in improved biomass of majority of C_3 and C_4 rainfed crops, however the response of C_3 crops was much higher than that of C_4 crops. Among the C_3 crops, the pulse crops recorded higher response to eCO_2 for biomass, yield and Harvest index. The performance of C_4 crop under eCO_2 was higher under moisture deficit stress than at well-watered condition revealing its ameliorative impact. With increased photosynthetic rate (Pn) and decreased transpiration rate (Tr), eCO_2 improved water use efficiency of rainfed crops. Earliness in flower initiation and improved seed yield with increase in female flowers were recorded with eCO_2 in monoecious castor bean. Elevated CO_2 improved total biomass and seed yield in sunflower, when the crop experienced maximum temperature below 35°C during anthesis and grain filling stages. In groundnut, eCO_2 led to improved oil quality, with increased oleic acid (omega-9) and linoleic acid (omega-6) content. It can be concluded that increasing atmospheric CO_2 will have an ameliorative capability against moisture deficit stress and increased temperatures to certain extent for majority of the rainfed crops for both biomass and yield.*

Keywords: Growth Response, Biomass, Phenology, Quality, Interaction Effect.

Atmospheric CO_2 level is predicted to reach 700 ppm by the end of this century associated with an increased temperature by 5.8°C coupled with substantial shifts in precipitation patterns (IPCC 2007). To meet food, fuel and fodder demand for the burgeoning population, we need to develop suitable cultivars with better growth and improved yields under predicted conditions. Increase in atmospheric carbon dioxide has a fertilization effect on crops and can promote their growth and productivity, however response is governed by plant type and other environmental conditions such as temperature, availability of soil moisture and nutrition (Kimball, 2011). High atmospheric CO_2 concentration can alleviate the negative effect of water deficit on plants due to decreased transpiration and conductance (Ewert *et al.*, 2002, Madhu and Hatfield, 2015). Warmer temperatures expected with climate change will impact plant productivity especially when crops at key reproductive stages are exposed to temperature above their optimum levels (Jagadish *et al.*, 2016). The growth and yield response of agricultural crops to high temperature differ, crops such as millets are considered tolerant as they have high temperature optima of 40°C (Prasad *et al.*, 2017). Under reduced transpiration rates with eCO_2, the canopy temperature is invariably higher and their combined effect on crop performance needs systematic evaluation.

Major rainfed crops grown during monsoon season such as sorghum (*Sorghum bicolor* L.), maize (*Zea mays* L.), pearl millet *(Pennisetum glaucum)*, blackgram (*Vigna mungo* L. Hepper), pigeon pea (*Cajanus cajan* L. Millsp.), groundnut (*Arachis hypogaea* L.), sunflower (*Helianthus annus* L.) and castor bean (*Ricinus communis* L.) were evaluated for their biomass and yield response at two elevated CO_2 (eCO_2) levels—550 ppm and 700 ppm in an open top chamber (OTCs) facility (Vanaja *et al.*, 2006). The studies with elevated temperature (+3°C) and its interaction with elevated CO_2 (550 ppm) were conducted in Free Air Temperature Elevation (FATE) facility at CRIDA. The FATE facility consists of nine 8 m dia rings fitted with 24 array of 2000W IR heaters above the canopy to increase the canopy temperature by 3°C above the ambient plot. Three FATE rings were fitted with perforated PU tubing for CO_2 supply which facilitated to quantify the impact of elevated temperature and CO_2 interaction.

Table 1: Improvement (%) in Total Biomass, Seed Yield and HI of Cereals, Pulses and Oilseed Crops at Two Elevated CO_2 Levels—OTC Studies

Crop	Variety	Total Biomass		Seed Yield		HI (%)	
		550 ppm	*700 ppm*	*550 ppm*	*700 ppm*	*550 ppm*	*700 ppm*
Cereal crops							
Sorghum	SPV–1616	1.3	14	1.2	17	Nil	2.7
Maize	Harsha	17	34	2.0	13	–13	–16
Pearl millet	ICTP 8203	26	32	25	26	0.6	–3.6
Pulse crops							
Blackgram	T-9	39	65	89	129	36	38
Pigeon pea	ICPL- 88039	27	91	48	128	14	19
Oilseed crops							
Groundnut	JL-24	35	15	37	16	2.2	0.72
Sunflower	KBSH-1	19	45	21	46	1.8	0.57
Castor bean	DCS-9	11	22	155	167	136	125

Studies with different rainfed cereals, pulses and oilseed crops in open top chambers at two elevated CO_2 levels—550 ppm and 700 ppm, revealed that the variability among the crops is significant and the response was highest with pulses for biomass, yield and HI, indicating that elevated CO_2 impacted even partitioning of biomass. The selected C_4 cereals are less responsive to elevated CO_2 and the improvement was more with vegetative biomass (Vanaja *et al.*, 2012) and this clearly indicated that the implications of increased atmospheric CO_2 are different with different crops (Table 1).

Flowering is a critical milestone in the life cycle of plants, and changes in the timing of flowering may alter processes at the species, community and ecosystem levels. Therefore, understanding flowering responses to global change drivers, such as elevated CO_2, is necessary to predict the impacts of global change on natural and agricultural ecosystems. Castor bean is a monoecious plant bearing both male and female flowers on the same spike. The flowering in castor is sensitive to temperature. Elevated CO_2 levels significantly reduced the duration of days to initiation of flowering and increased the number of female flowers of the spike (Figure 1) and thereby improved the seed yield (Vanaja *et al.*, 2008).

Fig. 1: Flowering Behaviour of Castor Bean at Two Elevated CO_2 Levels

As the atmospheric CO_2 content rises, plant water use efficiency or the amount of carbon gained per unit of water loss should increase dramatically due to improved photosynthetic rate coupled with low transpiration rate. Elevated CO_2 improved WUE by 44% in groundnut genotypes as photosynthetic rate increased by 36% and transpiration rate decreased by 21% (Vaidya *et al.*, 2014). It was observed that lesser reduction or even improvement of different components due to moisture deficit stress with eCO_2 (Table 2) both in C_4 (maize) and C_3 (sunflower) crops revealing its ameliorative impact (Vanaja *et al.*, 2011).

Majority of the crops tend to respond negatively, when the temperature exceeds optimum range and leads to reduced yield. However, the optimum temperature of different crops and their vulnerability to high temperatures varies with the genotype, developmental stage and duration of exposure. High temperature during the reproductive stages causes deleterious effects on the yield and quality (Prasad *et al.*, 2015). Exceeding crop-specific high temperature thresholds may result in a significantly higher risk of crop failure (Singh *et al.*, 2016).

Table 2: Decrease (%) in Root and Shoot Characters under Drought Stressed Conditions at Elevated (eCO_2) and Ambient CO_2 (aCO_2) in Maize (C_4) and Sunflower (C_3)

	Maize		Sunflower	
	[aCO₂]	*[eCO₂]*	*[aCO₂]*	*[eCO₂]*
Root Characters				
Root length	13.1	−39.3 (400)	30.5	−1.16 (104)
Root volume	69.7	12.0 (83)	58.6	24.7 (58)
Root dry weight	78.2	63.2 (19)	47.8	39.5 (17)
Shoot Characters				
Shoot length	31.5	28.4 (10)	39.2	23.8 (39)
Stem dry weight	50.4	26.5 (47)	48.4	38.1 (21)
Leaf area	71.1	49.8 (30)	63.6	44.0 (31)
Leaf dry weight	58.3	43.4 (26)	53.4	37.4 (30)
Root and Shoot Characters				
Total dry weight	64.3	45.9 (29)	50.7	38.1 (25)
Root shoot ratio	39.9	39.6 (1)	−6.6	3.1 (147)

* The values in parenthesis are the % improvement due to [eCO_2] under moisture stress.

Since the vegetative and reproductive processes have different responses to temperature, they show different stimulations by CO_2 at elevated temperature, hence beneficial effects of elevated CO_2 on photosynthesis, carbohydrate metabolism and vegetative growth are not always reflected in seed yield (Vu *et al.*, 2007). The biomass and yield response of sunflower hybrid KBSH-1 to elevated

CO_2 of 550 ppm differed with two temperature regimes. When the crop experienced maximum temperature below 35°C during the anthesis and grain filling stages, the eCO_2 (550 ppm) improved total biomass (19%), seed yield (21%), test weight (23%), HI (2%) and oil content (0.4%) and when it was exposed to more than 40°C the response was reduced for the biomass (2%), seed yield (9%), test weight (0.7%) and oil content (–5.5%) except HI (Vanaja *et al.*, 2013). The FATE studies with groundnut cv. Dharani at elevated canopy temperature (eT) recorded reduction in total biomass (20.4%), pod number (9.4%), seed yield (34.3%), test weight (28.5%) and HI (25.5%), while the eCO_2 (550 ppm) reduced the ill effects of high temperature (Figure 2) showing the protective effect of eCO_2 against high temperature.

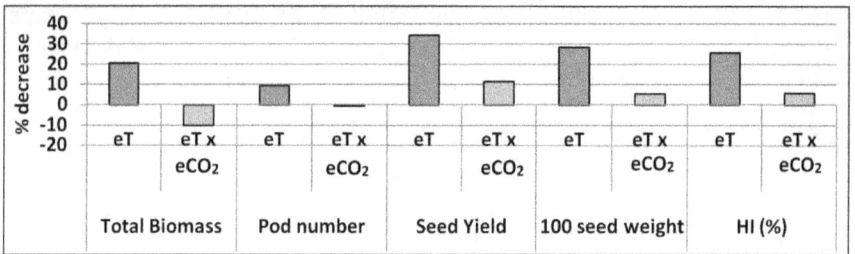

Fig. 2: The Impact of Elevated Canopy Temperature (+3°C) and its Interaction with Elevated CO_2 (550 ppm) on Biomass and Yield of Groundnut cv Dharani

Even small increases in temperature can have a larger effect than elevated CO_2, on grain quality. High temperatures enhance the grain protein content. Hence, it is expected to partially balance the CO_2 effect. Similarly, a 5°C increase in the temperature during seed development of soybean can decrease the isoflavone content in seeds by about 65%, thereby reducing the nutritive quality.

In edible oils, the component of unsaturated fatty acids, *viz.* oleic, linoleic and linolenic acids play a vital role for the human nutritional point of view while in the non edible oil, *viz.* castor oil it is the ricinoleic acid content that paves a way for its industrial utility. The experiments conducted with Open Top Chambers revealed that there was no significant change in the content and quality of castor bean oil at elevated CO_2 level of 550 ppm, however, the total oil yield obtained under elevated CO_2 conditions was significantly higher owing to the higher seed yields (Vanaja *et al.*, 2008).

The absolute values for oil content and protein content of groundnut cultivar JL-24 did not change much with 550 ppm of CO_2 level. The saturated fatty acids (palmitic and stearic) showed a decreasing trend at elevated CO_2 level while oleic acid (omega-9) and linoleic acid (omega-6) content registered an increasing trend (Yadav *et al.*, 2011). The increase in component of unsaturated fatty acids, *viz.* oleic and linoleic acids with increased atmospheric carbon dioxide concentration could prove beneficial for the human nutritional point of view under changed climatic conditions.

The impact of the future predicted increased CO_2 and temperature coupled with changes in precipitation patterns appear to be not that alarming for majority of rainfed crops. However to counter the adverse effects of climate change on agriculture, it is desirable to select the crops and their cultivars thereof, that can better utilize the increased concentration of CO_2 and perform better under high temperature and moderate moisture deficit conditions.

REFERENCES

Ewert. F., Rodriguez, D., Jamieson, P., Semenov. M.A., Mitchell, R.A.C., Goudriaan, J., Porter, J.R., Kimball, B.A., Pinter Jr, P.J., Manderscheid, R., Weigel, H.J., Fangmeier, A., Fereres, E. and Villalobos, F. (2002). Effects of elevated CO_2 and drought on wheat: Testing crop simulation models for different experimental and climatic conditions. *Agriculture, Ecosystems and Environment*, 93: 249–266.

IPCC (2007). Climate Change 2007: *The Physical Science Basis*. Contribution of Working Group I to the Fourth Assessment Report of the Intergovernmental Panel on Climate Change. Cambridge University Press, Cambridge.

Jagadish, S.V.K., Bahuguna, R.N., Djanaguiraman, M., Gamuyao, R., Prasad, P.V.V. and Craufurd, P.Q. (2016). Implications of high temperature and elevated CO_2 on flowering time in plants. *Frontiers in Plant Science*, 7: 913. (doi: 10.3389/fpls.2016.00913).

Kimball, B.A. (2011). Lessons from FACE: CO_2 Effects and Interactions with Water, Nitrogen, and Temperature. In: Hillel, D. and Rosenzweig, C. (Eds.), *Handbook of Climate Change and Agroecosystems: Impacts, Adaptation, and Mitigation,*(Eds.) Imperial College Press, London, 87–107.

Madhu, M. and Hatfield, J.L. (2015). Elevated Carbon dioxide and soil moisture on early growth response of soybean. *Agricultural Sciences*, 6: 263–278.

Prasad, P.V.V., Bheemanahalli, R. and Jagadish, S.V.K. (2017). Field crops and the fear of heat stress—Opportunities, challenges and future directions. *Field Crops Research*, 200: 1–8.

Prasad, P.V.V., Djanaguiraman, M., Perumal, R. and Ciampitti, I.A. (2015). Impact of high temperature stress on floret fertility and individual grain weight of grains orghum: Sensitive stages and thresholds for temperature and duration. *Frontiers in Plant Science*, 6: 820. (doi: 10.3389/fpls.2015.00820).

Singh, D., Singh, C.K., Tomar, R.S.S., Chaturvedi, A.K., Shah, D., Kumar, A. and Pal, M. (2016). Exploring genetic diversity for heat tolerance among lentil (*Lens culinaris* Medik.) genotypes of variant habitats by simple sequence repeat markers. *Plant Breeding*, 135: 215–223.

Vaidya, S., Vanaja. M., Sathish, P., Vagheera, P., Anitha, Y., Sowmya, P., Jainender and Jyothi Lakshmi, N. (2014). Impact of elevated CO_2 on growth and physiological parameters of groundnut (*Arachis hypogaea* L.) genotypes. *Journal of Plant Physiology & Pathology*, 3: 1–6: (*doi:http://dx.doi.org/10.4172/2329-955X.1000 138)*.

Vanaja, M., Jyothi, M., Ratnakumar, P., Vagheera, P., Raghuram Reddy, P., Jyothi Lakshmi, N., Yadav, S.K., Maheshwari, M. and Venkateswarlu, B. (2008).

Growth and yield responses of castor bean (*Ricinus communis* L.) to two enhanced CO_2 levels. *Plant, Soil and Environment,* 54: 38–46.

Vanaja, M., Maheswari, M., Ratnakumar, P. and Ramakrishna, Y.S. (2006). Monitoring and controlling of CO_2 concentrations in open top chambers for better understanding of plants response to elevated CO_2 levels. *Indian Journal of Radio & Space Physics,* 35: 193–197.

Vanaja, M., MaruthiSankar, G.R., Maheswari, M., Raghu Ram Reddy, P., Jyothi Lakshmi, N., Yadav, S.K., Archana, G. and Venkateswarlu, B. (2013). Response of sunflower traits to elevated CO_2 levels under cool and warm seasons. *Helia,* 36 (59): 85–98.

Vanaja, M., Raghu Ram Reddy, P., Yadav, S.K., Jyothi Lakshmi, N., Narasimha Reddy, A., Reddy, B.M.K., Maheswari, M. and Venkateswarlu B. (2012). Increasing CO_2 concentration and its probable influence on rainfed food and oilseed crop productivity in SAT. In: T. Bhattacharyya, Pal, D.K., Dipak Sarkar and Wani, S.P. (Eds.), *Impact of Climate Change in Soils and Rainfed Agriculture of Tropical Ecosystem.* Studium Press LLC, USA.

Vanaja, M., Yadav, S.K., Archana, G., Jyothi Lakshmi, N., Raghu Ram Reddy, P., Vagheera, P., Abdul Razak, S.K., Maheswari, M. and Venkateswarlu, B. (2011). Response of C_4 (Maize) and C_3 (Sunflower) crop plants to drought stress and enhanced carbon dioxide concentration. *Plant Soil and Environment,* 57: 207–215.

Vu, J.C., Allen, Jr, L.H. and Widodo, W. (2007). Leaf photosynthesis and Rubisco activity and kinetics of soybean, peanut, and rice grown under elevated atmospheric CO_2, supraoptimal air temperature, and soil water deficit. *Current Topics in Plant Biology,* 7: 27–41.

Yadav, S.K., Vanaja, M., Reddy, P.R., Jyothilakshmi, N., Maheswari, M., Sharma, K.L. and Venkateswarlu B. (2011). Effect of elevated CO_2 levels on some growth parameters and seed quality of groundnut (*Arachis hypogaea* L.). *Indian Journal of Agricultural Biochemistry,* 24(2): 158–160.

Carbon Sequestration and Climate Smart Agriculture

Ch. Srinivasarao

ICAR-Central Research Institute for Dryland Agriculture, Santoshnagar, Hyderabad 500 059, India
E-mail: cherukumalli2011@gmail.com

ABSTRACT: *The era of Anthropocene has been endorsed by both climatic and non-climatic factors. It may increase the risk of food and water security across the globe. Among all, carbon (C) sequestration is one of the important strategies that could mitigate the effect of climate change to some extent by transferring atmospheric CO_2-C into the long-lived natural pools such as soils and perennial green biomass. Soil organic carbon (SOC) has relation in plant growth and agronomic productivity. Most of the soils of India have been depleted their SOC stocks. Further, changes in climate are exacerbating the risk of soil erosion and alter nutrient cycling, and thereby impacting optimization of crop production. Therefore, conservation of soil organic matter (SOM) is essential to prevent soils from further degradation and for overall soil health maintenance. This article focuses climate change impact on agriculture, C sequestration and their role on vice-versa. The strategies for C sink (efficient use of crop residue, conservation agriculture, integrated nutrient management, agroforestry, cover crops, use of organic and other bio-solids etc.) along with soil water conservation measures can create a positive sink to soil C which will have marked influence on sustainable crop production and impart climate resilience.*

Keywords: Climate Change, Carbon Sequestration, Conservation Agriculture, Emission of GHGs, Climate Resilient Agriculture.

Climate change is becoming a serious global environmental concern. The signatures of climate change, have been seen in every sphere of life on the earth which ignited the issues towards set back free and ecofriendly environment. The climate has been changing naturally as volcanoes, tectonic plate movements since the beginning of time. Recently it is in the news due to anthropogenic emissions and man made disturbances. The IPCC AR5 on physical science basis report confirmed human intervention from 1750 AD. Concentrations of GHGs in the atmosphere exceeded the pre-industrial levels by about 40, 150 and 20% in 2011 CO_2, CH_4, N_2O, respectively (IPCC, 2013). The current total global anthropogenic emission is estimated at 400 ppm (WMO, 2013).

It was expected that increase in atmospheric CO_2 has a fertilization effect on crops through carbon (C) assimilation pathway and thus may encourage crop growth and productivity (Srinivasa Rao *et al.*, 2016). On the other hand, increase in temperature could reduce crop growth periods, alter photosynthetic processes, and affect pest populations, nutrient loss in soils. The transfer of C in the green biomass and its sink into the soils is one of the most important strategies to address the problem of land degradation and climate change mitigation. Secondly, CO_2 concentration in the atmosphere can be minimised by reducing the global energy use or through the substitution of fossil fuels by biomass and sequestering CO_2 in soils and biomass (Schrag, 2007). SOC refers to containing C in SOM; hence, conservation and maintenance of SOM is essential for increasing the resilient soils. Most agricultural soils have lost 30 to 75% of their antecedent SOC pool (Lal *et al.*, 2007). Agricultural, forest and other land-use change (AFOLU) contributes less than 25% (~10–12 Gt CO_2 eq/year) of anthropogenic GHG emission.

Soil samples were collected from eight ongoing long-term integrated nutrient management experiments under different rainfed agro ecologies under long-term manurial experiments in various locations, representing a wide range of climatic conditions in tropical India. Climate varied from arid, semi-arid, to sub-humid, with mean annual rainfall and temperature. Soils were alluvial, red, yellow, and black, and arid. Among the soil types, Inceptisols/Entisols, Vertisols and Vertic subgroups, and Aridisols were neutral to alkaline in reaction, and Alfisols/ Oxisols were acidic. Depth-wise sampling of soils (0.20 m intervals up to 1.00 m deep) was undertaken on the locations. The Walkley and Black (1934) method was used to estimate SOC. Bulk density of each horizon was determined by weight by volume. The size of C stock in each profile was calculated following the method described by Batjes (1996). It involved calculation of organic C by multiplying OC content (g C g^{-1} soil), bulk density (Mg m^3) of each layer, and thickness of this layer (m) for each horizon. This information is relevant in terms of comparing the soil C stocks among soil types, production systems, and climate, and accordingly suitable management practices could be identified for better C sequestration in dryland soils.

Long-term experiments are important for assessing long-term changes in SOC and crop yields and estimating C sequestration potential of agricultural lands (Srinivasa Rao *et al.*, 2013). The C sequestration potential in corresponding experiments ranged from 2.1 to 4.8 Mg C ha^{-1}. Integrated nutrient management (INM) treatments (using manure/crop residue of soybean, sorghum/green leaf manures, *Leucaena* etc.) increased the rate of SOC sequestration in semi-arid regions of India. SOC sequestration rates (Mg C $ha^{-1}year^{-1}$) were 0.57 for 50% recommended dose of fertilizer (RDF) + 4 Mg ha^{-1} groundnut shell, 0.82 for Farm Yard Manure (FYM) 10 Mg ha^{-1}+ 100% NPK, 1.64 for 10 Mg ha^{-1} FYM + 100% NPK, 0.89 for 25 kg N ha^{-1}(sorghum residue) + 25 kg N (*Leucaena* clippings), 0.42 for 50% recommended dose of N (RDN) (fertilizer) + 50% RDN (FYM), 1.26 for FYM 6 Mg ha^{-1} + 20 kg N + 13 kg P, and 0.32 for 100% organic (FYM) and in 28-year long-term pearl millet experiment at Agra, 13.97 for 50% N fertilizer + 50% N FYM and 0.73 at Arjia (unpublished), critical C input (Mg C ha^{-1}year^{-1}) was calculated for zero change and required organic source rate (Mg C ha^{-1}) under best management treatments indicated that FYM (or CR) in conjunction with mineral fertilizers was the most efficient system in building C pools, and is thus imperative for maintaining and enhancing the SOC stock in the major cropping systems in drylands.

CLIMATE SMART AGRICULTURE AND CARBON SEQUESTRATION

Climate smart agriculture practices include mulching, incorporation of crop residue, intercropping, conservation agriculture, crop rotation, integrated crop-livestock management, agro-forestry, improved grazing and improved water management. Through these resource conservation practices can reduce soil C losses by decreasing erosion, reducing oxidation of SOM and providing C inputs. Restoration of soil biota and their ecological processes breaks down organic matter into SOC fractions and stable organo-mineral complexes. In addition, such practices contribute to improve soil fertility and productivity. The global potential of SOC sequestration rate is estimated at 0.6 to 1.2 Gt C year^{-1}, comprising 0.4 to 0.8 Gt C year^{-1} through adoption of recommended management practices on cropland soils, 0.01 to 0.03 Gt C year^{-1} on irrigated soils and 0.01 to 0.3 Gt C year^{-1} through improvement of rangelands and grasslands (Lal *et al.,* 2007). Long-term manure application increases the soil C pools and the effects may persist for longer periods. Although both organic and inorganic forms of C are found in soils, land use management typically has a larger impact on SOC (Srinivasarao *et al.,* 2014).

CLIMATE CHANGE ADAPTATION AND MITIGATION IN AGRICULTURE

1. Need to adopt resource conservation technologies such as no-tillage, direct seeding of rice and crop diversification.
2. Crop residue management without affecting crop–livestock systems.
3. Use of short duration leguminous pulses and oilseed crops which can cope with intermittent short dry spells.
4. Efficient water use such as frequent but shallow irrigation, drip and sprinkler irrigation for high value crops, irrigation at critical stages.
5. Soil and water conservation practices such as ridges and furrows, land grading and leveling, field bunds, graded line bunds, contour trenching, con-servation furrows, mulching and judicious application of Farm Yard Manure (FYM).

6. Nutrient management practices should be followed for increasing the use efficiency of nutrients such as INM, optimum use of fertilizer doses, split application of nitrogenous and potassic fertilizers, deep placement, use of nitrification inhibitors for preventing loss through volatilization and leaching.

CONCLUSION

Coping with the adverse impact of climate change on agriculture will require careful management of natural resources like soil, water and biodiversity. Agriculture is the most sensitive sector to the climate change. Soil C sequestration is a win-win approach which can deal both adaptation and mitigation. Therefore, restoration of waste and degraded lands and ecosystems through adoption of improved management practices can enhance SOC and improve soil quality and health. Improving organic matter in soil will increase water holding capacity of soil which helps to cope up with intermittent dry spell, on the other hand it will release the nutrient in more judicious way which ultimately improve nutrient use efficiency and lower emission of GHGs. Adopting climate smart agriculture practices such as conservation agriculture, inter cropping, crop rotations, improve cultivars, INM, soil and water conservation can increase C sequestration in soils beyond the threshold limits. C sequestration potential of agroforestry systems have been established hypothetically; however it requires field measurements for validating these concepts.

REFERENCES

Batjes, N.H. (1996). Total carbon and nitrogen in the soils of the world. *European Journal of Soil Science,* 47: 151–163.

IPCC (2013). "Climate Change 2013: The Physical Science Basis." In: *Contribution of Working Group I to the Fifth Assessment Report of the Intergovernmental Panel on Climate Change.* (Stocker, T.F., Qin, D., Plattner, G.K., Tignor, M., Allen, S.K., Boschung, J., Nauels, A., Xia, Y., Bex, V. and Midgley, P.M. Eds.), pp. 1535, Cambridge University Press, Cambridge, United Kingdom and New York, NY, USA.

Jackson, M.L. (1973). Soil chemical analysis. New Delhi, India: Prentice Hall of India.

Lal, R., Follett, R.F., Stewart, B.A. and Kimble, J.M. (2007). Soil carbon sequestration to mitigate climate change and advance food security. *Soil Sci.,* 172(12): 943–956.

Schrag, D.P. (2007). Preparing to capture carbon. *Science,* 315: 812–813.

Srinivasarao, Ch., Lal, R., Kundu, S., Prasad Babu, M.B.B., Venkateswarlu, B. and Singh, A.K. (2014). Soil carbon sequestration in rainfed production systems in the semiarid tropics of India. *Sci. Tot. Environ.,* 487: 587–603.

Srinivasarao, Ch., Lal, R., Rao, A.S., Kund, S., Sahrawat, K.L., Chary, G.R., Thakur, P.B. and Srinivas, K. (2016). Carbon management as key to climate resilient agriculture. (Eds. Venkateswarlu B) Climate Resilient Agronomy, *Indian Society of Agronomy,* 182–202.

Srinivasarao, Ch., Venkateswarlu, B., Lal, R., Singh, A.K. and Kundu, S. (2013). Sustainable management of soils of dryland ecosystems of India for enhancing agronomic productivity, sequestering carbon (Sparks, D.L., Eds.). *Adv. Agron.*, 253–329.

Walkley, A. and Black, C.A. (1934). Estimation of organic carbon by chromic acid titration method. *Soil Science,* 37: 29–38.

WMO (2013). World Meteorological Organization Greenhouse Gas Bulletin. Geneva, p. 4.

Genetic Enhancement for Drought and Heat Resilience in Wheat

Harikrishna[1], A. Bellundagi[1], K.T. Ramya[1], L. Todkar[1],
K.C. Prashant Kumar[1], K.V. Prabhu[*1], Neha Rai[1], N. Jain[1],
P.K. Singh[1], G.P. Singh[1], N. Sinha[1], N.K. Singh[2],
P.C. Mishra[3], S.C. Misra[4] and P. Chhuneja[5]

[1]Indian Agricultural Research Institute, New Delhi 110 012, India
[2]National Research Center on Plant Biotechnology, New Delhi 110 012, India
[3]Jawaharlal Nehru Krishi Vishwa Vidyalaya, Jabalpur 482 004, India
[4]Agharkar Research Institute, Pune 411 004, India
[5]Punjab Agricultural University, Ludhiana 141 004, India
*E-mail: kvinodprabhu@rediffmail.com

ABSTRACT: *Due to complexity in phenotyping and inheritance of the traits, breeding for drought and heat tolerance is subject to complex environment, genotype × environment interactions which need to be accounted for while effecting selection on well-defined set of genotypes and environments. An International Core set for drought and heat tolerance consisting of 146 germplasm lines was characterized for physiological and morphological traits and traits-wise superior genotypes were identified. In the genetic background of GW322 and HD2733, 1–4 QTL (Meta QTLs) regions were transferred using Marker Assisted Backcross Breeding (MABB) strategy for heat and drought tolerance. To combine QTLs for drought and heat adaptive traits, Marker Assisted Recurrent Selection (MARS) was conducted on F_5 families in four different bi-parental populations where, in each population, the source parents carried positive alleles for drought and heat tolerance. These parents were diverse based on polymorphic 120 SSRs and 802 SNPs. Among 25 different traits phenotyped, NDVI at boot and grain filling canopy temperature at grain filling and milky stage, leaf area index (LAI), Fv/Fm, stomatal conductance, grain weight, biomass and harvest index showed significant correlation and regression values with yield under heat and drought stress. Based on validated QTLs, information in base population and multi-location (four location three year) testing of base populations 25 best individuals were selected in each population for inter mating at F_6 generation. Carefully chosen pair-wise inter crosses were attempted to combine QTLs using limited population size. Based on validated markers wide range of intercrossed progenies in advanced generation with range of 6–14 positive alleles were obtained after second round of recombination. Genotypes with high molecular score and more number of QTLs with MABB and MARS strategies out performed over checks in multi-location testing.*

Keywords: Molecular Markers, Climate Change, Validation, MABB and MARS.

W heat (*Triticum aestivum* L.) is one of the important crops consumed by human beings and it is cultivated in varying environmental conditions in India. Water and heat stress are the major constraint in sustaining yield under irrigated, rainfed and late-sown agro ecosystems. The terms 'drought' and 'heat' hide networks of complex phenomenon and pathways in it. Because of its complex nature in terms of networks of pathways involved, complex inheritance and difficulties in phenotyping; breeding for drought and heat-tolerance is difficult. Therefore, it is vitally important to identify drought and heat-tolerant wheat cultivars under high yield background in changing climatic conditions. Precise imposition of drought and heat stress is a challenging task. As a premier institute, IARI has made efforts to develop heat and drought tolerant varieties, state of art facility for phenotyping drought and heat tolerance traits and mapping populations at the Division of Genetics. Details on QTLs mapping and gene expression under stress environments are available in recent literature, however further utilization in breeding programme needs validation and recombination of positive alleles of these loci in superior yield backgrounds.

A total of 146 germplasm lines received under the Generation challenge programme and 17 popular varities were evaluated under irrigated timely sown, restricted irrigated, rainfed and late sown condition. Four MARS populations DBW43/HI1500 (160 families), HUW510/HI1500 (160 families), VL755/ UP2338//BAU/KAUZ/2/KS1/PBW562 (180 families) and PBW442/INQ*3/ TUKURU// DBW18 (180 families) were evaluated under restricted irrigated and rainfed conditions in Delhi, Pune, Powerkheda and Ludhiana in two replications. Six different MABB populations in the genetic background of HD2733 and GW322 were developed and used.

TRAITS FOR DROUGHT AND HEAT STRESS

Drought and heat tolerance are complex traits to phenotype in the field. But, dissecting it into component traits such as canopy temperature, NDVI, chlorophyll fluorescence, MSI, leaf area and agronomic traits such as thousand grain weight, grain weight per spike, biomass has been an option since these traits show significant association with grain yield under stress. Therefore use of these traits is important to select genotypes superior for drought and heat tolerance. Among 25 different traits phenotyped for drought stress NDVI at boot and grain filling stage, canopy temperature at grain filling and milky stage, leaf area index (LAI), Fv/Fm, grain weight per spike, thousand karnal weight, biomass and harvest index showed significant correlation and regression values with yield under heat and drought-stress. Strategy to increase the grain yield in wheat crop under moisture and heat stress is through the genetic manipulation of physiological traits is an option. However the trait yield under stress is an important trait to be considered for these abiotic stresses.

Physiological and molecular characterization of wheat working collection has brought out Indian tall genotypes C306 and its derivatives as distinct group from other genotypes. From the core set, 163 germplasms were characterized for physiological and morphological traits and traits-wise superior genotypes were identified. Popular mega varieties like HD2967, HD2733, GW322, GW366, PBW550, HD3086 and WH1105 were sensitive to heat and drought stress. Genotypes HD3043, HD2987, HD2781, RAJ3037, DBW43, DBW59, HI1500, C-306, HD2888, HD2932, Sokulu and Babax performed better under moisture-stress condition while HD3059, DBW14, WH730, Halna, WR544, HD3118, HD2985, Hindi 62, Chirya 7, Kauz//AA//Kauz, Excalibur, RAC875 and Pastor can be used as donor for heat tolerance. However genotypes DBW43, HD2932, WH1021 and RAJ3765 were discovered to be better donors for both drought and heat stress.

MARKER ASSISTED BACKCROSS BREEDING FOR HEAT AND DROUGHT TOLERANCE

In the genetic background of GW322 and HD2733 1–4 QTLs at Meta QTLs regions were transferred using six different marker assisted backcross populations for heat and drought tolerance. In all the six backcross populations two back-crosses were attempted. QTLs linked to component traits like Qtlyld(s).5D, Qtlyld(s).4A, Qtlda.5A, Qtlyld.4B, QtlCt.3B, Qtlbiomass.4A, Qtl.Stay-green.2D and Qtl.NDVI.2D were validated and transferred into the popular varieties GW322 and HD2733. A total of 25 QTL NILs are available at Division of Genetics in different combinations with background recovery ranges from 80.2% to 97.4%. The backcross derived lines of six MABB (marker-assisted backcross breeding) populations were evaluated under rainfed and late sown conditions for their performance with respect to yields of checks and parents. Five best lines were selected from each MABB population for their advancement to station trials and large plot trials.

MARKER ASSISTED RECURRENT SELECTION (MARS) FOR HEAT AND DROUGHT TOLERANCE

Lande and Thompson (1990) introduced the theory for optimizing weights given to each marker and demonstrated that this index is as efficient as the phenotypic score alone. This approach of marker assisted recurrent selection used only markers which have been identified as being significantly associated to QTL. The efficiency of MAS/phenotype selection is higher when the trait has a low heritability, the population size is large and the detected QTLs explain a large proportion of the trait variation. Hospital *et al.* (1997) showed that the use of marker index only allows early selection, without trait evaluation, thereby shortening selection cycles and accelerating genetic gain per cycle. Recombination of desirable alleles at QTLs loci in generation after generation using validated molecular marker information to increase the frequency of favourable alleles is MARS. In this approach phenotyping will be done for component traits and molecular marker scores will be used to select in segregating generations. Here off season summer nursery/glass houses will be used effectively to advance genetations with selection step based on marker score without evaluation. In crop species like wheat where recombination frequencies are less, selection and recombination steps of MARS will boost genetic enhancement.

In marker assisted recurrent selection (MARS), the information generated on QTLs in the mapping population of interest to develop superior lines with an optimum combination of favourable alleles originating from both parents were used. Reported QTLs and QTL alleles impacting the major traits of interest to the breeders are identified within breeding populations and accumulated through successive intercrossing only using molecular markers. Short listed recombined lines were then subjected to a final phenotypic screen to select the best lines. The approach was earlier used in sweet corn, leading to an increase in the frequency of favourable alleles from 0.50 to 0.80 at 18 out of 31 targeted loci in one F2 population and at 11 out of 35 loci in another F2 population (Edwards and Johnson, 1994). The approach has also been effectively used by Monsanto for improvement of several traits in corn, soybean and sunflower (Eathington *et al.*, 2007).

To combine QTLs for drought and heat adaptive traits using four bi-parental population MARS was conducted; in each case both parents carrying positive alleles for drought and heat tolerance. These parents were diverse based on physiological and molecular markers at 120 SSRs and 802 SNPs. Physiological and molecular characterization of wheat working collection has brought out C306 and its derivatives as distinct group from other genotypes. Based on validated QTLs information in the present population and multi-location (four location three year) testing of base populations 25 best individuals were selected in each population for intermating at F_6 generation based on the phenotypic data after AMMI analysis and validated linked molecular markers at QTLs regions top ranking individuals were selected. Carefully chosen pair-wise inter crosses were attempted to combine maximum number of QTLs using limited population size. Based on validated markers, wide range of a inter crossed progenies in advanced generation ($F_{2:6}$) with 6–14 positive alleles were obtained after second round of recombination (Figure 1).

Improved MABB and MARS lines were evaluated for drought adaptive traits in restricted irrigated condition along with check varieties and parents HI1500; HD3043; HD2987; DBW43 and C306 in four different agro-climatic conditions New Delhi, Pune, Powarkheda and Dindori. The experiment was conducted in lattice square design with two replications in 7.25 cm 2 plots size in four locations. Different morphological, physiological and agronomical traits were recorded in all the advanced lines.

Fig. 1: Advanced Line Tolerant to Drought Stress in
$F_{2:6}$ Generation in Comparison with Sensitive Genotype

Comparison biplot (Total – 94.80%)

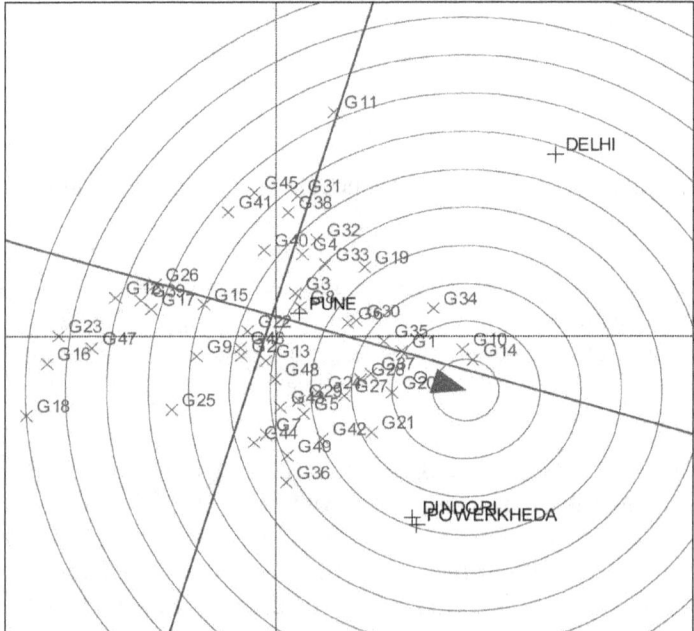

Fig. 2: Ranking of Advanced Lines and Checks Relative to
an Ideal Genotype (represented by green arrow)

Analysis of variance revealed presence of highly significant variations among the genotypes for yield and yield related traits. Ranking of advanced lines and checks relative to an ideal genotype were done by comparison with GGE biplot analysis. Genotypes G14, G10, G34, G21, G1, G20, G35, G28, G27, G37, G6, G30, G5 and G36 out performed over the check varieties in yield *per se* and G14, G10 and G34 were nearer to the ideal genotype (Figures 1 and 2). For thousand karnal weight, genotypes G14, G45, G23 and G28 were near ideal. As expected, the genotypes which contained more number QTLs out performed over checks. Genotypes with high molecular score and more number of QTLs with MABB and MARS strategies out performed over checks in multi-location testing.

DEVELOPMENT OF MULTI-PARENT ADVANCED GENERATION INTERCROSSES (MAGIC) POPULATION FOR DROUGHT AND HEAT TOLERANCE

An initiative has been taken to develop MAGIC populations, which present novel options and opportunities because of complex pedigree structure in the component lines. These offer great potential both for identifying genomic regions and for varietal development. In wheat, hybrid necrosis is the premature gradual death of the foliage in hybrids, it is based on 2 complementary genes $Ne_1Ne_1ne_2ne_2$ and $ne_1ne_1Ne_2Ne_2$; for each of these loci a multiple allelic series has been found. Genotypes were categorized for necrotic genes. Crosses with $Ne_1Ne_1ne_2ne_2$ and $ne_1ne_1Ne_2Ne_2$ were lethal while crosses like $Ne_1Ne_1ne_2ne_2/Ne_1Ne_1ne_2ne_2$, $Ne_1Ne_1ne_2ne_2/ne_1ne_1ne_2ne_2$, $ne_1ne_1ne_2ne_2/ne_1ne_1ne_2ne_2$ were not lethal; in this category the following founder lines HD3086, GW322, HI1544, HI1563, HD2932, HD2985, HD3043 and VL907 were used to avoid the lethality in the MAGIC population. Parent were selected based on their yield potential, combining ability, resistance/tolerance to biotic and abiotic stresses. A total of 70 double crosses were attempted from 28 single crosses; 35 paired 8 way crosses will be generated following the population size of 3000 individuals with 35 nodes in MAGIC population.

ACKNOWLEDGEMENTS

Authors thank funding source from Generation Challenge Programme (GCP) and National Innovation (originally, Initiative) for Climate Resilient Agriculture (NICRA) projects.

REFERENCES

Eathington, S.R., Crosbie, T.M., Edwards, M.D., Reiter, R.S. and Bull, J.K. (2007). Molecular markers in a commercial breeding program. *Crop Sci.* 47(S3): S154–S163.

Edwards, M.D. and Johnson, L. (1994). RFLPs for rapid recurrent selection. p. 33–40. *In Proc. Joint Plant Breeding Symposium Series of CSSA and ASHA, Corvallis, O.R.* 5–6 Aug. 1994. Am. Soc. Of Hort. Sci., Alexandria, V.A.

Harikrishna Singh, G.P., Neelu Jain, Singh, P.K., Sai Prasad, S.V., Divya Ambati, Das, T.R., Arun Kumar, Javiad Akther, Amasiddha Bhat, B., Priyanka Vijay, Nivedita Sinha, Mishra, P.C., Misra, S.C. and Prabhu, K.V. (2016). Physiological characterization and grain yield stability analysis of RILs under different moisture stress conditions in wheat (*Triticum aestivum* L.) *Ind J Plant Physiol.* DOI 10.1007/s40502-016-0257-9.

Hospital, F., Moreau, L., Lacoudre, F., Charcosset, A. and Gallais, A. (1997). More on the efficiency of marker-assisted selection. Theoretical and Applied Genetics 95, 1181 ± 1189.

Lande, R. and Thompson, R. (1990). Efficiency of Marker-Assisted Selection in the Improvement of Quantitative Traits. Genetics 124: 743–756.

Neelu Jain, Singh, G.P., Singh, P.K., Ramya, P., Harikrishna, Ramya, K.T., Leena Todkar, Amasiddha, B., Prashant Kumar, K.C., Priyanka Vijay, Vasudha Jadon, Sutapa Dutta, Neha Rai, Nivedita Sinha and Prabhu, K.V. (2014). Molecular approaches for wheat improvement under drought and heat stress, *Indian J. Genet.*, 74(4): 578–583.

Ramya, P., Singh, G.P., Neelu Jain, Singh, P.K., Pandey, M.K., Sharma, K., Arun Kumar, Harikrishna and Prabhu, K.V. (2016). Effect of recurrent selection on drought tolerance and related morpho-physiological traits in bread wheat. *PLOS ONE|* DOI:10.1371/journal.pone.0156869.

Adaptations to Climate Change

Enhancing Tolerance to Climatic Stresses in Rainfed Crops: The Road Ahead

M. Maheswari

ICAR-Central Research Institute for Dryland Agriculture, Santoshnagar, Hyderabad 500 059, India
E-mail: mmaheswari@crida.in

ABSTRACT: *Climate change impacts crop production through high temperature, increased incidence of droughts and floods due to altered precipitation patterns and other associated factors such as water logging, chilling, hail storms, and altered disease and pest dynamics. Agricultural systems already plagued by a plethora of challenges are under greater stress due to the accentuation of negative impacts of climate variability and climate change. The partially ameliorating effects of elevated CO_2 as demonstrated till now are not completely reassuring and concerns about over estimation of the positive effects of increased CO_2 do exist. In this context, developing resilient cultivars with consistently higher yields as well as tolerance to climatic stresses remains the key strategy for mitigating negative impacts of climate change on farm productivity. In this paper I review the various approaches to achieve the goals of this strategy.*

Keywords: Climatic Stresses, Reactive Oxygen Species, Abscisic Acid, Drought Susceptibility Index, Effective Use of Water.

Climate change impacts crop production through high temperature, increased incidence of droughts and floods due to altered precipitation patterns along with other associated factors including water logging, chilling, hail storms, disease and pest dynamics (Porter *et al.*, 2014; McKersie, 2015). Agricultural systems are already affected by several challenges while climate variability and climate change are accentuating further these negative impacts on food production. Although crop responses to CO_2 are suggested to partially ameliorate the negative impacts of climate change and water deficit on yields, concerns about over estimation of the positive effects of increased CO_2 do exist (Long *et al.*, 2006; IPCC 2013; Hatfield *et al.*, 2014). In this context, developing resilient cultivars with consistently higher yields as well as tolerance to climatic stresses is a key strategy.

Plants are exposed to a variety of climatic stresses in the field which occur either singly or in combination (Mittler, 2006). Several responses are elicited at morphological, physiological and molecular levels in crop plants under abiotic stress. Functional mechanisms and activation of a number of signaling pathways act in consort resulting in restoration of homeostasis and stress tolerance. Dissection of the complex traits into component traits, the association of traits with molecular markers and the knowledge of the candidate genes associated with functional traits and regulatory components involved in stress tolerance, will aid in developing tolerant genotypes (Shanker et al., 2014; McKersie, 2015). The genetic resources, especially land races and wild relatives from areas where past climates mimicked the projected future climates for agriculturally prime areas, could serve as donor genotypes for genetic enhancement of tolerance, maturity and yield attributes. In this context, the climate 'hotspot' areas which are more prone to risks due to weather extremes have to be identified and the possible stress environments are to be quantified.

CLIMATIC STRESS IN PLANTS

Oxidative stress leading to generation of reactive oxygen species (ROS) is central to most of the climatic stresses and at high concentrations ROS cause exacerbating damage to cellular components. Plants have mechanisms to detoxify ROS by the activation of anti-oxidative enzymes and accumulation of compatible solutes. However, if the levels of ROS exceed the ability of the plants to detoxify it, oxidative stress symptoms appear leading eventually to severe damage. These ROS also act as signal molecules in the activation of downstream signal transduction pathways. In plants, the mitogen-activated protein kinase (MAPK) cascade is activated in various abiotic stresses and in hormone responses that include ROS signaling (Moon et al., 2003; Hasanuzzaman et al., 2012).

GENE PYRAMIDING FOR STRESS TOLERANCE

The mechanism of adaptation to various abiotic stresses such as drought, heat and salinity either individually or in combination is unique and different. Since the plants are exposed to combinations of stresses in field, there is a need to pyramid tolerance genes to achieve satisfactory levels of benefits. To develop crop genotypes which can perform better under the predicted climate change, it is essential to understand the plant traits that are linked to adaptation. The pattern of plant adaptation is determined by the time duration and intensity of environmental conditions. Plant traits which favour yield and with a direct impact on mechanism of tolerance need to be considered in crop improvement programme for enhanced stress tolerance. A number of physiological traits are associated with the adaptability of the plant to stressed environments (Reynolds et al., 2016; Maheswari et al., 2012; 2016). One of the most important traits for stress tolerance is better root system and higher biomass production.

It is well known that abiotic stresses such as drought and heat elicit multigenic responses and it is important to pyramid tolerance genes to achieve satisfactory levels of benefits. A number of adaptive characteristics have been studied and used to identify tolerance to water limited environments including matching the phenology to the available water supply, early vigour, osmotic adjustment, carbon partitioning, transpiration efficiency, grain growth, membrane integrity, leaf senescence and others. Abscisic acid (ABA) plays a central role in root-to-shoot and cellular signaling in drought stress and initiates a signaling cascade to close stomata and reduce water loss. It also results in changes in leaf water potential, decreased photosynthesis and solute accumulation. Osmotic adjustment is another important mechanism facilitating plants under drought to continue water absorption and maintenance of tissue turgor thus contributing to higher photosynthetic rate and expansion growth. Proline and glycine betaine are the important compatible solutes accumulating under drought.

IMPROVING WATER-USE EFFICIENCY

One of the major approaches to crop improvement has been to assess drought tolerance on the basis of yield stability or drought susceptibility index (DSI) (Sinha *et al.*, 1986). There is a close relation between the available water in soil profile in post flowering period and yield. Significant progress has also been made in using the variability in water-use efficiency for breeding for drought tolerance (Udaykumar *et al.*, 1998). Since biomass production is tightly linked to transpiration, the effective use of water (EUW) and not the high water use efficiency is proposed to be important determinant of plant production under stress (Blum, 2009). Considerable progress has been made in the genetic dissection of flowering time, inflorescence architecture, and temperature and drought tolerance in certain model plant systems and by comparative genomics in crop plants.

HEAT TOLERANT VARIETIES

Temperature is a primary environmental factor controlling growth, development and adaptation of plants (Halford, 2009), and also determines phenology. There is a plethora of information on variation for heat tolerance in different crops species (Wheeler *et al.,* 2008; Rattunde *et al.*, 2012; Cairns *et al.,* 2013). Different physiological mechanisms may contribute to heat tolerance in the field, for example, metabolic tolerance as indicated by higher photosynthetic rates, stay-green, and membrane thermo stability, or heat avoidance as indicated manifested through canopy temperature depression and stomatal conductance. The development of heat-tolerant varieties is a feasible and cost-effective way to reduce the impacts of climate change. Regulation of stomatal function may be a target for optimal crop performance in different and potentially overlapping stresses such as drought and heat. Low canopy temperature is correlated with high stomatal transpiration and is used as a proxy for yield performance under optimal and stress conditions, facilitating drought and heat stress avoidance. The advantages and disadvantages for tolerance to different stress factors and environments need to be balanced, with incorporation of alleles conferring adaptive plasticity. Innovative solutions such as engineering ABA receptors to accept chemicals not related to ABA as agonists can contribute to the inexpensive field-scale manipulation of stomatal functions (Kissoudis *et al.,* 2016). Stay green trait is relevant in several crops such as maize, rice and sorghum can be used for yield improvement in dry, warm environments caused by climate change (Mir *et al.*, 2012).

IMPROVING TOLERANCE TO SALINITY

The degree of salt tolerance varies at different phases of life cycle of most plant species (Munns and Tester, 2008). In the simplest analysis of response of a plant to salinity stress, the reduction in shoot growth occurs in two phases; a rapid response to the increase in external osmotic potential and a slower response due to the accumulation of Na^+ in leaves. Of a number of plant responses to salt stress, over production of different types of compatible organic solutes, Na^+/K^+ ratio, ROS, stomatal closure, are the most common ones (Flowers, 2004; Munns and Tester, 2008). The challenge is to gain quantitative data for the role of specific salt tolerance mechanisms at the genetic level through to single cells and whole plants to develop models that predict the pathways leading to energy gains which will facilitate selection of crops with lower energy costs and greater yields (Munns and Gilliham, 2015).

Improving crop tolerance to climatic stresses is a highly complex challenge primarily due to genotype–environment interactions. Genetic variation of crops at intra-specific, inter-specific and inter-generic levels needs to be efficiently utilized in producing stress-tolerant genotypes. Availability of genetic variation in most of the crop species is another problem encountered for utilization for improvement through conventional approaches.

METABOLIC ENGINEERING, QTLs AND PHENOMICS

Metabolic engineering efforts for abiotic stress tolerance have largely focused on the expression of functional genes involved in osmolyte biosynthesis, enzymes coding for scavenging ROS, genes coding for LEA proteins, molecular chaperones, proteins involved in ion homeostasis. With the availability of information regarding the upstream elements and signaling components involved in various abiotic stresses and their cross talk they have been targeted to achieve multiple stress tolerance. Regulatory modules like MAPKs-based pathways, CDPK pathways and core hormone signaling modules which control the expression of a vast number of genes are key elements (Qin *et al.*, 2011). Characterization of the mechanism of action of the candidate TFs involved in stress crosstalk is of great importance.

Identification of QTLs for various traits controlling stress tolerance and pyramiding of gene/QTLs that independently confer tolerance to abiotic stress can be a robust and practical route towards increased resilience to abiotic stress combinations. Another molecular technology which gained considerable importance in developing climatic stress tolerance is marker assisted selection (MAS) as it would improve the efficiency of plant breeding through precise transfer of genomic region of interest (foreground selection) and accelerate recovery of the recurrent parent genome (background selection).

To enhance climate resilience in rainfed crops, a systematic analysis of trait-gene relations is essential to identify genes that regulate functional traits involved in stress tolerance. In this context, phenomics aided by image based, nondestructive high throughput phenotyping has emerged as a next generation tool for effectively dissecting quantitative traits related to yield and stress tolerance. The high throughput plant phenomics facility allows quantitative, non-destructive imaging allowing time series measurements which are crucial to follow the progression of growth and stress response (Fahlgren *et al.*, 2015). It is expected that high throughput plant phenotyping along with systems biology and crop modeling tools will immensely aid in improving crop adaptation to climatic stresses.

SUMMARY

Future strategies should take into account several species combinations and the wealth of genetic diversity existing in the land races and wild relatives. Also, high quality field phenotyping at sites representative of target environments needs to be addressed. An effective integration of conventional, molecular and genetic engineering as well as genomics and phenomics approaches seems to be the most promising strategy for enhancing tolerance to climatic stresses in rainfed crops.

REFERENCES

Blum, A. (2009). Effective Use of Water (EUW) and not Water Use Efficiency (WUE) is the target of crop yield improvement under drought stress. *Field Crops Research*, 112: 119–123.

Cairns, J.E., Crossa, J., Zaidi, P.H., Grudloyma, P., Sanchez, C., Araus, J.L., Thaitad, S., Makumbi, D., Magorokosho, C., Bänziger, M., Menkir, A., Hearne, S. and Atlin, G.N. (2013). Identification of drought, heat, and combined drought and heat tolerant donors in maize. *Crop Science*, 53: 1335–1346. (http: //dx.doi.org/10.2135/ cropsci2012.09.0545).

Fahlgren, N., Gehan, M. and Baxter, I. (2015). Lights, camera, action: high-throughput plant phenotyping is ready for a close up. *Current opinion in Plant biology*, 24: 93–99.

Flowers, T.J. (2004). Improving crop salt tolerance. *Journal of Experimental Botany*. 55: 307–319.

Halford, N.G. (2009). New insights on the effects of heat stress on crops. *Journal of Experimental Botany*, 60: 4215–4216.

Hassanuzzaman, M., Hossain, M.A., Teixeira da Silva, J.A. and Fujita, M. (2012). Plant response and tolerance to abiotic oxidative stress: Antioxidant defense is a key factor, In: Venkateswarlu, B., Shanker, A.K., Shanker, C., Maheswari, M. (Eds.), *Crop Stress and Its Management: Perspectives and Strategies*. Springer, Dordrecht, Heidelberg, London, New York, pp. 261–316. (doi10.1007/978-94-007-2220-0_8).

Hatfield, J., Takle, G., Grotjahn, R., Holden, P., Izaurralde, R.C., Mader, T., Marshall, E. and Liverman, D. (2014). Agriculture. In: Melillo, J.M., Richmond, T.C., Yohe, G.W. (Eds.) *Climate change impacts in the United States: The Third National Climate Assessment.* US Global Change Research Program, 6–1.

IPCC (2013). Climate change 2013: The physical science basis. Contribution of Working Group I to the Fifth Assessment Report of the Inter Governmental Panel on Climate Change. Stocker, T.F., Qin, D., Plattner, G.K., Tignor, M., Allen, S.K., Boschung, J., Nauels, A., Xia, Y., Bex, V. and Midgley, P.M. (Eds.) Cambridge, UK/Newyork: Cambridge University Press.

Kissoudis, C., Van de Wiel, C., Visser, R.G. and Van der Linden, G. (2016). Future-proof crops: Challenges and strategies for climate resilience improvement. *Current opinion in plant biology*, 30: 47–56.

Long, S.P., Ainsworth, E.A., Leakey, A.D., Nosberger, J. and Ort, D.R. (2006). Food for thought: Lower-than-expected crop yield stimulation with rising CO_2 concentrations. *Science,* 312: 1918–1921.

Maheswari, M., Yadav, S.K., Shanker, A.K., Anil Kumar, M. and Venkateswarlu, B. (2012). Overview of plant stresses: Mechanisms, adaptations and research pursuit. In: Venkateswarlu, B., Shanker, A.K. and Maheswari, M. (Eds.), *Crop Stress and Its Management: Perspectives and Strategies.* Springer, Dordrecht, Heidelberg, London, New York, pp. 1–18.

Maheswari, M., Vijaya Lakshmi, T., Varalaxmi, Y., Basudeb Sarkar, Yadav, S.K., Jainender Singh, Seshu Babu, G., Ashish Kumar, Sushma, A., Jyothilakshmi, N. and Vanaja, M. (2016). Functional mechanisms of drought tolerance in maize through phenotyping and genotyping under well watered and water stressed conditions. *European Journal of Agronomy,* 79: 43–57.

McKersie, B. (2015). Planning for food security in a changing climate. *Journal of Experimental Botany,* 66: 3435–3450.

Mir, R.R., Zaman-Allah, M., Sreenivasulu, N., Trethowan, R. and Varshney, R.K. (2012). Integrated Genomics, Physiology and Breeding Approaches for Improving Drought Tolerance in Crops. *Theoretical and Applied Genetics,* 125: 625–645.

Mittler, R. (2006). Abiotic stress, the field environment and stress combination. *Trends in Plant Science,* 11: 15–19.

Moon, H., Lee, B., Choi, G., Shin, D., Prasad, D.T., Lee, O., Kwak, S.S., Kim, D.H., Man, J., Bahk, J., Hong, J.C., Lee, S.Y., Cho, M.J., Lim, C.O. and Yun, D.J. (2003). NDP kinase 2 interacts with two oxidative stress activated MAPKs to regulate cellular redox state and enhances multiple stress tolerance in transgenic plants. *Nature,* 356: 710–713.

Munns, R. and Gilliham, M. (2015). Salinity tolerance of crops – what is the cost? *New Phytology,* 208: 668–673. doi: 10.1111/nph.13519

Munns, R. and Tester, M. (2008). Mechanisms of salinity tolerance, *Annual Review of Plant Biology,* 59: 651–681.

Porter, J.R., Xie, L., Challinor, A.J., Cochrane, K., Howden, S.M., Iqbal, M.M., Lobell, D.B., Travass, M.I., Netra Chhetri, N.C., Garrett, K., Ingram, J., Lipper, L., McCarthy, N., McGrath, J., Smith, D., Thornton, P., Watson, J. and Ziska, L. (2014). "Food security and food production systems", pp. 485–533.

Qin, F., Shinozaki, K. and Shinozaki, K.Y. (2011). Achievements and challenges in understanding plant abiotic stress responses and tolerance. *Plant and Cell Physiology*, 9: 1569–1582.

Rattunde, F., Haussmann, B.I.G., Rattunde, H.F., Weltzien-Rattunde, E. and Pierre, S.C. (2012). Breeding strategies for adaptation of pearl millet and sorghum to climate variability and change in West Africa. *Journal of Agronomy and Crop Science*, pp. 1–37.

Reynolds, M.P., Quilligan, E., Aggarwal, P.K. (2016). "An integrated approach to maintaining cereal productivity under climate change", *Global Food Security*, Vol. VII, pp. 9–18.

Shanker, A.K., Maheswari, M., Yadav, S.K., Bhanu, D., Attal., N.B. and Venkateswarlu, B. (2014). Drought stress responses in crops. *Functional Integrative Genomics*, 14: 11–22.

Sinha, S.K., Aggarwal, P.K., Chaturvedi, G.S. and Singh, A.K. (1986). Performance of wheat cultivars in a variable environment. *Grain Yeild Stability, Field Crops Research*, 13: 289–299.

Udaya Kumar, M., Seshasayee, M.S., Nataraj, K.N., Bindumadhava, H., Devendra, R., Aftab Hussain, I.S. and Prasad, T.G. (1998). Why has breeding for water use efficiency not been successful? An analysis and alternate approach to exploit this trait for crop improvement, *Current Science*, Vol. 74, pp. 994–1000.

Wheeler, T. Jagadish Krishna, Craufurd, P., Challinor, A. and Singh, M.P. (2008). "Effects of high temperature stress on grain crops in current and future climates", *Aspects of Applied Biology*, Vol. 88, pp. 22–28.

Rice Doubled Haploids for Aerobic Planting and Drought Situation

Farhad Kahani and Shailaja Hittalmani*

University of Agricultural Sciences, Bengaluru 560 065, India
*E-mail: shailajah_maslab@rediffmail.com

ABSTRACT: *Crop improvement through conventional plant breeding approach is time consuming as it requires six to seven generations of selfing to attain the desirable level of homozygosity. In the current experiment Double haploid (DH) breeding approach was used to generate homozygous lines, which reduced breeding cycles to two generations. Anthers were cultured in N6 medium supplemented with three colchicine concentrations (250, 500 and 750 mg/L) with three incubation durations (24, 48 and 72 hours). Highest rate of green plants regeneration (14.28%) was observed in 500 mg/L colchicine treatment for 48 hours of incubation with no albinos. Eight SSR markers out of 30 were found polymorphic between two parents, and these markers confirmed the true DH lines. Results from SSR markers showed either maternal or paternal banding pattern, representing the contribution of the alleles from corresponding parents. Genome size estimation by flow cytometry results indicated range of genome size from 453.2 Mb (DH6) to 468.02 Mb (DH3 and DH4). Remaining five DH lines had genome size similar to that of either parents. Experimental results from field evaluation indicated significant differences (5% level) for most of the traits under aerobic condition except for root volume, root length and panicle length. DH lines showed high grain yield (43.21 gl per plant), seed fertility (83.18%), number of panicles per plant (25.41), and number of grains per plant (1998.00) compared to their parents. Higher water use efficiency was exhibited by DH7 plants (6.98%), while DH2 plant had the least (4.79%). In brief, DH plant generation was quick and also produced superior recombinant plants for grain yield under water limited situation in just two generations. Confirmation of doubled haploids using co-dominant SSR markers and flow cytometry was accurate in detection of DH plants.*

Keywords: Homozygosity, Colchicine, SSR Markers, Flow Cytometry, Aerobic Planting.

The prime objective of any crop improvement program is the development of desirable homozygous lines after crossing parental lines with favourable traits. Conventional plant breeding methods require six to seven generations of selfing to attain the desirable level of homozygosity, which is a time consuming process. Whereas, development of new varieties through Double Haploid (DH) approach proves to be an advantage as homozygous lines could be obtained within two generations, by selecting haploids in first generation from a desired cross combination and subsequently doubling their genome in second generation. Haploid production has been reported in many crops including rice (Khush and Virmani, 1996), however its wide adoption is lacking.

Confirmation of true double haplois is an important factor before a plant is selected. To achieve this, a suitable co-dominant DNA marker for homozygosity testing is needed in breeding programme. Now a days simple sequence repeat (SSR) markers were successfully used in wheat (Muranty *et al.*, 2002) and maize (Aulinger *et al.*, 2003) DH line development, because of their quickness. In addition to DNA markers, flow cytometry has been proven as an efficient method for the rapid detection of DNA content in doubled haploids (Brito *et al.*, 2010).

Rice production frequently faces moisture stresses in rainfed ecosystem, so aerobic rice has been considered a promising cultivation method with limited use of irrigation water (Gandhi *et al.*, 2012). Therefore, development of suitable varieties that respond to less water with reasonably higher grain yield is the challenge that is essential to keep up the rice production for food security.

The major objective of this investigation was to develop rice genotypes through anther culture, confirmation of true double haploid lines by using both SSR markers and flow cytometry, and subsequently their field evaluation under aerobic situation.

Two rice genotypes IM 108 and IM 128 (derived from IR 50, an *indica* and Moroberecan, a *japonica* parents) were crossed and anthers from these F_1 hybrids were used to generate haploids and subsequently double haploid plants. For anther culture, basal medium (N6AK) consisted of N6 salts, iron sources, vitamins (Chu, 1978) and colchicine. Response of callus induction and green plant regeneration was evaluated using three colchicine concentrations at 250, 500 or 750 mg/L and three different incubation time of 24, 48 and 72 hours in the dark at 26°C. Subsequently, anthers were transferred to colchicines-free N6AK medium for 4–5 weeks and plants were regenerated from calluses by following the protocol of Alonanno and Guiderdoni (1994).

Eight polymorphic rice simple sequence repeat (SSR) markers were used for the detection of homozygous loci of parental types and confirmation of true double haploid plants. The banding pattern on agarose gel was designated as 1 for IM-108 and 2 for IM-128 specific bands, respectively. Fresh leaf samples from eight doubled haploid lines and their parents were used for estimation nuclear DNA content and ploidy by flow cytomerty, using the procedures of Arumuganathan and Earle (1991).

A set of eight doubled haploid lines confirmed by both SSR markers and flow cytometry to be doubled haploids were evaluated under aerobic condition during three *dry seasons of* 2014, 2015 and 2016 at UAS, GKVK, Bangalore, India. Data was recorded on twenty two different characters from shoot, root, and grain yield parameters. One way ANOVA was performed for DH lines and parents using SAS 9.1 program and significance differences (5% and 1%) were calculated.

CALLUS INDUCTION AND REGENERATION

High response to callus induction and regeneration was observed in 500 mg/L concentration of colchicine coupled with 48 hours incubation on MS medium. As high as 14.28% of green plants were obtained in this treatment. The regeneration rate of callus from the transferred callus varied from 1.83% in 500 mg/L colchicine concentration at 48 hours to 0.15 per cent in 750 mg/L colchicine concentration at 72 hours colchicine treatments (Table 1). Similar promotional effect of colchicine on androgenesis has also been observed by Khatun *et al.* (2003), Piyachai *et al.* (2011) and Würschum *et al.* (2013) in *indica* rice hybrids. The results also indicate that, colchicine increased the percentage of green plants at 24 and 48 hours periods whereas it decreased green plants formation at 72 hours colchicine treatments, which infers that extended incubation time has resulted in more albino plants (16.66%). The results show that application of N6 medium supplemented with 2, 4-D was advantageous to improve overall efficiency of callus induction and plant regeneration.

Table 1: Details of Regeneration Percentage of Callus from the Cross IM108 × IM128

Sl. No.	Colchicine Treatment	No. of Anthers Cultured	Transferred Callus	Regeneration Rate of Callus (%)	Green Plants (%)	Albino Plants (%)	Survival Rate (%)	
							Green Plants (%)	Albino (%)
1	24 h – 250 mg/l	1800	10	0.55	1	0	12.50	0
2	24 h – 500 mg/l	2130	11	0.56	1	0	9.09	0
3	24 h – 750 mg/l	2000	18	0.93	0	1	0	5.55
4	48 h – 250 mg/l	1990	17	0.85	2	0	11.76	0
5	48 h – 500 mg/l	1530	28	1.83	4	0	14.28	0
6	48 h – 750 mg/l	1700	10	0.51	0	1	0	10.00
7	72 h – 250 mg/l	2200	6	0.27	0	1	0	16.66
8	72 h – 500 mg/l	1750	7	0.36	0	0	0	0
9	72 h – 750 mg/l	2300	3	0.15	0	0	0	0

HOMOZYGOSITY BY SSR MARKERS AND FLOW CYTOMETRY

Eight polymorphic markers out of 30 SSR markers surveyed for polymorphism detected the presence of homozygous loci confirming that the DH lines were true double haploid (Figure 1). Four SSR markers (RM219, RM7, RM318 and RM324) exhibited maternal parent (IM108) type of banding pattern, while other marker loci (RM296, RM317 and RM21) exhibited paternal parent (IM128) type banding pattern (Figure 1). Likewise different maternal and paternal banding patterns for remaining doubled haploid lines were exhibited (Figure 1). Similar to these results, the findings in wheat (Muranty *et al.*, 2002) and maize (Aulinger *et al.*, 2003) also showed that the microsatellite locus was valuable for homozygosity testing at early stages.

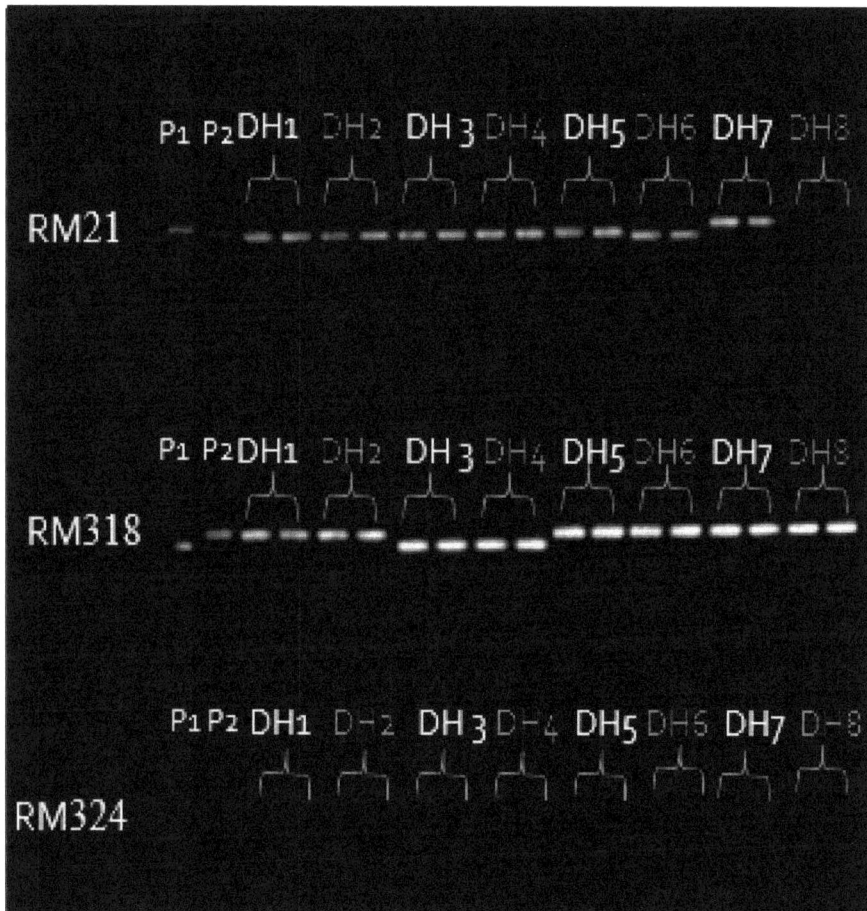

Fig. 1: Confirmation of True Doubled Haploid Lines Using SSR Markers

Genome size estimation results exhibited largest genome size in case of DH3 and DH4 plants (468.02 Mb) and smallest in DH6 (453.2). DH lines DH2, DH7 and DH8 had 458.3 Mb, which was similar to IM-128, male parent. Whereas, DH1 and DH5 had 463.2 Mb genome size, which was similar to female parent IM-108. These variations in genome of di-haploid were ascribed to the mitotic behaviour in each individual pollen grains from which it was doubled and their original ancestors (IR 50 and Moroberekan). Its robustness and reliability were exemplified by Brito *et al.* (2010).

FIELD EVALUATION OF DH LINES UNDER DIRECT SEEDED AEROBIC CONDITION

A total of 243 individual DH plants derived from eight DH lines were evaluated under aerobic field conditions and their agronomic traits were statistically analysed. Most of the characters were significant at 5% level except for root volume, root length and panicle length, which were found to be non-significant. Mean grain yield of DH5 (36.71 g), DH6 (35.00 g), DH7 (43.41 g) and DH8 (37.26 g) was superior to both parents (IM108–33.44 g, IM128–31.73 g), while reduction of grain yield plant^{-1} was observed for DH1, DH2 and DH3. These DH lines also showed high seed fertility (83.18%), number of panicles per plant (25.41), and number of grains per plant (1998.00) compared to their parents (Table 2). Similar reports were reported by Tadesse *et al.* (2013). Water use efficiency results indicated the identification of promising line for water limited environmental conditions in our study. These results showed that highest water use efficiency was present in DH7 (6.98%) while DH2 had least (4.79%) under aerobic condition. The parents, IM-108 and IM-128 exhibited 6.19% and 5.87% of WUE, respectively, with mean value of 5.94%.

Table 2: Morphological Characters of Eight Doubled Haploid Lines Grown Under Aerobic Conditions

SL No.	FGPP	WFG	UFGPP	NGPP	% F	% S
IM108 (P1)	720.00	90.00	300.00	1020.00	70.58	29.42
IM128(P2)	830.00	78.20	370.00	1200.00	69.16	30.84
F$_1$	980.00	85.25	321.00	1301.00	75.32	24.67
DH1	601.00	84.22	748.00	1349.00	44.55	55.45
DH2	1662.00	69.56	336.00	1998.00	83.18	16.82
DH3	976.00	89.21	1021.00	1997.00	48.87	51.13
DH4	686.00	85.44	376.00	1062.00	64.59	35.41
DH5	1100.00	100.52	600.00	1700.00	64.70	35.30
DH6	400.00	75.87	300.00	700.00	57.14	42.86
DH7	563.00	78.56	596.00	1159.00	48.57	51.43
DH8	1223.00	102.86	376.00	1599.00	76.48	23.52
$\bar{x}$ DH lines	901.38	85.78	544.13	1445.50	61.01	38.99
Max.	1662.00	102.86	1021.00	1998.00	83.18	55.45
Min.	400.00	69.56	300.00	700.00	44.55	16.82

FGPP = Fertile grains per plant; WFG = Weight of fertile grains; UFGPP = Unfertile grains per plant; NGPP = Number of grains per plant; F = Fertility%; S = Sterility%.

In conclusion, the results indicate that anther culture technique shortened the breeding cycle as we have generated and developed elite breeding lines within three growing seasons while conventionally, it takes six to seven seasons for selecting inbreds. Moreover, it was possible to develop plants that are homozygous and useful for aerobic cultivation in a short time.

REFERENCES

Alonanno, L. and Guidordoni, E. (1994). Increased double haploid plant regeneration from rice (*Oryza sativa* L.) anthers cultured on colchicine-supplemented media. *Plant Cell Rep.,* 13(8): 432–436.

Arumuganathan, K. and Earle, E.D. (1991). Estimation of nuclear DNA content of plants by flow cytometry. *Plant Molecular Biology Reporter*, 9: 229–241.

Aulinger, I.E., Peter, S.O., Schmid, J.E. and Stamp, P., 2003. Rapid attainment of a doubled haploid line from transgenic maize (*Zea mays* L.) plants by means of anther culture. In Vitro Cell. Dev., 39(2): 165–170.

Brito, G., Lopes, T., Loureiro, J., Rodriguez, E. and Santos, C., (2010). Assessment of genetic stability of two micro-propagated wild olive species using flow cytometry and micro-satellite markers. *Trees*, 24: 723–732.

Chu, C.C. (1978). The N6 medium and its applications to anther culture of cereal crops. In: *Proceedings of* Symposium on Plant Tissue Culture, 25–30 May, Science Press, Peking, 45–50.

Gandhi, Venkatesh, R., Rudresh, N.S., Shivamurthy, M. and Hittalmani, S.H. (2012). Performance and adoption of new aerobic rice variety MAS 946-1 (Sharada) in southern Karnataka. *Karnataka J. Agric. Sci.*, 25(1): 5–8.

Khatun, M.M., Ali, M.H. and Desamero, N.V. (2003). Effect of genotype and culture media on callus formation and plant regeneration from mature seed scutella culture in rice. *Plant Tissue Cult.*, 13: 99–107.

Khush, G.S. and Virmani, S.S. (1996). Haploids in plant breeding. In: Jain, S.M., Sopory, S.K. and Veilleux, R.E. (Eds.). In Vitro Haploid Production in Higher Plants. Vol. 1, Fundamental aspects and methods. Kluwer Academic Publishers. Dordrecht, pp. 11–33.

Muranty, H., Sourdille, P., Bernard, S. and Bernard, M. (2002). Genetic characterization of spontaneous diploid and rogenetic wheat and triticale plants. *Plant Breed.,* 121: 470–474.

Piyachai, P., Vearasilp, S., Sa-nguansak, T., Karladee, D. and Gorinstein, S. (2011). *In vitro* studies to produce double haploid in *indica* hybrid rice. *Biologia.*, 66: 1074–1081.

Tadesse, W., Tawkaz, S., Inagaki, M.N., Picard, E. and Baum, M. (2013). A Technical Manual Methods and Applications of Doubled Haploid Technology in Wheat Breeding. ICARDA Guide manual, pp. 1–16.

Würschum, T., Matthew, R., Tucker, M.R., Reif, J.C. and Maurer, H.P. (2013). Improved efficiency of doubled haploid generation in hexaploid triticale by *in vitro* chromosome doubling. *BMC Plant Biology,* 12: 109.

Glyoxalase Pathway: Characterization and Manipulation towards Developing Plant Stress Tolerance

Charanpreet Kaur, Sneh Lata Singla-Pareek and
Sudhir K. Sopory*
International Centre for Genetic Engineering and Biotechnology,
New Delhi 110 067, India
*E-mail: sopory@mail.jnu.ac.in

ABSTRACT: *Abiotic stresses limit plant growth and productivity. Crop plants encounter more than one stress at a time and to sustain agricultural productivity under such conditions, it is desirable to evolve crop types which are resilient to variable climatic conditions and stresses. Both molecular breeding and genetic engineering techniques have been undertaken by the researchers to improve stress tolerance levels of plants. Abiotic stress tolerance is a multigenic trait and hence, there is a need to identify the candidate gene(s) of specific tolerance pathways. Plants have different mechanisms to minimize the levels of toxic compounds produced during stress. Glyoxalase system constitutes one of the important metabolic pathways which can metabolize methylglyoxal (MG), a toxic byproduct of glycolysis to D-lactate. Under stress conditions, levels of MG rise to lethal concentrations and hence MG has to be detoxified. Glyoxalase system comprises two enzymes, glyoxalase I (GLYI) and glyoxalase II (GLYII). We demonstrated that transgenic plants expressing glyoxalase have better stress tolerance. Many researchers have tested the efficacy of glyoxalase pathway genes in improving stress tolerance in different plant systems. Research findings indicate that glyoxalases can be promising candidates for genetic manipulation in order to improve stress tolerance in crop plants.*

Keywords: Abiotic Stress, Methylglyoxal, Glyoxalase.

Abiotic stresses such as, salinity, drought and extreme temperatures frequently limit plant growth and productivity. In field conditions, plants generally encounter more than one stress at a time. In order to sustain agricultural productivity under such conditions, it is much desirable to generate crops which are resilient to variable climatic conditions and stresses. As a part of such attempts, both molecular breeding and genetic engineering techniques have been undertaken by the researchers to develop stress tolerance in plants but the progress is slow due to various technical and regulatory issues.

Abiotic stress tolerance, whether it be towards salinity, drought, extreme temperature or heavy metals, is a multigenic trait and hence, there is a need to identify appropriate candidate gene(s) which can show promising results, such that altering its levels has ample effect on a number of other processes related to tolerance mechanisms. For this, an inclusive knowledge of all metabolic processes related directly or indirectly to mechanism of stress tolerance is vital for generating stress-tolerant plants.

Glyoxalase system constitutes an important metabolic pathway which can metabolize methylglyoxal (MG), a toxic byproduct of glycolysis, to D-lactate and thus, plays an important role in controlling MG levels in the system. Under stress conditions, levels of MG rise to lethal concentrations and this has been observed as a general consequence of all abiotic stresses in plants (Yadav *et al.,* 2005; Hossain *et al.,* 2009). The toxicity of MG stems from the fact that it possesses two reactive groups, aldehyde and ketone, which can readily modify proteins, DNA and phospholipid moieties, rendering them inactive (Thornalley, 2008). Thus, being a strong glycating agent, elevated MG levels can cause extensive damage in the cellular milieu of living systems. The glyoxalase pathway can detoxify this cytotoxin and protect plants from its deleterious effects. Considering their important role in the biological systems, this pathway seems to be a good candidate for raising stress tolerance in plants (Singla-Pareek *et al.,* 2003).

Glyoxalase system is ubiquitous in nature and comprises two enzymes, glyoxalase I (GLYI) and glyoxalase II (GLYII). The glyoxalase I (GLYI) enzyme catalyes the first step of the glyoxalase pathway, converting MG into S-D-lactoyl-glutathione (SLG) and the second enzyme, glyoxalase II (GLYII) then converts SLG to D-lactate as shown in Figure 1 (Mustafiz *et al.,* 2014; Ghosh *et al.,* 2014). In this process, a molecule of reduced glutathione (GSH) is used initially to form hemithioacetal in spontaneous combination with MG, which then serves as the substrate of GLYI. The GSH is released in the second step catalysed by GLYII, along with the product D-lactate, thereby recycling GSH in the system. Glyoxalase pathway has been extensively characterized from the microbial and animal systems, however detailed work on plant glyoxalases is lacking.

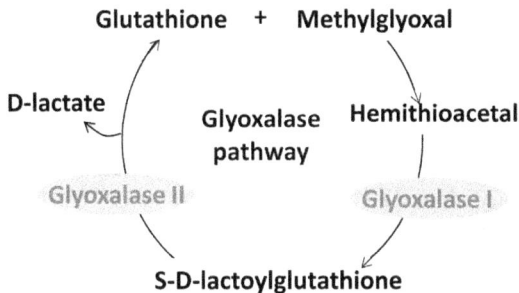

Fig. 1: Glyoxalase Pathway of the Living Systems

PRESENCE OF MULTIPLE GLYOXALASE ISOFORMS IN PLANTS

In plants, glyoxalases have been found to be present as multiple members (Mustafiz *et al.*, 2011; Kaur *et al.*, 2013). This is in sharp contrast to other eukaryotes and prokaryotes which mostly possess a single copy of glyoxalase genes. GLYI enzymes have been grouped into two categories based on the metal ion requirement for its catalytic activity viz. Ni(II) and Zn(II) dependent GLYI. Previously, Ni(II)-dependence of GLYI was linked to prokaryotes and the Zn(II)-dependence to eukaryotes. But characterization of newer GLYI candidates has led to rejection of this classification. In plants, both types of GLYI are present (Mustafiz *et al.*, 2014; Kaur *et al.*, 2016a; Jain *et al.*, 2016). Like GLYI, GLYII enzymes are also metalloenzymes but show no such metal-based preferences and usually possess binuclear metal centers binding Fe(II) and Zn(II) (Marasinghe *et al.*, 2005; Limphong *et al.*, 2010; Ghosh *et al.*, 2014).

We have previously reported that both rice and *Arabidopsis* genome possess 11 GLYI genes, with rice genome possessing four Ni(II)-dependent and one Zn(II)-dependent form and, *Arabidopsis* genome encoding three Ni(II)-dependent and one Zn(II)-dependent form (Mustafiz *et al.*, 2011). Rest of the 7 GLYI genes in rice and 8 GLYI in *Arabidopsis* possibly encode inactive forms of the enzyme. Recently, *G. max* genome has also been shown to possess multiple GLYI isoforms having eight Ni(II)-dependent and three Zn(II)-dependent GLYI and rest 13 possibly encoding inactive forms of the enzyme (Ghosh and Islam 2016). Likewise, multiple GLYII genes are also present in these species, with rice genome possessing three GLYII genes, *Arabidopsis* genome possessing five GLYII genes and *G. max* genome encoding 12 GLYII genes (Mustafiz *et al.*, 2011; Ghosh and Islam 2016; Kaur *et al.*, 2016b). Similar to GLYI family, some genes in even GLYII family of plants encode inactive forms of the enzyme and have been shown to adopt other functions, such as AtETHE1 and OsGLYII-3 (OsETHE1) proteins, which lack GLYII activity but possess sulfur dioxygenase activity (Holdorf *et al.*, 2012; Kaur *et al.*, 2014). Interestingly, these GLYI and GLYII isoforms have been found to be differentially induced under stress conditions (Mustafiz *et al.*, 2011; Kaur *et al.*, 2013; Ghosh and Islam, 2016).

We have recently characterized two GLYI candidates from rice, one being a Ni(II)-dependent GLYI, OsGLYI-11.2 enzyme, which is a monomeric enzyme possessing only one functionally active site (Mustafiz *et al.*, 2014), and the other being a homodimeric Zn(II)-stimulated GLYI, OsGLYI-8, which shows an apparently metal ion independent but stimulated nature and exhibits higher catalytic activity than OsGLYI-11.2 (Kaur *et al.*, 2016a). The most striking feature of this enzyme is its nuclear localization, previously not reported for any glyoxalase enzyme. This enzyme is important for MG detoxification in the nucleus and loss of its homolog in *Arabidopsis* leads to MG and salinity sensitivity in the system (Kaur *et al.*, 2016a). However, it is the *OsGLYI-11.2* gene which has been found to be highly stress-inducible and not OsGLYI-8

(Mustafiz *et al.*, 2011), indicating a likely housekeeping function of OsGLYI-8 which is indispensable for proper growth and development. Besides, there are many GLYI-like enzymes as well, which have highly stress-inducible nature such as, OsGLYI-6 from rice, but their functional role is still unknown.

ROLE OF GLYOXALASE PATHWAY IN PLANT STRESS TOLERANCE

Glyoxalases have been found to play an important role in raising stress tolerance in plants. These enzymes have been over-expressed in various plant systems, either individually (GLYI or GLYII) or as a pathway, and are known to impart tolerance against salinity, drought and heavy metal stresses in rice and tobacco. In first report of ectopic expression of a *GLYI* gene in the plant system, we reported that heterologous expression of *Brassica juncea BjGLYI* in tobacco could confer stress tolerance in plants (Veena *et al.*, 1999). It was found that the transgenic tobacco plants expressing *BjGLYI* could tolerate high NaCl (800 mM) and MG (20 mM) conditions much better than the untransformed plants (Veena *et al.*, 1999). Later, the same *GLYI* gene was transformed in *Vigna mungo*, and similar to tobacco, over-expression of *GLYI* in blackgram also imparted salinity tolerance (Bhomkar *et al.*, 2008). Further, *GLYI* gene from sugar beet has also been functionally validated and similar to other *GLYI*, over-expression of this gene in tobacco confers significant tolerance to MG, salt, mannitol and H_2O_2 treatments (Wu *et al.*, 2013). Furthermore, TaGLYI from wheat when overexpressed in tobacco results in enhanced tolerance to zinc (Lin *et al.*, 2010).

Like GLYI, GLYII gene has also been transformed in various plant species and the resulting transgenic tobacco, rice and *B. juncea* plants, harbouring rice GLYII gene, show improved tolerance towards high MG and NaCl concentrations (Singla-Pareek *et al.*, 2003; 2008; Wani and Gosal, 2011; Saxena *et al.*, 2011; Ghosh *et al.*, 2014). The transgenic tobacco plants have been shown to be able to grow, flower and set normal viable seeds under continuous salinity stress conditions (Singla-Pareek *et al.*, 2003). Likewise, the transgenic rice plants can also sustain growth and retain more favourable ion balance compared to the untransformed plants, leading to improved salt tolerance (Singla-Pareek *et al.*, 2008).

Importantly under salinity stress, double transgenics obtained from the transformation of both GLYI and GLYII genes perform even better than the single transgenic plants harbouring either GLYI or GLYII constructs, that too with only 5% yield penalty (Singla-Pareek *et al.*, 2003). The penalty on yield in the single construct-transformed plants was higher (Singla-Pareek *et al.*, 2003). Interestingly, expressing GLYI under the control of a stress-inducible rd29A promoter also abolishes any such yield penalty in GLYI overexpressing *B. juncea* lines (Rajwanshi *et al.*, 2016). Transgenic tomato and citrus plants transformed with the glyoxalase pathway genes also show enhanced salinity stress tolerance (Alvarez Viveros *et al.*, 2013; Alvarez-Gerding *et al.*, 2015). In tomato, this has been explained to be due to decrease in oxidative stress (Alvarez-Viveros *et al.,* 2013). In transgenic citrus plants, less yellowing, marginal burn in lower leaves and less than 40% of leaf damage is observed in comparison to the wild type plants (Alvarez-Gerding *et al.,* 2015). Further, double transgenic tobacco plants grow normally even in the presence of 5 mM $ZnCl_2$ without any yield penalty (Singla-Pareek *et al.,* 2006) and can tolerate toxic concentrations of other heavy metals, as cadmium and lead. Our recent work (unpublished) has shown that overexpression of GLYI and GLYII can confer multiple stress tolerance in rice.

Table 1: Genetic Manipulation of Glyoxalase Genes
for Improving Stress Tolerance in Plants

Source Species	Gene Name	Host Species	Stress	References
Brassica juncea	*BjGLYI*	Tobacco	MG, NaCl	Veena *et al.,* 1999
Oryza sativa	*OsGLYII*	Tobacco	MG, NaCl	Singla-Pareek *et al.,* 2003
Brassica juncea	*rd29A-BjGLYI, Cm-BjGLYI*	*Vigna mungo*	MG, NaCl	Bhalla-Sarin *et al.,* 2004
Brassica juncea	*BjGLYI*	Rice	NaCl	Verma *et al.,* 2005
Brassica juncea + *Oryza sativa*	*BjGLYI* + *OsGLYII*	Tobacco	NaCl, $ZnCl_2$	Singla-Pareek *et al.,* 2003; 2006
Brassica juncea	*BjGLYI*	*Vigna mungo*	NaCl	Bhomkar *et al.,* 2008
Brassica juncea	*BjGLYI*	Arabidopsis	NaCl	Roy *et al.,* 2008
Triticum aestivum	*TaGLYI*	Tobacco	Zn	Lin *et al.,* 2010
Oryza sativa	*OsGLYII*	Rice	MG, NaCl	Singla-Pareek *et al.,* 2008; Wani and Gosal 2011
Oryza sativa	*OsGLYII*	*Brassica juncea*	NaCl	Saxena *et al.,* 2011
Brassica juncea + *Pennisetum glaucum*	*BjGLYI* + *PgGLYII*	Tomato	NaCl	Alvarez Viveros *et al.,* 2013
B. vulgaris	*BvM14-GLYI*	Tobacco	MG, NaCl, mannitol and H_2O_2	Wu *et al.,* 2013
Oryza sativa	*OsGLYI-11.2*	Tobacco, *E. coli*	MG, NaCl, Mannitol, H_2O_2	Mustafiz *et al.,* 2014
Oryza sativa	*OsGLYII-2*	Tobacco, *E. coli*	MG, NaCl	Ghosh *et al.,* 2014
Brassica juncea + *Pennisetum glaucum*	*BjGLYI* + *PgGLYII*	Carrizo Citrange	NaCl	Alvarez-Gerding *et al.,* 2015
Arabidopsis thaliana	*AtGLYI2, AtGLYI3, AtGLYI6*	*E. coli*	MG, NaCl, Mannitol, Heat, H_2O_2	Jain *et al.,* 2016
Brassica napus	*BnGLYI-2, BnGLYI-3*	*Pichia pastoris*	Heat, cold	Yan *et al.,* 2016
Oryza sativa	*OsGLYI*	Rice	MG, NaCl, $ZnCl_2$	Zeng *et al.,* 2016
Brassica juncea	*rd29A-BjGLYI*	*Brassica juncea*	MG, NaCl, Mannitol, $ZnCl_2$	Rajwanshi *et al.,* 2016

Recently, we have checked the efficacy of the Ni(II)-dependent OsGLYI-11.2 in conferring stress tolerance in *Escherichia coli* as well as the model plant tobacco. Heterologous expression of OsGLYI-11.2 in tobacco leads to improved adaptation against various abiotic stresses as a result of increased MG scavenging, lower Na^+/K^+ ratio and maintenance of reduced GSH levels (Mustafiz *et al.,* 2014). Similarly, ectopic expression of a rice GLYII, OsGLYII-2, in *E. coli* and tobacco also provides improved stress tolerance possibly by maintaining levels of MG and GSH as well as better photosynthesis rate and reduced oxidative damage in transgenic plants under stress conditions (Ghosh *et al.,* 2014). Heterologous expression of Ni(II) and Zn(II) dependent GLYI enzymes from *Arabidopsis* also leads to better tolerance against various stress conditions in *E. coli* (Jain *et al.,* 2016). Recently, a *BnGLYI3* from *B. napus*, has been shown to be involved in seed thermotolerance as over expression of this gene imparts thermo- and cold-tolerance in yeast (Yan *et al.,* 2016). Table 1 summarizes the available reports describeing the role of glyoxalase pathway genes in stress tolerance in various plants.

To conclude, various groups have by now tested the efficacy of glyoxalase pathway genes in improving stress tolerance in different plant systems and their findings indicate that glyoxalases can be promising candidates for genetic manipulation in order to raise stress tolerance in plants. We believe that this pathway by detoxifying MG indirectly affects a number of metabolic and developmental processes in plants thereby leading to such promising results in stress response.

REFERENCES

Alvarez Viveros, M.F., Inostroza-Blancheteau, C., Timmermann, T., Gonzalez, M. and Arce-Johnson, P. (2013). Overexpression of GlyI and GlyII genes in transgenic tomato (*Solanum lycopersicum* Mill.) plants confers salt tolerance by decreasing oxidative stress. *Mol. Biol. Rep.* 40, 3281–3290.

Alvarez-Gerding, X., Cortés-Bullemore, R., Medina, C., Romero-Romero, J.L., Inostroza-Blancheteau, C., Aquea, F. and Arce-Johnson, P. (2015). Improved salinity tolerance in *Carrizo citrange* rootstock through overexpression of glyoxalase system genes. *Biomed Res Int.* 2015, 827951.

Bhalla-Sarin, N., Bhomkar, P., Debroy, S., Sharma, N., Saxena, M., Upadhyaya, C.P., Muthusamy, A., Pooggin M. and Hohn, T. (2004). Transformation of *Vigna mungo* (blackgram) for abiotic stress tolerance using marker free approach. *Australian Agronomy Conference.*

Bhomkar, P., Upadhyay, C.P., Saxena, M., Muthusamy, A., Prakash, N.S., Poggin, K., Hohn, T. and Sarin, N.B. (2008). Salt stress alleviation in transgenic *Vigna mungo* L. Hepper (blackgram) by overexpression of the glyoxalase I gene using a novel Cestrum yellow leaf curling virus (CmYLCV) promoter. *Mol. Breeding*, 22, 169–181.

Ghosh, A. and Islam, T. (2016). Genome-wide analysis and expression profiling of glyoxalase gene families in soybean (*Glycine max*) indicate their development and abiotic stress specific response. *BMC Plant Biol.*, 16, 87.

Ghosh, A., Pareek, A., Sopory, S.K. and Singla-Pareek, S.L. (2014). A glutathione responsive rice glyoxalase II, OsGLYII-2, functions in salinity adaptation by maintaining better photosynthesis efficiency and anti-oxidant pool. *Plant J.* , 80, 93–105

Holdorf, M.M., Owen, H.A., Lieber, S.R., Yuan, L., Adams, N., Dabney-Smith, C. and Makaroff, C.A. (2012). *Arabidopsis* ETHE1 encodes a sulfur dioxygenase that is essential for embryo and endosperm development. *Plant Physiol.*, 160, 226–236.

Hossain, M.A., Hossain, M.Z. and Fujita, M. (2009). Stress-induced changes of methylglyoxal level and glyoxalase I activity in pumpkin seedlings and cDNA cloning of glyoxalase I gene. *Aust. J. Crop. Sci.*, 3, 53–64.

Jain, M., Batth, R., Kumari, S. and Mustafiz, A. (2016). *Arabidopsis thaliana* contains both Ni^{2+} and Zn^{2+} dependent Glyoxalase I enzymes and ectopic expression of the latter contributes more towards abiotic stress tolerance in *E. coli. PLoS One*, 11, e0159348.

Kaur, C., Mustafiz, A., Sarkar, A.K., Ariyadasa, T.U., Singla-Pareek, S.L. and Sopory, S.K. (2014). Expression of abiotic stress inducible ETHE1 like protein from rice is higher in roots and is regulated by calcium. *Physiol. Plant.*, 152, 1–16.

Kaur, C., Sharma, S., Singla-Pareek, S.L. and Sopory, S.K. (2016a). Methylglyoxal detoxification in plants: Role of glyoxalase pathway. *Ind. J. Plant Physiol.*, 21, 377–390.

Kaur, C., Tripathi, A.K., Nutan, K.K., Sharma, S., Ghosh, A., Tripathi, J.K., Pareek, A., Singla-Pareek, S.L. and Sopory, S.K. (2016b). A nucleus-localized rice glyoxalase I enzyme, OsGLYI-8 functions in the detoxification of methylglyoxal in the nucleus. *Plant J.*, DOI: 10.1111/tpj.13407.

Kaur, C., Vishnoi, A., Ariyadasa, T.U., Bhattacharya, A., Singla-Pareek, S.L. and Sopory, S.K. (2013). Episodes of horizontal gene-transfer and gene-fusion led to co-existence of different metal-ion specific glyoxalase I. *Sci. Rep.*, 3, 3076.

Limphong, P., McKinney, R., Adams, N., Makaroff, C., Bennett, B. and Crowder, M. (2010). The metal ion requirements of *Arabidopsis thaliana* Glx2-2 for catalytic activity. *J. Biol. Inorg. Chem.*, 15, 249–258.

Lin, F., Xu, J., Shi, J., Li, H. and Li, B. (2010). Molecular cloning and characterization of a novel glyoxalase I gene *TaGly* I in wheat (*Triticum aestivum* L.). *Mol. Biol. Rep.*, 37, 729–735.

Marasinghe, G.P., Sander, I.M., Bennett, B., Periyannan, G., Yang, K.W., Makaroff, C.A. and Crowder, M.W. (2005). Structural studies on a mitochondrial glyoxalase II. *J. Biol. Chem.*, 280, 40668–40675.

Mustafiz, A., Ghosh, A., Tripathi, A.K., Kaur, C., Ganguly, A.K., Bhavesh, N.S., Tripathi, J.K., Pareek, A., Sopory, S.K. and Singla-Pareek, S.L. (2014). A unique Ni^{2+}-dependent and methylglyoxal-inducible rice glyoxalase I possesses a single active site and functions in abiotic stress response. *Plant J.*, 78, 951–963.

Mustafiz, A., Singh, A.K., Pareek, A., Sopory, S.K. and Singla-Pareek, S.L. (2011). Genome-wide analysis of rice and *Arabidopsis* identifies two glyoxalase genes that are highly expressed in abiotic stresses. *Funct. Integr. Genom.*, 11, 293–305.

Rajwanshi, R., Kumar, D., Yusuf, M.A., DebRoy, S. and Bhalla-Sarin, N. (2016). Stress-inducible overexpression of glyoxalase I is preferable to its constitutive overexpression for abiotic stress tolerance in transgenic *Brassica juncea. Mol. Breeding*, 36, 1–15.

Roy, S.D., Saxena, M., Bhomkar, P.S., Pooggin, M., Hohn, T. and Bhalla-Sarin, N. (2008). Generation of marker free salt tolerant transgenic plants of *Arabidopsis thaliana* using the *gly I* gene and *cre* gene under inducible promoters. *Plant Cell Tissue Org. Cult.*, 95, 1–11

Saxena, M., Roy, S.B., Singla-Pareek, S.L., Sopory, S.K. and Bhalla-Sarin, N. (2011). Overexpression of the Glyoxalase II gene leads to enhanced salinity tolerance in *Brassica juncea. The Open Plant Sci. J.*, 5, 23–28.

Singla-Pareek, S.L., Reddy, M.K. and Sopory, S.K. (2003). Genetic engineering of the glyoxalase pathway in tobacco leads to enhanced salinity tolerance. *Proc. Natl. Acad. Sci. USA*, 100, 14672–14677.

Singla-Pareek, S.L., Yadav, S.K., Pareek, A., Reddy, M.K. and Sopory, S.K. (2006). Transgenic tobacco overexpressing glyoxalase pathway enzymes grow and set viable seeds in zinc-spiked soils. *Plant Physiol.*, 140, 613–623.

Singla-Pareek, S.L., Yadav, S.K., Pareek, A., Reddy, M.K. and Sopory, S.K. (2008). Enhancing salt tolerance in a crop plant by overexpression of glyoxalase II. *Transgenic Res.*, 17, 171–180.

Thornalley, P.J. (2008). Protein and nucleotide damage by glyoxal and methylglyoxal in physiological systems – role in ageing and disease. *Drug Metabol. Drug Interact.*, 23, 125–150.

Verma, M., Verma, D., Jain, R.K., Sopory, S.K. and Wu, R. (2005). Over expression of glyoxalase I gene confers salinity tolerance in transgenic *japonica* and *indica* rice plants. *Rice Genet. Newslett.*, 22, 58–62.

Veena, Reddy, V.S. and Sopory, S.K. (1999). Glyoxalase I from Brassica juncea: Molecular cloning, regulation and its over-expression confer tolerance in transgenic tobacco under stress. *Plant J.*, 17, 385–395.

Wani, S.H. and Gosal, S.S. (2011). Introduction of *OsglyII* gene into *Oryza sativa* for increasing salinity tolerance. *Biol. Plantarum*, 55, 536–540.

Wu, C., Ma, C., Pan, Y., Gong, S., Zhao, C., Chen, S. and Li, H. (2013). Sugar beet M14 glyoxalase I gene can enhance plant tolerance to abiotic stresses. *J. Plant Res.*, 126, 415–425.

Yadav, S.K., Singla-Pareek, S.L., Ray M., Reddy M.K. and Sopory S.K. (2005). Methylglyoxal levels in plants under salinity stress are dependent on glyoxalase I and glutathione. *Biochem. Biophys. Res. Commun.*, 337, 61–67.

Yan, G., Lv, X., Gao, G., Li, F., Li, J., Qiao, J., Xu, K., Chen, B., Wang, L., Xiao, X. and Wu, X. (2016). Identification and characterization of a glyoxalase I gene in a rapeseed cultivar with seed thermotolerance. *Front. Plant. Sci.*, 7.

Zeng, Z., Xiong, F., Yu, X., Gong, X., Luo, J., Jiang, Y., Kuang, H., Gao, B., Niu, X. and Liu, Y. (2016). Overexpression of a glyoxalase gene, OsGly I, improves abiotic stress tolerance and grain yield in rice (*Oryza sativa* L.). *Plant Physiol. Biochem.*, 109, 62–71.

Adaptation of Crops to Climate Change: Relevant Mechanisms and their Genetic Enhancement

M. Udayakumar*, Ramu S. Vemanna, D. Prathibha, Gowsiya Shaik, S. Mahesh, G.J. Pavithra and M.S. Sheshshayee

Department of Crop Physiology, University of Agricultural Sciences, GKVK, Bengaluru 560 065, India
*E-mail: udayakumar_m@yahoo.com

ABSTRACT: *The predicted changes in climatic factors especially temperature and VPD substantially increase water requirement of the crops signifying the need to improve water productivity. Improving transpiration efficiency of the crops has phenomenal relevance. There are now options for genetic enhancement of transpiration efficiency of crops. High transpiration would affect specific physiological processes and reproductive parameters, carbon use efficiency, respiratory efficiency and several factors of cell metabolism. Many adaptive mechanism/traits have been identified in crops and also the molecular mechanisms that reprogram the genetic machinery for adaptation are being evaluated. Significant progress made in this area provides leads in genetic enhancement of these traits. Transgenic crops tested so far provided evidences on the relevance of genes regulating specific mechanisms. These leads will complement the ongoing programme for genetic enhancement of heat stress tolerance. Molecular breeding approaches are emerging as potential option to identify QTL's associated for some of the relevant mechanisms, though restricted to integrated traits. Accurate phenotyping for specific traits is the major limitation for exploiting genetic variability. Yet another limitation is narrow genetic variability, therefore identifying wild relatives from warmer climate and introgressing superior alleles will be rewarding.*

Keywords: Green House Gases, Vapour Pressure Deficit, Multi-Parent Advanced Generation Inter Cross.

Several strategies, ranging from observation of the ice cores, floral and faunal distribution records, stable isotopes etc., form the basis of examining the past climate and also to some degree, provide evidence to predict the future. Several models and scenarios predict mean change in the specific environmental factors and their impacts on crop productivity. Unlike the natural factors (solar variability, volcanic dust levels and geological changes) which historically have been causative factors for climate change, the predicted climate change is predominantly due to the green house gases (GHG) driven by anthropogenic activities. Even with restriction in the emission of GHG the likely scenario of location specific changes in the critical environmental variables are well predicted and their impacts are being assessed (UNFCC, Climate change). Based on the emission scenario of GHG, several models have predicted significant increase in CO_2, leading to rise in temperature and Vapour pressure deficit (VPD). These changes will have a direct effect on the precipitation pattern and evaporation.

CLIMATE CHANGE AND WATER PRODUCTIVITY—
NEED TO IMPROVE TRANSPIRATION EFFICIENCY

The most significant factor directly related to crop productivity is change in water availability and water demand patterns. The temperature driven indirect changes in precipitation pattern and increased evapo-transpiration (ET) affects water resources/availability and crop water demand. The water scarcity thus emerges as a single most determining factor that affects global ecosystem and crop yields. The other dimension to this water crisis is increasing "virtual water" requirements to meet the increasing food needs which is population driven. The present water availability in terms of surface water (684 km^3) and ground water (423 km^3) together is 1107 km^3 in India. The per-capita water availability decreased from 3008 m^3/yr/person in 1951 to 938 m^3 in 2010 and predicted to be 814 m^3 and 687 m^3 in 2025 and 2050 respectively (Rakesh-Kumar *et al.,* 2005; Chaterjee, 2011).

It is crucial to assess the water footprint of the available water for the different consumption needs of the people. The water foot print for the average consumer is around 3800 lit/day, out of which 3.8% of the water relates to home water use and the remaining 96.2% of the water footprint is invisible related to the agricultural and industrial products of which 91.5% is agricultural products. This signifies that the bulk of the water requirement is for food production (Hoekstra and Mekonnen, 2012).

With significant advancement in genetic enhancement of crops and production technologies the water required to produce one kilo calorie energy through food is 0.67 litres/vegetarian diet and 1.05 litres/meat diet (Hoekstra, 2010). Given the food consumption patterns and suggested food calorie requirement of 3400 kcal/day/person, the percapita water requirement is 840 m^3/yr/person. However, the available water to agriculture is 729 m^3. With diminishing allocation of water to agriculture which is likely to decrease from existing 80 to 65% by 2050, the predicted per capita availability of water for agricultural production would be 420 m^3 in 2030 and 240 m^3 in 2050. Since 90% of water footprint comes from the agricultural products, the choice of products/crops with less hidden virtual water like coarse cereals, tuber crops assumes significance when compared to the major food crops like rice and wheat. The choice of the climate smart crops has to be viewed from this perspective.

Improving water productivity would be an option to sustain crop productivity. However, the predicted increase in temperature will have far reaching effects not only on water availability but also on water productivity. One of the major reasons is high temperature induced increase in Evapotranspiration (ET). Every one degree increase in temperature is likely to increase the potential ET by 16–25% between 20–35°C based on E-pan measurements (Figure 1a; source—http://onlinehydro.sdsu.edu/onlinethornthwaite.php). Productivity per unit water in-put substantially decreases by 2080, which is predominantly due to increase in ET and decrease in grain filling period (Figure 1b).

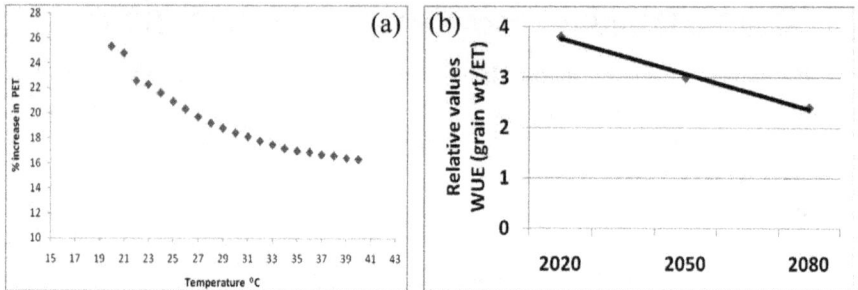

Fig. 1: The predicted increase in potential evapotranspiration (PET) for every 1°C increase in temperature. a) At lower mean temperature the increase in PET is more compared to high temperature. b) WUE at crop level at predicted high temperature in 2050 and 2080.

The water requirement of the crops is determined based on ET crops and losses due to conveyance, application and the distribution. This signifies the importance of improving irrigation efficiency to minimise the losses which account for 30–40% of water requirement. The need is to develop more cost effective novel technologies to improve irrigation efficiency and their adoption. Further 15–30% gap exists between water requirement based on ETC and actual crop water requirement considering crop transpiration dynamics. Therefore, optimization of irrigation needs based on soil and plant water status using sensor and non invasive imaging technologies has significance (Junker *et al.*, 2014). The agronomic mitigation options have significant relevance to improve water productivity. However, from the genetic enhancement point of view enhancing Transpiration Efficiency (TE) has paramount importance to improve the water productivity. The major component of ET is transpiration and it contributes up to 70% of the ET. Therefore, improving TE under the conditions of climate change is one of the crucial aspects. High temperature driven increase in VPD would substantially enhance the transpiration without being associated with carbon gain thus, reducing the TE. Though increase in CO_2 is likely to improve the carbon gain there is contradictory effects of CO_2 and temperature on carbon gain. The accepted concept of CO_2-physiological forcing further enhances the leaf temperature, enhancing the VPD driven transpiration and also decreases the carbon use efficiency (Cao *et al.*, 2010). Improving TE has been one of the major thrust and recent scientific discoveries provided leads in developing tools to capture genetic variability and assessing genetic and physiological basis for variability and finally to identify relevant mechanisms that need genetic enhancement.

Existing genetic variability in specific plant traits/mechanisms that maintain lower leaf temperature (water mining and waxes) and improve mesophyll efficiency by higher carboxylation activity of RUBISCO and its activation can be exploited to improve TE even at elevated temperature. Improving the TE has been one of the major research thrust. Our earlier studies showed that the TE is a relevant trait across the genotypes with adequate canopy cover and total transpiretion. We demonstrated the importance of identifying genotypes with TE achieved by improving mesophyll efficiency (Udayakumar *et al.*, 1998, Sheshshayee *et al.*, 2013). Our recent studies provided convincing evidence that improving water use and water use efficiency by trait based breeding substantially improved water productivity (Prathiba, 2016). By introgressing these two traits the productivity of rice under semi irrigated aerobic conditions has been substantially improved (Figures 2 and 3).

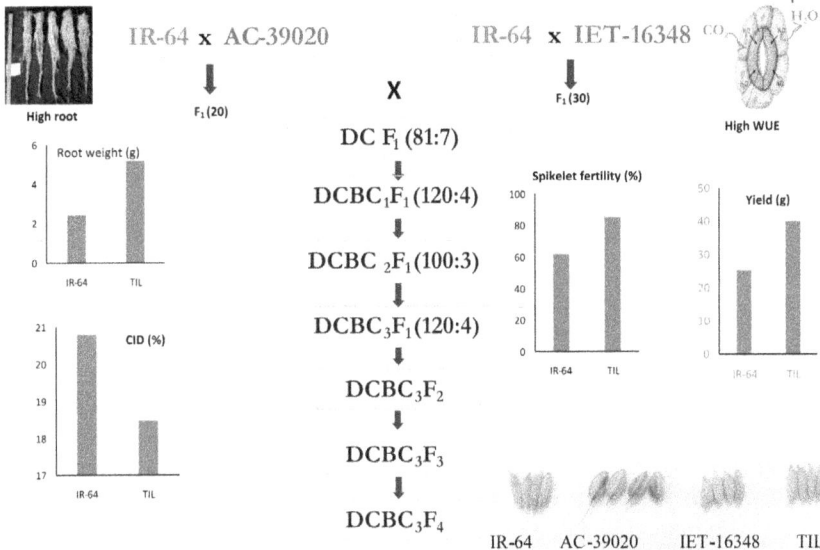

Fig. 2: The QTLs associated with water use efficiency (WUE) and roots have been identified from our earlier studies were introgressed in to IR64 by adapting a marker assisted backcross strategy. Some of the trait introgressed lines which are in BC3F5 generation showed improved root characteristics, improved WUE and productivity. The approach not only provides evidences about the relevance of QTLs but also signifies the importance of introgressing the traits by this approach.

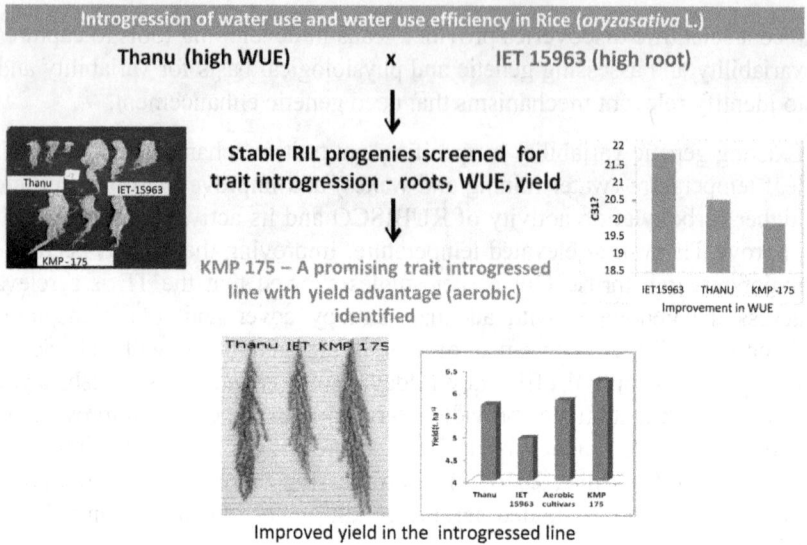

Fig. 3: Conventional breeding strategy adapted to intogress root and WUE traits in rice by identifying specific donors. The RIL population were extensively phenotyped and one of the transgenerational segregant KMP-175 showed substantial improvement in both root character and WUE, besides higher productivity under aerobic rice cultivation conditions. This genotypes is now being released for cultivation.

Therefore one of the research foci should be enhancing the TE of crops to reduce the water footprint and to increase the water productivity. In the rainfed eco-system, with the increased demand of water requirement, crops experience increased moisture stress. A combined water and high temperature stress is the predicted scenario in rainfed conditions, which substantially affects the productivity. Adaptive mechanisms/traits to sustain and improve growth under this complex abiotic scenario warrant a comprehensive approach.

PHENOLOGY AND PHYSIOLOGICAL TRAITS

Besides the elevated temperature driven diminishing effect on TE, the temperature has direct effects on several plant processes which affect plant growth and productivity. The factors that largely affecting the productivity are decrease in crop duration especially reproductive phase, substantial decrease in carbon use efficiency despite of high CO_2, decrease in respiratory efficiency, and deleterious effect on the reproductive parameters finally large impact on structural integrity and cell metabolism (Yunlai Tang *et al.*, 2007; Driedonks *et al.*, 2016).

Phenology–Crop Duration

The increase in air temperature resulting in faster crop development and shorter crop duration (Challinor *et al.*, 2005), has been shown to decrease the yield (Islam *et al.*, 2012) (Figure 4a). Several convincing proofs indicate a reduction in the duration of growth stage II (GSII) and GSIII, which subsequently affect grain characters like grain number and weight (Figure 4b and c).

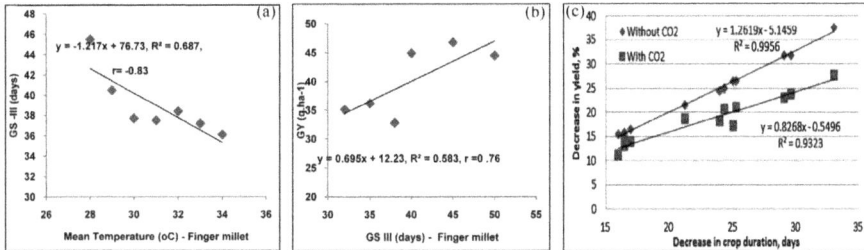

Fig. 4: Effect of temperature on grain yield. a) Response of GSIII to temperature finger millet (UAS, B), b) Effect of GSIII duration on grain yield (UAS, B), c) Relation between crop duration and yield (Islam, 2012).

The basic molecular mechanisms involved to bring change in duration of crops especially advancement of reproduction are not well elucidated. However, more recent studies implicate the high temperature induced changes chromatin remodulation and phytochrome B in regulating the genes associated with flower initiation and cell division (Jung *et al.*, 2016). In addition to the reduction in crop duration at elevated temperature, pollen viability, pollen tube growth and spikelet fertility decreases, and fruit-set is affected leading to poor crop productivity (Rang *et al.*, 2011). In many crops including warm season annuals such as legumes, rice, even in maize the reproductive stage is highly vulnerable to temperature increase. High temperature leads to flower abortion and reduced fertility. The male reproductive organs especially pollen development are temperature sensitive compared to maternal tissue of the pistil and female gametophyte (Driedonks *et al.*, 2016). In an elegant experiment Rang *et al.* (2011). Demonstrated substantial decrease in pollen development and its germination at high temperature compared to water stress suggesting that high temperature has much more deleterious effect (Rang *et al.*, 2011). At high temperature, failure in tapetum differentiation and deleterious effects on microsporogenesis process leads to improper pollen development, pollen viability and growth which ultimately affects spikelet fertility, thus substantially decrease the crop productivity (Sakata and Higashitani, 2008).

Despite increase in CO_2 the net carbon gain decreases mainly by enhanced dark and photo respiratory processes. Increase in temperature demands more maintenance respiration for cell metabolism. Inspite of increase in respiration, respiratory efficiency is shown to be decreased (Figure 5). Several reports showed at higher Q_{10} values respiratory efficiency substantially decreases at a given temperature. High Q_{10} is related to lower respiration efficiency and it results in lower relative growth rates (RGR, Figure 6) (Van-Iersel, 2003; Wan and Luo, 2003). The basic molecular mechanisms regulating the mitochondrial respiration efficiency will be the key issue to identify the climate resilient crop species/

varieties. Genotypes from warm adapted crops show less increase in respiration with increase in temperature signifying the possibility of arriving at molecular mechanisms regulating these processes (Loka and Oosterhuis, 2010).

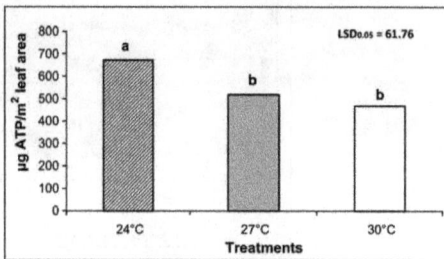

Fig. 5: At Higher Temperature ATP Levels Decreases Inferring Lower Respiration Efficiency

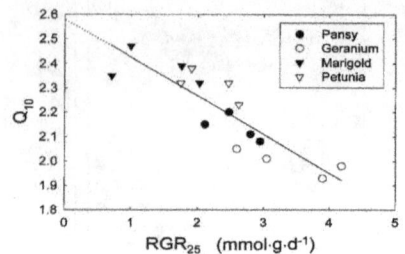

Fig. 6: Q_{10} and RGR Relation in Different Species

(*Source:* Van-Iersel, 2003)

The climate change driven increase in temperature has several effects on the above discussed physiological parameters thus affecting the crop productivity. The physiological parameters which are affected by elevated temperature have in fact emerged as important traits to assess genetic variability and subsequently for their genetic enhancement. However, the response to higher temperature effects shows significant variations across the species and even across the genotypes. Mainly such variability is predominantly due to cellular tolerance mechanisms which are often upregulated upon exposure to temperature.

IMPROVING THERMO-TOLERANCE–MOLECULAR MECHANISMS

Cellular level response to higher temperature effect by reprogramming their genetic machinery is the key for adaptation. Besides the basal thermo-tolerant mechanisms, the upstream and downstream genes that governs the acquired tolerance is the major focus of the recent studies. The flow chart summarises the present understanding and provides the leads for genetic enhancement by molecular approaches (Figure 7).

At least five putative sensors have been proposed to trigger heat shock response. They include, a plasma membrane channel that initiates Ca flux, Histone sensors in the nucleus and two unfolded protein sensors in the ER and Cytosol (Mittler *et al.*, 2011). PhyB has been implicated in temperature sensing and initiating the cascade of events to regulate flowering time and other processes (Legris *et al.*, 2016; Jung *et al.*, 2016). The histone sensing processes are predominantly associated in recruiting the alternate variant H2A.z in place of normal H2A and the process is mediated by a group of proteins such as actin related proteins 6 (ARP6) (Kumar and Wiggie, 2010). Sensing the temperature induced unfolded protein response is another upstream response associated with endoplasmic reticulum membrane localized IRE1 (Mittler *et al.*, 2011). The involvement of specific kinases like Calmodulin is crucial in bringing about the heat shock activation of

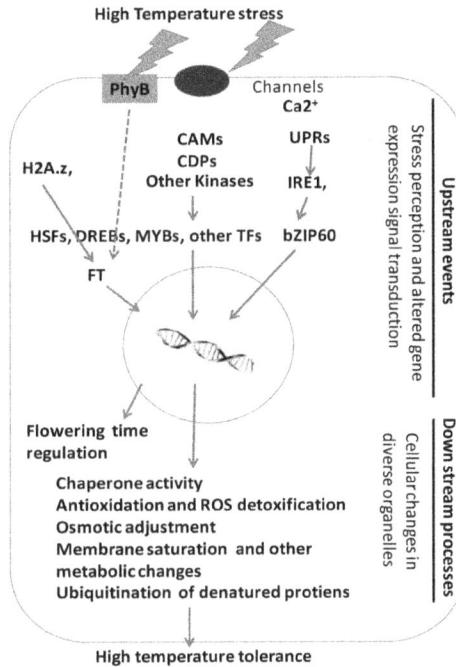

Fig. 7: Model depicting temperature stress perception, signal transduction and altered gene expression to regulate relevant cellular level processes for adaptation (UAS, B; modified from Grover *et al.,* 2013).

the most important group of transcription factors, HSFs has been implicated as a crucial to bring about the expression of several HSPs thus improving thermo tolerance. Besides, HSFs, other transcription factors like DREBS, WRKYs, MYBs, bZIPs, etc., are considered to be relevant and important in upregulating several downstream genes that improve high temperature tolerance (Grover *et al.*, 2013).

Temperature induced changes in transcriptome and proteome has been extensively studied and the information provided leads to decipher the major metabolic processes and cellular level changes that have relevance in improving cellular tolerance. Six major cellular tolerance mechanisms have been implicated has relevance under high temperature such as chaperone activity, detoxification of cytotoxic ROS, osmotic adjustment, membrane saturation for stability, ubiquitination of denatured proteins and flowering time regulation. Network of the genes associated with these six cellular level tolerance mechanisms and their upstream regulators will provide new leads and options.

Fairly good leads exist to predict the likely changes in the temperature and its likely impact on physiological processes and also the cellular level mechanisms that are affected by the high temperature. Further the important mechanisms/traits which are relevant to improve tolerance and productivity have been well elucidated. Besides, the molecular cues that brings altered mechanisms to tolerance have also been well elucidated. This basic information in fact forms the basis for genetic enhancement of the relevant traits to sustain and improve the productivity under predicted climate change scenario.

IMPROVING CROP ADAPTATION: GENETIC ENHANCEMENT APPROACHES

Several translation programs mainly transgenic or molecular breeding approaches provided some leads in validating the relevance of specific functional or regulatory genes associated with high temperature tolerance and also in identifying several QTL's associated with temperature tolerance.

The thermo-tolerance genes that are being chosen to develop transgenics predominantly addressed to improve three important mechanisms; a) improving the fluidity of the membranes upon heat shock, (b) regaining the redox homoeostasis by improving the activity of ROS scavenging machinery and (c) prevention of deleterious proteins confirmation and elimination of non-native aggregations formed during stress by over expressing proteins that function as chaperons. Several transgenics expressing genes regulating many of the above processes have been developed and their importance has been validated. The major emphasis has been to identify genes that related to the chaperons and ROS scavenging pathways (Figure 8).

35S: HSF4 transgenic groundnut plants

Wild type   Transgenic

Fig. 8: Transgenic Groundnut Plants Expressing HSF4
Transcription Factor Improves Thermotolerance

Enhanced thermo-tolerance shown to be achieved by overexpression of HSP/HSFs in several species (Grover *et al.,* 2013). In one of our recent studies, we have demonstrated that overexpression of HSF4 significantly improved thermo-tolerance in Groundnut (Figure 8) (Shankar, 2011). Maintenance of cooler canopy by water mining avoids high temperature effect. Therefore co-expression of relevant genes improving the root characters, protein turnover and stability has relevance for improved adaptation under climate change scenario. In a recent study we have demonstrated that the groundnut transgenics co-expressing HSF, PDH45 and Alfin1 showed higher temperature tolerance and water mining by increased root growth leading to cooler canopy (Ramu *et al.,* 2016). These transgenics showed improved tolerance to both elevated temperature and moisture stress (Figure 9).

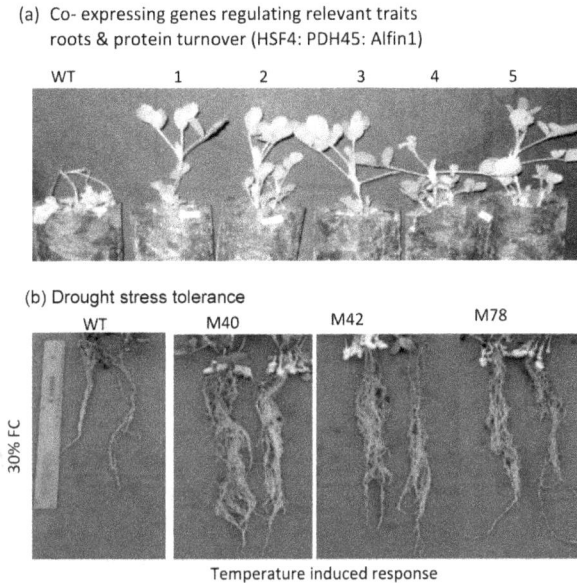

(a) Co- expressing genes regulating relevant traits roots & protein turnover (HSF4: PDH45: Alfin1)

(b) Drought stress tolerance

Temperature induced response

Fig. 9: Improved thermotolerance and root growth in the transgenics co-expressing HSF4:PDH45:Alfin1. a) Temperature induced tolerance in multigene expressing plants, b) improved root growth under 30% field condition.

Inspite of the genomic information in several crops and refined genotyping option, the inadequate progress in identifying the QTL linked to stress tolerance is mainly because of three important reasons. The major reason is phenotyping for specific adoptive mechanisms/traits under defined temperature regimes. The other lacuna is narrow genetic diversity for important traits. Therefore, the identification of superior wild alleles that are lacking in cultivated germplasm is of great importance. Search for wild relatives from warmer climates has relevance. Inadequate attempts to identify the casual genes in major QTL's is another limitation. The present emphasis in molecular breeding has been in identifying QTL's in mapping population developed from contrasts that largely differ in growth and productivity under high temperature because of limitation to phenotype specific adaptive traits. With this approach, QTLs for several heat tolerance traits has been identified and most significant amongst them are those associated with canopy temperature and chlorophyll florescence. QTL's associated with necrosis, pollen viability and growth, spikelet fertility have also been identified. Recently, a team lead by Li *et al.* (2015), identified a major QTL from *Oryza glaberima* thermo-tolerance 1 (TT1) (Li *et al.,* 2015) which substantially improves the thermo-tolerance. This QTL is associated with unique gene TT1 that encodes a α2 subunit of 26S proteosome that involved in degradation of ubiquitinated proteins that were denatured due to heat stress. This study signifies the importance of identifying QTLs from wild relatives from warmer climates.

Besides the polygenic basis of heat tolerance, the crops are likely to experience both water and high temperature stress. Therefore several adaptive mechanisms have to be introgressed to achieve desired levels of tolerance. From this context, more advanced approach is Multi-Parent Advanced Generation Inter Cross (MAGIC) population which might provide requisite leads in identifying the genes involved in thermo tolerance and subsequently their introgression.

In summary, the climate change scenario predicts increase in the temperature and VPD which substantially enhances demand for water and also affects crop TE. High temperature would alter specific physiological processes such as crop duration (especially GS-II and GS-III), reproductive fertility parameters, carbon use efficiency, respiratory efficiency and several factors of cell metabolism. Good leads exist in the terms of relevant adaptive mechanism/traits. Major focus is on the discovery of thermotolerance genes regulating the fluidity of membranes upon heat stress, and the genes involved in hyper activation of ROS scavenging mechanisms and prevention of deterioration of protein confirmation and degradation of ubiquitinated denatured proteins. Several transgenics developed provided evidences about the relevance of genes regulating specific mechanisms. These leads will complement the ongoing programme for genetic enhancement of heat stress tolerance. The knowledge of heat tolerance mechanisms aids to design of strategies to screen the germplasm for heat tolerance traits. Molecular breeding approaches are emerging as potential option to identify QTL's associated for some of the relevant mechanisms, though restricted to integrated traits. The major limitation is inabilities to develop trait specific mapping population. Accurate phenotyping for specific traits is the major limitation for exploiting genetic variability. Besides the other lacuna is narrow genetic diversity. A comprehensive exploration to identify wild relatives from warmer climate will be rewarding.

ACKNOWLEDGEMENT

We thank Dr. Thimmegowda, Associate professor, Department of Dry land Agriculture, University of Agricultural Sciences, GKVK, Bangalore for providing PET values.

REFERENCES

Cao, L., Bala, G., Caldeira, K., Nemanid, R. and Ban-Weiss, G. (2010). Importance of carbon dioxide physiological forcing to future climate change. *Proceedings of the National Academy of Science*, 107(21), pp. 9513–9518.

Challinor, Wheeler, T.R., Craufurd, P.Q., Slingo, J.M. (2005). Simulation of the impact of high temperature stress on annual crop yields. *Agricultural and Forest Meteorology,* 135, pp.180–189.

Chatterjee (2011). Water resources of India. http://climatechangecentre.net/pdf/water resources.pdf.

Driedonks, N., Rieu, I. and Vriezen, W.H. (2016). Breeding for plant heat tolerance at vegetative and reproductive stages. *Plant Reproduction*, 29(1), pp. 67–79.

Grover, A., Mittal, D., Negi, M. and Lavania, D. (2013). Generating high temperature tolerant transgenic plants: achievements and challenges. *Plant Science*, 205, pp. 38–47.

Hoekstra, A.Y. (2010). A global and high-resolution assessment of the green, blue and grey water footprint of wheat. Value of water research report series no. 47.

Hoekstra, A.Y. and Mekonnen, M.M. (2012). The water footprint of humanity, Proceedings of National Academy of Sciences, 109(9), pp. 3232–3237.

Islam, A., Lajpat, R., Ahuja, Garciab, Luis A., Maa, Liwang, Saseendrana, Anapalli S., Thomas, C. and Troutd, J. (2012). Modelling the impacts of climate change on irrigated corn production in the central great plains. *Agricultural and water management*, 110, pp.98–108.

Jung, J.H., Domijan, M., Klose, C., Biswas, S., Ezer, D., Gao, M., Khattak, K., Box, M., Charoensawan, V., Cortijo, S., Kumar, M., Grant, A., James C.W., Locke, Jaeger, Wigge, P.A. (2016). Phytochromes function as thermosensors in Arabidopsis. *Science* 10.1126/science.aaf6005.

Junker, A., Moses, M., Muraya, Weigelt-Fischer, K., Arana-Ceballos, F., Klukas, C., Albrecht, E., Melchinger, Rhonda, C., Meyer, Riewe, D., and Altmann, T. (2014). Optimizing experimental procedures for quantitative evaluation of crop plant performance in high throughput phenotyping systems. *Frontiers in Plant Science*. 5, 770.

Kumar, S.V., and Wigge, P.A. (2010). H2A.Z-containing nucleosomes mediate the thermosensory response in Arabidopsis. *Cell*, 140, pp.136–147.

Legris, M., Klose, C., Burgie3, S., Costigliolo, C., Neme, S., Hiltbrunner, A., Philip, A., Wigge, Schafer, E., Richard, D., Vierstra, Jorge, J., Casal (2016). Phytochrome B integrates light and temperature signals in Arabidopsis. *Science, DOI: 10.1126/science.aaf5656.*

Li, X.M., Chao, D.Y., Wu, Y., Huang, X., Chen, K., Cui, L.G., Su, L., Ye, W.Y., Chen, H., Chen, H.C., Dong, N.Q., Guo, T., Shi, M., Feng, Q., Zhang, P., Han, P., Shan, J.X., Gao, J.P. and Lin, H.X., 2015. Natural alleles of a proteasome α2 subunit gene contribute to thermotolerance and adaptation of African rice. *Nature Genetics,* 47(7), pp. 827–833.

Loka and Oosterhuis (2010). Effect of high night temperatures on cotton respiration, ATP levels and carbohydrate content. *Environmental and Experimental Botany*, 68, pp. 258–263.

Mittler, R., Finka, A., and Goloubinoff, P. (2011). How do plants feel the heat? *Trends in Biochemical Sciences*, 37, 3.

Prathiba (2016). Introgression of root and water use efficiency traits by marker assisted backcross strategy in rice (Oryza sativa L.) and validation of progeny through physiological characterization. Ph.D. Thesis, University of Agricultural Sciences, GKVK, Bangalore, India.

Rakesh Kumar, R.D., Singh and K.D. and Sharma (2005). Water resources of India. *Current Science,* 89, pp. 5–102.

Ramu, V.S., Swetha, T.N., Sheela, S.H., Babitha, C.K., Rohini, S., Reddy, M.K., Tuteja, N., Reddy, C.P., Prasad, T.G. and Udayakumar, M., 2016. Simultaneous expression of regulatory genes associated with specific drought-adaptive traits improves drought adaptation in peanut. *Plant Biotechnology Journal*, 14(3), pp. 1008–1020.

Rang, Z.W., Jagadish, S.V.K., Zhou, Q.M., Craufurd, P.Q. and Heuer, S. (2011). Effect of high temperature and water stress on pollen germination and spikelet fertility in rice. *Environmental and Experimental Botany*, 70(1), pp. 58–65.

Sakata, T. and Higashitani, A. (2008). Male sterility accompanied with abnormal anther development in plants – genes and environmental stresses with special reference to high temperature injury. *International Journal of Plant Developmental Biology*, 2(1), pp. 42–51.

Shankar (2010). Development of multiple gene construct and overexpression of multigene cassette for abiotic stress tolerance in plants. M.Sc. Thesis, University of Agricultural Sciences, GKVK, Bangalore, India.

Sheshshyee *et al.* (2013). Enhancing water use efficiency and effective use of water as a potential strategy to develop rice cultivars suitable for semi-irrigated aerobic cultivation. International Dialogue on perception and prospects of designer rice. *Directorate of rice research*, Hyderabad, 261–272.

Tang, Y., Wen, X., Lu, Q., Yang, Z., Cheng, Z., and Lu, C. (2007). Heat Stress Induces an Aggregation of the Light-Harvesting Complex of Photosystem II in Spinach Plants. *Plant Physiology*, 143(2), pp. 629–638.

Udayakumar, M. and Sheshshayee, M.S., Nataraj, K.N., Bindumadhava, H., Aftab Hussain, I.S., Devendra, R., and Prasad, T.G. (1998). Why has breeding for water use efficiency not been successful? An analysis and alternate approach to exploit this trait for crop improvement. *Current Science*, 74(11), pp. 994–1000.

UNFCC—United Nations Framework Convention on Climate Change (2007). Climate Change: Impacts, Vulnerabilities and Adaptation In Developing Countries.

Van Iersel, M.W., 2003. Short term temperature change affects the carbon exchange characteristics and growth of four bedding plant species. *Journal of American Society and Horticultural Science.* 128, pp. 100–106.

Wan, S., and Luo, Y. (2003). Substrate regulation of soil respiration in a tallgrass prairie: Results of a clipping and shading experiment. *Global Biogeochemistry Cycles*, 17(2), pp. 1054.

Genetic Engineering of Component Traits of Drought Tolerance in Rice

V.V. Santosh Kumar, S.K. Yadav, R.K. Verma, S. Shrivastava and C. Chinnusamy*

Division of Plant Physiology, ICAR-Indian Agricultural Research Institute, New Delhi 110 012, India
*E-mail: viswanathan@iari.res.in

ABSTRACT: *Rice productivity must be doubled to sustain food security in near future. Currently rice productivity under rainfed ecosystem is very low due to water deficit stress. Rice cultivation uses about 50% of irrigation water, and is highly sensitive to water deficit stress. Hence understanding and improving water use efficiency and drought tolerance of rice is necessary to sustain rice cultivation in the global climate change scenario with increasing fresh water scarcity and frequency of drought stress. This study was initiated to engineer ABA signalling, cytokinin metabolism and ICE1 pathway in rice to manipulate various component traits of drought tolerance, source-sink relationship and yield in rice. ABA regulates both dehydration avoidance and tolerance mechanisms under drought stress in plants. Genome-wide analysis revealed that rice genome encodes12 ABAR homologs. Real-time qRT-PCR expression analysis of ABARs in different tissues and stress conditions revealed differential expression patterns suggesting non-redundant roles of ABARs in development and stress responses. Analysis of drought tolerance of RD29A: ABAR6 transgenic rice lines in pot culture under greenhouse conditions revealed that RD29A: ABAR6 transgenic rice lines use less water, and are more tolerant than non-transgenic lines. ABAR6 enhances drought tolerance of rice by enhancing dehydration avoidance through transpiration minimization, and cellular tolerance to dehydration. The transgenic rice lines over-expressing AtICE1 showed enhanced tolerance to drought, salt and cold stress. Physiological analysis showed that AtICE1 transgenic plants maintained better RWC and chlorophyll and membrane stability under drought, salt and cold stresses. AtICE1 is also known to regulate stomatal development in Arabidopsis. We found that over-expression of AtICE1 in rice enhanced the stomatal density but reduced the stomatal size in transgenic plants as compared to NT wild-type plants. The results demonstrate that ICE1 function in stomatal development and abiotic stress tolerance is conserved across evolutionarily diverse plant species.*

Keywords: ABA Receptors, ICE1, Stomata, Water Use, Transpiration.

Rice, the major food crop of the world, has very low Water Use Efficiency (WUE) and is highly sensitive to water deficit (drought) stress. Water requirement for rice crop is expected to increase due to the increase in evapo-transpiration under global warming scenario in future. Fresh water scarcity is a looming threat for agriculture production and specifically rice cultivation in near future. Rice accounts for about 50% of irrigation water, and thus rice cultivation is expected to be unsustainable in future as the per capita water availability is expected to decline by 15 to 54% in most river basins by the year 2025 (Guerra *et al.,* 1998). Besides, rainfed lowland and upland rice is cultivated in about 45% of the rice grown area in the country, which are subjected to intermittent soil moisture deficit that causes severe yield loss. The yield of about 40% of rice grown area which experiences drought stress is less than one ton per ha. There-fore, genetic improvement in Water Use Efficiency (WUE) and drought toler-ance of rice is a priority research target for food security as well as minimizing agricultural water use. Research work presented here focuses on functional validation of genes involved in various component mechanisms of drought tolerance in rice.

PRESENT STATUS

Drought tolerance mechanisms of plants can be partitioned into dehydration avoidance and dehydration tolerance mechanisms. Plants can avoid dehydration by maximization of water uptake from the soil to support photosynthetic rates and minimizing water loss through stomata, while dehydration tolerance mechanisms include metabolic adjustments, protection of cellular components and maintenance of cellular integrity. Plant stress hormone Abscisic Acid (ABA) positively regulates both dehydration avoidance and dehydration tolerance mechanisms (Chinnusamy *et al.,* 2008). Since the late 1960, significant efforts have been made to understand the ABA signalling mechanisms with an aim to utilize the genes involved in ABA signaling for improving stress tolerance of crops. Only recently, the START domain proteins PYRABACTIN RESISTANCE1 (PYR1)/PYR1-like (PYL)/Regulatory Components of ABA Receptors (RCARs) have been identified as *bona fide* ABA receptors (ABARs) (Ma *et al.,* 2009, Park *et al.,* 2009). The core ABA signaling pathway for stress responsive gene expression and stomatal closure consists of PYR1/PYL/RCAR ABARs, clade A protein Phosphatase 2Cs (PP2Cs) and SnRK2 family ser/thr kinases (Fujii, *et al.,* 2009, Melcher *et al.,* 2009). The SnRK2 kinases phosphorylate effectors such as ion channels, enzymes and DNA binding proteins, and thus mediate ABA responses (Cutler *et al.* (2010). Among the hormone receptor families in plants, ABAR family is the largest with 14 members. The functional specificity/ redundancy of ARARs needs to be established in development and biotic and abiotic stress tolerance of plants. Inducer of CBF Expression 1, a MYC-type basic helix-loop-helix (bHLH) transcription factor, is a master regulator of cold acclimation and freezing tolerance in Arabidopsis (Chinnusamy *et al.,* 2003). Later studies showed that ICE1-CBF cold response pathway is conserved in diverse plant species. Transgenic analysis in different plant species revealed that tolerance to cold and other abiotic stresses is significantly enhanced by genetic engineering ICE1-CBF pathway (Chinnusamy *et al.,* 2010). Besides regulating stress tolerance, ICE1 is also involved in regulation of stomatal density in both dicot model plant Arabidopsis (Kanaoka *et al.,* 2008) and monocot model plant Brachypodium (Raissig *et al.,* 2016), and flowering in Arabidopsis (Lee *et al.,* 2015). Hence this study was initiated to engineer ABA signaling and ICE1 pathway in rice to manipulate various component traits of drought tolerance in rice.

Genome-Wide Identification of ABARs in Rice

Arabidopsis PYR/PYL family of ABARs protein sequences were used as query to search rice genome database (http://rice.plantbiology.msu.edu) to identify ABAR family members in rice.

Expression Analysis of Gene by Real Time qRT-PCR

Rice non-transgenic (NT) and transgenic lines were subjected to various abiotic stresses. Tissues samples were harvested and used for total RNA isolation.

Gene Cloning

Coding sequences of ABARs were cloned from drought tolerant rice cv. Nagina 22. AtICE1 coding sequence and stress-inducible RD29A promoter were cloned from *Arabidopsis* ecotype Columbia.

Gene Constructs for Plant Transformation

pC1300_*AtRD29A::OsABARs*—Rice ABAR genes were clonedin a modified pCAMBIA1300 under stress inducible *AtRD29A* promoter.

pANDA_*UQ::ABAR6-RNAi*—To silence ABAR6, partial CDS of ABAR6 was cloned in pANDA vector.

pC1300_*AtRD29A::AtICE1*—Arabidopsis *AtICE1* gene was cloned in a modified pCAMBIA1300 under stress inducible *AtRD29A* promoter.

All these gene constructs were confirmed by restriction analysis and sequencing, and then were transformed in to *Agrobacterium tumefaciens* EHA105.

Agrobacterium Mediated Genetic Transformation of Rice

Embryogeniccalli derived from mature seeds of rice cv. PusaSugandh 2 or MTU1010 were transformed with *Agrobacterium* EHA105 stains harbouring above gene constructs. Putative transgenics were confirmed by with *hptII* as well as T-DNA specific primers, qRT-PCR and Southern blot analysis.

Physiological Analysis of Drought Tolerance of Transgenics

Confirmed transgenic rice events and non-transgenic vector control plants were subjected to osmotic (20% PEG), salt (200 mMNaCl) and cold (6°C) stresses for indicated duration at seedling stage. Plants were grown in soil filled pots and drought stress was imposed by withholding water till soil matric potential reached to –80 kPa. Data of SMC, RWC, chlorophyll and membrane stability were measured following the standard protocols. Stomatal density was measured in microscope (EVOS®FL) at 40X and stomatal size was measured by SEM (Zeiss Neon-40) in WT and transgenic events. After end of the stress period, plants were recovered with regular irrigation, and were grown to full maturity to record grain yield. All the experiments were conducted in the transgenic greenhouse conditions.

Abscisic acid (ABA) is an essential phytohormone which is involved in a host of biological processes, including plant development and responses to biotic and abiotic stresses. The specific functions of individual members of ABARs in developmental and stress responses are being currently investigated worldwide. Towards unravelling the functions of ABA receptors in rice, we carried out genome wide analysis and identified 12 ABAR genes in rice genome. Phylo-genetic analysis grouped these ABARs into three distinct sub-families: the monomeric receptors formed subfamily I and II, while the dimeric receptors formed subfamily III. Real time RT-PCR expression analysis revealed differen-tial regulation of *ABARs* in different tissues and stress conditions. For example, ABAR3 showed upregulation in both root and shoot under osmotic, salt and cold stress, while ABAR2 was downregulated by these stresses. ABAR9 was upregulated only in shoot while ABAR10 was upregulated only in root under NaCl stress. ABAR3 and ABAR4 showed very high expression in panicle during grain filling as compared with other receptors. These results suggest non-redundant roles of *ABARs* in development and stress responses.

For gene function validation, rice transgenics over-expressing ABAR6 and RNAi silencing lines were developed and confirmed by PCR, qRT-PCR and southern analyses. Phenotyping of transgenic rice lines (T1) over-expressing *ABAR6* gene under RD29A promoter was carried out in pot experiments in transgenic greenhouse. RD29A::ABAR6 transgenics showed better tolerance to drought stress. To further confirm the tolerance of ABAR6 transgenics, 5 events of ABAR6 at T2 generation were phenotyped for drought tolerance. Transgenic lines showed significantly higher cellular tolerance in terms of higher RWC and lower drought induced leaf death as compared with non-transgenics under drought. Daily water use of ABAR6 transgenics and NT plants were studied by gravimetric method. Plants were grown up to 60 days under non-stress conditions, and were subjected to 3 cycles of water-deficit stress (–80 kPa). Transgenic plants used about 30% less water as compared with NT control plants. Excised Leaf Water Loss (ELWL) assay confirmed that transgenic leaves loose considerably less water as compared with the leaves of NT plants.

Scanning Electron Microscopy also showed that ABAR6 plants close stomata early as compared with NT plants under drought stress conditions. These analyses revealed that ABAR6 confers tolerance to rice plants through both dehydration avoidance and tolerance mechanisms.

Transgenic rice plants over-expressing AtICE1 showed enhanced tolerance to cold, drought and salt stressesat seedling, vegetative and flowering stages. Transgenic ICE1 plants exhibited enhanced tolerance to these stresses in terms of RWC, chlorophyll and membrane stability. Stomatal density was observed higher while stomatal size was smaller in transgenic plants as compared to non-transgenics. The results suggest that the *Arabidopsis* ICE1 is functional in rice and the over-expression of ICE1 improves the tolerance of rice to abiotic stresses.

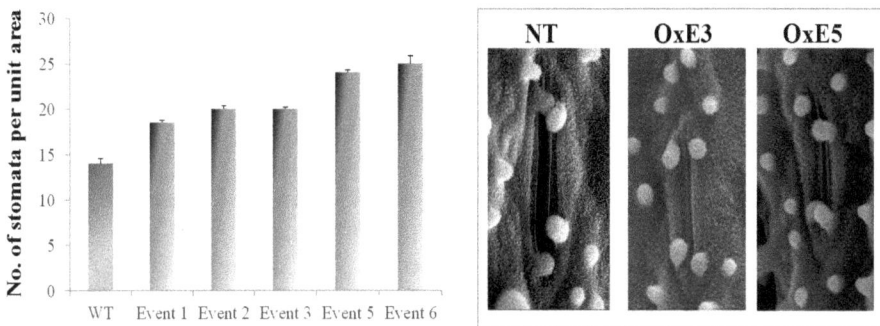

Fig. 1: AtICE1 Over-Expression Alters Stomatal Number and Size in Transgenic Rice. NT, Non-transgenic plants, OxE3 and OxE5 are rice transgenic over-expressing AtICE1 under stress inducible AtRD29A promoter.

ACKNOWLEDGEMENTS

The work presented here is supported by grants from NASF, ICAR (Phen 2015/2011-12), DBT, Gove of India (BT/Bio-CARe/02/863/2011-12; BT/PR8787/ AGR/36/750/2013) and ICAR-IARI in-house projects.

REFERENCES

Chinnusamy, V., Gong, Z. and Zhu, J.K. (2008). ABA-mediated epigenetic processes in plant development and stress responses. *J. Intg Plant Biol.,* 50: 1187–1195.

Chinnusamy, V., Ohta, M., Kanrar. S., Lee, B.-h., Hong. X., Agarwal, M. and Zhu, J.K. (2003). ICE1, A regulator of cold induced transcriptome and freezing tolerance in *Arabidopsis. Genes Dev.,* 17: 1043–1054.

Chinnusamy, V., Zhu, J.K. and Sunkar, R. (2010). Gene regulation during cold stress acclimation in plants. *Methods Mol Biol.,* 639: 39–55.

Cutler, S.R., Rodriguez, P.L., Finkelstein, R.R. and Abrams, S.R. (2010). Abscisic acid: Emergence of a core signaling network. *Annu Rev Plant Biol.,* 61: 651–679.

Fujii, H., Chinnusamy, V., Rodrigues, A., Rubio, S. Antoni, R., Park, S.Y., Cutler, S.R., Sheen, J. Rodriguez, P.L. and Zhu, J.K. (2009). In vitro reconstitution of an abscisic acid signalling pathway. *Nature,* 462: 660–664.

Guerra, L.C., Bhuiyan, S.I., Tuong, T.P. and Barker, R. (1998). Producing more rice with less water from irrigated systems. SWIM Paper No. 5, Colombo, International Water Management Institute.

Kanaoka, M.M., Pillitteri, L.J., Fujii, H., Yoshida, Y., Bogenschutz, N.L., Takabayashi J., Zhu, J.K. and Torii, K.U. (2008). SCREAM/ICE1 and SCREAM2 specify three cell-state transitional steps leading to *Arabidopsis* stomatal differentiation. *Plant-Cell,* 20: 1775–85.

Lee, J.H., Jung, J.H. and Park, C.M. (2015). Inducer of CBF Expression 1 integrates cold signals into Flowering Locus C-mediated flowering pathways in Arabidopsis. *Plant J.,* 84: 29–40.

Ma, Y., Szostkiewicz, I., Korte, A., Moes, D., Yang, Y., Christmann, A. and Grill, E. (2009). Regulators of PP2C phosphatase activity function as abscisic acid sensors. *Science,* 324: 1064–1068.

Melcher, K., Ng, L.M., Zhou, X.E., Soon, F.F., Xu, Y., Suino-Powell, K.M., Park, S.Y., Weiner, J.J., Fujii, H., Chinnusamy, V., Kovach, A., Li, J., Wang, Y. Li., J., Peterson, F.C., Jensen, D.R., Yong, E.L., Volkman, B., F., Cutler, S.R., Zhu, J.K. and Xu, H.E. (2009). A gate-latch-lock mechanism for hormone signalling by abscisic acid receptors. *Nature,* 462: 602–608.

Park, S.Y., Fung, P., Nishimura, N., Jensen, D.R., Fujii, H., Zhao, Y., Lumba, S., Santiago, J., Rodrigues, A., Chow, T.F.F., Alfred, S.E., Bonetta, D., Finkelstein, R., Provart, N.J., Desveaux, D., Rodriguez P.L., McCourt, P., Zhu, J.K., Schroeder, J.I., Volkman, B.F. and Cutler, S.R. (2009). Abscisic acid inhibits type 2C protein phosphatases via the PYR/PYL family of start proteins. *Science,* 324: 1068–1071.

Raissig, M.T., Abrash, E., Bettadapur, A., Vogel, J.P. and Bergmann, D.C. (2016). Grasses use an alternatively wired bHLH transcription factor network to establish stomatal identity. *Proc NatlAcadSci. USA*, 113: 8326–31.

Precision Agriculture—Challenging the Climate Change

Anil Kumar Singh

Rajmata Vijayaraje Scindia Agriculture University, Gwalior 474 002 (M.P.), India

E-mail: vcruskvvgwl@gmail.com

ABSTRACT: India has been able to produce enough to feed its 1.25+ billion as well as export food grains to other countries. But the challenge to feed a projected population of 1.6 billion in 2050 and provide them with nutrition security, is a formidable one because of the continuously degrading natural resource base and shrinking of good agriculture land for cultivation. It has been estimated that to meet the demands of that vast population, land productivity has to be enhanced four times along with three fold increase is water productivity and concomitant increase of six times in labour productivity.

Water and agrochemicals (fertilizers, pesticides, herbicides etc.) are vital inputs in the agricultural production system. Water, the most vital of them, is becoming scarcer each passing day. The country became 'water stressed' in 2007 and if the 'business as usual scenario' continues, the entire country may became 'water scarce' by 2050. It is estimated that only a handful of basins will have per capita water availability above the critical level by 2025. The irrigation projects developed at a very huge cost to the exchequer, function at an efficiency level of 38–40 per cent. Mismanagement of water ultimately leads to degradation of natural resources. In spite of a plethora of technologies available like drip and sprinkler having water application efficiency between 60–90%, water use efficiency continues to be low. The latest estimates put the area under drip and sprinkler irrigation systems around 7.8 million ha out of a potential of 69.5 m ha. It has been established that increase in water use efficiency automatically results in increase in nutrient use efficiency. A typical example is drip fertigation which leads to a quantum saving of at least 40% fertilizers applied.

Use efficiency of fertilizer nitrogen, which constitutes more than 60% of total plant nutrients consumed in India is abysmally low; 30–40% in rice and 40–60% in other crops. Excessive use of nitrogenous fertilizers is leading to ground water pollution as well as increased N_2O emission. Nitrogen use efficiency can be increased by treating urea with nitrification inhibitors or coating with some hydrophobic substances to retard the release of urea in soil solution or its microbial oxidation to nitrates, which leach down or are lost to the atmosphere as N_2 or NO_2 gases. The Global Warming Potential (GWP) of NO_2 is about 310 times that of CO_2. Phosphorus use efficiency is even lower at 15–20% while that of micro nutrients is dismally low varying between 2–5%. Such low use efficiencies lead to considerable financial losses to the exchequer and cause serious environmental hazards. Development of crop-region specific customized fertilizers is needed to maximize fertilizer use efficiency. It is important to note that inclusion of nutrients like boron, copper, manganese and molybdenum has to be done cautiously because of a narrow margin between the deficiency and toxicity thresholds of these nutrients. We have to do away with the conventional form of agriculture that focuses on Recommended Dosage of Fertilizes with emphasis on NPK only. Soil Health Card Scheme provides ample opportunity to switch over to site specific nutrient management immediately. Precision farming is tailored to provide the exactly needed amount of a specific input at a given location.

The high production levels have to be not only sustained but also achieved with emphasis on energy savings and low emission technologies considering climate change impacts. These aspects have not been factored in the projected production estimates of various commodities. The year 2015 had been the warmest year on record with 2016 aiming to break this record. Annual CO_2 concentrations crossed the 400 ppm level in 2015 and rainfall is becoming more erratic with higher variability. Obviously higher amount of inputs would be needed to obtain the same levels of productivity. Technology-driven farming is the only option for increasing farmers' income and reducing cultivation costs with the ultimate aim of shifting to precision farming. Unfortunately, more than 80% of the farmers fall in the small and marginal category. Small sized farms coupled with scattered holdings will be a major hindrance in switching over to precision farming as it involves a seamless merging of remote sensing, GIS, GPS, sophisticated machinery and sensors for application of various inputs in measured quantities varying over space and times.

The Government's decision to use drone technology for estimating crop yields and damage is an indication of its intention to use modern technologies in agriculture. Precision farming is already fairly common in developed nations like USA, Australia and several European Countries where the focus is on cost cutting, minimal environmental damage and high quality produce. It is now possible to differentiate between stresses caused by different nutrients through hyper spectral signatures. Smart phone controlled drip system is no larger a distant goal. Flying drones have already entered the domain. Can self-driving tractors, robotic harvesters, integration of data technology into day to day operations in the digital era, field level data collection by sensors attached to tractors or installed in the field, be far behind? The sooner we shift to personalized farming, the faster will be reduction in environmental foot prints, productivity enhancement in a sustainable manner and moderation of climate change impacts.

Climate Smart Agriculture in Intensive Cereal Based Systems: Scalable Evidence from Indo-Gangetic Plains

M.L. Jat

International Maize and Wheat Improvement Center (CIMMYT),
NASC Complex, New Delhi 110 012, India
E-mail: M.Jat@cgiar.org; www.cimmyt.org

ABSTRACT: *The Indo-Gangetic plains, one of the highly populated regions of the world having extreme density of poverty and malnourished people is most vulnerable region in terms of climate change induced variability and hence has serious implications for food security and economy. Since the impact of climate change induced variability is complex and has local level causes and effects, the solutions are not simple and require locally-adapted practices with real time decision making by the farming households/community. The practices, technologies or services which simultaneously help in improving adaptive capacity, improve food security and reduce environmental foot prints of food production defined as 'Climate Smart Agriculture' needs a system and community based approach. Therefore, evidence on such Climate Smart Agricultural Practices (CSAPs) adapted to local situations are needed for their targeting to relevant farm typologies and investment priorities for impact at scale. The Climate Smart Village (CSV) concept, a community based approach to sustainable agriculture provides common platform to researchers, extension agents, local governments, farmers groups, private sectors and the service sector to collaborate and identify most appropriate CSAPs to tackle local challenges related to climate change. In this article we provide the provide a synthesis on scalable CSAPs in intensive cereal based systems of IGP.*

Keywords: Adaptation, Business Models, Conservation Agriculture, Climate Smart Villages, GHG Mitigation, Sustainable Intensification.

South Asia, home to about 1.5 billion people, over 30 per cent of whom are still living in poverty, faces a major challenge in achieving rapid economic growth to reduce poverty and attaining other Millennium Development Goals under emerging challenges of natural resource degradation, energy crisis, volatile markets and risks associated with global climate change (Jat *et al.*, 2016; Lal., 2016). During past half century (1965–2015); in process of achieving multi-fold increase in crop production in the region, inefficient use and inappropriate management of non-climate production resources (water, energy, agro-chemicals) have vastly impacted the quality of the natural resources and also contributed to climatic variability affecting farming adversely. The natural resources in South Asia specially in Indo-Gangetic plains (IGP) are 3–5 times more stressed due to population, economic and political pressures compared to the rest of the world and can potentially add to adversity of climatic risks, making a large number of people in the region vulnerable to climate change. Increasing climatic variability affects most of the biological, physical and chemical processes that drive productivity of agricultural systems including livestock and fisheries (Easterling *et al.*, 2007).

With no scope for horizontal expansion of farming we need to produce 70% more food to feed the projected world population of 9.7 billion by 2050. Nonetheless, having high risks of climate change induced extreme weather events, the crop yields in the region are predicted to decrease by 10 to 40% by 2050 with risks of crop failure in several highly vulnerable areas. Increase in mean temperature, increased variability both in temperature and rainfall patterns, changes in water availability, shift in growing season, rising frequency of extreme events such as terminal heat, floods, storms, droughts, sea level rise, salinization and perturbations in ecosystems have already affected the livelihood of millions of people. Studies (Sivakumar and Stefanski., 2011) show that there would be at least 10% increase in irrigation water demand in arid and semi-arid region of Asia with a 1°C rise in temperature. Thus, climate change could result in the increased demand for irrigation water, further aggravating resource scarcity. Moreover, climate change on the one hand, can intensify the degradation process of natural resources which are central to meet the increased food demand, while on the other hand, changing land use pattern, natural resource degradation (especially land and water), urbanization and increasing pollution could affect the ecosystem in this region directly and also indirectly through their impacts on climatic variables (Lal., 2016). For example, about 51% of the Indo-Gangetic Plains may become unsuitable for wheat crop, a major food security crop for India, due to increased heat-stress by 2050 (Lobell *et al.*, 2012; Ortiz *et al.*, 2008). Therefore, adaptation to climate change is no longer an option, but a compulsion to minimize the loss due to adverse impacts of climate change and reduce vulnerability (IPCC, 2014). Moreover, while maintaining a steady pace of development, the region would also need to reduce its environmental footprint from agriculture. Considering these multiple challenges, agricultural technologies that promote sustainable intensification and adapting to emerging climatic variability yet mitigating GHG emissions (climate smart agricultural practices) are scientific research and development priorities in the region (Dinesh *et al.*, 2015). There are a wide range of agricultural practices that have the potential to increase adaptive capacity of production system, reduce emissions or enhance carbon storage yet increasing food production. In this paper, we provide scalable evidence on Climate Smart Agriculture Practices (CSAPs) in intensive cereal based systems of IGP.

Under the aegis of CGIAR research program on Climate Change, Agriculture and Food Security (CCAFS) and other related projects, CIMMYT and other organizations in collaboration with range of partnerships including ICAR, SAUs, CGIAR Institutions, ARIs, NGOs, development departments, farmer organizations, service provider and community based organizations have been engaged in mainstreaming CSAPs through generating science backed scalable hard evidence across the range of farm typologies in the Indo-Gangetic plains (CIMMYT-CCAFS, 2014; Dinesh *et al.*, 2015). Climate-Smart Villages (key components are given in Figure 1) are the sites where researchers from national and international organizations, farmers' cooperatives, local government leaders, private sector organizations and key policy planners come together to identify which CSAPs are most appropriate to tackle the climate and agriculture challenges in the village. In CSVs, a portfolio of CSAPs adapted to local farming system is adopted by the community for multiple benefits of increased productivity, income and resilience to climatic variability. The idea is to integrate climate-smart agriculture into village development plans, using local knowledge and expertise and supported by local institutions. We piloted over 60 CSVs across Indo-Gangetic plains and the evidence from these community participatory pilots have been generated. Across various CSV sites, we evaluated and demonstrated the performance of individual CSAPs over conventional practices and generated evidence on various indicators. In various CSVs; the sites of participatory learning on CSA, we generated hard evidence on *"portfolios of climate smart agriculture interventions"* through participatory strategic research in intensive rice-wheat system so as to demonstrate what combination of practices (portfolio) have synergistic and multiplier effects on various indicators. The protocols of the portfolio used in these strategic research trials are given in Table 1.

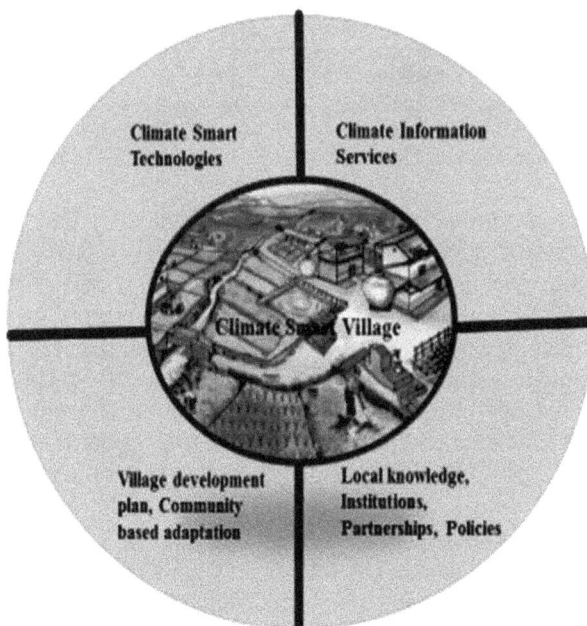

Fig. 1: Key Components of a Climate Smart Village (CSV)

Table 1: Description of Practice Portfolios under Different Scenarios
in Intensive Rice-Wheat Rotation in Western IGP

	Scenario	Tillage	Crop Establish-ment	Laser Land levelling	Cultivars	Residue Management	Water Manage-ment	Nutrient Manage-ment
1.	Business as usual	FP	TPR–CTW	No	Pusa44 - PBW343	FP (Burning or removal of rice and wheat residues)	FP	FFP
2.	Improved with Low intensity	FP	TPR–CTW	No	Pusa44 - PBW343	Incorporation of wheat stubble (1.5 to 2 t/ha) - 100% of rice	FP	FFP
3.	Improved with High intensity	Reduced till (RT)	DSR–RTW	No	Pusa44 - HD2967	Incorporation of wheat stubble (1.5 to 2 t/ha) - 100% of rice Retention	FP	SR
4.	CSA with Low intensity	RT –Zero till (ZT)	RT DSR–ZTW	Yes	PR114 - HD2967	Incorporation of wheat stubble (1.5 to 2 t/ha) - 100% of rice Retention	SR	SR + NCU
5.	CSA with medium intensity¹	ZT	ZT DSR–ZTW	Yes	PR114 - HD2967	Retention wheat stubble (1.5 to 2 t/ha) - 100% of rice	Tensio-meter based	SR + GS guided N + NCU
6.	CSA with high intensity¹	ZT	ZT DSR–ZTW	Yes	PR114 - HD2967	Retention wheat stubble (1.5 to 2 t/ha) -100% of rice	Tensio-meter based	SR + GS guided N + NCU

¹Scenario 5 and 6 had additional layering of ICTs and Index Insurance.

FP—Farmers practice, TPR—Puddled transplanted rice, CTW—conventional till wheat, RTW—reduced till wheat, DSR—direct seeded rice, ZTW—Zero till wheat, FFP—farmers' fertilizer practice, SR—State recommendation (Ah-hoc), GS—Green seeker, NCU—Neem coated urea.

Analysis of data from large number of field trials for their performance on key climate smart indicators (yield, income, water, energy, NUE, GHG emission) were performed to generate evidence across IGP. The results (Table 2) revealed that conservation agriculture (CA) based management (zero tillage, residue retention, direct seeded rice), precision water and nutrient management and laser land leveling practices qualifies as CSAPs (Jat *et al.*, 2015). CA based management practices reduces production cost, increase yields and economic benefits. Analysis of forty farmers' participatory field trials conducted for three consecutive years in Haryana demonstrated that the total cost of wheat production in zero tillage using turbo happy seeder technology (Sidhu *et al.*, 2015) was on an average, 23% less than that of conventional tillage (CT) whereas the net income was significantly higher in CA (Aryal *et al.*, 2015). Similarly, through long-term trial on tillage and crop establishment methods in rice-wheat system of eastern IGP, we found that the productivity of rice-wheat was higher under CA-based system (ZT rice-ZT wheat with and without residue retention) as compared to CT systems irrespective of the climate risks. The results from the another set of participatory trials in eastern IGP also showed the complementarity of various practices, for example layering of no-till maize-wheat rotation with residue recycling, inclusion of legume and site-specific nutrient management enhances yield (2.35 t ha^{-1} yr^{-1}) and income (USD 941) along with resource conservation benefits. However, higher yields and income does not provide the evidence of technology to be called as climate smart. We therefore, evaluated these practices for their role in adapting to climate risks as well as their mitigation potential to identify their multiple wins and verify as climate smart practices.

Table 2: Field Scale Evidence on Performance Indicators of Various Climate Smart Agriculture Practices

CSA Intervention	Performance Indicators of CSA Intervention Compared to Business as Usual Intervention					
	Yield Gain/Loss (kg ha^{-1})	*Economic Gains/Loss (US$ ha^{-1})*	*Water Saving (m^3 ha^{-1})*	*Energy Saving (MJ ha^{-1})*	*Increase in NUE (as kg/kg)*	*Reduction in GHG (kg CO$_2$eq ha^{-1})*
Zero tillage in wheat (without residue)	342	131	414	3040	1.44	1507[¶]
Zero tillage with residue in wheat	468	190	550	2650	1.61	–
Permanent beds in maize/wheat	195	289	1650	–	1.33	–
Direct seeded rice	+/–150	136	3000	–	–	420[§]
Improved water management	375	97.51	405	–	1.40	–
Nutrient Expert based SSNM in wheat	500	104	–	–	10	200
Laser leveling (RW system)	600	130	2500	–	–	330

[¶]Based on life cycle analysis, [§]Based on soil flux.

GHG—Greenhouse gases, NUE-Nitrogen use efficiency.

Source: Synthesized by the author from various CSA trials and related publications.

We evaluated the role of CA and other sustainable intensification practices for their adaptive capacity to reduce climate risksin the intensive cereal based systems in IGP. Our studies on CA in rice-wheat system in western IGP (Sapkota *et al.*, 2015), showed that retention of rice residue on soil surface lowered the canopy temperature in wheat by 1–2°C at grain filling period (between 138–153 days after sowing). Surface retention of crop residues (no-till systems) is strategically located at soil-atmosphere interface and offers profound water conserving effect by reducing run-off and evaporative losses which buffers the abiotic stresses. Adoption of ZT in cereal cropping system in IGP has also been reported to advance the planting time thereby increasing the thermal window for wheat and thus escaping from terminal heat effect specially in eastern IGP. We collected data from 208 farmers in Haryana for the 2 contrasting wheat seasons; 2013–14 (a period with normal rainfall i.e., normal year) and 2014–15 (a period with untimely excess rainfall i.e., bad year). Our analysis shows that whilst average wheat yield was greater under CA than CT during both bad and normal years, the difference was two-fold greater during the bad year (16% vs. 8%). This provides new hard evidence that CA can cope better with the climatic extremes, in this case untimely excess rainfall, compared to CT. Absolute yield of the CA and CT was 10% and 16% lower in the bad year compared to the normal year, respectively. The Govt had to pay huge compensation to farmers for this yield loss in wheat during 2014–15. Whereas our study revealed that if, as targeted by the Haryana government in 2011, one million ha of wheat was brought under CA, the state would have produced an additional 0.66 million tonnes of wheat in 2014–15, equivalent to US$ 153 million (Aryal *et al.*, 2016).

These practices not only help in adaptation to climate risks but also contribute to reduction in environmental foot prints. Reduced power and energy requirements in CA translates into less fuel consumption, lower working time and slower depreciation rates of equipment, all leading to mitigation from farm operations as well as from the machinery manufacturing processes. On an average, by adopting of ZT in rice-wheat system of IGP, farmers could save 36-liter diesel ha^{-1} equivalent to a reduction in 93 kg CO_2 emission ha^{-1} yr^{-1}. Through our continuous monitoring of GHGs by using static chamber method in a rice-wheat production system of western IGP, we found much higher emission of CH_4 from rice production in puddled transplanted field with continuous flooding (50–250 mg CH_4 m^{-2} day^{-1}) compared to Direct Seeded Rice (DSR) production system (< 50 mg CH_4 m^{-2} day^{-1}). In this study, total cumulative GHGs emission (soil flux of CO_2, N_2O and CH_4) in terms of CO_2-equivalent was about 27% higher in the conventional tillage than in CA-based rice-wheat system. Through life-cycle analysis (using Cool Farm Tool) of wheat production in western IGP, we found that global warming potential of conventional till wheat with ad-hoc nutrient management was significantly higher than in ZT- with precision nutrient management (Sapkota *et al.*, 2014).

Above results provides the evidence on food security (yield, income), adaptive capacity and mitigation; the 3 key components of CSA but in isolation. However, our results from synthesis of the participatory strategic research on portfolio of CSAPs in intensive rice-wheat rotation of western IGP indicates that system yield, net income, water use, energy use efficiency & GHGs mitigation varies greatly with layering (portfolio) of various CSAPs. Results (Table 3) revealed that improved management with low intensity practices (residue incorporation) does not lead to immediate gains (yield, income, water, energy) over business as usual (residue burning) except marginal (7%) reduction in GHGs. However, CSA practices with varying degree of intensity (layering of various practices) have led to progressive gains in yield (5–9%); income (15–25%), water savings (21–28%), energy efficiency (3.6–5.7%) with 13–28% lower environmental foot prints.

Table 3: Portfolio of Climate Smart Agriculture Practice Influence on System Yield, Net Income, Water Use, Energy Use Efficiency and GHGs Mitigation in Intensive Rice-Wheat System of Western IGP (pooled analysis of 2014–15 and 2015–16)

	Scenario	RW System Yield (t ha⁻¹ yr⁻¹)	Net Income (INR ha⁻¹ yr⁻¹)	Irrigation Water Use (cm ha⁻¹ yr⁻¹)	Energy Use Efficiency	Global Warming Potential (kg CO₂ eq ha⁻¹ yr⁻¹)
1.	Business as usual	11.61^E	122879^E	234^A	11.21^E	5207^B
2.	Improved with Low intensity	11.71^{DE} (0.09%)	124539^E (1.4%)	234^A (0 %)	11.31^E	5574^A (−7%)
3.	Improved with High intensity	11.98^{CD} (3.2%)	134613^D (10%)	201^B (−14 %)	14.05^D (2.53%)	4504^C (−14%)
4.	CSA with Low intensity	12.19^{BC} (5.0%)	141591^C (15%)	184^C (−21%)	15.29^C (3.64%)	4510^C (−13%)
5.	CSA with medium intensity¶	12.38^{AB} (6.6%)	147606^B (20%)	169^D (−28%)	16.70^B (4.90%)	3956^D (−24%)
6.	CSA with high intensity¶	12.66^A (9.0%)	153328^A (25%)	169^D (−28%)	17.66^A (5.75%)	3765^E (−28%)

¶Scenario 5 and 6 had additional layering of ICTs and Index Insurance.

The evidence from our research has led to stakeholder buy-in for mainstreaming CSAPs and scaling through CSV approach. The CSV model demonstrates strong scalability through unique and interrelated elements of CSA-led business models, innovation platforms, knowledge networks, ICTs, gender and youth empowerment; thereby facilitates convergence of AR4D programs under the umbrella of National Action Plan on Climate Change (NAPCC).

REFERENCES

Aryal, J.P., Sapkota, T.B., Stirling, C.M., Jat, M.L., Jat, H.S., Rai, M., Mittal, S. and Sutaliya, J.M. (2016). Conservation agriculture-based wheat production better copes with extreme climate events than conventional tillage-based systems: A case of untimely excess rainfall in Haryana, India. *Agriculture, Ecosystems and Environment,* 233: 325–335.

Aryal, J.P., Sapkota, T.B., Jat, M.L. and Bishnoi, D. (2015). On-farm economic and environmental impact of zero-tillage wheat: A case of north-west India. *Experimental Agriculture,* 51: 1–16

CIMMYT-CCAFS (2014). Climate Smart Villages in Haryana. International Maize and Wheat Improvement Center (CIMMYT), CGIAR Research Program on Climate Change, Agriculture & Food Security (CCAFS), CIMMYT India, NASC Complex, Pusa New Delhi, India, P 12.

Dinesh, D., Frid-Nielsen, S., Norman, J., Mutamba, M Loboguerrero Rodriguez, A., Campbell, B. (2015). Is Climate-Smart Agriculture effective? A review of selected cases. *CCAFS Working Paper No.* 129.

Easterling, W., Aggarwal, P., Batima, P., Brander, K., Erda, L., Howden, S., Kirilenko, A., Morton, J., Soussana J.F., Schmidhuber J, T.F. (2007). Food, fibre and forest products. Climate Change 2007: impacts, adaptation and vulnerability, in: Parry, M., Canziani, O., Palutikof, J. (Eds.), Contribution of Working Group II to the Fourth Assessment Report of the Intergovernmental Panel on Climate Change. Cambridge University Press, Cambridge, pp. 273–313.

IPCC (2014). Climate change 2014, synthesis report. pp. 151 WMO, Geneva, Switzerland.

Jat, M.L., Dagar, J.C., Sapkota, T.B., Yadvinder-Singh, Govaerts, B., Ridaura, S.L., Saharawat, Y.S., Sharma, R.K., Tetarwal, J.P., Hobbs, H. and Stirling, C. (2016). Climate Change and Agriculture: Adaptation Strategies and Mitigation Opportunities for Food Security in South Asia and Latin America. *Advances in Agronomy*, 137: 127–236.

Jat, M.L., Yadvinder-Singh., Gill, G., Sidhu, H.S., Aryal, J.P., Stirling, C. and Gerard, B. (2015). Laser-Assisted Precision Land Leveling Impacts in Irrigated Intensive Production Systems of South Asia. *Advances in Soil Science*, Soil Specific Farming: Precision Agriculture, Vol. 22, eds R. Lal and B.A. Stewart, CRC Press. pp. 323–352. doi/pdf/10.1201/b18759-14.

Lal, R. (2016). Feeding 11 billion on 0.5 billion hectare of area under cereal crops. *Food and Energy Security,* 5(4): 239–251.

Lobell, D.B., Sibley, A. and Ivan Ortiz-Monasterio, J. (2012). Extreme heat effects on wheat senescence in India. *Nature Climate Change.* 2, 186–189. doi:10.1038/nclimate1356.

Ortiz, R., Sayre, K.D., Govaerts, B., Gupta, R., Subbarao, G.V., Ban, T., Hodson, D., Dixon, J.M., Iván Ortiz-Monasterio, J. and Reynolds, M. (2008). Climate change: Can wheat beat the heat? *Agriculture, Ecosystems and Environment.*126, 46–58.

Sapkota, T.B., Jat, M.L., Aryal, J.P., Jat, R.K. and Arun, K.C. (2015). Climate change adaptation, greenhouse gas mitigation and economic profitability of conservation agriculture: Some examples from cereal systems of Indo-Gangetic Plains. *Journal of Integrative Agriculture,* 14(8): 1524–1533.

Sapkota, T.B., Majumdar, K., Jat, M.L., Kumar, A., Bishnoi, D.K., McDonald, A.J. and Pampolino, M. (2014). Precision nutrient management in conservation agriculture based wheat production of Northwest India: Profitability, nutrient use efficiency and environmental footprint. *Field Crops Research,* 155: 233–244.

Sidhu, H.S., Singh, M., Yadvinder-Singh, Blackwell, J., Lohan, S.K., Humphreys, E., Jat, M.L., Singh, V. and Sarabjeet-Singh (2014). Development and evaluation of Turbo Happy Seeder to enable efficient sowing of wheat into heavy crop rice residue in rice-wheat rotation in the IGP of NW India. *Field Crops Research,* 184: 201–212.

Adaptation to Climate Change:
A Fishery Technology Perspective

C.N. Ravishankar

ICAR-Central Institute of Fisheries Technology, Matsyapuri, Willingdon Island, Cochin 682 029, India
E-mail: directorcift@gmail.com

ABSTRACT: *Fisheries play an important role in food supply, food security and income generation for millions of people. Issues related to climate change will further impact these resources which are already stressed with issues like overfishing and diminishing returns. There are reports regarding far-fetching impacts on the safety and quality of seafood, especially the increase in bioburden on algal toxins, organic chemical residues and toxic metals which are related to climate change. Technologies that can reduce the contributing factors, like the emission of GHGs, together with responsible fishing operations and reduction in effort for fishing, can make a significant dent in reducing emissions related to fishing. Adaptation is always involved for benefitting from these issues and an interdisciplinary approach cutting across different disciplines is a pre-requisite for developing mitigation measures.*

Keywords: Fishing Vessel, Fuel Consumption, Green House Gases, Algal Toxins, Shell Fish Poisoning.

UNFCC (1992) defines climate change as the "Change of climate that is attributed directly or indirectly to human activity that alters the composition of the global atmosphere and that is in addition to natural climate variability observed over comparable time periods".

Fisheries play an important role in food supply, food security and income generation for more than 1.5 million people in India. Being in the open access to a large extent, the intense competition among the different stakeholders to share the limited resources, has resulted in plateauing of catches in the capture fisheries, both from the Inland and marine sectors. While the production has stabilized, there has been a drastic change in the fluctuation of catches of major species and stocks. Despite recent reductions of fleet sizes in many developed countries, the size of the global fishing fleet has doubled in the last 30 years and is now estimated at 4.6 million vessels (FAO, 2016), of which about 64% are engine powered.

The global warming expected to occur over this century will have its reper-
cussions in the sea also and it is predicted that the over the next 40 years, the
average temperature of the Indian seas will rise by 1 to 3°C. Changes in the current
system, acidification, and sea level rise are expected as a result of increasing
temperature and these changes will affect the fish living in these waters. Though
there are a large number of ways in which the changes in climate would affect
fishes, research to ascertain the far reaching consequences is just progressing.

Increase in temperature may have both positive and negative impacts on fishes.
Commercial fishing in the lower latitudes can gain due to the higher temperatures and hence an increase in the abundance of selected species would be noticed.
Changes in the geographical extent of the species as seen in case of Indian oil
sardine (*Sardinella longiceps*) and Indian mackerel (*Rastrilliger kanagurta*) with
their extended geographical ranges in the recent years are examples.Though
many of the early trends show the negative impacts due to climate change,
improvements in the fishing technology and the increase in effort over the years
would have also affected the fluctuation in the fish stocks.

Out of the 4.6 million vessels globally, 3.5 million are in Asia, and 80% of these
vessel are powered by some engines. With dwindling fishery resources, the
capacity and the installed engine power of these vessels are on the increase, to
enable them to fish in deeper and distant waters. Fuel costs for the fishing
industry have risen substantially over the years. FAO and World Bank reported
that the global fishing fleet consumes approximately 41 million tonnes of fuel
per annum at a cost of $22.5 billion (World Bank and FAO, 2009), generating
approximately 130 million tonnes of CO_2. Though there is a paucity of detailed
data on Green House Gases (GHGs) emissions from fishing vessels (Buhaug *et al.*,
2009), the conservative estimate for global fisheries is at 1.7 t of CO_2 generated for
capture of one tonne of fish. This average shows that capture fisheries is one of the
most energy intensive food production methods. Though passive methods of
fishing like seining and lining have much lower energy requirements. The average
quantity of CO_2 generated for capture of one tonne of fish in India is estimated at
1.18 t CO_2/t of fish (Figure 1).

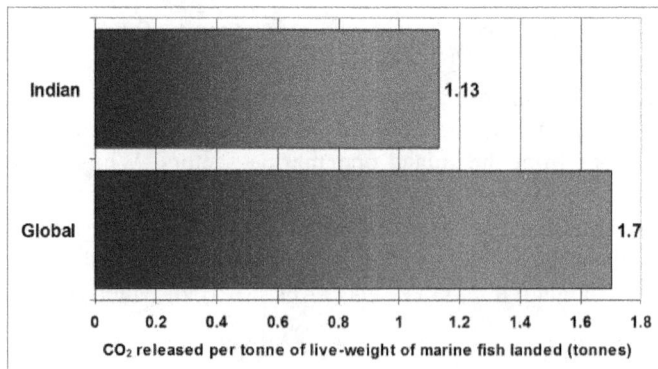

Fig. 1: GHG Emissions of Indian and World Fishing Fleet

Some of the preventive methods to reduce the impact of GHGs would be to improve the performance of fishing craft and gear in terms of saving energy. This would involve the use of efficient vessel designs with reduced drag and energy saving features that would cut down on the fuel usage (Figure 2).

Fig. 2: Energy Efficient Fishing Vessel Developed by CIFT

Modified gears like semi-pelagic trawl systems with reduced bottom contact and higher filtration capabilities can significantly reduce fuel consumption. Use of thinner twines for fabrication of gears has found to significantly reduce drag offered by the gear. Right sizing of engine power with the vessel design and gear used and in optimum speed for fishing will also significantly reduce fuel usage. Use of by catch reduction technologies in trawls (most energy intensive fishing), is found to reduce fuel consumption by 0.1–0.2%, by reduction in drag due to reduced by catch and will help also help in conservation in addition to reducing impacts of GHGs emissions.

Tapping of alternate renewable energy sources like solar energy for propulsion and use of alternate materials for construction of fishing canoes/vessels will reduce the dependence on forests for traditional wood. Development of solar fishing boat (Figure 3) and innovations like Rubber wood and Coconut wood canoes, which are often a byproduct in the production cycle, are some steps towards this direction. Products developed using remote sensing technologies like the Potential Fishing Zones (PFZ) developed by Indian National Centre for Ocean Information Services (INCOIS), when used by fishermen is found to reduce the fishing time and hence equivalent reduction in the fuel consumption (Subramanian *et al.*, 2014).

Climate change with concomitant change in ocean water temperature also has a direct impact on the safety of fishery products. During recent years an unusual escalation of harmful algal blooms has been observed in many marine and coastal regions of the world. An increasing number of outbreaks due to consumption of seafood with natural toxins produced by HAB organisms leading to Amnesic Shellfish Poisoning (ASP), Diarrheic Shellfish Poisoning (DSP), Neurotoxic Shellfish Poisoning (NSP), Azaspiracid Shellfish Poisoning (AZP), Paralytic Shellfish Poisoning (PSP), and ciguatera fish poisoning is reported (Marques *et al.*, 2010). Abrupt changes in sea surface temperature has resulted in increase in growth of HAB forming dinoflagellates like *Prorocentrum* and *Dinophysis* (Peperzak, 2003), increase in the temporal period of *Alexandrium catenella* and range expansion of Ciguatoxic dinoflagellate *Gambierdiscus toxicus* (Moore *et al.*, 2008).

Fig. 3: CIFT-SUNBOAT, a Solar Powered Fishing Boat for Inland Waters

In recent years, seafood exported from India has faced a number of rejections from European Union due to presence of ciguatoxin in Red Snapper and Barracuda. Climate change related extreme events are also attributed to increased residual load of chemical hazards like polychlorinated dibenzo-p-dioxins and dibenzofurans (PCDD/Fs), methyl mercury (Booth and Zeller, 2005) as well as increased bioburden of pesticides and veterinary drug residues in fishery environments (Shriner and Street, 2007). Similarly, emergence of pathogenic strains of *Vibrio, Aeromonas and Clostridium* in seafood has been attributed to climate change phenomena like sea level elevation, flooding and salinity changes (FAO, 2008).

As fisheries sector would be disproportionately affected by climate change, it is essential to devise a comprehensive management strategy for fishing and seafood industry, to ameliorate the stress from climate related changes and other environmental stressors like overfishing, over capacity and bioburden of veterinary drugs and pesticides to provide sustainable and safe food for the coming generation.

REFERENCES

Booth, S. and Zeller, D. (2005). Mercury, food webs, and marine mammals: Implications of diet and climate change for human health. *Environmental Health Perspectives*, 113, 521–526.

Buhaug, Ø., Corbett, J.J., Endresen, Ø., Eyring, V., Faber, J., Hanayama, S., Lee, D.S., Lee, D., Lindstad, H., Markowska, A.Z., Mjelde, A., Nelissen, D., Nilsen, J., Pålsson, C., Winebrake, J.J., Wu, W.-Q. and Yoshida, K. (2009). Second IMO GHG Study 2009. International Maritime Organization (IMO), London, UK, pp. 220.

FAO (2008). Food safety and climate change. FAO conference on food security and the challenges of climate change and bioenergy. <http://www.fao.org/ag/agn/agns/files/HLC1_Climate_Change_and_Food_Safety.pdf>.

FAO (2016). The State of World Fisheries and Aquaculture. FAO, Rome.

Marques, A., Nunes, M.L., Moore, S.K. and Strom, M.S. (2010). Climate change and seafood safety: Human health implications. *Food Research International*, 43(7), 1766-1779.

Moore, Stephanie K., Trainer, Vera L., Mantua, Nathan J., Parker, Micaela S., Laws, Edward A. and Backer, Lorraine C. (2008). Impacts of climate variability and future climate change on harmful algal blooms and human health. *Environmental Health*, 7 (Suppl. 2), S4.doi:10.1186/1476-069X-7-S2-S4.

Peperzak, L. (2003). Climate change and harmful algal blooms in the North Sea. *Acta Oecologia*, 24, 139–144.

Shriner, D.S. and Street, R.B. (2007). North America. Climate change 2007: Impacts, adaptation and vulnerability. In M.L. Parry, O.F. Canziani, J.P. Palutikof, P.J. vander Linden and C.E. Hanson (Eds.), Contribution of working group II to the fourth assessment report of the inter-governmental panel on climate change. Cambridge, UK: Cambridge University Press.

Subramanian, S., Sreekanth G.B., Manjulekshmi N., Singh, N.P., Kolwalkar, J., Patil, T. and Fernandes, P.M. (2014). Manual on the Use of Potential Fishing Zone (PFZ) forecast. Technical bulletinNo. 40, ICAR Research Complex for Goa (Indian Council of Agricultural Research), Old Goa–403402, Goa, India.

UNFCC (1992). United Nations Framework Convention on Climate Change, 1992. FCCC/INFORMAL/84 GE.05-62220 (E) 200705.

World Bank and FAO (2009). The Sunken Billions. The Economic Justification for Fisheries Reform. Agriculture and Rural Development Department/The World Bank,Washington, DC, pp. 100.

Nanotechnology for Management of Infestations and Grain Quality Monitoring

N.L. Naveena[1], Amruta Ranjan Behera[1], Rudra Pratap[*1] and Krishna Chaitanya[2]

[1]Center for Nanoscience and Engineering, Indian Institute of Science, Bengaluru 560 012, India

[2]Department of Chemical Engineering, Indian Institute of Technology, Chennai 600 036, India

*E-mail: pratap@cense.iisc.ernet.in

ABSTRACT: *Pest infestation in agricultural produce is a major issue both in the field and in the storage environment. Its detection is primarily done by manual observation which makes human intervention inevitable. Further-more, traditional methods are used to control pests which involve use of harmful pesticides. For better pest management, it is desirable to have a sensor network for automated, real time, and remote monitoring of pests. This requires development of low-cost sensors for field deployment. Application of nanotechnology has already resulted in cost reduction and miniaturization of various sensing systems. Here, we propose spectroscopy based sensing methods for detection of pests. Results from initial trials are presented. Also, towards management of pests in grain-storage, we report some results on efficacy of silver nano-particles synthesized from plant extracts using a scalable production process. The quality of stored grains also changes with aging due to time dependent internal chemical changes. At present, the age of grains is inferred from their physical appearance, which is very subjective and depends on individual's experience. In this paper we also report explora-tion of NIR spectroscopy to deduce age of grains.*

Keywords: IR Spectroscopy, Pest Detection, Nanosensors, Nanoparticles, Pest Management.

Climate change and associated problems such as erratic rainfall, frequent occurrence of drought, etc., are leading to new challenges in food security that require novel solutions for boosting productivity and safe storage of agricultural produce. Recent advances in nanotechnology provide several tools for potential technological interventions in agriculture. While production of various kinds of nanoparticles has led to a lot of experimentation for their usage in fertilizers and pesticides, the strides made in micro and nano sensor technology are just about beginning to find applications in agriculture. The goal of our work is to explore the use of nanotechnology for grain storage and post harvest produce monitoring.

Pest infestation is a major concern in field and storage conditions. Timely detection of pests and accurate assessment of their population densities are fundamental to Integrated Pest Management (IPM) approaches. So far the key methods for gathering such data are scouting and periodic survey of fields, and visual inspection of grains in storage. These methods are very subjective, time consuming, and often ineffective as the infestation is frequently identified at a stage which is quite late leading to significant damage. Such damages can be minimized if the infestation can be detected at an early stage. Thus it is desirable to have scientifically sound and robust methods of detection which can provide actionable information. In our proposed solution a handheld NIR (Near-infrared) spectrometer is used to acquire reflectance spectra of stored food grains, which is subsequently analyzed by machine learning algorithms to classify healthy and infested grains. For field conditions we propose a technique for detection of insects by sensing the unique Volatile Organic Compounds (VOCs) released by them using MIR spectroscopy.

We also report the synthesis of silver nanoparticles from karanjin, extracted from *Pongamia pinnata* seeds, and its efficacy as a safe fungicide. We are studying the use of these karanjin coated nanoparticles as safe pesticides for grains in storage. In addition, we are also developing techniques for sensing the quality of stored grains, especially in terms of ascertaining their age. We are exploring the use of NIR spectroscopy as a feasible technique for classifying grains based on their storage duration.

INFRARED SPECTROSCOPY FOR MONITORING INFESTATION

Infrared spectroscopy is a well-known technique for chemical analysis and is routinely used in laboratories. However the equipment used for such analyses are bulky and expensive (e.g., FTIR) and are not amenable to field applications. Recent developments have resulted in handheld sensing systems that enable measurement in field conditions (Anon, 2016). Efforts are being made at our center for development of similar miniaturized systems that exploit nanotech-enabled components.

NIR Spectroscopy for Detection of Infestation in Storage

Near-infrared reflectance spectroscopy is a well-established technique for obtaining chemical signature of substances. Here, the measurements are carried out using SCiO – a handheld NIR spectrometer developed by Consumer Physics Inc. It includes a light source that illuminates the sample and a spectrometer that measures intensity of the reflected light in the wavelength range of 759–1052 nm. It communicates the measured spectral data to a smartphone wirelessly through an application, which in turn, uploads it to a cloud-based server. Machine learning algorithms are used to establish correlations between various types of grains (grains with different levels of infestation) and their corresponding NIR reflectance spectra. This enables creation of classification models and hence a standard database that can be subsequently used for detecting levels of infestation in grains. In the current study, we have used rice grains infested with rice weevil (*Sitophilus oryzae*) as a representative sample.

The rice (var: Sona Musoorie) sample was purchased from APMC, Yeshwanth-pur, Bangalore with approximately 15% moisture content (wet basis) for the study. Moisture content was determined using a standard oven method (ASAC Standards, 2003). The insect species, *Sitophilus oryzae*, were obtained from Storage Entomology Lab, AICRP on Post-Harvest Technology, University of Agricultural Sciences, GKVK, Bengaluru. Adult insects were released in 100 gram of rice and named as infested and another fresh sample was used as uninfested. All the samples were kept at room temperature (27–30°C) and 70% relative humidity. For imaging, 50% of the samples were used randomly.

Figure 1(a) shows the acquired reflectance spectra from rice samples. In this case three kinds of samples have been scanned: (i) uninfested rice, (ii) infested rice with insects in pupa stage, and (iii) infested rice with insects in adult stage. Multiple samples were scanned from each category. The curves denote the reflected intensity at various wavelengths. It can be observed that there is significant difference in the pattern of spectral variation among the three categories. This variation is due to the difference in morphology and characteristics of the grain surfaces, which can be attributed to the chemicals released by insects as well as the physical damage caused by them. It shows that near-infrared spectroscopy can be used to classify infested grains from healthy ones. These scan data were used as the training data-set to create a classification model. Robustness of this model was tested by applying it on the training data-set

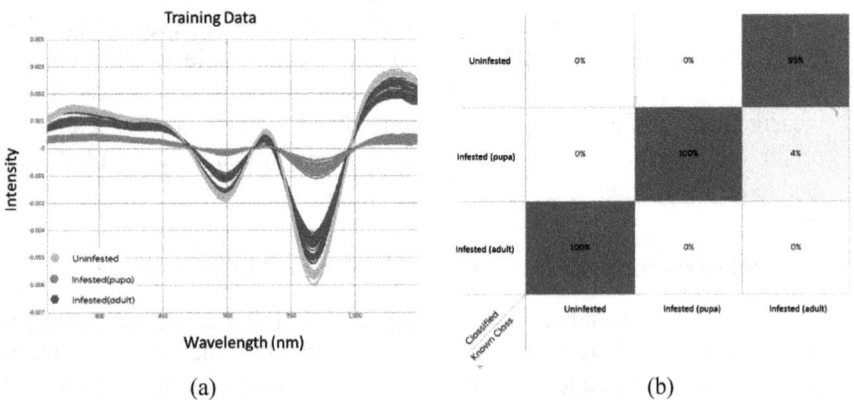

(a)   (b)

Fig. 1: (a) Reflectance spectra from 3 types of rice samples: uninfested, infested rice–insect in pupa stage, infested rice—insect in adult stage. (b) Classification ability of the model when applied to the training dataset.

itself to classify them. Figure 1(b) shows the model's performance. Here the X-axis denotes the actual grain category and the Y-axis denotes the classified category as predicted by the model. The numbers in each square block represent how much percentage of a particular rice type (X-value) is classified as Y-value (corresponding to the block) type. Thus in an ideal case, all the diagonal boxes should show 100% values. Here, it can be observed that the model's performance is very close to ideal.

MIR Spectroscopy for Detection of Infestation in Field Conditions

Mid-IR spectroscopy is well known for its selectivity as the absorption peaks of gases in this regime are sharp which helps in distinguishing them. The proposed solution is based on Non-Dispersive Infrared (NDIR) spectroscopy. The goal of this technique is to detect the trace chemicals used by insects for intra-species communication to ascertain their presence. The schematic of the proposed system is shown in Figure 2.

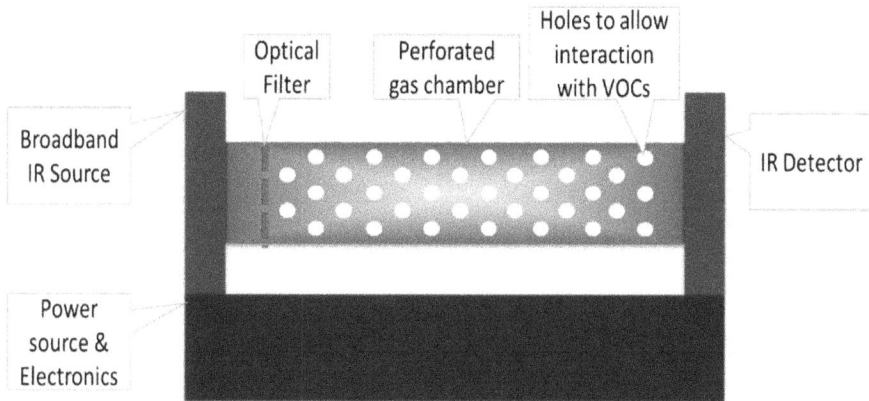

Fig. 2: A Schematic of the Insect Detector

The radiations emitted by the broadband IR source is detected by the IR detector. An optical filter is used to allow radiation with only desirable wavelength to pass through. VOCs enter the tube through the holes and absorb radiation at their characteristic specific wavelengths. Thus the presence of a VOC alters the amount of radiation that falls on the detector and from the detector output its quantity can be inferred. A survey has been carried out to identify important crops and the corresponding major infestations. For those insects, the signature VOCs have been identified. This data in compiled form is presented in Table 1. Experiments are in progress to validate the infrared absorption spectra of these VOCs. For this purpose artificially synthesized VOCs are being used. From the spectra of each VOC, specific wavelengths will be identified which can be used as markers to recognize them unambiguously. Interference from atmospheric gases will be taken into consideration while making this identification.

Table 1: List of Pests, the Crops Infested by Them and Associated VOCs

Pest	Crops Infested	Associated VOC	Absorption Peaks (Wave-Number in cm⁻¹)
Spodoptera litura	Pulses, Brinjal, Cauliflower, Cabbage	(Z, E)-9,11-Tetradecadienyl acetate	1730, 985, 950
Tuta absoluta	Tomato	(3E, 8Z, 11Z)-tetradeca-3,8,11- trienyl acetate	3009, 2962, 2929, 2856, 1741, 1456, 1364, 1233, 1033, 968
Scirophaga incertulas	Rice (in the field)	(Z, Z)-7,11-Hexadecadienal	2927, 2858, 2715, 1728, 1654, 1458, 1099, 1011, 914

NIR SPECTROSCOPY FOR DETECTION OF AGEING IN STORED GRAINS

Ageing of grains leads to chemical changes in them. We are developing a NIR reflectance spectroscopy based technique to ascertain the age of grains and classify them based on the chemical content.

Five different varieties of paddy were collected from farms. The collected grains were dried to reduce the moisture level (12 per cent) and stored separately in air-tight plastic containers. The observations (scanning of grains) were recorded at weekly intervals using SCiO. During each observation, a total of 15 scans were recorded from each variety. Measured spectra for paddy variety Jaya from day 1

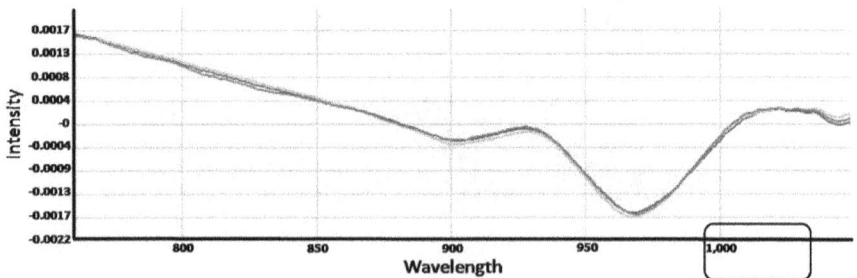

Day 1

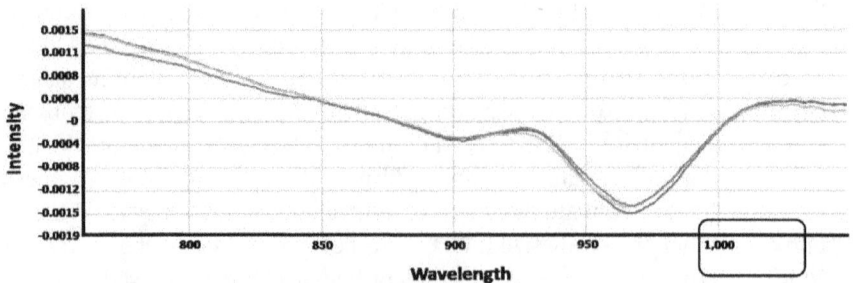

Day 179

Fig. 3: Spectral Variations Observed in the Paddy Variety, Jaya

and day 179 are shown in Figure 3. From these two images it can be observed that in the wavelength range 950–1000 nm (marked by the rectangle) there is a slight shift that represents the change in chemical composition of the grain. This technique requires further analysis and refinement for building a good predictive model.

SYNTHESIS OF SILVER NANOPARTICLES FROM *PONGAMIA PINNATA* SEEDS

Pongamia pinnata seeds (2 kg) collected from the forest area of Indian Institute of Science campus, Bangalore, were dried under shade, powdered, and used for extraction using Soxhlet apparatus with methanol. After evaporation of methanol, the obtained semisolid compound was fractionated by column chromatography over silica gel (60 mm) and the presence of the desired compound (karanjin) was confirmed by spectroscopic techniques (LC-MS, NMR). The extracted compound was used for synthesis of silver nanoparticles.

About 10 ml of 1 mm solution of silver nitrate was prepared using double distilled water. Then 1 ml of karanjin was added and the reaction mixture was sonicated (3 minutes) and exposed to microwave irradiation for two minutes at 850 W. The resultant reaction mixture was cooled to room temperature and centrifuged at 15,000 rpm for 10 minutes. The obtained pellets were re-suspended in deionized water and centrifuged for 5 min at 15,000 rpm. The resulting material was dried in hot-air oven at 55°C for 24 hrs and used for characterization.

The AgNPs were characterized for the presence of pure silver with UV-Visible Spectrophotometer showing a surface plasmon resonance peak at 424 nm (Figure 4(a)) indicating complete reduction of silver nitrate (Edison *et al.*, 2016). The crystalline nature of Ag was confirmed by XRD characterization (Figure 4(b)). The presence of various functional groups of karanjin as reducing and capping agents on the surface of AgNPs were confirmed by FT-IR analysis (Gannimani *et al.*, 2014). SEM, TEM (Figure 4(c)), and AFM characterizations were carried out to ascertain the topology, size distribution (20-30 nm), and poly-dispersed nature of the karanjin coated AgNPs.

Antifungal activity of the AgNPs was tested against saprophytic and pathogenic fungi, *Aspergillus flavus* and *A. niger* by the standard agar disc diffusion method (Anon, 1997). The fungal strains were grown in a broth medium containing potato dextrose for 72 hrs and used for the study. The negative (distilled water) and positive (amphotericin B) controls, along with karanjin were included for the antifungal activity assay. The petriplates were then incubated at 25°C for 72 hrs in an incubator. The zone of inhibition (mm) around the disc was observed and recorded. The results indicate that the karanjin capped silver nanoparticles showed the maximum zone of inhibition against *A. flavus* (15 mm) and *A. niger* (14 mm) compared to karanjin. On the other hand, the negative control (distilled water) did not exhibit any zone of inhibition. The positive control showed antifungal activity against both the fungi (Figure 4(d)).

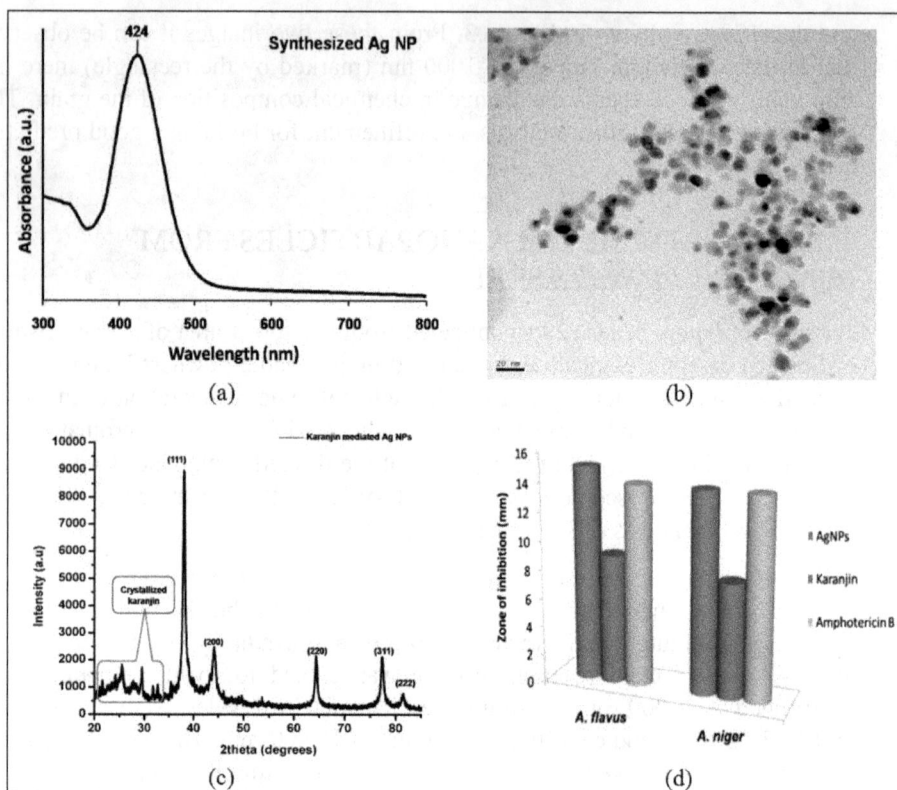

Fig. 4: (a) UV spectroscopy (b) XRD pattern (c) TEM image (d) Antifungal activity of karanjin mediated silver nanoparticles synthesized by microwave heating.

CONCLUSIONS

We have carried out tests with handheld NIR spectrometer and showed that it can be used for detecting pests in grains in storage conditions as well as for detecting and eventually ascertaining the age of stored grains. Some preliminary work is also carried out to develop a handheld MIR spectrometer that will be key for detecting VOCs for effective monitoring of grain storage. We have also reported initial work on the synthesis and use of karanjin capped biosynthesized AgNPs as fungicides for grain storage.

REFERENCES

Anon (1997). National Committee for Clinical Laboratory Standards. Performance Standards for Antimicrobial Disc Susceptibility Test, 6th ed. 1997; Approved Standard, M2-A6; Wayne, PA, USA.

Anon (2016). The first molecular sensor that fits in the palm of your hand. https://www.consumerphysics.com/myscio/scio.

ASAE Standards, 50th ed. (2003). S352.2 Moisture Measurement—Unground Grain and Seeds. St. Joseph, Mich.: ASAE.

Edison, T.N.J.I., Lee, Y.R. and Sethuraman, M.G. (2016). Green synthesis of silver nanoparticles using *Terminalia cuneata* and its catalytic action in reduction of direct yellow-12 dye, Spectroch. *Acta Part A: Mol. and Biomol. Spectrosc.* 161: 122–129.

Gannimani, R., Perumal, A., Krishna, S.B., Sershen, Muthusamy, K., Mishra, A. and Govender, P. (2014). Synthesis and antibacterial activity of silver and gold nanoparticles produced using aqueous seed extract of *Protorhus longifolia* as a reducing agent. *Journal of Nanomaterials and Biostructures,* 9(4): 1669–1679.

Climate Change Impacts on Water Resources and Adaptations

A.K. Gosain

Civil Engineering Department, Indian Institute of Technology, Delhi,
New Delhi 110 016, India
E-mail: gosain@civil.iitd.ac.in

ABSTRACT: *One of the key sectors that are going to be adversely impacted by climate change is that of water resources. Throughout the World, countries are engaged in formulating policies to cope with the implications of climate change to the water resources. India has also responded by formulating the National Water Mission under the National Action Plan for Climate Change (NAPCC). India has also made two National Communications to the United Nations Framework Convention on Climate Change (UNFCCC) in 2004 and 2012 respectively. Impacts of climate change on the water resources of various river basins have been evaluated by using the SWAT (Soil and Water Assessment Tool), a distributed hydrological model under the present condition and the future conditions. Projections made by the Regional Climate Models (RCMs) on the future weather conditions have been used as input in the SWAT model. Implications on account of the changing weather conditions have been evaluated in terms of changes from the present levels. Extremes in terms of floods and droughts have also been evaluated on temporal as well as spatial scales of the river basins. This analysis has been done by making various assumptions in the absence of required data on utilization.*

Keywords: IWRM, GIS, National Water Mission, SWAT.

Quantification of the implications of climate change on water resources of Indian river systems is more complex on account of the poor understanding of these basins. Presently multiple organizations/ministries have various programmes ranging from local watershed management that intends empowering the village level communities, to large scale projects in the form of big dams, but all targeting the common water resources. All these programmes despite using the common water resources hardly interact with each other thereby violating the Integrated Water Resources Management (IWRM) philosophy.

A common framework that can incorporate all the development of water resources at various scales in a dynamic manner shall be key to integrated water resources development and management in a sustainable manner. The IIT Delhi has taken an initiative to put such a framework together (see http://gisserver. civil.iitd.ac.in/natcom). This is a GIS based information system and has been populated with very extensive information on the present status of water resources as well as implications of climate change under various future scenarios. This information base is also very useful for evaluating the various adaptation options relevant for specific areas.

The study evaluates the impacts of climate change on water resources of the river basins of India. The Figure 1 shows river systems of the country modeled using SWAT hydrological model (refer http://gisserver.civil.iitd.ac.in/natcom).

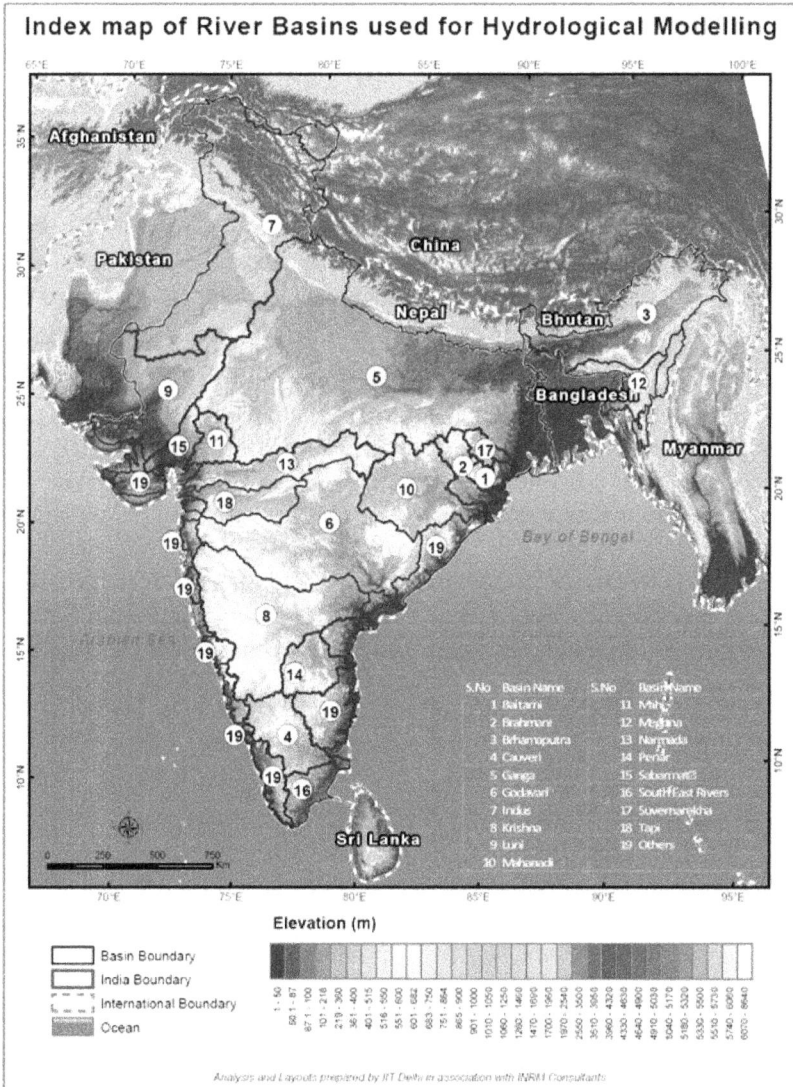

Fig. 1: River Basins of India Used for Hydrological Modelling

Impacts of climate change on the water resources are likely to have many implications such as effect on irrigated agriculture, hydro power, environmental flows in the dry season, higher flows during the wet season and thereby causing severe droughts and flood problems. The hydrological modeling has been used to first calibrate and validate each river basin using the past historical hydrometeorological data and then used the validated model for assessing climate change impacts on water resources under future scenarios. Deployment of the simulation models is the only option for climate change impact assessment. In the present study the SWAT hydrological model as described in the following paragraphs has been used.

The Soil and Water Assessment Tool (SWAT) model (Arnold *et al.*, 1998; Neitsch *et al.*, 2002) is a distributed parameter and continuous time simulation model. The SWAT model has been developed to predict the hydrological response of catchments to natural inputs as well as the manmade interventions. Water and sediment yields can be assessed. The model (a) is physically based; (b) uses readily available inputs; (c) is computationally efficient to operate and (d) is continuous time and capable of simulating long periods for computing the effects of management changes. The major advantage of the SWAT model is that unlike the other conventional conceptual simulation models it does not require much calibration and therefore can be used on un-gauged watersheds (in fact the usual situation).

In SWAT, a watershed is divided into multiple sub-watersheds, which are then further sub-divided into unique soil/land-use characteristics called Hydrologic Response Units (HRUs). The water balance of each HRU in SWAT is represented by four storage volumes: snow, soil profile (0–2 m), shallow aquifer (typically 2–20 m), and deep aquifer (> 20 m). Flow generation, sediment yield, and non-point-source loadings from each HRU in a sub-watershed are summed, and the resulting loads are routed through channels, ponds, and/or reservoirs to the watershed outlet. Hydrologic processes are based on the following water balance equation:

$$SW_t = SW + \sum_{i=1}^{t}(R_{it} - Q_i - ET_i - P_i - QR_i)$$

where SW is the soil water content minus the wilting-point water content, and R, Q, ET, P, and QR are the daily amounts (in mm) of precipitation, runoff, evapotranspiration, percolation, and groundwater flow, respectively. The soil profile is subdivided into multiple layers that support soil water processes, including infiltration, evaporation, plant uptake, lateral flow, and percolation to lower layers. The soil percolation component of SWAT uses a storage routing technique to predict flow through each soil layer in the root zone. Downward flow occurs when field capacity of a soil layer is exceeded and the layer below is not saturated. Percolation from the bottom of the soil profile recharges the shallow aquifer. If the temperature in a particular layer is 0°C or below, no percolation is allowed from that layer. Lateral subsurface flow in the soil profile is calculated simultaneously with percolation. The contribution of groundwater flow to the total stream flow is simulated by routing shallow aquifer storage component to the stream (Arnold *et al.*, 1993).

The study determines the present water availability in space and time. The same framework is then used to predict the impact of climate change on the water resources with the assumption that the land use shall not change over time. A total of 90 years of simulation have been conducted; 30 years belonging to IPCC SRES A1B Baseline (BL), 30 years belong to IPCC SRES A1B near term or Mid-Century (MC) climate scenario and 30 years belong to IPCC SRES A1B long term or End-Century (EC) climate scenario.

While modeling, each river basin has been further subdivided into reasonable sized sub-basins so as to account for spatial variability of inputs under the baseline and GHG scenarios. Detailed analyses have been performed to quantify the possible impacts on account of the climate change.

DATA USED

Spatial data used in the modeling of the study areas and their source include:

- Digital Elevation Model: SRTM, of 90 m resolution, http://srtm.csi.cgiar.org
- Drainage Network–Hydroshed: http://hydrosheds.cr.usgs.gov
- Soil maps and associated soil characteristics (source: FAO Global soil)–http://www.lib.berkeley.edu/EART/fao.html
- Land use (source: Global landuse): http://glcfapp.glcf.umd.edu:8080/esdi/index.jsp

The Meteorological data pertaining to the river basins are also required for modeling. These data elements include daily rainfall, maximum and minimum temperature, solar radiation, relative humidity and wind speed. These weather data as per following details were made available by the IITM Pune.

- *Climate Change:* PRECIS Regional Climate Model outputs for Baseline (1961–1990, BL), near term (2021–2050, MC) and long term or end-century (2071–2098, EC) for A1B IPCC SRES scenario (Q14 QUMP ensemble).

Having performed the SWAT modeling, detailed outputs have been analyzed with respect to the two various water balance components such as water yield, actual evapotranspiration, ground water recharge that are highly influenced by the weather conditions dictated by temperature and allied parameters. Furthermore, the analysis has also been extended to the detection of extreme events of droughts and floods that may be triggered on account of the climate change (Gosain *et al.*, 2011) and are of major concern to the local societies. One sample output depicting the change in water yield in various basins of India is depicted in Figure 2.

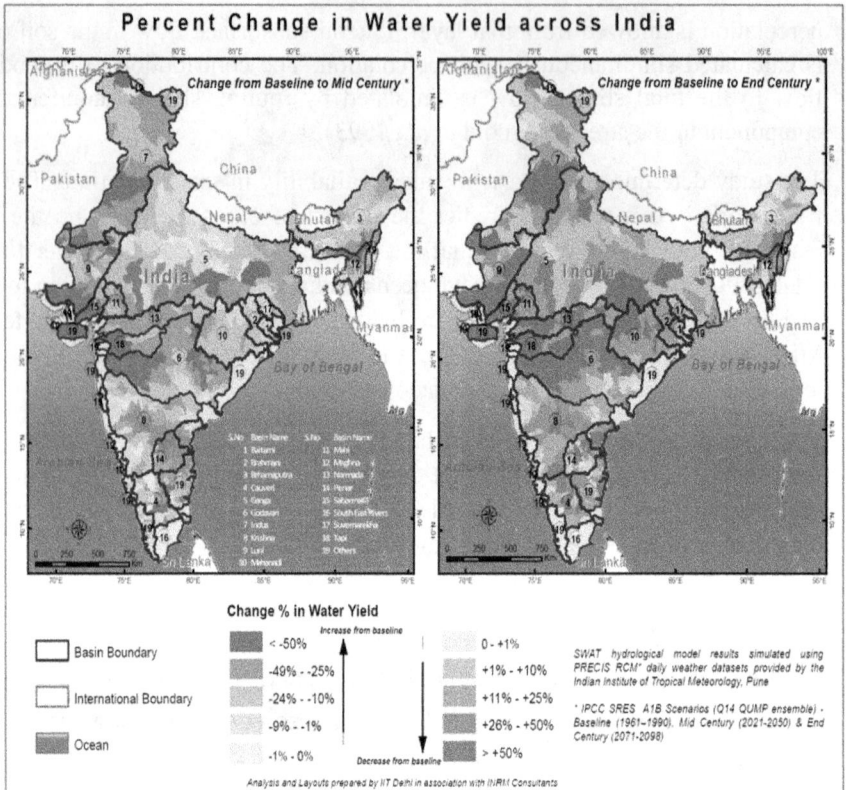

Fig. 2: Change in Water Yield (water availability)
Towards 2030s and 2080s with Respect to 1970s

All the analyses have been performed by aggregating the inputs/outputs at the sub-basin level that are the natural boundaries controlling the hydrological processes and have been depicted accordingly using the GIS. Since it is not possible to put all the results covering the spatial and temporal scales in a physical report, a GIS framework has been created and all the results have been uploaded in this frame-work which is accessible to every user (http://gisserver.civil.iitd.ac.in/natcom). This framework can also be very effective in the integration of various sectors for implementing integrated water resources management strategy.

ADAPTATIONS

The adaptation to impacts of climate change to water resources shall revolve around the water resources practitioner/manager whose responsibility is to balance supply of water (be it from rivers, groundwater, impoundments, return flows or water transfers) with demand for water so that allocations can be made in a sustainable manner using IWRM. The climate change shall impose the additional challenges that would generally revolve around engineering issues of changes in supply/demand and limits to the design of hydraulic structures in regard to system failure, as well as around environmental consequences of changes in hydrological regimes, including in-stream flow requirements (IFRs).

There is the transitional component of the hydrological system, which represents the changes happening in the wetlands, the riparian zones and estuaries. Another important aspect is the ecosystem and its interaction with other sub-systems.

A framework is required that has a model base (suit of models) to support the IWRM activities. The challenges for the future lie in the improvement of representations of hydrological and hydraulic processes in the models because they remain simplifications of complex dynamic interactions. These modelling systems can also be used for providing improved early warning lead times in accordance with increased reliability of weather forecasts (Kabat *et al.*, 2003).

Forecasts of climate change may remain debatable for some time; evidence of increased climate variability is incontestable, and the severity of that variability demands urgent responses from water managers. These measures include the conventional technological elements of water infrastructure, like storage reservoirs, recharge wells, etc., but with an emphasis on techniques for boosting the yield of available resources (water recycling/reuse, desalination). Risk sharing and access to credit for affected families are among the financial mechanisms that are being adapted to combat floods and droughts. Additionally, modification of land-use patterns, crop selection and tillage practices can also be considered and modeling framework can be used to generate scenarios through simulation for the purpose.

REFERENCES

Arnold, J.G., Allen, P.M. and Bernhardt, G.T. (1993). A comprehensive surface groundwater flow model. *Journal of Hydrology*, 142: 47–69.

Arnold, J.G., Srinivasan, R., Muttiah, R.S. and Williams, J.R. (1998). Large-area hydrologic modeling and assessment: Part I. Model development. *J. American Water Res. Assoc,* 34(1): 73–89.

Gosain, A.K., Sandhya Rao and Anamika Arora (2011). Climate change impact assessment of Water Resources of India, *Current Science*, 101(3), pp. 356–371.

Gosain, A.K., Sandhya Rao and Debajit Basuray (2006). Climate change impact assessment on hydrology of Indian River basins, *Current Science*, 90(3), pp. 346–353.

Kabat, P., Schulze, R.E., Hellmuth, M.E. and Veraart, J.A. (editors), (2003). Coping with impacts of climate variability and climate change in water management: A scoping paper. DWC-Report no. DWCSSO-01, *International Secretariat of the Dialogue on Water and Climate,* Wageningen, Netherlands.

Neitsch, S.L., Arnold, J.G., Kiniry, J.R., Williams, J.R. and King, K.W. (2002). Soil and Water Assessment Tool-Theoretical Documentation (version 2000). Temple, Texas: Grassland, Soil and Water Research Laboratory, Agricultural Research Service, Blackland Research Center, Texas Agricultural Experiment Station.

Agro-Hydro-Technologies and Policies for Adaptation to Climate Change: An Assessment

N.K. Tyagi[1] and P.K. Joshi[2]

[1]Agricultural Scientists Recruitment Board, New Delhi 110 012, India
[2]International Food Policy Research Institute–South Asia, New Delhi 110 012, India
[1]E-mail: nktyagi1947@gmail.com; [2]p.joshi@cgiar.org

ABSTRACT: *Adaptations are considered to be very potent mechanisms to modify the negative impacts and magnify the positive impacts of climate change in agriculture. A conceptual framework of hierarchical adaptations and the corresponding agro-technologies which bring resilience in the water resource system are discussed. The effectiveness of a few water smart technologies such as micro irrigation, zero tillage, laser levelling etc, which have been introduced in sizeable area in India, is reviewed to highlight the close nexus between water, energy and food production systems. A first order assessment of the role of supportive policies of the Green Revolution Technologies (GRTs) in adaptation to climate change in India has been made. These adaptations decreased the energy foot prints of food grain production (tCO_{2e}/t_{FG}) from 1.196 in 1990 to 0.907 in 2010. The performance of GRTs policies in respect of sustainability of both surface as well as ground water systems was low. Considering the dynamic nature of climate change, a set of technology and policy research issues are indicated to keep pace with the rapidly changing climate scenario.*

Keywords: Hydro-Technology, Productivity, Virtual Mitigation, Climate Change, Policies, Mainstreaming.

Universal food security is on top of the agenda of sustainable development as underlined in the 'Future We Want' (UN, 2012). A major challenge for achieving this goal is the climate change which is projected to adversely impact agriculture, because it is a highly weather dependent enterprise. The key impacts of climate change on agriculture will be transmitted through water in terms of increased irrigation demands in response to increased temperatures and decreased rainfall, particularly during winter season; degradation of water quality; and increased flooding risk (GOI, 2010). Mitigation and adaptation are two major mechanisms to turn down the heat to keep operating in safespace of resource boundaries.

In agriculture, adaptation is considered to be a major rational policy response to climate change for reducing vulnerability, increasing resilience and moderating the risks of climate impacts (Howden *et al.*, 2007). The key issues addressed herein are: intensity of climate change and the nature of adaptations; agro-technologies which would help ensuring the food and nutrition security with minimum trade-offs between increased production and the environment (health of soil and water resources systems); and the role of climate centric agriculture development policies in promoting climate smart technologies. Finally, the technology and policy research areas to cope with the continuously changing climate are identified.

ADAPTATION TO CLIMATE CHANGE: CONCEPTUAL FRAMEWORK

The impact of rise in green house gases (GHGs) under climate change is addressed through two distinct but complementary approaches—mitigation and adaptation. In agriculture the opportunities for adaptation, which connotes adjustment to moderate the impacts of climate change, are higher than mitigation. The adaptation requirements depend on the vulnerability in terms of loss in production and/or income from agriculture (Howden *et al.*, 2010) under the given set of biophysical as and socio-economic factors. As the degree of climate change increases, the efficacy of the adaptation measures goes down and so do the benefits requiring change from incremental adaptations, to systems adaptations and finally transformational adaptations like the ones indicated in Figure 1. Further, there are limits on effectiveness of the measures arising from

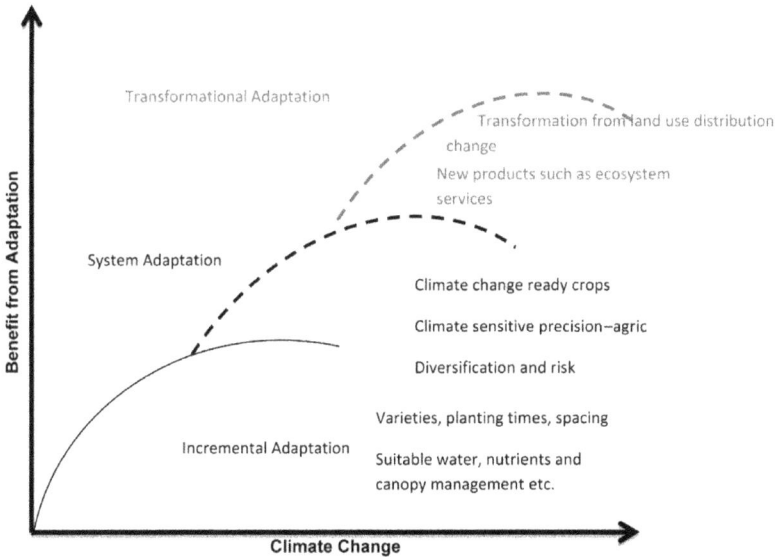

Fig. 1: Levels of Adaptation in Relation to Benefits from Adaptation Actions and Degree of Climate Change (Howden *et al.*, 2010)

biophysical factors (the ecological tipping points) which create absolute limits for adaption, social limits (how much is acceptable) and economic limits (how much is affordable) (Shipper, 2009). The guiding principles for building resilience in water resources systems are based on: limiting water as renewable supply, adaptive allocation, transparent water markets and maintenance of environmental flows. Minimizing economic and environmental tradeoffs often remains an issue in observance of these principles.

Hydro-Agro-Technologies for Building Resilience in Water Resources Systems

There are a number of technological, economic, regulatory and policy based options which may be used to increase the resilience of water resources (2030 WRG, 2009). There is very strong empirical evidence to show that, increasing land and water productivity though various agro-technological interventions and their mainstreaming in public development policies is the key to minimizing the projected water demand and supply gaps (Schipper, 2009; 2030 WRG, 2009; Chakraborty *et al.*, 2012; Iglesias and Garrote, 2015).

Enhancement of agricultural productivity without emitting excessive additional green house gasses into the atmosphere is one of the keys to adaptation led mitigation.There being a very close nexus between water, energy and food production systems, adaption of different technologies, gives rise to differential GHG emissions. Collectively, these can be termed as climate smart agricultural technologies (Agarwal, 2008). The important water smart technologies include improved irrigation techniques (irrigation scheduling, laser levelling, micro-irrigation, system of rice intensification, alternate wetting and drying (AWD), deficit irrigation etc). Some of these technologies such as laser levelling, micro-irrigation and reduced tillage have been out scaled in sizeable areas. Laser levelling, which has been extensively promoted in Indo-Gangetic Plain, was found to save water to the extent of 20–30%, increase yields by 15–20% and the reduction in energy used in pumping was a bonus (Jat *et al.*, 2006). Similarly micro irrigation which has so far been extended over 4 Mha, proved to be a 'triple wins' intervention as it was estimated to have increased production by 3.483 Mt, reduced water use by 0.73 Mham, and effected GHG reduction of 5.555 CO_2e, Mt (Table 1) at average efficiency of 30% (Joshi *et al.*, 2015).

Introduction of zero-till drill has made a revolutionary change in seed bed preparation and seeding of crops by reducing the cost and time required for sowing. A special feature of this technology, which is hugely significant for climate change adaptation, is its energy saving. The water productivity of zero tillage system in rice wheat could be higher by 15–37%, while net global warming potential lower by 26–31% as compared to conventional tillage systems (Pathak *et al.*, 2011). An increase of 28% in water productivity in wheat has been reported from Bihar (Upadhyaya and Sikka, 2016).

Some other adaptations like adjustment in crop areas, reallocation of water or introduction of tolerant cultivars have been found to be useful (Howden *et al.*, 2010; Iglesias and Garrote, 2015), but may generate conflict between productivity (income) and environmental sustainability goals, as is happening in northwest India's rice-wheat system and groundwater decline.

No single technology can reduce the water demand supply gap, and therefore adaptation to climate change requires adoption of multiple technologies. The optimal technology mix varies with location and socioeconomic situations of the adaptors. Decision making prioritization tools like cost curve, payback period curve and quantitative modelling, which now have become available may help in deciding the portfolio of actions (2030 WRG, 2009 Ahmed and Suphachalasai, 2014).

Table 1: Water Saving, Production Increase, Food Grain Increase and Emission Reduction from due to Existing 3.87 Mha Area under Micro Irrigation (Tyagi *et al.*, 2014)

Parameters	Increase in Water Saving Productivity/Food Grain Availability, and Reduction in Emissions at Different Efficiencies		
	20%	*30%*	*40%*
Saving in water, Mha-m	0.488	0.733	1.47
Increase in production (Mt)	2.522	3.483	4.644
Increase in food grain availability, (kg/cap/yr)	2.08	3.13	4.16
Reduction in GHG emission, CO_2e, Mt	3.704	5.555	7.605

Table 2: Simulated Yield, Irrigation, Global Warming Potential and Net Benefits Resource-Conserving Technologies in Modipuram (Pathak *et al.*, 2011)

Treatment	Rice + Wheat Yield (t/ha)	Rice + Wheat Irrigation (cm)	Irrigation WP (kg/m³)	GWP (CO₂ ekg ha⁻¹)	Net Benefit (USD/ha)
Puddling + TP rice and CT in wheat	12.2	271.4	0.449	5853	563
DS rice after ZT and DS after ZT in wheat	11.1	188.7	0.588	4408	651
TP rice after ZT and DS + NT in wheat	11.6	229.9	0.505	4752	629

CT—Conventional tillage, DS—Drill seeded, TP—Transplanted, ZT—No till, WP—Water productivity, GWP—Global warming potential.

MAINSTREAMING DEVELOPMENT MEASURES AND POLICIES

Most mitigation and adaptation interventions require policy support and have to be mainstreamed into development programmes of the governments for large scale implementation. In recent years, the government policies have often been expressed through action programmes. A large number of actions programmes aimed at conservation of natural resources, improving water use efficiency, and transfer of risk caused by weather aberrations, have been initiated under National Mission for Sustainable Agriculture (GOI, 2010). Several other such missions/programmes were launched in the last quarter of 20th century to promote green revolution technologies (GRTs). These action programmes, to name a few, included irrigation command area development, accelerated irrigation benefit in agriculture (AIBA), National mission on micro-irrigation, National Innovations on Climate Resilient Agriculture (NICRA), crop insurance schemes, weather advisory service etc. (Joshi *et al.*, 2015).

IMPACT ASSESSMENT OF GRTs AND DEVELOPMENT POLICIES ON MITIGATION AND ADAPTATION TO CLIMATE CHANGE

The Government of India has promoted development of agriculture by incentivising GRTs (improved seeds, irrigation and fertilizers) with policy focus on subsidies on water, electricity, fertilizers and implements. The role of these development policies in combating climate change has been recently evaluated by Joshi *et al.* (2015). The first order assessment indicated that these policies were highly successful in reducing the potential GHGs intensification, which has been termed as virtual mitigation (Figure 2). The virtual mitigation was of the order

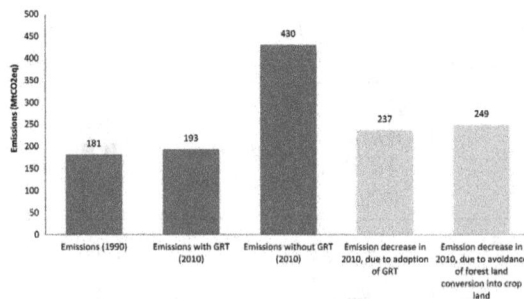

Fig. 2: Estimated Annual GHG Emissions from Land under Food Grains with without Adoption of GRTs

of 237 MtCO$_2$e. The incremental adoption of GRTs between 1990–2010 resulted in production and productivity increases, and lowered the food grain production foot print (tCO$_2$e/t$_{FG}$) from 1.196 in 1990 to 0.907 in 2010. Additionally, it avoided the deforestation of 56.6 Mha of additional land, which otherwise would have been put under the plough to reach food grain production of 232 Mt, achieved in 2010.

Further, the study established that the adaptation capacity in terms of food grain availability improved by 26 percent during 1990–2010. But in so far as sustainability is concerned, with degree of water abstraction between 0.60 to 0.66 range by 2010, these policies resulted in creating high stress in both, surface as well as ground water resources (Joshi *et al.*, 2015).

SUMMARY AND WAY FORWARD

Adaptations have been, and will continue as the mainstay of climate smart water resources management. The incremental adaptations through GRTs were useful in improving agricultural productivity and consequent adaptation led virtual mitigation, but had environmental cost. Current agro-technical fixes may become ineffective due to continued heating of earth, and in such a situation higher level agro-hydro-technologies will be needed to ensure food and environmental security. Models like "My Climate" would be useful to set up technology generation research programmes; and unlike the past, more climate centric policies and institutions would be required.

The farm, irrigation projects or basin level improvements would have to be achieved through a combination of technologies in the form of precision agriculture which may be based on real-time data on weather, soil and water quality, crop maturity and equipment etc. Technology development alone will not be sufficient, as there are economic, social and cultural barriers to adaptations. It would, therefore be important to assess the effectiveness of adaptation options under different operating policy regimes, geographical differentiations and risk transfer programmes.

Agriculture is gradually getting commercialized, and therefore, increasing productivity alone will not remain the targeted objective, but will be replaced by profitability. Generally, only the adaptations of greater economic value for the farmers would be adopted. The use of decision support tools like cost curve, pay back curve, and the models like IMPACT will be needed to charter policy pathways (Rosegrant *et al.*, 2012; 2030 WRG, 2009).

Water resources management is highly political in nature, and the climate induced scarcity will make it still more political. The most effective way to motivate policy makers to integrate climate smart water resources management into development plans will need generating research based strong empirical evidence and establishing effective communication channels with them.

REFERENCES

Aggarwal, P.K. (2008). Global Climate Change and Indian Agriculture: Impacts, Adaptation and Mitigation. *The Indian J. agric. Sci.*, 178(11): 911–919.

Ahmed, M. and Suphachalasai, S. (2014). Assessing the Costs of Climate Change and Adaptation in South Asia. Mandaluyong City, Philippines: Asian Development Bank

Chakraborty, S.S., Mohan, S., Tyagi, N.K. and Chander, S. (2012). Water Resources Management. New Delhi, Indian National Academy of Engineers (INAE), p. 170.

GoI (Government of India) (2010). National Mission for Sustainable Agriculture-Strategies for Meeting the Challenges of Climate Change. Department of Agriculture and Cooperation, Ministry of Agriculture, Government of India, New Delhi http://agritech.tnau.ac.in/nmsa/pdf/climate_change (Last retrieved November 12, 2016).

Howden, S.M., Crimp, S. and Nelson, R.N. (2010). Australian Agriculture in a Climate of Change. In: '*Managing Climate Change: Papers from GREENHOUSE 2009 Conference*'. (Eds, I. Jubb, P. Holper, W. Cai). Melbourne CSIRO Publishing, pp. 101–112.

Iglesias, A. and Garrote, L. (2015). Adaptation Strategies for Agricultural Water Management under Climate Change in Europe, *Agricultural Water Management*, 155: 113–124.

Jat, M.L., Chandana, P., Sharma, S.K., Gill, M.A. and Gupta, R.K. (2006). Laser Land Leveling-A Precursor Technology for Resource Conservation, Rice-Wheat Consortium Technical Bulletin Series 7, New Delhi, India: Rice-Wheat Consortium for the Indo-Gangetic Plains, p. 48.

Joshi, P.K., Aggarwal, P.K., Tyagi, N.K. and Pandey, D. (2015). Role of Development Policies in Combating Climate Change Issues in Indian Agriculture: A First Order Assessment of Irrigation and Fertilizer Policies. In: 29th ICAE Congress-Agriculture in an Interconnected World, August 8–14, 2015 Milan, Italy.

Pathak, H., Saharawat, Y.S., Gathala, M. and Ladha, J.K. (2011). Impact of resource-conserving technologies on productivity and greenhouse gas emissions in the rice-wheat system. *Greenhouse Gas Sci Technol.* 1: 1–17. http://libcatalog.cimmyt.org/download/reprints/96198.pdf.

Rosegrant, M. and the IMPACT development team (2012). International Model for Policy Analysis of Agricultural Commodities and Trade (IMPACT*): Model Description, Washington DC International Food Policy Research Institute, www.ifpri.org/sites/default/files/publications/impactwater2012.pdf.

Tyagi N.K., Joshi, P.K., Aggarwal, P.K. and Pandey, D. (2014). Micro-Irrigation for Enhancing Energy and Water Use Efficiency: Analysis of Climate Change Mitigation and Adaptations Impacts in India, In: Energy and Water for Development (Abstract Volume), World Water Week Conference August 31, September 5, 2014, SIWI, Stockholm, p. 100. Available at: http://www.worldwaterweek.org/wp-content/uploads/2014/08/2014_Abstract_Volume_web.pdf (Last accessed 12 Nov, 2016).

2030 Water Resource Group (2009). Charting Our Water Future: Economic Frameworks to Informed Decision-Making. McKinsey & Company, p. 185.

Upadhyaya, A. and Sikka, A.K. (2016). Concept of Water, Land and Energy Productivity in Agriculture and Pathways for Improvement, *Irrigation and Drainage Systems Engineering*, 5(1): 1–10 .

United Nations (2012). The Future We Want. Final document of the Rio + 20 Conference, http://rio20.net/en/iniciativas/the-future-we-want-final-document-of-the-rio20-conference.

Water Resources Management for Climate Change Adaptation and Sustainable Water Security

N.H. Rao

National Academy of Agricultural Research Management, Rajendranagar, Hyderabad 500 030, Telangana, India
E-mail: nhrao@naarm.ernet.in

ABSTRACT: *Water is the key medium through which climate change impacts agriculture, food security and economy. Rising average and extreme temperatures and changing rainfall patterns caused by climate change alter the hydrologic cycle. Together, they intensify the risk of water insecurity by: (i) altering soil moisture, surface water flows and groundwater recharge patterns, (ii) increasing the variability of available supplies and (iii) increasing the risk of extreme hydrologic events like floods and droughts (IPCC, 2014). Small changes in temperature and rainfall patterns are amplified significantly in surface flows (World Bank, 2014). Climate change can therefore make water systems more stressed and vulnerable. Water stress has been consistently flagged among the top three risks with greatest potential impact on global economy (World Economic Forum, 2016). This paper reviews the current status of water security in India and the state-of-art in science of climate change impacts on water resources, to develop a unified framework for managing water security.*

Keywords: Climate Change Adaptation, Water Security,

Water security is defined as the availability of an *acceptable* quantity and quality of water for health, agricultural production, livelihoods and ecosystems, coupled *with an acceptable level of water-related risks* to people, environment and economies (UN, 2015). The definition combines both access risk for productive water services and destructive risk to human and natural systems. Extreme floods and droughts are most obvious manifestations of the latter. Rainfall, temperature and hydrologic variabilities that can have chronic impacts on production are sources of access risks. Water security is about managing both types of risks at a *tolerable* level for society, across local (farm) to regional and global scales (Grey *et al.*, 2013). In addition to climate change, water security is also threatened by growing population, heavy dependence on agriculture, urbanization, water pollution, ground water depletion and deteriorating water infrastructure. Climate change amplifies the risks for water security from all these sources.

This paper reviews the current status of water security in India and the state-of-art in science of climate change impact on water resources, to develop a unifield framework for managing water security. The framework for managing water security is scalable and provides for an anticipatory, and structured quantitative analysis that integrates: (i) uncertainties and extremes in future water availability and use, (ii) vulnerabilities of water use (e.g. agricultural) systems and (iii) evaluation of alternate water development, management and technology pathways for chosen risk levels. The framework also enables informed decisions across different scales on how science, innovation and institutions can be leveraged to ensure water security.

WATER SECURITY STATUS

India's annual average renewable fresh water resource is 1869 billion cubic meters (BCM), as determined by its national water balance (Table 1). Per capita availability of water in India has declined from 5831 m^3 in 1950 to 1608 m^3 in 2010. By 2050 it is expected to decline further to 1139 m^3, or to only about 22% of the 1950 level. The present surface water storage per capita is only about 200 m^3, compared to 2500 m^3 in China, 6000 m^3 in USA, and a world average of 800 m^3. Per capita water availability also varies spatially by river basins from over 14000 m^3 in Ganga-Brahmaputra-Meghna basin to about 300 m^3 in Sabarmati basin. Per capita availability < 1000 m^3 is classified as water scarcity by UN. Many river basins in India are water scarce.

Table 1: India's National Water Resources and Use
(adapted from CWC, 2015; CGWB, 2014)

Resource	*Quantity in BCM* *(billion cubic metres or million km³)*
Annual precipitation	4,000 (100% of precipitation)
Evaporation + soil water and groundwater storage	2,131 (53% of precipitation)
Runoff (including about 35 BCM groundwater outflow into rivers-base flow)	1,869 (47% of precipitation)
Storage capacity:	
Completed projects (2015)	253 (13.5% of average annual runoff)
Under construction	51 (2.7% of average annual runoff)
Proposed	109
Total (by 2050?)	414 (22% of average annual runoff)
Utilizable water resources	1123 (28% of precipitation; 60% of runoff)
Surface water	690 (37% of average annual runoff)
Ground water	433 (11% of precipitation)

Water demand (BCM)	Low			High		
	Total	Irriga-tion	Municipal/ Industry Energy	Total	Irriga-tion	Municipal/ Industry/ Energy
2010	694	543 (78%)	97	710	557 (78%)	99
2025	784	561 (72%)	153	843	611 (72%)	162
2050	973	628 (65%)	234	1180	807 (68%)	262

Per Capita availability (cubic metres)		
1950	5831	Average per capita water availability range by river basins 300–14000 M^3 (water stress:1000–1700 M^3; water scarce 500–1000 M^3)
2010	1608	
2050	1139	

Though surface water availability is higher, ground water is more easily accessible locally, less vulnerable to climate fluctuations, and stabilizes agricultural production and livelihoods during droughts. Groundwater meets 85% of rural water needs and 80% of urban water needs. Over 60% of total water and 88% of groundwater is used for irrigation. The level of ground water development is very high (100–145% of recharge; critical) in north west India, where ground water is depleting by 40 mm/year. Ground water development is high (75–85% of recharge—semi-critical) in Tamil Nadu, Uttar Pradesh, Gujarat and Himachal Pradesh. In other states, ground water development is below safe limit of 70% of net recharge (CGWB, 2014). A recent satellite based ground water assessment of 37 largest aquifer systems of the world found that ground water depletion in the north-west India aquifer system is the highest (Richey *et al.,* 2015). Nearly 60% of districts in the country are vulnerable with respect to groundwater security.

Overall, India's water security is at far greater risk than currently acknowledged. Even without climate change, vast regions in India will become water stressed by 2025 and water scarce by 2050.

CLIMATE CHANGE AND WATER SECURITY

Global and national assessments (IPCC, 2014; Chaturvedi *et al.,* 2012; World Bank 2014) have provided projections of climate change (temperature, precipitation, including their extremes), river flows, and hydrometeorological events like droughts and floods (Table 2). The direct effects of climate change on water resources are most directly manifest in changes in river flows and extremes. The effects on soil water, groundwater, and water quality are less direct, and less analyzed. Taken together, the likelihood of water scarcity, driven by climate change alone, is as high as 30–50 percent for India (World Bank, 2014). The broad assessments in Table 2 are based on global climate models (GCMs) of 100–300 km resolutions downscaled to 50/25 km resolutions using

Table 2: Observed Trends and Projections of Key Hydrologic Variables
(adapted from World Bank, 2009; Chaturvedi *et al.*, 2012)

Key Variable	Observed Trends	Projections for 21ˢᵗ Century (base line 1961–90)
Precipitation	• Trend for mean annual precipitation is unclear • Decrease in mean seasonal rainfall • Increase in number of monsoon break days • Decrease in frequency of rainy days in central, north and east India; Increase in rainy days in AP, Karnataka • Decrease in frequency of extreme rainfall days • Increase in frequency of extreme precipitation events in eastern India; decline in some parts	• *Mean annual precipitation:* increase of 4–5% by 2030s; 6–14% by 2080s; for all over India except for a few regions in short term projections (2030s); increase in inter-annual variability • Both, frequency of years with above normal monsoon rainfall and extremely deficit rainfall is expected to increase • *Extreme precipitation:* consistent positive trend in frequency of extreme precipitation (> 40 mm/day) days for 2060s and beyond; 30–40% increase in frequency of > 100 mm/day events (a 1-in-20 year annual maximum daily precipitation amount is *likely* to become a 1-in-5 to 1-in-15-year event by the end of the 21st century in many regions)
Temperature	• Increasing trend; 0.68°C/ 100 years in annual mean temperature • Increase more pronounced in post monsoon and winter • Greater increase in night temperature and in post monsoon months	• Mean warming in the range 1.7–2°C by 2030s and 3.3–4.8°C by 2080s compared to pre-industrial times • More increase in night temperature • Increase in frequency of extreme temperatures (a 1-in-20 year hottest day is *likely* to become a 1-in-2 year event by the end of the 21st century in most regions) • consecutive day warm spells beyond the 90ᵗʰ percentile, will lengthen to 150–200 days under RCP 8.5, but only to 30–45 days under RCP 2.6 • Projections more reliable than for precipitation
Droughts	• Increasing frequency of dry days, intensity and, area affected by droughts (IG Plains, coastal south India, central Maharashtra) • Increasing frequency of multiyear droughts	• Increasing in lower latitudes; decreasing in higher latitudes; complex patterns • More frequent in some areas, especially in north-western India, Jharkhand, Orissa and Chhattisgarh
Floods	No significant trend	Increasing trend because of increasing trends in extreme rainfall
Cyclones	Increases in intensity	Increase in intensity; uncertain changes in frequency and track; average cyclone maximum wind speed is *likely* to increase; it is *likely* that global frequency of tropical cyclones will either decrease or remain unchanged
River flows	No long term trends in annual flows—attenuated by increasing storages in peninsular rivers	Increase in annual mean inflows; but skewed seasonally—high increase in monsoon season; decrease in non-monsoon season; higher risk of floods in monsoon, drought in non-monsoon seasons (e.g.: for Ganges-Brahmaputra system, the mean flow increases by only 4 percent, whereas the low flow decreases by 13 percent and the high flow increases by 5 percent)
Ground-water	Rapidly declining groundwater levels (40 mm/year) in northwest India and urban concentrations; no significant trend in other regions	Declining trends can accelerate in view of skewed distributions of rainfall and extreme rainfall

local observations and Regional Climate Models (RCMs). RCMs at 10/4/1/0.8 km resolution, of significance to extreme event/flood hydrology and small watershed management are becoming increasingly available (Westra *et al.*, 2015). More recently statistical downscaling has been used to generate historical and future climate datasets to scale-free point estimates (Wang *et al.*, 2016). Most studies on climate change impacts on river basin flows and agricultural impacts combine downscaled climate models data at these resolutions with hydrologic basin models and crop models (World Bank, 2014; Mathison *et al.*, 2015; Elliott *et al.*, 2014). With the advent of big data analytics, and based on local observations (including radar and crowd sourced data), downscaling climate model projections to hyper local resolutions for weather forecasting, agricultural advisories, and hydrologic design has become even more feasible (Bell *et al.*, 2016). Much of this data is available on demand, as a service to users.

The challenge for water security is to leverage high resolution historical climate and model data, hydrologic and crop models, and big data analytics in a unified framework to generate better insights into climate change related uncertainties and extremes for improved strategic and tactical management of water systems.

FRAMEWORK FOR OPERATIONALIZING CLIMATE SMART SUSTAINABLE WATER SECURITY IN INDIA

The typology and design for climate smart water security management in India must keep in view the following:

- A one-size-fits-all approach will not work as climate, water resource, land use and risk situations are unique to locations/regions.
- Predominance of agricultural water use in India (68% even by 2050) and ground water use for irrigation (88%).
- Agricultural water security in India is equally concerned with green water (soil water), and blue water (surface and groundwater) as over 60 mha will be rainfed even after development of full irrigation.
- Groundwater is the more widespread and dependable source, particularly during droughts. Its recharge management is critical for water security.
- Understanding changes in surface flow hydrology is important for managing river flows and reservoir storages.

The digital watershed atlas of India classifies the nation's water systems into river basins, catchments, sub-catchments, watershed, sub-watershed and microwatersheds for water related decisions. This provides an effective scalable natural spatial unit typology for water security assessment that meets the above criteria (Figure 1).

It permits relatively independent modelling of climate, hydrologic, land and agricultural production processes over each typological unit and scale the processes successively from micro-watersheds to higher levels for integration. The most fundamental spatial unit is the micro-watershed (500 to 1500 ha).

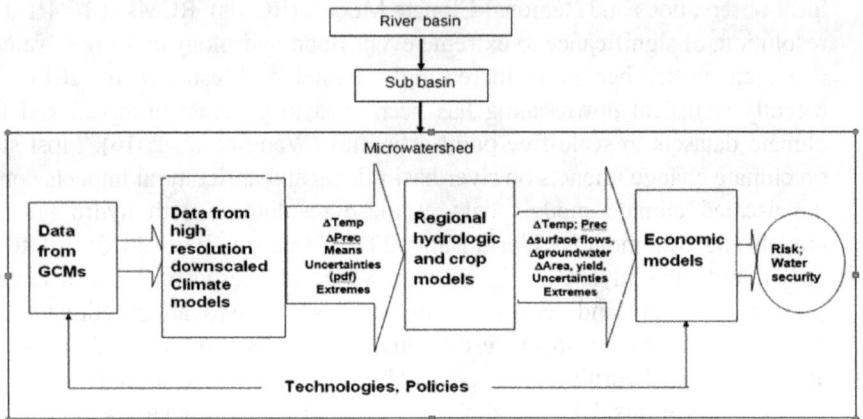

Fig. 1: Scheme for Climate Smart Water Security

Model outputs for each level can include: (i) risk assessments for water and agricultural production by anticipating climatic extremes over the long term (decadal, 30–100 years) and (ii) changes to green and blue water regimes and impacts on agriculture in the short term (3–10 years). The outputs will help to devise integrated water resources management (green and blue water) plans. The micro-watershed analysis can be integrated across scales to larger typology units of sub-basins and river basins to estimate climate change risks on river flows, groundwater recharge, soil moisture and agricultural production. Based on acceptable risk levels, the same framework can be applied to evaluate the costs and effects of measures to mitigate climate change risks for green water and blue water (surface flows and groundwater) to increase future water security. Examples of such measures include soil-water conservation technologies, water saving technologies in irrigation, water harvesting structures, large reservoirs, groundwater recharge systems, conjunctive use of groundwater and surface water, improved monitoring of water use and sources, increasing crop water productivity, changing cropping systems, and developing alternate sources of water (desalination, municipal waste water).

The climate and water science data essential to address questions of sustainable water security across scales in the face of climate change is increasingly accessible. But the effectiveness of the science to deliver water security for the nation depends on the scientific and institutional capacities to absorb the science, create the data networks to apply the science and develop new inter-disciplinary and innovation competencies.

REFERENCES

Bell, David E., Reinhardt, M. and Shelman, M. (2016). Harvard Business School Case 516–060.

Central Ground Water Board (2014). Dynamic Groundwater Resources, p. 295.

Central Water Commission (2015). Water and Related Statistics, p. 165.

Chaturvedi, R.K., Joshi, J., Jayaraman, M, Bala, G. and Ravindranath, N.H. (2012). Multi-model climate change projections for India under representative concentration pathways. *Current Science*, 103: 792–802.

Elliott, J., Deryng, D., Müller, C., Frieler, K., Konzmann, M., Gerten, D., Glotter, M., Flörke, M., Wada, Y., Best, N., Eisner, S., Fekete, B.M., Folberth, C., Foster, I., Gosling, S.N., Haddenland, I., Khabarov, N., Ludwig, F., Masaki, Y., Olin, S., Rosenzweig, C., Ruane, A.C., Satoh, Y., Schmid, E., Stacke, T., Tang, Q. and Wisser, D. (2014). Constraints and potentials of future irrigation water availability on agricultural production under climate change. *Proceedings of National Academy of Sciences USA*, pp. 3239–44.

Grey, D., Garrick, D., Blackmore, D., Kelman, J., Muller, M. and Sadoff, C. (2013). Water security in one blue planet: twenty-first century policy challenges for science. *Phil. Trans. R. Soc. A* 371 (doi:10.1098/rsta.2012.0406).

IPCC (2014). Climate Change 2014: The Physical Science Basis. Contribution of Working Group I to the Fifth Assessment Report of the Intergovernmental Panel on Climate Change. Cambridge, United Kingdom and New York, NY, USA: Cambridge University Press; 2014.

Mathison, C., Wiltshire, A.J., Falloon, P. and Challinor, J. (2015). South Asia river-flow projections and their implications for water resources. *Hydrology and Earth System Science.*, 19, 4783–4810.

Richey, A.S., Thomas, B.F., Lo, M., Reager, J.T., Famiglietti, J.S., Voss, K., Swenson, S. and Rodell, M. (2015). Quantifying renewable groundwater stress with GRACE. *Water Resources Research*, 51, 5217–5238.

United Nations (2015). Water for a Sustainable World. Paris, UNESCO, 138 pp.

Wang, T., Hamann, A., Spittlehouse, D. and Carroll, C. (2016). Locally downscaled and spatially customizable climate data for historical and future periods for North America. *PLoS ONE* 11(6):e0156720. doi:10.1371/ journal. pone.0156720

Westra, S., Fowler, H.J., Evans, J.P., Alexander, L.V., Berg, P., Johnson, F., Kendon, E.J., Lenderink, G. and Roberts, N.M. (2014). Future changes to the intensity and frequency of short duration extreme rainfall. *Rev. Geophys.*, 52, 522–555, doi:10. 1002/2014RG000464.

World Bank (2014). Turn Down the Heat, Climate Extremes, Regional Impacts, and The Case For Resilience, p. 254.

World Economic Forum (2016). The Global Risks Report, 11th Edition, The Global Competitiveness and Risks Team, pp. 103.

Adaptation to Climate Change: Water Management in Agriculture

T.B.S. Rajput

Water Technology Centre, Indian Agricultural Research Institute,
New Delhi 110 012, India
E-mail: tbsraj@iari.res.in

ABSTRACT: *Agriculture is the largest consumer of India's limited water resources developed at a high cost. Changing climate is adversely affecting the spatial and temporal availability of water. In view of the increasing demand for water in agriculture, industries and domestic needs, the available water resources need to be used efficiently. Mismatch between available water supplies and crop water requirements, in terms of quantity and timing are a major cause of low water use efficiency at field level particularly, in canal command areas. Optimal utilization of ground water, preferably in conjunction with surface water, wherever applicable, offers a better insurance against climatic variability in agricultural production system. Adoption of appropriate water distribution policies, maintenance of conveyance and water distribution network and regulated water application are the keys for achieving efficient utilization of available irrigation water. Field rectangulation, land leveling, selection of cropping pattern and irrigation method also play important role in enhancing water use efficiencies at field level. Automation, adequate water application as per temporal and spatial crop needs through modern techniques and tools help in maximizing water productivity to achieve the famous objective of "more crop per drop". This paper attempts to highlight different strategies, technologies and practices for efficient water distribution, application and its use in agricultural production system.*

Keywords: Water Use Efficiency, Modernization of Irrigation, Conjunctive Use, On-Farm Water Management.

Climate change is likely to influence political and economic stability in many countries as tensions fuelled by rising food insecurity. By 2050 climate change is expected to impact negatively on more than half of all food crops in sub-Saharan Africa, and at least 22% of the area cultivated by the World's most important crops (Campbell *et al.*, 2011). Against a backdrop of enhanced climatic risks and associated agro-ecological and socio-economic threats, preserving and enhancing food security requires higher agricultural productivity. More productive and resilient agriculture requires changes in the management of natural resources including water to attain higher efficiency in their use.

As population growth and economic expansion accelerated the use and abuse of water resources over the past few decades has resulted in a greater mismatch between water availability and water demand. This mismatch has brought water scarcity and water quality crises in many regions of the world and has also resulted in the destruction of some fresh water resources (World Bank, 2005).

Innovative technologies, including the improvements in the indigenous technologies, are needed to fully utilize limited water resources and to safeguard these resources against pollution. Water is a very important and valuable community resource and irrigation sector being the main user of the resource, it is essential to modernize the irrigation system for optimal use of water resources by economizing water consumption per unit yield of agricultural products.

Water for agriculture is becoming increasingly scarce. The causes include decreasing its physical availability as a consequence of silting of reservoirs, malfunctioning of irrigation systems, falling groundwater tables and increased competition from urban and industrial sectors. Under the circumstances, it is essential to develop strategies and technologies for efficient agricultural water management for surface as well as for groundwater to improve productivity. Studies of pumped irrigation schemes in Central Asia, China and elsewhere have shown that improvements in water use efficiency of irrigation scheme translate into energy savings and so can be an effective way of reducing emissions (Zou *et al.*, 2012). Options to improve irrigation scheme performance include 1. Changes in cropping pattern, 2. Changes in soil tillage practices, 3. Conjunctive use of ground water and surface water, 4. Optimizing the time of pump use, 5. Within scheme water storage and 6. Recycling of water.

INDIA'S WATER RESOURCES

The annual rainfall varies from about 310 mm in western Rajasthan to over 11400 mm in Meghalaya with an average value of 1170 mm for the entire country which is the highest anywhere in the world for a country of comparable size (Lal, 2001). About 85 percent of this rainfall occurs during four to five months of the year. The annual precipitation occurring over the geographical area of 329 mha of the country amounts to 4000 Billion Cubic Meters (BCM).

Irrigation development in India gained momentum after Independence. The net irrigated area in 1950–51 was 20.85 million ha, with 1.71 million ha irrigated during more than one crop season. The gross irrigated area was 22.6 million ha (Sharma and Paul, 1999). The planners assigned a very high priority to irrigation in the five-year plans. As a result of the sustained efforts for development of water resources India had achieved 80 mha irrigated area out of a total estimated potential of 140 mha. Projected water demand in the country is presented in Table 1.

Table 1: All-India Projected Water Demand by Different Uses (2010, 2025 and 2050)

Type of Use	Water Demand in BCM								
	Standing Sub-Committee of Ministry of Water Resources			*National Commission on Integrated Water Resources Development*					
	2010	*2025*	*2050*	*2010*		*2025*		*2050*	
				Low	*High*	*Low*	*High*	*Low*	*High*
Irrigation	688	910	1,072	543	557	561	611	628	807
Drinking water	56	73	102	42	43	55	62	90	111
Industry	12	23	63	37	37	67	67	81	81
Energy	5	15	130	18	19	31	33	63	70
Other	52	72	80	54	54	70	70	111	111
Total	813	1,093	1,447	694	710	784	843	973	1,180

ADAPTATION TO CLIMATE CHANGE: WATER MANAGEMENT INTERVENTIONS

Depending on local context, needs and interests, there are opportunities for improving water management that can significantly contribute to people's livelihood and make them more resilient to the adverse impacts of climate change. Many of the challenges in adapting to climate change relate to 1. capturing rainfall, 2. managing water resources, 3. enhancing soil moisture retention and 4. improving water use efficiency.

Surface Water Utilization

The most obvious and most common adaptation to variable rainfall is the establishment of irrigation. In past few decades this is the intervention commonly favoured by government investing considerable capital investments. Most irrigation supports the monsoon crop (i.e. supplements rainfall) and typically only 15–35% is used for double cropping. Dry season crops are rarely fully irrigated, but are planted to take advantage of the beginning or end of the wet season rains or utilize residual soil moisture.

There is a need for canal systems to respond quickly to flow changes. Computer technology is being recognised as a special tool for monitoring and analysing the data and development of decision support system (DSS). Special efforts by computer engineers and communication groups have helped in establishing management information system (MIS) for operation of reservoir and canals which helped in improvement in flood control, irrigation and hydropower. Through an effective MIS and DSS, precious water can be saved and efficiencies improved by quickly responding to sudden change in demand. In future, from manual operation of canal system, one could ultimately shift to automatic regulation as precise discharge measurements and better communication facilities are available.

Improving canal water utilization requires not merely improvement of engineering parameters like lining of canals/improvement of structures/providing additional field channels and equity in water distribution, but also application of a complex combination of field disciplines (agronomic/management/field measurement and hydro-sociological aspects) to the irrigated agriculture sector. To ensure sustainability, focused efforts are required in the area of adequate infrastructure such as power supply and communication network, adequate training of staff, implementation of available technology and security against vandalism with active participation of all the stakeholders in its implementation and operation.

Ground Water Utilization

Where reliable ground water supplies are available, the advantages over surface water for irrigation can be very significant. Aquifers provide both storage as well as transmission of water, reducing the need for large scale infrastructure and irrigation can be developed quickly and with low capital cost. In contrast to most large canal systems, water is available directly on demand, allowing farmers flexibility and control over timing and quantity of supply.

Groundwater irrigation thus offers opportunities for adaptation to climate change through stabilization of water supply and intensification of cropping. However two important aspects must be kept in mind before promoting expansion of ground water use. First, the largely unregulated groundwater exploitation poses questions of sustainability in long term (Taylor, 2012). Some of the suggestions for making sustainable groundwater use include: 1. Demand reduction through crop selection, 2. Adoption of water saving technologies, 3. Supply augmentation through managed aquifer recharge and 4. Energy rationing/pricing. Second, the potential impacts of climate change on ground water availability and demand must be considered. For example, change in rainfall pattern may impact aquifer recharge and rise in sea levels may result in salinization of aquifers in deltaic regions. (Holman *et al.*, 2012).

Although the buffering capacity of groundwater to climate change is typically higher than that of surface waters drawn from rivers and ponds, groundwater is not exempt from draught and water table declines and can undermine irrigation performance. Streams may be depleted during the dry season, or located too far from where the water is needed. In this situation, small on-farm water storage can be beneficial to deal with the risks of draught at the onset or during the monsoon season, or to allow limited irrigation at the beginning of the dry season.

Due to over draft, substantial ground water level declines are being witnessed both in hard rocks and alluvial areas. Ground water quality in coastal areas has also been affected by sea water intrusion due to excessive ground water development. Pollution of ground water due to increased industrial activity and sewage disposal also rising. The ground water development in such areas therefore needs to be augmented through suitable measures to provide sustainability and protection of ground water reservoirs. These involve 1. construction of check dams, 2. percolation tanks, 3. recharge wells, 4. subsurface dykes, 5. roof water harvesting and 6. other innovative artificial recharge works and monitoring of the impact of these structures.

Conjunctive Water Use

Conjunctive use of surface and ground water offers major opportunities for irrigated agriculture, and is realistic adaptation strategy for climate change (GWP, 2012). Ground water development of large alluvial aquifers, closely linked to rivers and replenished annually by the monsoon, can potentially replace surface storage (resulting in reduced costs and evaporative losses). Appropriate physical conditions exist in the large alluvial tracts. In other places, managed aquifer recharge, a technology applied to capture and store wet season flows, could be used to both mitigate the impacts of flooding and enhance water security, offsetting potential groundwater overuse by enhancing aquifer storage. Informal conjunctive use also provides opportunities for overall performance of formal irrigation scheme. For example farmers have invested in shallow tube wells and diesel/electric pumps in large parts of major canal systems in north India to supplement water provided by the canal system. The informal irrigation taps shallow aquifers, which are at least partly recharged through canal leakage/ seepage from the main scheme (IWMI, 2015).

Micro Irrigation Technologies

India, already stands first in area coverage under micro irrigation and is likely to surge ahead in its adoption in future in view of the positive push from the Government. Integration of micro-irrigation with major irrigation projects, particularly in their tail end reaches, are likely to become a reality soon to bring at least 10 percent command under micro-irrigation as proposed by the Government. Integration of micro-irrigation with watershed projects particularly for utilization of harvested water too are likely to result in efficient utilization of available water resources in agriculture with significant savings of water required for extending the effective irrigation command area. Micro-irrigation, conventionally has been considered useful only for a limited number of widely spaced crops. During the last decade micro-irrigation systems have been evolved to efficiently irrigate close growing vegetables, cereals, pulses and other crops too.

Agronomic Measures

The primary adaptation by farmers to climate change will be through improved farming practices. A wide range of agronomic measures at field to farm level are available to increase water productivity and reduce risks in cropping systems. Agronomic measures are related to soil management (erosion prevention and control, improving fertility, structure, organic matter), soil cover, crop varieties, mixed cropping and crop rotation. Many agronomic measures are encapsulated within the principles of conservation agriculture (minimum soil disturbance through no till or minimum tillage systems, permanent soil cover using crop residues, plastic mulches or cover crops). Conservation agriculture also promotes precision placement of inputs to reduce use of agricultural chemicals (FAO, 2011).

Adoption of Scientific Water Management Practices

Productivity of irrigated areas in India leaves much to be desired. Major reason for this situation is unscientific practices of water management (Rajput and Patel, 2005). The scientific water management practices may include, 1. Efficient Conveyance and distribution network, 2. Irrigation Scheduling, 3. Canal Automation, 4. Precision land levelling and 5. Efficint Water use at the farm. Efficinet water use at the farm may be realised through, *i.* maintenance of water courses, *ii.* Use of Water Efficient Crops, *iii.* Choice of irrigation method, *iv.* Choice of cropping pattern and *v.* Use of Micro Irrigation.

CONCLUSION

Climate change is occurring in, and adding to, a highly dynamic and uncertain agricultural context. Improved water management is the key to the significant transformation that agriculture must undergo in order to meet the interlinked challenges of increasing food demand and climate change. Managing current climate variability is the best indicator of the ability to manage future variability and, effective technologies and practices–some of which already exist and are highly innovative. Though there are some simple generic solutions to climate change, water smart interventions can reduce the risk associated with climate variability, improve yields and increase income, in both irrigated and rain fed systems.

To be effective, multi-faceted interventions must be tailored to local conditions and all costs and benefits fully evaluated. Experience from developing agrarian economies shows that improvement in water management is an important first step for increasing agricultural production and reducing poverty. Better water management reduces the risk of crop failure, enables the cultivation of more than one crop a year, and facilitates farmer's investment in improved varieties and fertilizers.

REFERENCES

Campbell, B., Mann, W., Melendz-Oritz, R., Streek C. and Tennigkeit, T. (2011). Addressing agriculture in climate change–A scoping report. Available at http: cgspace.cgiar.org/handle/10568/10306 (accessed 23 November, 2015).

Food and Agriculture Organization of the United Nations (FAO) (2011). Save and Grow–a policy makers guide to the sustainable intensification of smallholder crop production, Rome, FAO, p. 112.

Holman, I.P., Allen, D.M., Cuthbert, M.O. and Goderniaux, P. (2012). Towards best practice for assessing the impacts of climate change on ground water, *Hydrology Journal*, 20: 1–4.

Taylor, R.G. (2012). Groundwater and climate change, *Nature Climate Change*, 3: 322–329.

International Water Management Institute (IWMI), 2015. Improving waste management in Myanmar's dry dry zone for food security, livelihoods and healt, Colombo, Sri Lanka: IWMI, doi:10.5337/2015.213.

Global Water Partnership (GWP) (2012). Groundwater resources and irrigated agriculture. *Global Water Partnership Perspectives Paper*, Stockholm: Global Water Partnership, p. 19.

Zou, X., Li, Y., Gao, Q. and Wan, Y. (2012). How water saving irrigation contributes to climate change resilience—A case study of practices in China, *Mitigation and Adaptaion Strategies for global change,* 17(2): 111–12.

World Bank (2005). India's Water Economy: Bracing for a Turbulent Future, World Bank, Washington.

Lal, M. (2001). Climate change-Implications for India's water resources, *J. India Water Res. Soc.*, 101–119.

Sharma, B.R. and Paul, D.K. (1999). Water Resource of India. *In:* 50 Years of Natural Resource Management Research', ICAR. New Delhi, pp. 31–48.

Rajput, T.B.S. and Neelam Patel (2005). Enhancement of field water use efficiency in the Indo-Gangetic plain of India, *Irrigation and Drainage (ICID),* 54(2), 189–203.

Climate-Smart Technologies for Drought Vulnerability—NICRA Experiences

B.K. Ramachandrappa* and M.N. Thimmegowda

University of Agricultural Sciences, GKVK, Bengaluru 560 065, India
*E-mail: bkr_agron@yahoo.co.in

ABSTRACT: *Drought is a predominant cause of low yields world-wide. In arid and semi-arid tracts that include Eastern dry-zone of Karnataka, drought is the major factor contributing to climatic vulnerability. Reduced rainy days, increased rainfall intensity and intermittent dry spells are major climatic constraints in the region. Improved agricultural practices, viz. selection of appropriate varieties, method of crop establishment, in-situ moisture conservation, ex-situ rainwater harvesting and its efficient utilization and effectiveness of real time agro advisory services were demonstrated in Chikkamaranahalli cluster, Bengaluru Rural district of Karnataka during 2011 to 2015. Long duration variety (MR-1) for July sowing, medium duration (GPU-28) for August first fortnight and short duration (GPU-48) for August second fortnight sowing performed better. In-situ moisture conservation through opening conservation furrows between paired rows of pigeon pea under finger millet + pigeon pea (8:2) and groundnut + pigeon pea (8:2) recorded significantly higher yields compared to farmers' practice. Establishment of finger millet through transplanting and direct sowing using modified bullock drawn seed drill recorded higher yield compared to conventional broadcasting. The excess run-off water harvested in lined ponds, used for diversified purposes, viz. protective irrigation at critical stages, nourishing plantation crops, kitchen garden, pisciculture and Azolla cultivation has brought economic sustainability and livelihood security of farmers in rainfed situation. Real time agro advisory services helped decision-making on timeliness of operation. The climate resilient practices helped in bringing sustainability under vagaries of monsoon.*

Keywords: Climate Resilience, Drought, Farm Pond, Intercropping, Agro Advisory.

Agriculture is the source of livelihood for nearly two-thirds of the population in India. The fate of Indian agriculture is gambling around monsoon and market. Eastern dry-zone of Karnataka is traditionally a rainfed agrarian eco-system with predominance of finger millet, groundnut and pulses. Changing climatic scenarios in recent past underscore the necessity of climate-smart agriculture. The impact of climate change and vulnerability in the domain on agricultural production is quite evident in the recent years. Rainfall and its behaviour fluctuates widely among the climatic factors in the domain. Reduced rainy days, increased rainfall intensity, delayed onset of monsoon, excessive rains and prolonged dry spells are the major rainfall behaviour affecting the crop growth, yield and quality of the produce. Further, the soils of the domain are light textured *alfisols* with low to medium soil fertility, and poor water holding capacity.

Climate-smart agriculture gains momentum considering all the issues of climate and soil in bringing sustainability of farmers in the region. A project on climate-smart agriculture was conceptualized to study the impact of climate and sustainable practices on crop productivity during 2011 under National Innovations in Climate resilient agriculture. The study area chosen in the dry tracts of Karnataka have finger millet, groundnut, bajra, sorghum, chickpea, castor, sesame, pigeon pea, cowpea, field bean and horse gram as the major crops. All these crops are prone to water stress owing to deficit rainfall and rapid loss of soil water from profile (Mallareddy *et al.,* 2015).

Action research called "National Innovations in Climate Resilient Agriculture (NICRA)" funded by ICAR–Central Research Institute for Dryland Agriculture (CRIDA) was conceptualized for Chikkamaranahalli cluster (Chikkamaranahalli, Chikkamaranahalli colony, Chikkaputtayyanapalya, Mudalapalya and Hosapalya), Nelamangala Taluk, Bengaluru Rural district since 2011. Farming constraints were identified through participatory rural appraisal (PRA) and the interventions for resilience were planned based on the constraints. Fields were selected based on the willingness of farmers to engage in participatory research to evaluate the science-based strategies. Capacity building of selected farmers was undertaken through trainings in multi-disciplinary approach. Farmers were involved in all research intervention, *viz.* soil sampling, input application and yield estimation. The climate resilient agriculture practices, *viz.* selection of crops and cropping systems, improved varieties, rainwater harvesting and timely agriculture operations through farm mechanization were demonstrated. Major soils in the domain area are sandy loam to sandy clay loam with acidic to neutral reaction (pH 4.3 to 6.5). Bimodal rainfall prevailed in the domain with peaks during May and Sept.–Oct. The cluster receives a normal annual rainfall of 750 mm (Figure 1) and its major share during *Kharif* season. The total rainfall was above normal during 2014 (+23.0%) and 2015 (+20.8%), negative normal during 2011 (–7.7%) and 2013 (–12.4%) while deficit during 2012 (–41.2%).

Delayed onset of monsoon and intermittent dry spells prevailed in the domain. Finger millet is the major staple food crop grown traditionally in the domain. Farmers resort to a common variety and dry spell hampers its productivity. In order to bring resilience, climate-smart varieties were demonstrated according to the sowing window. For the farmers having specificity in longer duration varieties, transplanting was implemented. Moisture conservation furrow in groundnut and finger millet was encouraged involving sowing of groundnut or finger millet with pigeon pea in 8:2 row proportions with 60 cm spacing between the paired rows of pigeon pea and opening of conservation furrow between the paired rows of pigeon pea was adopted as a strategy for soil and moisture conservation. Modified bullock drawn seed drill was introduced as a strategy to maintain optimum plant population in finger millet. *Ex-situ* rainwater harvesting with farm pond of 250 m^3 capacity and multiple use of harvested water was emphasized. Real time agro advisory services were circulated twice weekly through posting the messages on the notice boards of milk collection centers.

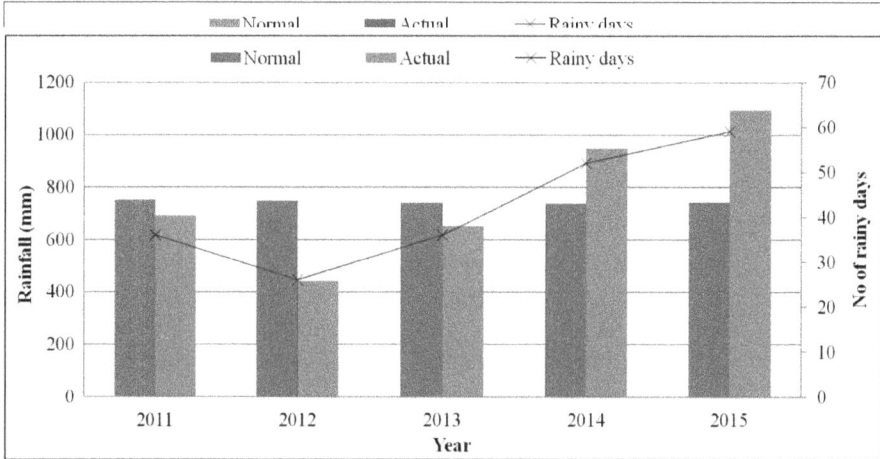

Finger millet varieties: Regular onset of monsoon during 2011, 2013 and 2014 resulted in higher grain yield, net returns and B:C ratio with long duration finger millet variety (MR-1) sown during July. Delayed onset of monsoon during 2012, the agro-advisory suggested the farmers to adopt a medium duration variety (GPU-28) for late sowing in August. Farmers adopting the medium duration variety (GPU-28) realized higher grain yield, net returns and B:C ratio (1720 kg/ha, ₹ 18096/ha and 2.31, respectively) compared to the long duration variety (MR-1). Delayed sowing of long duration variety pose end-season moisture stress on plants and getting caught under low temperature during flowering and results in yield loss. In a situation with normal onset of monsoon, direct sowing of finger millet variety MR-1 was found to be more remunerative than direct seeding of GPU-28 (Hegde and Jayarama Reddy, 1983).

Transplanting of finger millet: Finger millet tolerates transplanting shock and works as a strategy for delayed onset of monsoon. Transplanting of finger millet recorded significantly higher finger millet grain yield (2564 kg/ha) compared to direct sowing (2281 kg/ha). Maintenance of optimum plant population, timeliness in flowering and maturity are the causes for higher yield (Ramachandrappa *et al.*, 2013).

Modified bullock drawn seed drill for finger millet: Finger millet crop sown using modified bullock drawn seed drill recorded significantly higher grain yield (2417 kg/ha), net returns (₹ 35965/ha) and B:C ratio (3.21) compared to farmer's practice (1770 kg/ha, ₹ 22173/ha and 2.36 respectively). The modified seed drill facilitated timely sowing, optimum plant population, ease of inter-cultivation and ultimately increased the yield (Ramachandrappa *et al.*, 2014).

Moisture conservation furrow: Intercropping finger millet + pigeon pea (8:2) with conservation furrow between paired rows of pigeon pea recorded significantly higher finger millet grain equivalent yield (3774 kg/ha) and B:C ratio (2.56) compared to finger millet + *akkadi*. Groundnut + pigeon pea (8:2) with conservation furrow recorded significantly higher groundnut equivalent yield (1623 kg/ha) and B:C ratio (2.70) compared to sole groundnut (566 kg/ha and 1.90 respectively). The increased yield and economic benefit was associated with increased soil profile moisture as a result of conservation furrow (Raikwar and Srivastva, 2013). Finger millet/groundnut and pigeon pea are crops with contrasting plant-architecture both above and below ground. The root systems with different zones of nourishment exhibit complementarity; also pigeon pea being a legume helped the companion crop through biological nitrogen fixation.

Efficient utilization of farm pond water: It was estimated that only about 10% of the annual rainfall on the drylands is beneficially used for supporting vegetation cover, replenishing the ground water and other purposes (Oweis and Taimeh, 1996). Mr. Ravikumar S/o Ramaiah, Mudalapalya used harvested water for nourishing plantation crops, leafy vegetables, banana, drumstick and curry leaf during dry spell and realized a total profit of ₹ 10,300. Mr. Gubbanna S/o Shivanna, Chikkamaranahalli used harvested water for nourishing finger millet crop, vegetables, besides mango, drumstick and chilli, pisciculture, *Azolla* cultivation and realized an overall benefit of ₹ 64,850. Mr. Krishnappa S/o Huchaiah, Chikkaputtaiahnapalya realized overall benefit of ₹ 5,500 through cultivation of tomato and leafy vegetables by utilizing harvested water in farm pond.

Real time agro-advisory services: Economic impact studies indicated considerable benefit to farmers who adopted the agro-met advisories. The per cent gain in income due to adoption of suggested contingency cropping systems ranged from 22 to 379 over traditional cropping systems. The yield increase in cropping systems, *viz.* finger millet + pigeon pea, groundnut + pigeon pea, pigeon pea + cowpea and pigeon pea + field bean was to the tune of 1040, 741, 481 and 594 kg ha^{-1}, respectively. Rajegowda *et al.* (2008) had earlier reported that, in the eastern dry zone of Karnataka the farmers adopting the agromet advisories realized an average additional benefit of 31.4, 24.7, 16.2 and 20.6% in finger millet, pigeon pea, field bean and tomato, respectively.

Contour bunding with Khus and Nase grass as live bunds: Contour bunds were erected at 15 m interval. Grasses like *Khus* and *Nase* on bunds were established to reduce runoff (11–12%), soil loss by erosion (36%) and conserve moisture for longer period and facilitate crop to survive in dry periods and increased yield.

Contour cultivation practices, viz. sowing, inter-cultivation, etc: Contour farming was implemented in all the demonstrations, whose effect is 15–20% higher yield across the crops.

Tied-ridging and mulching: Tied-ridging is practiced in widely spaced crop by earthing up soil from inter-row and placing to the base of crop row with staggered bunds. Mulching is the practice of covering soil. Tied-ridges were formed at 20–25 DAT and organic mulches available in the farm, *viz.* thinned excess plants, weeds before seed setting were mulched in chilli to curtail evaporation losses, increase infiltration, reduce soil loss and improve the soil health.

Ridges and furrows in widely spaced crops: Ridges and furrows were demonstrated in widely spaced crops, *viz.* maize, sunflower, chilli, castor, pigeon pea etc. with an intension of anchorage to the crop, conservation of rain water and safe disposal of excess rain water.

Soil health management with double cropping/green manuring/INM: Bi-modal distribution of rainfall in the southern part of the state encourages double cropping. Maintenance of soil health was another thrust to be addressed under dryland csituation considering the proverb 'dryland soils are not only thirsty but also hungry'. Legumes as green manures or in crop rotation help for maintenance of soil health and fertility. Double cropping of cowpea finger millet, green manuring horse gram-finger millet, *Ex-situ* incorporation of glyricidia-finger millet and integrated nutrient management helped in maintaining soil health and sustainability.

Contingent crops for delayed sowing: Early season drought due to delay in onset of monsoon is directly responsible for shortfalls in area sown under major crops compared to normal situation. Loss of food grain, livelihood and shortage of fodder for animals are the critical issues to be addressed under early season drought. Hence, appropriate contingency crop planning combining possible cereals, pulses, oilseeds and fodder crops for late sowing situation (August end to September) was developed. Small millets, *viz.* foxtail, barnyard, little, proso and kodo millet, short duration pulses, *viz.* rice-bean, horse gram, field bean, cowpea, oilseeds, *viz.* sunflower and niger and fodder crops, *viz.* fodder maize, fodder sorghum, fodder bajra, horsegram, *etc.* form good choice even during late September sowing.

Drought mitigation practices: Midseason/intermittent drought is common under dryland tracts of domain districts. It is well established that, potassium is known to induce drought tolerance through osmo regulation. Application of 1% KCl (MOP) during drought when the plant starts expressing wilting symptom, thiourea @ 2% after retrieving drought in comparison with water spray (control) was demonstrated in finger millet. Plant tolerance up to 15–20 days under potassium spray was better compared to thiourea and water spray (Table 1).

Alternate land use: Agro-forestry with silver oak on bunds and field crops in the catchment-storage-command relationship as an inter-terrace management strategy, agri-horti systems with pomelo, custard apple, amla-based intercropping systems were promising as possible long-term strategy for climatic aberrations.

Table 1: Performance of Crops to Real Time Climate Resilient Agricultural Practices

Climate Resilient Practices	Improved Practices	Traditional Practice	Increase in Yield (%)
Finger millet planting technique	Transplanting	Direct sowing	12.4
	Modified bullock drawn seed drill	Farmer's practice	36.6
Moisture conservation furrow	Intercropping finger millet + pigeon pea (8:2) with conservation furrow	Finger millet + *akkadi*	68.5
	Groundnut + pigeon pea (8:2) with conservation furrow	Sole groundnut	186.0
Drought mitigation chemicals	Thiourea	Without foliar spray	45.0
	2% KCl		35.0
Live bunds	Khus grass	Control	18.9
	Nase grass		36.3
Organic mulch in chilli	Mulching and tied ridging	Without mulch	75.0

REFERENCES

Hegde, B.R. and Jayarama Reddy, M. (1983). For late establishment, transplanting is better than drilling in ragi. *Current Research*, 12: 23.

Mallareddy, Thimmegowda, M.N., Shankaralingappa, B.C., Narayan Hebbal and Fakeerappa Arabhanvi (2015). Effect of moisture conservation practices on growth and yield of finger millet + pigeon pea intercropping system. *Green Farming*, 6(6): 1273–1275.

Oweis, T. and Taimeh, A. (1996). Evaluation of a small basin water harvesting system in the arid region of Jordan. *Water Resources Management*, 10, 21–34.

Raikwar and Srivastva P. (2013). Productivity enhancement of sesame (*Sesamum indicum* L.) through improved production technologies. *African Journal of Agricultural Research*, 8(47): 6073–6078.

Rajegowda, M.B., Janardhanagowda, N.A., Jagadeesha, N. and Ravindrababu, B.T. (2008). Influence of agromet advisory services on economic impact of crops. *Journal of Agrometeorology*, 10: 215–218.

Ramachandrappa, B.K., Shankar, M.A., Dhanapal, G.N., Sathish, A., Jagadeesh, B.N., Indrakaumar, N., Balakrishna Reddy, P.C., Thimmegowda, M.N., Maruthi Sankar, G.R., Srinivasa Rao, Ch. and Murukanappa, 2013, Four Decades of Dryland Agricultural Research for *Alfisols* of Southern Karnataka, Directorate of Research publication, UAS, Bangalore. 308 p.

Ramachandrappa, B.K., Shankar, M.A., Sathish, A., Alagundagi, S.C., Surakod, V.S., Thimmegowda, M.N., Shirahatti, M.S., Guled, M.B., Khadi, B.M., Jagadeesh, B.N. and Srinivasa Rao, Ch., 2014. Rainfed Technologies for Karnataka. All India Co-ordinated Research Project for Dryland Agriculture, University of Agricultural Sciences, Bangalore and University of Agricultural Sciences, Dharwad Karnataka, India, 105 p.

Adaptations to Climate Change: Water Management and Security

Jagdish Krishnaswamy

Ashoka Trust for Research in Ecology and the Environment (ATREE),
Bengaluru 560 064, India
E-mail: jagdish@atree.org

ABSTRACT: *There are threats and concerns emerging from climate change, uncertainty about future climate and other global and regional change drivers on hydrology, forests, water resources and agriculture. Using the data from time-series of climate, and satellite based normalised difference vegetation index as well as recent high resolution data from instrumented catchments in the Western Ghats, and applying trend analyses and time-varying regression models, responses to climate change and predictions of future responses were examined. Declining Monsoon and increase in more intense rains could aggravate dry season stress on ecosystems and increase vulnerability under future climate change. In the Himalayas, browning of forests due to temperature induced moisture stress is a mechanism that needs further scrutiny. In central India, over 70% of the region experienced water-stress for four months or more annually. We need a concerted effort to fill gaps in our knowledge of the inter-action of land-use and land-cover with climate change and other drivers of ecosystem responses.*

Keywords: Rain Events, Land-Use and Land-Cover, Global Change, Indian Monsoon, Browning of Forests, Ecological Flows.

The role of climate change, other global change and its interaction with land-use and land-cover change is a relatively less understood for ecosystems in India. India's continued development depends on the availability of adequate water. The Indian monsoon has been declining in recent decades even as more intense rain events are increasing, with major consequences for ecosystems and well-being of over 1.2 billion people. Climate models have also been unable to simulate recent multi-decadal decline of the Indian Monsoon bringing into question uncertainty about future precipitation projections for India (Figure 1).

In this paper, recent work in India and elsewhere to assess what is known about the ongoing and future responses of diverse ecosystems to climate change and other global change drivers is discussed.

Time-series trend analysis was used to assess trends in climate dynamic models with time-dependent regression parameters was applied to study the time evolution

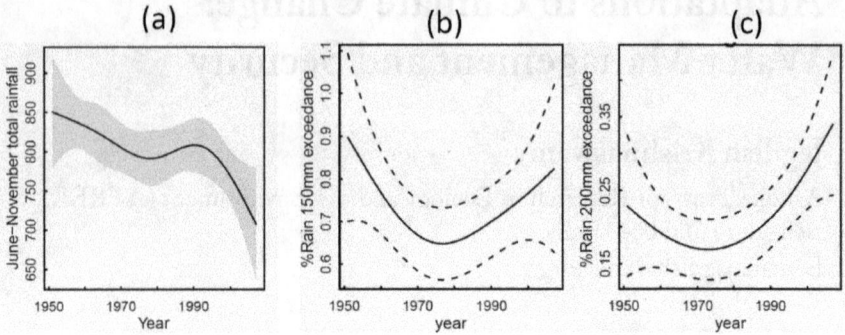

Fig. 1: Recent Decline in Precipitation in India (a) Even as More Proportion of Intense Rains Have Increased (b, c) (Krishnaswamy and Vaidyanathan, 2015)

of NDVI–climate relationships. High resolution data from instrumented catchments in the Western Ghats and data driven modeling approaches were used to examine hydrologic response of forests, grasslands and tree plantations to intense rain. Finally, for central India a physical hydrologic model and available spatially explicit data were used to estimate the intra-annual dynamics of water stress across the central Indian Highlands over the period 2002–2012. The spatial distribution of water demanding sectors (*e.g.* industry, domestic, irrigation, livestock and thermal power generation) was investigated and the vulnerability of urban centers to water stress was examined.

Although the impact of deforestation on enhancing flood risk is well known, the effects of forest degradation and reforestation on hydrological processes in the humid tropics are less well established, especially under scenarios of climate change and changes in rain intensity regimes. Certain combinations of land-cover, soil types and agro-ecosystems in the Western Ghats are already vulnerable to increased surface flows under current rainfall regimes, but the responses of these and other land-cover and soil types to future changes in rainfall regimes is less well understood. Although climate models have not performed well in terms of simulating the observed historical trends (*e.g.* decline in Monsoon since 1950s), their predictions of more frequent Extreme Rain Events (EREs) have been matched by data.

The impact of deforestation on enhancing flood risk is well known (van-Dijk *et al.*, 2008), however, the effects of forest degradation and reforestation on floods and the hydrological cycles in the humid tropics are less well established, especially under scenarios of climate change and increase in EREs. Certain combinations of land-cover, soil types and agro-ecosystems in the Western Ghats are already vulnerable to increased surface flows under current rainfall regimes, but the responses of these and other land-cover and soil types to future changes in rainfall regimes is less understood.

Recent evidence from basins dominated by sub-surface flows including two from the Western Ghats has suggested that time to peak in a basin or "flashiness" of a basin is a non-linear decreasing function of rainfall intensity (Figure 2). Under climate change influenced higher intensity rainfall regimes, flooding risk could be enhanced even in well vegetated basins such as in the Western Ghats (Chappell *et al.*, 2017).

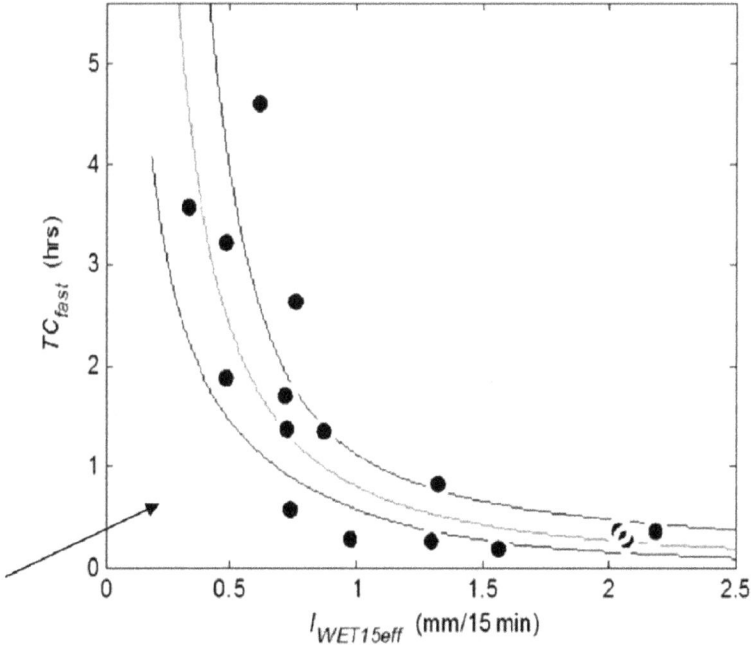

Fig. 2: Time to Peak of Rain Induced Storm Flow in Catchments ("flashiness") is a Non-Linear Decreasing Function of Rain Intensity (Chappell *et al.*, 2017).

In the Himalayas and in other tropical mountains, there is strong evidence that warming interacting with drier precipitation episodes can result in loss of "greenness" and that warming may not necessarily result in "greening" in cold mountain environments, where vegetation and growth periods are assumed to be limited by temperature. The time series of annual maximum NDVI for each of five continental regions shows mild greening trends followed by reversal to stronger browning trends around the mid-1990s (Figure 3). During the same period increasing trends in temperature were observed, but changes in precipitation was marginal. The amplitude of the annual greenness cycle increased with time, and was strongly associated with the observed increase in temperature amplitude. The relationship between vegetation greenness and temperature weakened over time or was negative. Such loss of positive temperature sensitivity has been documented in other regions as a response to temperature-induced moisture stress. Tropical mountain vegetation including in the Himalayas is considered sensitive to climatic changes, so these consistent vegetation responses across widespread regions indicate persistent global-scale effects of climate warming and associated moisture stresses.

(a)   (b)

$$NDVI_t = Level_t + B1_t(Temperature_t) + B2_t(Precipitation_t) + e_t$$

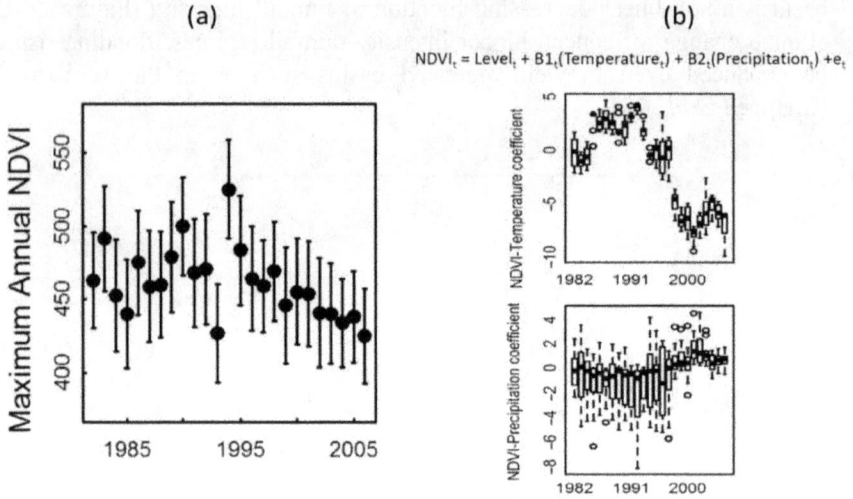

Fig. 3: (a) Himalayan forests showed a browning response starting in the mid 1990s (Krishnaswamy *et al.*, 2014). (b) Time-varying regression models showing the temperature induced moisture stress as a potential driver of "browning" in Himalayan forests.

About 74% of the area of the central Indian Highlands experienced water stress (defined as demand exceeding supply) for four or more months annually (Clark *et al.*, 2016). The rabi season irrigation drives the intra-annual water stress across the landscape. The Godavari basin experiences the most surface water stress while the Ganga and Narmada basins experience water stress due to groundwater deficits as a result of rabi irrigation. All urban centers experience water stress at some time during a year. Urban centers in the Godavari basin are considerably water stressed with some towns that experience water stress eight months annually. Irrigation dominates water use accounting for 95% of the total water demand, with substantial increases in irrigated land over the last decade.

CONCLUSIONS

land-use and land-cover change (LULC) and blue-water transformations (abstraction of water from rivers and more recently, ground-water) have been both intensive and extensive in India. Managing land use to promote hydrologic functions will become increasingly important as water stress increases. These, in combination with other dimensions of global change including climate change and climate variability, proliferation of alien and invasive species and urbanization have potentially generated complex responses from socio-ecological systems. In addition, the absence of long-term and proper gauging records and long-term monitoring sites have retarded ecohydrologic research.

The field of ecohydrology and its relevance for policy and management knowledge gaps needs urgent articulation for diverse audiences. Recent insights from the limited ecohydrological work in India illustrate the data and methodological challenges in understanding ecohydrologic change at various spatial and temporal scales. There is also concern about the broader canvas of uncertainty in hydro-climatology in the context of the inability of climate models to simulate historical trends. We need to prioritize gaps in training, knowledge and data that we urgently need to address if ecohydrology in India has to make meaningful contribution to science and society.

REFERENCES

Clark, B., DeFries, R. and Krishnaswamy, J. (2016). Intra-annual dynamics of water stress in the central Indian Highlands from 2002 to 2012. *Regional Environmental Change*, 16(1), 83–95.

Chappell, N.A., Jones, T.D., Tych, W. and Krishnaswamy, J. (2017). Role of rainstorm intensity under estimated by data-derived flood models: Emerging global evidence from subsurface-dominated watersheds. *Environmental Modelling & Software*, 88, 1–9.

Krishnaswamy, J. and Vaidyanathan, S. (2015). La Nina and Indian Ocean Dipole influence on distribution of daily rain intensities in India. Poster presentation in session A13A of the American Geophysical Union meeting *AGU Fall Meeting 2015* in San Francisco 14–18 December 2015.

Krishnaswamy, J., John, R. and Joseph, S. (2014). Consistent response of vegetation dynamics to recent climate change in tropical mountain regions. *Global change biology*, 20(1), 203–215.

van, Dijk, van Noordwijk, M., Calder, I.R., Bruijnzeel, S.L.A., Schellekens, J. and Chappell, N.A. (2008). Forest–flood relation still tenuous–comment on 'Global evidence that deforestation amplifies flood risk and severity in the developing world' by CJA Bradshaw, NS Sodi, KS H. Peh and BW Brook. *Global Change Biology,* 15(1), 110–115.

Climate Change: Adaptations of Livestock, Poultry and Fisheries

G.S.L.H.V. Prasada Rao

Centre for Animal Adaptation to Environment and Climate Change Studies, Kerala Veterinary and Animal Sciences University, Mannuthy, Thrissur 680 656, Kerala, India
E-mail: grao@kvasu.ac.in

ABSTRACT: *Climate change in India is true and in tune with global warming. Rise in temperature is likely to be around 2–3 °C by the end of this century with regional uncertainties in rainfall causing a threat to primary sectors, viz. agriculture, livestock, poultry, aquaculture, forestry, biodiversity (both land and ocean), land and water resources. Weather related disasters, such as cyclones, cloud bursts, hailstorms, floods, drought, cold and heat waves are frequent in recent years as a result of climate change and detrimental to the society linked sectors directly or indirectly to a considerable extent. Increase in sea surface temperature and sea level rise are the climate extremes. These weather and climate extremes are likely to be more frequent under the projected global warming and climate change scenario. Impacts of weather and climate changes and adaptation strategies in plantations, livestock, poultry and fisheries against the ill effects of climate variability/ climate change are discussed in brief.*

Keywords: Monsoon, Temperature, Rainfall, THI, SST.

The monsoon directory over the country (Kerala coast) since last 147 years (1870–2016) indicated that the onset of monsoon is on 1st June with +/– 7 days. It varies from 25th May to 8th June in majority of the years. The earliest monsoon was recorded on 11th May in 1918 while belated monsoon on 18th June in 1972. The mean onset of monsoon during the tri-decade of 1901–30 was on 4th June while in other tri-decades it revolved around 1st June. The trend analysis since 1870 also indicated that the onset of monsoon is stable and it tends to be around 1st June (Figure 1). The fact file of onset monsoon in 2011 was on 29th May while 5th June in 2012, 1st June in 2013, 6th June in 2014, 5th June in 2015 and 8th June in 2016. It indicated that inter-annual variations are expected within normal onset of monsoon of one standard deviation in majority of the years. It is also understood that the monsoon set may be early (before 25th May) or late (after 8th June) occasionally during which the monsoon rainfall is likely to be below normal or normal across Kerala, indicating that the chances of excess rainfall in such years (early or late monsoon years) are likely to be less (Rao and

Gopakumar, 2016). Crop seasons appear to be shortened in recent years due to erratic behavior of monsoon rainfall with late arrival and early withdrawal of monsoon in some pockets. In detail studies in this direction need to be taken up on shifting of crop seasons in relation to southwest and northeast monsoons.

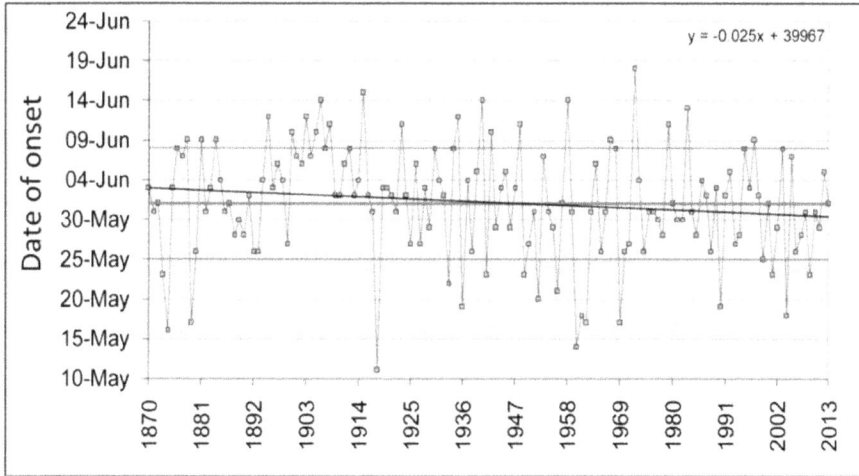

Fig. 1: Onset of Monsoon Over India (Kerala Coast)
from 1870 to 2016

RAINFALL

Decline in annual rainfall is noticed since last 40–60 years across the Country though rainfall increase is the trend over a long period of time. Two consecutive drought years were noticed in 2014 and 2015 identical to 1965 and 1966. 2002, 2009, 2012, 2014 and 2015 were declared as drought years in recent years. Decrease in monsoon rainfall while increase in post monsoon is the trend across the State of Kerala though cyclic trends of 40–60 years were noticed in annual and monsoon rainfall. Of course, the decline in annual rainfall is evident since last 50–60 years. It also reveals that the percentage contribution of rainfall in monsoon season to the annual rainfall was declining while increasing during the post monsoon season in Kerala. However, increase in post monsoon rainfall may not compensate rainfall decrease in monsoon season. Significant decline in monsoon rain fall was noticed in 2012 (26%), 2015 (26%) and 2016 (34%) across Kerala. Interestingly, there was a shift in climate as a result of changes in thermal and moisture regimes over the State of Kerala. The State of Kerala as a whole was moving from wetness to dryness within the Humid Climate from B_4 to B_3 since last 100 to 150 years. One of the major factors could be due to alarming deforestation and forest fires that took place during the above said period across the Western Ghats within and outside the State of Kerala.

TEMPERATURE

Temperature increase is widespread over the globe and is greater at higher northern latitudes. Land regions have warmed faster than the oceans, which has already started affecting the climatic phenomenon in different parts of the world and India is no exception. The mean annual temperature for India as a whole has risen by 0.56°C (Figure 2) while 0.76°C in maximum temperature and 0.22°C in minimum temperature. Increase in temperature is high across the West Coast and the North East among different zones of the Country. Increase in temperature is relatively high during post monsoon season and winter when compared to that of southwest monsoon and summer seasons. 2009, 2010, 2015 and 2016 were the recent warm years and summer heat wave is not uncommon across the Country. Warming Kerala is also real as the trend in temperature was increasing significantly since 1980s in tune with the global warming. Within the State, the rate of increase in temperature was high across the high ranges, followed by the low lands while moderate increase along the midlands. It could be attributed to alarming deforestation across the high ranges and the effect of increase in sea surface temperature along the Coast. At the current rate of increase in temperature, it is projected that increase in maximum temperature is likely to be around 1.5°C by 2100 A.D. while 0.3–0.4°C in the case of minimum temperature. Increase in mean surface air temperature is likely to be less than 1°C by 2100 A.D. The decade 1981–90 was the warmest and driest decade in Kerala during which the plantation crops' production was adversely affected to a considerable extent. The year 1987 was the warmest, followed by 1983. Increase in night temperature is noticed during winter in recent years and flowering of fruit crops is adversely affected. It is more so in the case of mango.

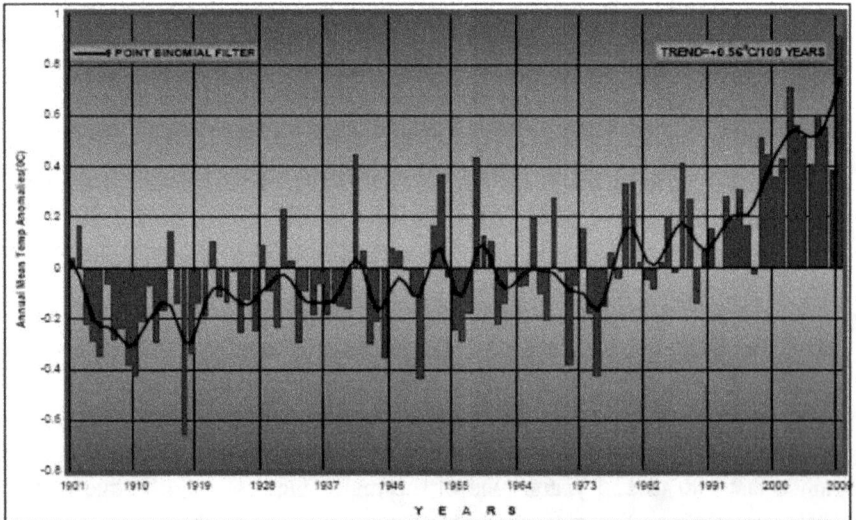

Fig. 2: All India Mean Annual Temperature Anomalies for the Period 1901–2009
(*Source:* IMD, 2010)

IMPACTS ON LIVESTOCK

India owns 57% of the world's buffalo population and 16% of the cattle population. It ranks first in the world in respect of cattle and buffalo population, third in sheep and second in goat population. The sector utilizes crop residues and agricultural by-products for animal feeding that are unfit for human consumption. Livestock sector has registered a compounded growth rate of more than 4.0% during the last decade, in spite of the fact that a majority of the animals are reared under sub-optimal conditions by marginal and small holders and milk productivity per animal is low. Increased heat stress associated with rising temperature may, however, cause distress to dairy animals and possibly impact milk production. A rise of 2 to 6°C in temperature is expected to negatively impact growth, puberty and maturation of crossbred cattle and buffaloes. The low producing indigenous cattle are found to have a high level of tolerance of these adverse impacts than high yielding crossbred cattle. Therefore, high producing crossbred cows and buffaloes will be affected more by climate change. Direct effects include multiple stresses due to temperature, humidity, radiation, low plane of nutrition, heavy rains, pests and diseases. The cyclical phenomenon of drought and floods has additional effects. The indirect effect of climate is the reduced availability of quality feed ingredients. Direct and indirect effects cause stress which cause depletion of body reserves and there by reduced production, growth and reproduction. Among all stresses that the climate offers, the thermal stress due to effective environmental temperature in the most important. Challenge before the scientific community of the tropical world is to find ways to enhance milk production in the prevailing climatic conditions. Historically the traditional livestock production largely depended on heat tolerant native breeds that produced less milk compared to temperate exotic breeds. The dairy sector now largely

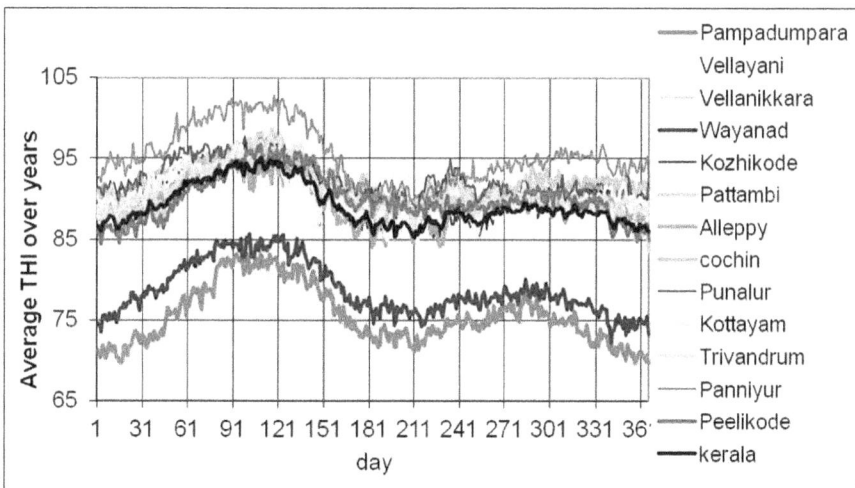

Fig. 3: Daily Mean Temperature and Humidity Index (THI)
at Selected Locations Across Kerala

comprises of extensive and expanding crossbred population in Kerala. For crossbreds, increased air temperature, and humidity measured as temperature humidity index (THI) above critical thresholds are related to low dry matter intake (DMI) and to reduced efficiency of milk production cause significant heat stress. The daily mean temperature humidity index (THI) at selected locations over Kerala is depicted in Figure 3. The results indicated that THI may not be conducive from February to May.

IMPACTS ON POULTRY

Poultry farming is very sensitive to heat stress. Egg production and size are likely to reduce in high summer temperature. It may not be that much severe in the humid tropics when compared to that of arid and semi-arid regions. However, increase in the egg production is hindered and it is reduced by 5% due to thermal stress even in the humid tropics. If there is no sprinkler, reduction in egg production could have been around 10%. The sprinkler reduces the poultry house temperature by 2–3°C. High summer temperature led to reduction in egg weight from 52 grams to 45 grams. This problem is reduced by providing more protein in the feed and also supplementation of amino acids like methionine. The fertility of the eggs received at Revolving fund hatchery was reduced from 95 to 75% due to the hot summer. The hatchability also reduced from 85 to 50% on total egg set in a Revolving fund hatchery, which led to the closure of the hatchery in the remaining period of summer.

DISEASE OUTBREAK AND MORTALITY IN POULTRY

A poultry farmer 'Basheer of Thrissur district' although he vaccinated all 8000 ducklings for duck plague, due to summer stress the immunity did not develop which led to the outbreak of duck plague and loss of 50% (4,000) of his ducklings (45 days old) due to the outbreak of duck viral hepatitis (duck plague) in April and May 2013. Another farmer 'Varghese of Thrissur' lost around 1000 ducklings (3 months old) out of 7000 due to pasteurellosis outbreak in April, 2013. There was a mortality of 8 pullets out of 700 pullets reared in cages due to heat stroke accompanied by AI (artificial insemination) stress at Poultry Farm, Mannuthy. The same thing had happened in a commercial broiler breeder farm with the mortality of 250 birds due to heat stroke and AI stress. The heat stress can be alleviated by providing cool water by installing sprinklers at roof top. The analysis of mortality data from 2004 to 2009 at the Project Directorate of Poultry, Hyderabad revealed that the overall mortality was increased as the ambient temperature rises in broiler, layer and native chickens. The mortality started increasing when the temperature reaches 32°C and the peak was observed at 38 to 39°C (13.5%). The mortality was highest in broiler type chickens followed by layers and native chicken. The mortality due to heat stress in broiler type birds was started appearing at the ambient temperature 30°C, while in the layer and native chicken the heat stress related mortality was observed at the ambient temperature of 31°C. The deaths due to heat stress were 10 times more in broiler type chickens as compared to layer and native type chickens. The mortality due to heat stress was negligible in native (desi type chickens) which may be due to low metabolic rate and natural heat tolerance. Another study was conducted to find the influence of high ambient temperature on the feed intake body temperature and respiratory rate in commercial layers for 13 weeks. The consumption which was 108 g/bird/day at 28°C was reduced to 68 g/bird/day at the shed temperature 37.8°C.

IMPACTS ON MARINE FISHERIES

A rise in temperature is detrimental and will have rapid effect on the mortality of fish and their geographical distributions. Oil sardine fishery did not exist before 1976 in the northern latitudes and along the East Coast of India as the resource was not available/and sea surface temperature (SST) was not congenial. With the warming of sea surface, the oil sardine is able to find the temperature to its preference especially in the northern latitudes and eastern longitudes, thereby extending the distributional boundaries and establishing fisheries in larger coastal areas (Vivekanandan *et al.,* 2009a). The dominant demersal fish, the threadfin breams have responded to increase in SST by shifting the spawning season off Chennai. During the past 30 year period, the spawning activity of *Nemipterus japonicus* reduced in summer months and shifted towards cooler months. A similar trend was observed in *Nemipterus mesoprion* too. Analysis of historical data showed that the Indian mackerel is able to adapt to rising in sea surface temperature by extending distribution towards northern latitudes, and by descending to depths (Vivekanandan *et al.,* 2009b).

Climate change seems to be imminent and it is a threat to livestock, poultry and fishery sectors. To mitigate the climate change effects a range of adaptive strategies need to be considered. Therefore, it is high time that the government agencies and policy makers to be proactive with short term and long term strategies to mitigate the ill effects of weather related disasters with the people's participation and minimize the losses against the weather related disasters. Need of the hour is expect the unexpected guests in the form of weather related disasters under the projected climate change scenario and tune to the system to minimize the losses with well preparedness at various levels. These issues are being tackled as part of National Initiative on Climate Resilient Agriculture (NICRA). At the state level, such initiatives need to be taken up on priority basis in developing technologies against the weather extremes, for which a definite state action plan on climate change (SAPCC) in animal agriculture and allied sectors need to be highlighted and tackled with concerted efforts. To me it appears, we need to have an action plan similar to that of NICRA at the State level with technical people in the form of 'State Initiative on Climate Resilient Agriculture', which includes animal agriculture.

REFERENCES

Rao, G.S.L.H.V.P. (2011). Climate Change Adaptation Strategies in Agriculture and Allied Sectors. Scientific Publishers, Jodhpur (India), p. 330.

Rao, G.S.L.H.V.P. and Gopakumar, C.S. (2016). Climate Change and Plantations in the Humid Tropics. New India Publishing Agency, New Delhi, p. 492.

Rao, G.S.L.H.V.P., Varma, G.G. and Beena, V. (2017). Livestock Meteorology. New India Publishing Agency, New Delhi, p. 492.

Vivekanandan, E., Rajagopalan, M. and Pillai, N.G.K. (2009a). Recent trends in sea surface temperature and its impact on oil sardine. *In:* Global Climate Change and Indian Agriculture: Case Studies from the ICAR Network Project (Ed. P.K. Aggarwal). ICAR Publication, pp. 89–92.

Vivekanandan, E., Hussain Ali, M. and Rajagopalan, M. (2009b). Impact of rise in seawater temperature on the pawning of threadfin beams. *In:* Global Climate Change and Indian Agriculture: Case Studies from the ICAR Network Project (Ed. P.K. Aggarwal). ICAR Publication, pp. 93–96.

An Overview of Climate Change Impacts on Marine Productivity and Fisheries

N. Ramaiah

CSIR-National Institute of Oceanography, Dona Paula, Goa 403 004, India
E-mail: ramaiah@nio.org

ABSTRACT: *Climate change affects us all severely. The future Oceans are expected to experience ocean acidification, rising temperature and sprawling hypoxic zones. The combined effect of this deadly trio is bound to adversely affect the intricately interconnected web of organisms (e.g. from microbes to phytoplankton to zooplankton to shellfish to fish to marine mammals) in the marine ecosystems. Infectious diseases, which are important drivers within ecosystems may become inevitable, and as a consequence of changing oceans, would impact marine biota. This presentation will cover many examples of researches done in the country and highlight the need to adapt to climate resilient fisheries, to utilize and to evolve conservation practices essential to safeguard natural resources.*

Keywords: Fish Harvests, Mariculture, Climate Resilient Fisheries, Indian Ocean Warming, Seas Around India.

Global warming and acidification are the two major climate-related changes being experienced in the ocean environment. Increases in sea surface temperature (SST) is understood to cause ocean stratification decreasing nutrient entrainment to the euphotic zone and making the surface layer lowly productive. The consequences of impeded nutrient entrainment are decreased primary productivity in the euphotic upper layers and likely shifts in phytoplankton type and abundance. Subsequent events of climate variability and change could have greater ramifications for marine ecosystems as elevated sea surface temperature influences water column stability, nutrient entrainment, and biodiversity and compositional abundance of plankton and their growth and productivity. It is but obvious that any change in the plankton diversity and abundance will have consequences on the marine food-web and on other trophic levels.

An alarming recent realization is that, on a global scale, a warmer climate could cause a rapid, overall reduction in marine life. This is attributable to increased temperatures which cause decreased marine phytoplankton production. Basically, climate response to marine autotrophs can be linked to increased stratification. When the ocean surface warms, it essentially becomes "lighter" than the cold, dense water below which is loaded with nutrients. This lightening process effectively separates phytoplankton in the surface layer from the nutrients below them. All these are altering the complex balance of biogeochemical cycles and climate feedback mechanisms. Corals, the key community in the coral reefs offering various services to mankind, bleach in response to warming seas. During coral bleaching, they lose their endosymbionts that causes starvation till recovery in a few weeks to several months or not at all. This recovery if any, and the duration for recovery render corals vulnerable for most of the secondary threats such as algal overgrowth and outbreaks of diseases. The impact might affect the community structures, trophic interactions, age distributions, among many other physical and chemical changes as has been observed in relation to diseases in seagrasses (e.g. eelgrass wasting disease; reviewed in Burge *et al.*, 2014), reef-building corals (Aronson and Precht, 2001), oysters (Mann *et al.*, 2009), and sea urchins (Lauzon-Guay *et al.*, 2009). This destruction has a cascading effect on other coral communities like reef fishes and so the reef fishery gets affected.

Optimal performance of an organism is affected by temperature, ocean carbonate chemistry and hypoxia. For animals, Oxygen and Capacity Limited Thermal Tolerance (OCLTT) govern their specialization to be inside specific, limited temperature ranges and, their sensitivity to its extremes. Optimal OCLTT allows integration of other stressors on a thermal matrix of performance. The biological functioning can thus be credited to such modulations. Any analyses investigating on a few such but vital parameters are central to response evaluations, for instance as a consequence of climate change. Diseases have had large impacts on both cultured and wild harvests of commercially important species, such as salmon [e.g. *Ichthyophonus* infection in marine and anadromous fish (reviewed in McVicar 2011) and viral infections in Atlantic and Pacific salmon (Kurath and Winton 2011)], abalone (e.g. withering syndrome; Friedman *et al.*, 2000), and crustaceans (e.g. protozoan infections of natural populations and viruses in aquacultured species; Stentiford *et al.*, 2012). Interactions between hosts, pathogens, and the environment are known to govern disease outbreaks. A balance toward or away from a high-intensity disease state can be due to change in any of these components and, climate change can alter disease outbreak likelihood (Garrett *et al.*, 2013). Globally, we are only beginning to understand the effects of infectious diseases in the ocean and how climate change will affect marine host-pathogen interactions (Harvell *et al.*, 2009), both of which are critical for informed conservation and management strategies.

Effects of climate change and ocean acidification are being documented in oceans around the world (IPCC 2014, Doney *et al.*, 2012). It is now recognized that rising levels of atmospheric carbon dioxide (CO_2) are leading to increased global atmospheric and ocean temperatures. Ocean warming is likely to continue if significant and sooner reductions in CO_2 levels are not in place. Increasing temperatures lead to physical impacts on ocean systems, including rising sea levels, increased ocean stratification, loss of sea ice, and altered oceanic circulation (Howard *et al.*, 2013). Warming temperatures have already affected the survival, growth, reproduction, health, and phenology of marine biota. In addition to being economically and ecologically valuable (Gilman *et al.*, 2008) marine biota are influenced by both direct effects of stressors on populations and species and indirectly by changes in species interactions, including competition, mutualism, parasitism and predation (Kordas *et al.*, 2011). Preserving the biota and managing them effectively require thorough knowledge of the ill effects of such stressors, including climate change. The strategy is to investigate as to how to minimize the recovery period so that it could save the corals from many of the secondary threats. With advancement in molecular techniques and our improved abilities to use them, precise measurements of all the relevant parameters are possible to address the resilience and vulnerability of the biota in the seas around us in the wake of changing climate.

Gross overview of marine ecosystem functioning of the seas around India can be summarized as follows. The western Indian Ocean is more productive during the SW monsoon and the eastern, during the NE monsoon. Although the up to 0.4°C raise in temperature is reported from the northern Indian Ocean (Roxy *et al.,* 2016), the biological response at different trophic levels is not yet studied. It is imperative to glean the community alterations, productivity aberrances and consequent shifts in food-web dynamics. The coastal waters of India support a rich and abundant planktonic life. Many researchers have related hydrographic differences such as temperature, salinity, upwelling and surface currents with seasonality in biological productivity characteristics. Our coastal ecosystems experience intra-annual/seasonal differences. During the summer months of April–June the temperature are the warmest (28–33°C), and generally low during monsoon months of June–October (24–27°C). There are also wide salinity gradients (2.9–34.5) and chlorophyll concentrations (1.4–18.6 µg l^{-1}). The overall mean abundance of phytoplankton cell counts are known to vary intra annually by ~4.3 folds and zooplankton by ~3.6 folds.

India has been harvesting annually on an average 3.5 Million Tones of marine fish and shellfish with a coastline of over 7500 km and an Exclusive Economic Zone (EEZ) of 2.01 Million Sq Km. We are also increasing our mariculture production efforts. Marine related activities employ semi- or non-skilled population of over three million Indians and earn a revenue of about ₹ 75000 Crores. Our endeavor therefore ought to be on safeguarding our natural treasuries and thwart climate change impacts by suitable policy promulgations and implementation. Therefore, an all inclusive study of microbial (bacterial, archaeal, mycotic), autotrophic (phytoplanktonic and picoplanktonic), zooplanktonic and nektonic communities and their biological processes need to be carried out in the context of elucidating the impacts of changing climate.

REFERENCES

Aronson, R.B. and Precht, W.F. (2001). White-band disease and the changing face of Caribbean coral reefs. In: The Ecology and Etiology of Newly Emerging Marine Diseases, *Dev. Hydrobiol.*, 159: 23–35.

Burge, C.A., Mark Eakin, C., Friedman, C.S., Froelich, B., Hershberger, P.K., Hofmann, E.E., Petes, L.E., Prager, K.C., Weil, E., Willis, B.L., Ford, S.E. and

Harvell, C.D. (2014). Climate change influences on marine infectious diseases: Implications for management and society. *Ann Rev Mar Sci.,* 6: 249–277.

Doney, S.C., Ruckelshaus, M., Emmett Duffy, J., Barry, J.P., Chan, F., English, C.A. and Polovina, J. (2012). Climate change impacts on marine ecosystems. *Annual review of marine science,* 4, 11–37.

Friedman, C.S., Andree, K.B., Beauchamp, K., Moore, J.D., Robbins, T.T., *et al.* (2000). "*CandidatusXenohaliotiscaliforniensis*", A newly described pathogen of abalone, *Haliotis* spp., along the west coast of North America. *Int. J. Syst. Evol. Microbiol.,* 50: 847–55.

Garrett, K.A., Dobson, A.D.M., Kroschel, J., Natarajan, B., Orlandini, S., *et al.* (2013). The effects of climate variability and the color of weather time series on agricultural diseases and pests, and on decisions on their management. *Agric. for. Meteorol.* 170: 216–27.

Gilman, E., Ellison, J., Duke, N. and Field, C. (2008). Threats to mangroves from climate change and adaptation options: A review, *Aquat. Bot,* 89: 237–25.

Harvell, C.D., Altizer, S., Cattadori, I.M., Harrington, L. and Weil, E. (2009). Climate change and wildlife diseases: When does the host matter the most? Ecology, 90: 912–20.

Howard, J., Babij, E., Griffis, R.B., Helmuth, A., Himes-Cornell, P., *et al.* (2013). Oceans and marine resources in a changing climate. Oceanogr. *Mar. Biol.,* 51: 71–192.

Kordas, R.L., Harley, C.D. and O'Connor, M.I. (2011). Community ecology in a warming world: The influence of temperature on interspecific interactions in marine systems. *Journal of Experimental Marine Biology and Ecology,* 400(1), 218–226.

Kurath, G. and Winton, J. (2011). Complex dynamics at the interface between wild and domestic viruses of finfish. *Curr. Opin. Virol.,* 1: 73–80.

Lauzon-Guay, J.S., Scheibling, R.E. and Barbeau, M.A. (2009). Modelling phase shifts in a rocky subtidal ecosystem. Mar. Ecol. Prog. Ser., 375: 25–39.

Mann, R., Harding, J.M. and Southworth, M.J. (2009). Reconstructing pre-colonial oyster demographics in the Chesapeake Bay, USA. Estuar. Coast. Shelf Sci., 85: 217–22.

McVicar, A.H. (2011). *Ichthyophonus.* In *Fish Diseases and Disorders,* Vol. 3: *Viral, Bacterial, and Fungal Infections,* ed. PTK Woo, DW Bruno, pp. 721–47. Cambridge, MA: CAB Int. 2nd ed.

Roxy, M.K., Modi, A., Murtugudde, R., Aalsala, V.V., Anickal, S.P., Prasanna Kumar, S., Ravichandran, M., Vichi, M. and Lévy, M. (2016). A reduction in marine primary productivity driven by rapid warming over the tropical Indian Ocean, Geophys. Res. Lett., 43, 826–833.

Stentiford, G.D., Neil, D.M., Peeler, E.J., Shields, J.D., Small, H.J., *et al.* (2012). Disease will limit future food supply from the global crustacean fishery and aquaculture sectors. *J. Invertebr. Pathol.,* 110: 141–57.

Adaptations of Livestock, Poultry and Fisheries to Climate Change

R.C. Upadhyay

National Dairy Research Institute, Karnal 132 001, Haryana, India
E-mail: upadhyay.ramesh@gmail.com

ABSTRACT: High environmental temperature and lack of prior condition-ing under climatic changes may result in a catastrophe and death of livestock and poultry. Heat waves in particular can push vulnerable species beyond their survival threshold limits. The loss in productivity may be observed up to 30% following extreme heat or cold wave events. Diversified milk produc-tion and value chain issues require attention at policy, legislation and imple-mentation levels. The milk hygiene, adulteration, process efficiency and lack of quality control need to be addressed on priority. Low productivity of livestock, inadequate feed and fodder, animal health care and vaccination, livestock carcass and waste disposal etc do need to be fixed for reducing carbon and water foot prints. Similarly poultry and fisheries production systems need to be adapted suitably for meeting climate change challenges and future nutritional needs of growing population.

Keywords: Milk Production, Egg Production, Fish Yield, Health Care, Livestock Waste Management.

With an estimated livestock of 512 million heads, India accounts for amongst the largest livestock population in the world, distributed over 100 million households in approximately 600,000 villages in varied agro-climatic conditions. The locally adapted livestock species and poultry are able to cope with normal variations in climatic conditions. Indigenous livestock and poultry breeds in different agro-climates sustain rural communities and their livelihoods. Zebu breeds (e.g. Sahiwal, Tharparkar, Sindhi, Kankrej, Gir, Ongole etc.) are well adapted to heat, scarce feed and water availability and some of the tropical diseases. The adaptive characteristics in the skin (Singh *et al.,* 2014; Uttarani *et al.,* 2014), coat, immune functions are not only unique to cope with environment challenges and variability but also able to reproduce and sustain reproductive functions. The livestock and poultry will be impacted severely by climate change both directly and indirectly. The negative impacts of climate change will be experienced more on livelihoods by vulnerable resource poor marginal and small holder population in different agro-climatic conditions even with best adaptive measures and use of Indigenous technical knowledge (Maiti *et al.,* 2014, 2015). Indian livestock and livestock production system have enough resilience and can cope with climate change. Warming of the Himalayas may affect the Yak and mule husbandry at higher altitudes, where animal feeds and fodder is less available, that may compel some of the nomadic communities to seek other species for production and explore new routes and pastures for animals.

Major likely impacts of temperature rise on livestock physiological functions have been presented in an earlier publication (Upadhyay *et al.,* 2008, 2010). The negative impacts will be experienced on milk production of crossbred cattle and buffaloes (Upadhyay *et al.,* 2010) more than zebu cattle. Animal growth and reproduction functions will also be affected and incidence of animal diseases is likely to increase. The crossbred population of cattle, sheep and other species will be vulnerable to heat stress and emerging diseases without adequate measures for adaptations. Buffaloes under intensive management will experience difficulty and an absence of wallowing opportunity and/or heat alleviation conditions may limit their production and reproduction efficiency. Existing rural animal shelters neither protect them from intense solar radiation nor provide adlib water access. Young animals suffer most due to their underdeveloped thermo-regulatory mechanisms in the improper animal housing. However, well adapted native breeds are able to sustain long duration sun exposure and water scarcity, therefore, may not be impacted much as they possess inherent genetic capacity to withstand large variations in temperature experienced in different agro-climatic zones. Studies on gene expression of Tharparkar a well adapted Zebu cattle (Mehla *et al.,* 2014) and Sahiwal (Vamsikrishna *et al.,* 2014) have shown that adapted cattle breeds exhibit differences in the gene expression that distinguishes them from other vulnerable cattle breed crosses with Taurine cattle breeds (Parva and Upadhyay, 2014). The zebu cattle also have differences in the skin colour and coat to cope with hot conditions (Singh *et al.,* 2014). The differences in antioxidants, metabolic parameters also explain their capacity to withstand large variations and adapt under tropical climatic conditions.

In this presentation some current and recent climate change adaptation issues or activities have been highlighted that may help better adaptation of the livestock, poultry and fisheries sector (Table 1). Comprehensive presentations on fisheries and poultry are available in this volume.

LIVESTOCK AND POULTRY

Adaptation to climate change with specific reference to livestock and poultry production is unlikely to be achieved with a single strategy. Therefore, different approaches need to be adopted at different resource poor farms and for different species. Up-gradation and modifications in animals' and poultry housing should be addressed on priority for reducing impact of temperature rise and variations. In general animal production functions need efficiency enhancement and improve-ment. Early onset of animal puberty and maturity (cattle and buffalo) in different breeds by adequate feeding, scientific management and health care can be achieved. Better feeding and management of growing animals will improve weight

Table 1: Climate Change Impacts and Adaptation Measures
in Livestock, Poultry and Fisheries

Climate Change Impact	Adaptation Measure
Livestock • Decreased Production (milk yield and growth). • Decreased reproduction. • Increased disease incidence (bacterial, viral, parasitic).	• Increase production efficiency by improved animal housing, ventilation, heat stress alleviation. • Balanced animal feeding, management and health care. • Enhanced reproduction efficiency by targeting early animal puberty, reduction in inter-calving interval and reducing dry period. • Selective breeding of adapted cattle and buffaloes for long lactation period. • Investment costs for disease investigation, prevention and control.
Poultry • Decreased growth rate and reproductive capacity. • Increased cold/heat stress and decreased poultry production. • Increased incidence of poultry diseases.	• Poultry house designs based on eco-friendly building material. Use of green energy equipments and/or eco-friendly natural system to maintain optimal temperature during different seasons and reduce the risk of heat stress on poultry. • Increased investment in poultry house ventilation and cooling systems. • Energy costs for warming poultry house during winter are likely to reduce due to relatively warm winter and shortened winter season. • Locally available feed stuffs may reduce feed costs. • Poultry meat and meat products may increase in price and with feed prices possibly decreasing • Poultry farming may become more profitable and viable.
Fisheries • Reduced fish yield yields • Yield variability	• Increase fishing effort avoiding over exploitation. • Selective breeding for increased resilience. • Shift to culture-based fisheries. • Encourage native aquaculture species to reduce impacts. • Modify or Change feeds and management system. Improve water-use efficiency. • Aquaculture infrastructure investments (e.g. nylon netting and raised dykes in flood-prone pond systems) • Diversify livelihood portfolio (e.g. algae cultivation for biofuels or engage in nonfishery economic activity such as ecotourism). • Ecosystem approach to fisheries/aquaculture and adaptive management.

gain and increase growth efficiency for early onset of puberty. Thus proper feeding of lactating, pregnant and growing animals will improve not only efficiency of animal production functions but also reduce costs of maintenance of livestock under climate change scenarios.

India is rich in animal and poultry biodiversity. Huge animal genetic resources and variation is likely to help coping and adaptation to climate change in different agro-climatic zones and coastal areas. Indian diets are supplemented both by animal and plant proteins. Animal proteins (milk, meat) are part of diet mainly due to its quality in terms of high densities of energy, protein, and other critical nutrients normally required. Demand for milk and milk products and livestock products is foreseen to increase significantly in the future, therefore, necessitate better animal management and health care. Diversified milk production and value chain is likely to be impacted by temperature rise. The issues of milk hygiene, adulteration, process efficiency and quality control need to be addressed on priority at policy, legislation and implementation levels. Livestock low productivity, inadequate feed and fodder, health care, animal vaccination, etc. do require attention equally and investments both at center and state level. The existing diverse breeds of livestock and poultry in different agro-climatic zones of India will help cope and adapt to climate change and climate variability. This will necessitate infrastructure development for livestock and poultry commensurate to production level and future development. The proper animal houses and development of climate resilient animal shelters for better animal comfort and stress reduction is a general requirement of resource poor livestock and poultry farmers. Livestock and poultry waste management and water use are other issues requiring attention. Present livestock manure disposal is unhygienic and makes people vulnerable to pathogens and is a health hazard due to zoonotic disease transmission.

Animal feed and fodder resource management has received little attention. In general, present animal feeding and feeds and fodder resource management is poor. Some of the feed resources (e.g. paddy straw, wheat straw, sugarcane tops, etc.) are burned in open leading wide spread pollution. These feeds and fodders may be suitably stored or processed for enriching nutrients for animal feeding. Therefore, a proper management system of animal feeds and fodder resources will be critical for climate change adaptation in India. In order to improve production efficiency of livestock both feeding and management should be improved to obtain higher weight gain (during late pregnancy, growing phase) and milk production during lactation. The infrastructure and human resource development for assisted or aided reproduction should also be a priority to improve reproductive efficiency of existing low maturing or performing livestock breeds and species.

The infrastructure for livestock disease prevention, vaccination, surveillance and health care is inadequate to cater to the need of more than 510 million livestock population. This should be developed on priority for long term use and improving efficiency and productivity of livestock. Indigenous birds are able to sustain under hot weather and are resistant to many tropical diseases. However poultry flocks based on exogenous genotypes are particularly vulnerable to temperature rise and variability due to their narrow zone of thermal comfort and to tolerate limited temperature variations. Therefore, poultry farmers need to consider making adaptations to help reduce cost, risk and concern in the future.

FISHERIES

The anticipated temperature change and variability will impact marine, inland, freshwater and other fishing opportunity due to vulnerability of water and ecosystems. The impacts on marine fisheries may be observed in the form of fish distribution, migration, regional extinction, change in phonological events, etc (Vivekanadan *et al.*, 2009). Similarly other fisheries systems are likely to be impacted in various ways by climatic changes. Adaptive measures in fisheries need to be addressed at different levels and as per nature of fisheries systems and vulnerability. Resource poor fishermen and communities vulnerable to climate change may change fish rearing pattern and/or are likely to shift to other options for sustaining livelihoods. Marine, freshwater, inland fisheries, etc will also be impacted by globalization simultaneously. Some of the farm communities may opt for alternatives for livelihood and rear other animal species. Often long association with livestock, breeds and other topographical conditions do restrict them from adopting new ventures, yet some of the entrepreneureal farmers may switch over.

REFERENCES

Maiti, S., Jha, S.K., Sanchita Garai, Arindam Nag, Chakravarty, R., Kadian, K.S., Chandel, B.S., Datta, K.K. and Upadhayay, R.C. (2014). Adapting to climate change: Traditional coping mechanism followed by the Brokpa pastoral nomads of Arunachal Pradesh, India. *Indian Journal of Traditional Knowledge*, 13(4), pp. 752–761.

Maiti, S., Jha, S.K., Sanchita Garai, Arindam Nag, Chakravarty, R., Kadian, K.S., Chandel, B.S., Datta, K.K. and Upadhyay, R.C. (2015). Assessment of social vulnerability to climate change in the eastern coast of India. *Climatic Change*. DOI: 10.1007/s10584-015-1379-1.

Mehla, K., Magotra, A., Choudhary, J., Singh, A.K., Mohanty, A.K., Upadhyay, R.C., Srinivasan, S., Gupta, P., Choudhary, N., Antony, B. and Khan, F. (2014). Genome-wide analysis of the heat stress response in Zebu (Sahiwal) cattle. *Gene*, 533(2): 500-7. DOI: 10.1016/j.gene.2013.09.051.

MoEF (2010). Climate change and India: A 4X4 assessment. A sectoral and regional analysis for 2030. November 2010. MoE & F, GOI.

Parva, M. and Upadhyay, R.C., (2014). Heat Shock Protein 72 Expression of Sahiwal and Karan-Fries during Thermal Stress. *Indian J. Dairy Sci.*, 67(2).

Singh A.K., Upadhyay, R.C., Malakar, D., Sudarshan Kumar and Singh, S.V. (2014). Effect of thermal stress on HSP70 expression in dermal fibroblast of zebu (Tharparkar) and crossbred (Karan-Fries) cattle. *Journal of Thermal Biology*, 43: 46-53. DOI: http://dx.doi.org/10.1016 /j.jtherbio.2014.04.006.

Upadhyay, R.C., Rani, R., Asharaf, S., Ashutosh, Singh, S.V., Somvanshi, S.P.S. and Anil Kumar (2010). Effect of climate changes on buffalo milk production. *Rev. Vet.*, 21: 256–258.

Upadhyay, R.C., Singh, S.V. and Ashutosh (2008). Impact of climate change on livestock. *Indian Dairyman*, 60(3): 98–102.

Uttarani, M., Sohan Vir Singh, Anil Kumar Singh, Suresh Kumar and Upadhyay, R.C. (2014). Expression of skin color genes in lymphocytes of Karan Fries cattle and seasonal relationship with tyrosinase and cortisol. *Tropical Animal Health and Production*, 46(6): DOI: 10.1007/s11250-014-0620-7.

Vamsikrishna Kolli, Upadhyay, R.C. and Dheer Singh (2014). Peripheral blood leukocytes transcriptomic signature highlights the altered metabolic pathways by heat stress in zebu cattle. *Res. Vet. Sci.*, 96(1): 102-10. DOI: 10.1016/j.rvsc. 2013.11.019

Vivekanadan, E., Hussain, A.M., Jasper, B. and Rajagopalan, M. (2009). Vulnerability of corals to warming of the Indian seas: a projection for the 21st Century. *Current Science,* 97, 11, 1654–1658.

Climate Change and its Resilient Measures to Augment Poultry Production

R.N. Chatterjee* and U. Rajkumar

ICAR-Directorate of Poultry Research, Rajendranagar, Hyderabad 500 030, Telangana, India
*E-mail: rnch65@gmail.com

ABSTRACT: *Climate change is posing many challenges to mankind through its adverse effects in different fields. Poultry contributes very less to carbon sequestration and GHG emissions; the challenge is increased temperature and its mitigation strategies for sustainable poultry productivity. Temperature is one of the important factors that exerts a negative influence on the performance of poultry and causes huge losses in terms of loss of productivity, reduced reproductive efficiency, increased stress, reduced immune competence and increased investment costs to mitigate the effects of climate change. Higher temperature has significant adverse effects on both broiler and layer production. Modern poultry genotypes produce more body heat, due to their greater metabolic activity and faster growth and higher productivity. The combination of different mitigation strategies like management, nutritional modulations, genetic and epigenetic approaches in a systematic way may result in reducing the adverse effects of climate change and help in sustainable production of poultry produce. The effective climate resilient practices with strong scientific backup are essential to minimize the adverse effects of climate change on poultry.*

Keywords: Green House Gases, Climate Resilient Strategies, Poultry Productivity, Climate Change.

INTRODUCTION

Climate change is happening globally posing challenges that need to be mitigated to avoid possible adverse effects. The major threat of climate change has been an increase in the carbon dioxide (CO_2), methane (NH_4), nitrogen oxide (N_2O) and other green house gases (Krishna, 2011) over a period of time. Climate change in many parts of the world adversely affects socioeconomic sectors which include water resources, agriculture, forestry, fisheries, animal husbandry and poultry. Farmers are facing several challenges due to climate variation and the extent of losses farmers are likely to incur may not be clear in empirical terms but it is projected to cause extensive harm to farm productivity (Chatterjee and Rajkumar, 2016). Livestock productionaccounts for 18 percent of global anthropogenic green house gases (GHG) emissions of which cattle contribute major share, while poultry contribution is marginal, only 8% to the livestock emissions (Steinfeld *et al.*, 2006).

Temperature is one of the important factors that exerts a negative influence on the performance of poultry and causes huge losses in terms of loss of productivity, reduced reproductive efficiency, increased stress, reduced immune competence and increased investment costs to mitigate the effects of climate change (Rajkumar *et al.*, 2015a and b). Climate change especially increased temperature affects poultry production by reducing poultry yield and nutritional quality of feeds, increasing disease and disease-spreading pests, reducing water availability and making it difficult for birds to survive. Poultry species are more vulnerable to heat stress due to increased temperature as birds can tolerate a narrow zone of temperature range; 18–24°C is the thermo-neutral zone for the birds. An increase in temperature beyond this range due to environment or other metabolic factors will lead to cascading effects on thermo-regulation that could be lethal to the birds.

In the 20th century, there was an increase of 0.65°C in the average global temperature and 0.2% to 0.3% increase of precipitation in the tropical region. The internal body temperature of chickens (41–42°C) is higher than that of mammalian livestock and humans (36–39°C). Poultry birds have considerably less threshold to heat stress as compared with other animals as they lack sweat glands. Though the impact is negligible in terms of carbon sequestration but it is significant in terms of productivity and mortality of the birds. The magnitude of environmental impacts is highly dependent on production practices and especially on manure management practices. The magnitude of GHG is very negligible, the only challenge the poultry industry face under global warming is predicted increased temperature. Therefore, there is an urgent need to develop sustainable production systems along with the proper mitigation strategies to combat increased heat stress in poultry.

IMPACT OF CLIMATE CHANGE ON POULTRY

Carbon footprint and green house gases from poultry: The carbon footprint is a measure of the exclusive total amount of CO_2 emissions that are directly or indirectly caused by an activity or is accumulated over the life stages of a product (Wiedmannand Minx, 2008). A carbon footprint involves not only CO_2 emissions but also includes N_2O and CH_4 emissions which are expressed in CO_2 equivalents (CO_2e). Most of the CO_2 generated from the poultry industry is primarily from the utilization of fossil fuels. Apart from this, these gases are also emitted from manure during handling and storage. Nitrous oxide and CH_4 emissions are dependent on management decisions about manure disposal and storage as these gases are formed in decomposing manures as a by-product of nitrification/de-nitrification and methanogenesis respectively.

In animal agriculture, the greatest contribution to CH_4 emissions is enteric fermentation (21%) and manure management (8%). The distribution of CH_4 emissions from enteric fermentation among animal types poultry had the lowest amount with 0.57 lbs (0.26 kg) of CH_4 per animal per year compared to dairy cattle with 185–271 lbs (84–123 kg) and swine with 10.5 lbs (4.8 kg) of CH_4 (Dunkley, 2011). Poultry industry is in advantageous position as one pound of chicken meat adds 7.05 lbs of emissions to GHG (Dunkley, 2011).

Growth and Production: Higher temperature has been shown to have significant adverse effects on both broiler and layer production (Ghazi *et al.*, 2012; Imik *et al.*, 2012; Rajkumar *et al.*, 2015a). Heat stress impairs overall poultry and egg production by modifying the bird's neuro-endocrine profile both by decreased feed intake and by activation of the HPA axisleading to reduced feed intake, excessive panting to maintain thermo regulation and diverting more energy towards homeostasis instead of growth and production. However, even though the detrimental effects of heat stress in broilers seem to be very consistent, it is important to consider that stocking density has a major role as a potential compounding factor (Estevez, 2007). The chronic heat exposure negatively affects fat deposition and meat quality in broilers, in a breed-dependent manner. In fact, recent studies demonstrated that heat stress is associated with depression of meat chemical composition and quality in broilers (Lu *et al.*, 2007).

Productivity of laying hens flocks was affected by many factors, including environmental stress, mostly heat stress, which has been probably one of the most commonly occurring challenges in many production systems around of the world. Decreased feed intake is very likely the starting point of most detrimental effects of heat stress on production, leading to decreased body weight, feed efficiency, egg production and quality (Deng *et al.*, 2012).

Physiological and biochemical: Poultry seems to be particularly sensitive to temperature-associated environmental challenges, especially heat stress. Modern poultry genotypes produce more body heat due to their greater metabolic activity and faster growth and higher productivity. Understanding and controlling environmental conditions is crucial to successful poultry production and welfare. Under high temperature conditions, birds alter their behavior and physiological homeostasis seeking thermo-regulation, thereby decreasing body temperature. Birds subjected to heat stress conditions spend less time for feeding, more time for drinking and panting, as well as more time with their wings elevated, less time moving or walking, and more time resting (Lucas and Rostagno, 2013; Mack *et al.*, 2013). Air sacs are very useful during panting, as they promote air circulation on surfaces contributing to increase gas exchanges with the air consequently, the evaporative loss of heat. High environmental temperatures alter the activity of the neuro-endocrine system of poultry, resulting in activation of the hypothalamic-pituitary-adrenal (HPA) axis, and elevated plasma corticosterone concentrations. Body temperature and metabolic activity are regulated by the thyroid hormones, triiodothyronine (T3) and thyroxine (T4), and their balance.

Immuno-suppressing effects of heat stress on broilers and laying hens were revealed in terms of lower relative weights of thymus and spleen in laying hens subjected to heat stress (Ghazi *et al.*, 2012); reduced lymphoid organ weights in broilers under heat stress conditions (Quinteiro-Filho *et al.*, 2010). As the result of stress, bird's body attempts to maintain its thermal homeostasis leading to increased production of reactive oxygen species (ROS) resulting in stage of oxidative stress, and starts producing and releasing Heat Shock Proteins (HSP) to try and protect itself from the deleterious cellular effects of ROS (Rajkumar *et al.*, 2015a; Vinoth *et al.*, 2015).

CLIMATE RESILIENT STRATEGIES

Management strategies: Maintaining the in house temperatures is very important for sustaining the productivity from the birds. In extreme summer management through spreading the paddy/wheat straw on roofs and sprinkling water on the roofs maintains the temperatures at comfortable level. Effective cross ventilation, use of coolers and foggers is also recommended. In winters, generally gunny bags are used to cover the sides especially in high raised poultry houses.

Foul smelling odours can be controlled by minimizing the surface of manure in contact with air, frequent collection of litter, closed storage (bags or closed sheds); cooling systems can be equipped with bio-filters and air scrubbers that trap odours from the ventilation airflow; lowering litter's water content achieved by the incorporation of hydrophilic products such as hashes, rice husk, peanut husk, dust or sawdust; applying deodorant products to feed or directly to animal houses and building wind protection structures. The proliferation of flies and mosquitoes can be controlled by minimizing the surface of manure in contact with air, frequent collection of litter at shorter intervals than the length of the larvae development cycle, lowering litter's moisture and applying insecticides.

Nutritional modulation: Nutritional management aims to reduce pollution load by limiting excess nutrient intake and/or improving the nutrient utilization efficacy of the bird. Nutrition management can also allow improvement to feed conversion ratios through optimal diet balancing and feeding regimes, and improvement to feed digestibility. Feeding the antioxidants like Vit E, plant extracts and trace minerals like selenium, chromium, Zinc etc. reduces the stress condition and improves the heat tolerance in birds. Many researchers formulating feeds that closely match the nutritional requirements of birds in their different production and growth stages to reduce the amount of nutrients excreted. Use of low-protein diets supplemented with amino acids, and low-phosphorus diets with highly digestible inorganic phosphates; using good quality, uncontaminated feed (e.g. in which concentrations of pesticides and dioxins are known and do not exceed acceptable levels) which contains no more copper, zinc, and other additives than is necessary for animal health.

Epigenetic adaptation: Adaptations acquired in the life of organisms other than genetic means are called epigenetic adaptations. Poultry species can tolerate a narrow zone of temperature, 18–24°C which is the thermo-neutral zone for the birds. Increase in temperature beyond this range due to environment will lead to cascading effects on thermo-regulation and could be lethal to the birds. Thermal manipulation during embryogenesis (pre-natal) induces physiological memory due to epigenetic adaptation to high temperature eliciting the improved thermo tolerance during the post natal life (Yahav, 2008). Pre-exposure of embryo to high or low temperatures during incubation improves the adaptability to hot and cold environments respectively, during the post natal life (Yahav *et al.*, 2004; Rajkumar *et al.*, 2015 a&b; Vinoth *et al.*, 2015). Changes in incubation temperatures during critical period of development of the thermoregulatory system can result in long-lasting modifications to the cellular and molecular neuronal mechanism of temperature regulation (Janke and Tzschentke, 2010). Daily cyclical higher incubation temperatures, depending on the length of exposure and the days of the temperature modification, appear to improve tolerance of chickens to higher ambient temperatures (Yahav *et al.*, 2009). Adaptation to higher temperature at early age either during embryogenesis or first week of life of chicks improves the thermal tolerance to high temperatures in post natal life.

Breeding for heat resistance: Breeding for heat resistance in poultry as a mitigation strategy focuses on selection for heat tolerance in poultry and includes utilization of major genes for traits that promote stress tolerance.

Selection for Heat Tolerance

The magnitude of the reduction in body weight and production at high temperatures is very high leading to considerable economic losses to the farmer. Although most standard breeding stocks are selected in temperate climates, the genotypes may respond differentially to high temperature even if they have similar performance in thermo-neutral environment. The commercial broilers perform better in winter than summer. Therefore, the broilers' genotype should be taken into account in broiler production in tropical and subtropical regions. Because fast-growing broilers produce more heat and have a higher heat load, the effect of heat stress is more pronounced in commercial broiler stocks and in broilers with high growth potential compared to the slower-growing chickens (Lin *et al.*, 2006). The genetic potential for rapid growth is not achieved under high temperature in fast-growing strains mainly due to reduced feed intake in hot conditions (Deeb and Cahaner, 2002). Heat adaptation in broilers can be improved by applying selection in a hot environment. However, such selection may lead to reduced growth potential at normal air temperatures. Therefore, the parameters used in selection should be specific to season. It is essential to optimize the heat tolerance and production based on the location and environment to overcome the adverse effects of the heat stress.

Major Genes

Major genes like Naked neck, Dwarf and Frizzle with proven abilities to combat heat stress may be utilized in tropical countries. The major genes can be utilized in developing crosses, either broiler or layer, to improve the heat tolerance in terminal crosses. Naked neck (Na) gene reduces feather mass by 20% and 40% (relative to body weight) in the heterozygous (Na/na) and homozygous (Na/Na) birds respectively, compared with the fully feathered birds. The lower feather mass increases the effective surface of heat dissipation and increases the sensible heat loss from the neck region in naked neck chickens (Rajkumar *et al.*, 2011). Dwarf (dw) gene results in a reduction of 30–40% of adult body size and leads to speculation about the inherent heat tolerance of dwarf broiler breeders. Frizzle (F) gene may reduce the heat insulation of feather by curling the feathers and reducing their size.

CONCLUSION

Climate change is posing many challenges to mankind through its adverse effects in different fields. Poultry contributes very little to carbon sequestration and GHG emissions; the challenge is the increased temperature and its mitigation strategies for sustainable poultry productivity. The effective climate resilient practices with strong scientific backup are the need of the hour to minimize the adverse effects of climate change on poultry.

REFERENCES

Alade, O.A. and Ademola, A.O. (2013). Perceived effect of climate variation on poultry production in OkeOgun area of Oyo state. *J. Agri. Sci.* 5(10). doi: 10.5539/jas.v5n9p176.

Attia, Y.A., Hassan, R.A., Tag El-Din, A.E. and Abou-Shehema, B.M. (2011). Effect of ascorbic acid or increasing metabolizable energy level with or without supplementation of some essential amino acids on productive and physiological traits of slow-growing chicks exposed to chronic heat stress. *J. Anim. Physiol. Anim. Nutr.*, 95: 744–755.

Chatterjee, R.N. and Rajkumar, U. (2016). An outlook on climate ready poultry production. In: Proceedings of National seminar on Integrating Agri-horticultural and allied research for food and nutritional security in the era of global climate disruption. ICAR RC for NEH Manipur Centre, Imphal, pp. 191–199.

Deeb, N. and Cahaner, A. (2002). Genotype-by-environment interaction with broiler genotypes differing in growth rate.3. Growth rate and water consumption of broiler progeny from weight-selected versus nonselected parents under normal and high ambient temperatures. *Poult. Sci.* 81: 293–301.

Deng, W., Dong, X.F., Tong, J.M. and Zhang, Q. (2012). The probiotic *Bacillus licheniformis* ameliorates heat stress-induced impairment of egg production, gut morphology, and intestinal mucosal immunity in laying hens. *Poult. Sci.* 91: 575–582.

Dunkley, C. (2011). Global worming: How does it relate to poultry? UGA extension bulletin 1382. 1–7.

Estevez, I. (2007). Density allowances for broilers: Where to set the limits? *Poult. Sci.* 86: 1265–1272.

Ghazi, S.H., Habibian, M., Moeini, M.M. and Abdolmohammadi, A.R. (2012). Effects of different levels of organic and inorganic chromium on growth performance and immune competence of broilers under heat stress. *Biol. Trace Elem. Res.,* 146: 309–317.

Imik, H., Ozlu, H., Gumus, R., Atasever, M.A., Urgar, S. and Atasever, M. (2012). Effects of ascorbic acid and alpha-lipoic acid on performance and meat quality of broilers subjected to heat stress. *Br. Poult. Sci.* 53: 800–808.

Janke O. and Tzschentke B. (2010). Long lasting effect of changes in incubation temperature on heat stress induced neuronal hypothalamic c-Fos expression in chicken. *Open Ornith. J.*3: 150–155.

Krishna, P.P. (2011). Economics of Climate variation for Smallholder Farmers in Nepal: A review. *J. Agri. Envi.*: 6–13.

Lin, H., Jiao, H.C., Byse, J. and Decuypre, E. (2006). Strategies for preventing heat stress in poultry. *World's Poult. Sci. J.* 62: 71–86.

Lu, Q., Wen, J. and Zhang, H. (2007). Effect of chronic heat exposure on fat deposition and meat quality in two genetic types of chicken. *Poult. Sci.* 86: 1059–1064.

Lucas, J.L. and Rostagno, M.H. (2013). Impact of Heat Stress on Poultry Production. *Animals*, 3: 356–369.

Mack, L.A., Felver-Gant, J.N., Dennis, R.L. and Cheng, H.W. (2013). Genetic variation alters production and behavioral responses following heat stress in 2 strains of laying hens. *Poult. Sci.*, 92: 285–294.

Niu, Z.Y., Liu, F.Z., Yan, Q.L. and Li, W.C. (2009). Effects of different levels of vitamin E on growth performance and immune responses of broilers under heat stress. *Poult. Sci.* 88: 2101–2107.

Quinteiro-Filho, W.M., Ribeiro, A., Ferraz-de-Paula, V., Pinheiro, M.L., Sakai, M., As, L.R., Ferreira, A.J. and Palermo-Neto, J. (2010). Heat stress impairs performance parameters, induces intestinal injury, and decreases macrophage activity in broiler chickens. *Poult. Sci.*, 89: 1905–1914.

Rajkumar, U. Reddy M.R., Rama Rao, S.V., Radhika, K. and Shanmugam, M. (2011). Evaluation of growth, carcass, immune competence, stress parametres in Naked Neck chicken and their normal siblings under tropical winter and summer temperatures. *Asian-Austral. J. Anim. Sci.*, 24: 509–516.

Rajkumar, U., Shanmugam, M., Rajaravindra, K.S., Vinoth, A. and Rama Rao, S.V. (2015b). Effect of increased incubation temperature on juvenile growth, immune and serum biochemical parameters in selected chicken populations. *Indian J. Anim. Sci.*, 85(12): 1328–1333.

Rajkumar, U., Vinoth, A., Shanmugam, M., Rajaravindra, K.S. and Rama Rao, S.V. (2015a). Effect of embryonic thermal exposure on Heat shock proteins (Hsps) gene expression and serum T3 concentration in coloured broiler populations. *Anim. Biotechn.* 26(4): 260–267.

Steinfeld, H., Gerber, P., Wassenaar, T., Castel, V., Rosales, M. and Hann, C.de (2006). L*ivestock's Long Shadow: Environmental Issues and Options.* Rome: FAO.

Vinoth, A., Thirunalasundari, T., Tharian, J.A., Shanmugam, M. and Rajkumar, U. (2015). Effect of thermal manipulation during embryogenesis on liver heat shock protein expression in chronic heat stressed coloured broiler chickens. *J. Therm. Biol.* 53: 162–171.

Wiedmann, T. and Minx, J. (2008). A Definition of 'Carbon Footprint'. In: C.C. Pertsova, Ecological Economics Research Trends: Chapter 1, pp. 1–11, Nova Science Publishers, Hauppauge NY, USA.https: //www.novapublishers.com/ catalog/product_info.php?products_id=5999.

Yahav, S. (2008). Thermal manipulation during the perinatal period. Does it improve thermo tolerance and performance of broiler chickens? In *proceedings of the 19th Australian Poultry Science Symposium, New South Wales, Australia.*

Yahav, S. (2009). Alleviating heat stress in domestic fowl: different strategies. *World's Poult. Sci. J.* 65: 719–732.

Yahav, S., Colin, A., Shinder, D. and Picard M. (2004). Thermal manipulations during broiler chick's embryogenesis–the effect of timing and temperature. *Poult. Sci.* 83: 1959–1963.

Unique Traits of Zebu Cattle under Tropical Climatic Conditions

Sohan Vir Singh* and Simson Soren

Climate Resilient Livestock Research Centre, NICRA, ICAR-NDRI, Karnal, Haryana 132 001, India
*E-mail: sohanvir2011@gmail.com

ABSTRACT: *The origin of the Bos indicus (Zebu) and Bos taurus is from the same ancestor. They have undergone separate evolution for thousands of years and this evolution made the Bos indicus thermo-tolerant. The adapt-ability of Zebu cattle was achieved by acquiring genes that confer the thermo-tolerance at physiological and cellular level. The morphological characteristics also play important role in adaptability. The peripheral circulation is superior in Zebu cattle. The expression of heat shock proteins during heat stress is lower in Zebu than the crossbred cattle. It has been revealed that the melanin pigmentation is higher in Zebu, which also plays role in adaptation. Some of the Zebu breeds (Tharparkar, Nagori and Sahiwal) are well adapted to hot dry desert conditions, able to reduce their metabolic requirements to minimum and conserve energy for diversion to production (milk and/or work). These mechanisms are rarely found in livestock species located in other areas. There are several unique characteris-tics in Zebu cattle which make them resilient to tropical climatic conditions.*

Keywords: Zebu Cattle, Heat Tolerant, Low Metabolic Rate, Coat Colour, Melanin.

Zebu cattle are exposed to heat and other climatic stress factors, which are common in tropical climatic conditions, leads to adaptability to tropical climate. Adaptability means the capacity or ability to survive and reproduce to a defined environment. India is one of the most diverse lands in the world. It has many descriptive and non-descriptive livestock found in different agro-climatic zones. It has been noticed that the livestock breeds which are adapted to stressful environments have unique adaptive traits e.g. heat tolerance, parasitic and disease resistance, ability to cope with water scarcity and poor quality feed etc. These traits enable them to survive and reproduce under stressful environmental conditions. The propagation of these adaptive traits without disturbing the productive traits through modern technologies might help to face the challenges of climate change.

MORPHOLOGICAL CHARACTERISTICS OF ZEBU CATTLE

Zebu cattle well adapted to soil, plant and climatic conditions that prevail in different agro-climatic zones were studied. Most of the breeds have been named on the basis of their habitat or location to which they have before well adapted. One or more specific regions have used these breeds for their improvement. Zebu breeds have small size and low body weight with small barrel shaped body and slender legs. They have a hump and a dewlap. The head is held high in most zebu breeds. Since most of these breeds have been developed for draught purpose long legs with articulate joints provide ample capacity to run and swiftly move even under moist soils. The balanced fore and hind body quarters help them in propelling body and moving forward with loads at moderate speeds. Balanced body is mainly due to small size and low volume of internal organs. Small sized rumen, reticulum, omasum and abomasum, do not distend down the belly of these Zebu draught breeds contrary to heavy bodied of Taurine breeds. Some of the Zebu breeds (Tharparkar, Nagori and Sahiwal) well adapted to hot dry desert conditions are able to reduce their metabolic requirements to minimum and conserve energy for diversion to production (milk and/or work) without extra energy expenditure.

THERMOREGULATORY MECHANISMS IN ZEBU CATTLE

Cattle have to regulate the internal body temperature within a set point by matching the amount of heat production and the heat flow from animal's internal parts to surface and to the environment. Heat flow occurs through conduction, convection and radiation (sensible heat loss). The sensible heat loss i.e., conduction and convection depend upon the surface area per unit body weight. It also depends upon the temperature gradient between the animals and the air. Heat loss through radiation depends upon the reflective properties of the hair as well as the surface area. Evaporation is one of the best ways of heat relief. Therefore, providing easily digestible diet is recommended during summer to the animals for energy requirement. When the air temperature reaches the skin temperature, evaporation becomes the major route for heat exchange with the environment. White colour reflects more than darker colour. The light coloured hair coat and the sleek and shiny reflect a greater portion than the dark coloured.

Low Metabolic Rate

Low metabolic rate is advantageous if the feed quality and/or quantity is low. The quality and quantity of feed is affected during extreme climatic stress, where Zebu cattle can survive, give some milk and reproduce due to their low metabolic activity and are adapted to such stressful conditions. However, the crossbred animals cannot maintain their production performance which is reduced drastically during extreme stress. The adapted animals recycle the nutrients more efficiently than temperate breeds (Bayer and Feldmann, 2003).

The performance of *Bos taurus* breed is better than Zebu cattle, however, there is loss of weight and they fail to survive when fed poor quality feed. These changes not significantly seen in adapted breeds. The mRNA expression of metabolism related genes (Dio2, TRIP11) was found lower in Tharparkar than Karan Fries heifers during hot humid season (Naidu, 2016). The thyroid hormone, skin temperature and rectal temperature were positively correlated with the expression level of deiodinase type 2 (Dio2) and Thyroid Hormone Receptor Interacting Protien 11 (TRIP 11) genes in PBMC. The magnitude of TRIP11 gene expression was higher in Karan Fries heifers than Tharparkar. These characteristics clearly indicate better adaptability of Zebu cattle under tropical climatic conditions than crossbred.

Superior Peripheral Blood Flow and Coat Colour

The blood flow in the periphery is not only important for nourishment but is also sufficient for exchange of heat dissipation from internal to the surface of the body and to the environment. The peripheral blood flow increases to release the heat via conduction and convection. The skin of Zebu cattle is soft, smooth and clean due to the superior skin blood circulation as compared to crossbred cattle. The blood flow was positively correlated with the temperature of the body parts and it varied in different seasons.

The hair and skin pigmentation is also one of the adaptive traits of heat tolerant animals. Melanin pigmentation helps in adaptive mechanism and act as an antioxidant. Studies also revealed that the skin pigmentation (melanin) is higher in Tharparkar than Karan Fries cattle (Maibam *et al.,* 2014, 2016). The basis of coat colour in mammals including cattle is the presence or the absence of melanin pigment (eumelanin and pheomelanin). Eumelanin is responsible for black and brown colours and pheomelanin for reddish brown. The rate limiting enzyme for melanin synthesis is tyrosinase. Eumelanin intensifies skin pigmentation and thus helps in photoprotection because of its efficiency in blocking ultraviolet rays (UV) and scavenging reactive oxygen species. The expression of skin colour related genes (MC1R and PMEL) in lymphocytes and plasma tyrosinase activity were higher in Tharparkar than Karan Fries cattle. It shows that the ability of Karan Fries cattle to protect themselves from the harmful UV radiation by melanisation was less compared to Tharparkar.

Sweating and Panting Rate

Evaporation involves in sweating rate and respiratory minute volume. Evaporative cooling by sweating and panting is the most important mechanism for body heat dissipation under hot climates. The sweat glands of Zebu cattle are not only baggy-shaped and higher in volume, but are also closer to the skin surface than those of *Bos taurus.* Comparative studies showed that Zebu cattle are more dependent upon sweating to dissipate excess body heat, while *Bos taurus* are more likely to utilize an increase in respiration rates.

THERMOTOLERANT GENES

Zebu cattle regulate their body temperature better under heat stress than crossbred and temperate cattle. Heat stress influence the synthesis and expression of heat shock proteins (HSPs). They are the important for cell survival and cellular functioning. In the hot environmental niche, a greater amount of constitutive HSP 70.8 (HSPA8) is found in Zebu cattle during non-stress conditions (Singh *et al.,* 2014). The HSPA8 assists in the day to day cell functions of protein folding and unfolding, prevention of polypeptide aggregation, disassembly of large protein complexes, and aid in the translocation of proteins between cellular compartments. Heat shock proteins (HSPs) act as a molecular chaperones. HSPs have been considered to play crucial roles in environmental stress tolerance and thermal adaptation. Several studies in bovine, mice and human cells gave evidence that constitutive elevation of the HSPs levels provides cyto-protection upon thermal stress (Collier *et al.,* 2006). However, continuous temperature rise does not protect cellular damage due to an imbalance between various physiological and cellular functions (Patir and Upadhyay, 2010). Among members of the HSP family, HSP70i (HSPA1A and HSPA2) is the most temperature sensitive and induced by various physiological, pathological, and environmental stressors (Kumar *et al.,* 2015). Microarray analysis revealed that a total 460 transcripts were differentially expressed with a fold change of P2 in peripheral blood leukocytes on exposure to heat (42°C, 4 hours) in Tharparkar cattle. Further analysis is required to understand their functional role in livestock (Kolli *et al.,* 2014). Heat stress (40°C) did not increase ROS formation and lower expression of HSP 70.1 and 70.2 in Tharparkar whereas reverse trend was observed in Karan Fries cattle (Singh *et al.,* 2014). It is clear that Tharparkar cattle are adapted to tropical climatic conditions. Similarly, the spermatozoa of Tharparkar bulls observed were not significantly different in mRNA expression of HSP70 during summer and winter (Rajoriya *et al.,* 2014), it further indicates that the sperm quality of heat tolerant breed might not be affected due to hot dry and hot humid environment under tropical climatic conditions.

ADAPTATIONS TO PARASITES

The infestation of parasites seemed to vary between individuals and breeds. Ectoparasite can directly affect the animals and it may carry diseases like anaplasmosis, babesiosis etc. Some of the breeds of Zebu cattle are resistance to ticks. Glass *et al.* (2005) reported that the Sahiwal cattle are more resistant than European (Holstein) dairy breed calves to tick-borne *Theileria annulata* infection. Therefore, the selection for tick resistant breeds can be an advantage for farming system.

FERTILITY

Fertility is a broad term. It involves various chains of physiological events and every event is critical and important to established pregnancy. The milk production and fertility seems to be negatively related. Fertility may be impaired with high yielding cattle. High metabolic rate with high productivity and the genetic makeup of the animals is inversely proportional to reproductive functions. Better reproductive performance of Zebu cattle might be due to low milk yield or low metabolic heat production. As Zebu cattle are low yielding as compared to crossbred cattle, but the total productive life is more in Zebu.

CONCLUSION

There are several molecular characteristics in Zebu cattle among descriptive or non-descriptive breeds which still have to be revealed. The climate change scenario is likely to affect the livestock production system and food security. Different area is expected to be affected by drought, flood, cyclone etc. in the near future. Crossbred cattle yield more milk than Zebu cattle, whereas, the management of crossbred cattle may not be economical under heat stress. The production and reproduction are negatively impacted under heat stress. The livestock production system which is a major source of livelihood is affected due to global warming. It is an important sector to be transferred into an economical enterprise which can alleviate poverty. Selection of adaptive breed and genetic adaptation of livestock to new production conditions may be an important factor to meet the challenges in future. Several traits of Zebu cattle can be transferred to non-adapted breeds through reproductive and biotechnological tools. Genetic selection of good productive Zebu cattle may be beneficial in near future. Faster genetic gains of these traits can be achieved with new technologies, including genomic selection and advanced reproductive technologies.

REFERENCES

Bayer, W. and Feldmann, A. (2003). Diversity of animals adapted to smallholder system. Conservation and Sustainable Use of Agricultural Biodiversity. http://www.eseap.cipotato.org/UPWARD/Agrobio-sourcebook.htm.

Collier, R.J., Dahl, G.E. and VanBaale, M.J. (2006). Major advances associated with environmental effects on dairy cattle. *J. Dairy Sci.*, 89: 1244–1253.

Glass, E.J., Preston, P.M., Springbett, A., Craigmile, S., Kirvar, E., Wilkie, G. and Duncan Brown, C.G. (2005). Bos taurus and Bos indicus (Sahiwal) calves respond differently to infection with Theileria annulata and produce markedly different levels of acute phase proteins. *Int. J. Parasitol.*, 35: 337–347.

Kolli, V., Upadhyay, R.C. and Singh, D. (2014). Peripheral blood leukocytes transcriptomic signature highlights the altered metabolic pathways by heat stress in zebu cattle. *Res Vet Sci.,* 96: 102–110.

Kumar, A., Ashraf, S., Sridhargoud, T., Grewal, A., Singh, S.V., Yadav, B.R. and Upadhyay, R.C. (2015). Expression profiling of major heat shock protein genes during different seasons in cattle (Bos indicus) and buffalo (Bubalus bubalis) under tropical climatic condition. *J. Thermal Bio.,* 51: 55–64.

Maibam, U., Hooda, O.K., Sharma, P.S., Mohanty, A.K., Singh, S.V. and Upadhyay, R.C. (2016). Expression of HSP70 genes in skin of zebu (Tharparkar) and crossbred (Karan Fries) cattle during different seasons under tropical climatic conditions', *J. Thermal Bio. (in press).*

Maibam, U., Singh, S.V., Singh, Kapoor, S.A.K. and Upadhyay, R.C. (2014). Expression of skin color genes in lymphocytes of Karan Fries cattle and seasonal relationship with tyrosinase and cortisol. *Trop. Anim. Health Prod.* Doi.10.1007/s11250-014-0620-7.

Naidu, C. (2016). Metabolic profile and expression pattern of some genes in Tharparkar and Karan Fries (Tharparkar × Holstein Friesian) heifers during different seasons. M.V.Sc. thesis submitted to ICAR-National Dairy Research Institute (Deemed University).

Patir, H. and Upadhyay, R.C. (2010). Purification, characterization and expression kinetics of heat shock protein 70 from Bubalus bubalis. *Res. Vet. Sci.*, 88(2): 258–262.

Rajoriyal, J.S., Prasad, J.K., Ghosh, S.K., Perumal, P., Kumar, Anuj and Shobhana Kaushal (2014). Studies on effect of different seasons on expression of HSP70 and HSP90 gene in sperm of Tharparkar bull semen. *Asian Pac. J. Reprod.*, 3(3): 192–199.

Singh, A.K., Upadhyay, R.C., Malakar, D., Kumar, S. and Singh, S.V. (2014). Effect of thermal stress on HSP70 expression in dermal fibroblast of zebu (Tharparkar) and crossbred (Karan-Fries) cattle. *J. Thermal Bio.*, 43: 46–53.

Potential for Adaptation of Fisheries to Climate Change

E. Vivekanandan

Central Marine Fisheries Research Institute, Kochi 682 018, India
E-mail: evivekanandan@hotmail.com

ABSTRACT: *Climate change poses a threat to the sustainability of fisheries. The impacts are likely to reduce fish abundance and thereby income to the fishers. The impact of climate change on fisheries can be reduced by undertaking several adaptation measures by addressing the issue at two levels: (i) improving the resilience of fish populations by adopting effective fisheries management measures and (ii) improving the resilience of fishing communities by adopting effective livelihood and life protection measures. The adaptations can occur by (i) improving fisheries management; (ii) increasing resilience of communities, and (iii) adopting long-term plans. For the fisheries sector, climate change notwithstanding, there are several issues to be addressed. Strategies to promote sustainability should be in place before the threat of climate change assumes greater proportion.*

Keywords: Fish Stocks, Fishing Communities, Improved Management.

INTRODUCTION

The rapid pace of anthropogenic climate change poses a threat to the food production systems including fisheries. In India, the fisheries production has reached about 10 million tonnes from the marine, freshwater and brackish water sub-sectors, with the GDP of ₹ 78,000 crores. About one million people are engaged in full-time fishing and another 2 million in part-time fishing and fishing-related activities. The annual export from fisheries amounts to ₹ 32,000 crores. In recent years, evidences are accumulating that climate change is impactting the fishes and fisheries (Vivekanandan, 2011).

The impact of climate and its seriousness can be reduced/modified by adaptation of various kinds. Identifying potential adaptation options is important for impact assessment (to identify adaptations that are likely to occur) and for policy development. Identifying adaptation options is based upon the following three considerations: (i) adapt to what? (ii) who or what adapts? and (iii) how does adaptation occur? (Smit *et al.*, 2000).

---

* Ocean Partnership Project, Bay of Bengal Programme, Intergovernmental Organisation, Chennai–600 018, India.

ADAPTATION TO WHAT?

Adaptation can be considered in the context of causes of impacts. In fisheries, most often, the causes are climate or weather conditions such as water temperature, acidity, intensity and amount of rainfall, magnitude and frequency of storms and cyclones, speed and direction of wind and current, sea level rise, etc. It is also expressed as ecological or human-driven causes like water availability, ecosystem impacts etc. In reality, these causes are not independent and act jointly on fish stocks.

Most fish species have a fairly narrow range of optimum temperatures related to their basic metabolism and availability of food organisms. Being poikilotherms, even a difference of 1°C or 0.1 unit pH in water may affect their distribution and life processes. The more mobile species should be able to adjust their ranges over time, but less mobile and sedentary species may not. Depending on the species, the area it occupies may expand, shrink or be relocated. This will induce increases, decreases and shifts in the distribution of fish, with some areas benefiting while others lose. From the recent investigations carried out by the Central Marine Fisheries Research Institute (CMFRI) and Central Inland Fisheries Research Institute (CIFRI), the following responses to climate change by different fish species are discernible in the Indian waters:

1. Shift/Extension of distributional boundary;
2. Shift/extension of depth of occurrence;
3. Shift in spawning grounds; and
4. Phenological changes.

While these changes are beneficial to a few fish species and geographical areas, the proportion of benefitted species and areas are likely to decrease with increasing intensity of climate change and climate variability. These changes are likely to negatively affect fish populations, and thereby, their abundance. In turn, the fish catches will reduce, and this will have great negative impacts on the social and economic conditions of fishers, traders and others in the supply chain.

Limits and barriers to adaptation restrict the ability of fishers, particularly the artisanal fishers to address the negative impacts of climate change (Islam *et al.*, 2014). The limits include physical characteristics of climate and sea, like higher frequency and duration of cyclones and sea level rise. Barriers include technologically poor boats, inaccurate weather forecast, poor radio signal, lack of access to credit, low incomes, lack of education, skills and livelihood alternatives, unfavourable credit schemes, lack of enforcement of fishing regulations and maritime laws, and lack of access to fish markets. These local and wider scale factors interact in complex ways and constrain completion of fishing trips, coping with cyclones at sea, safe return of boats from sea, timely responses to cyclones and livelihood diversification.

WHO OR WHAT ADAPTS?

The question of who or what adapts should focus on the spatial scale as well as the species or ecosystem or economics or social structure. IPCC (1996) notes that adaptation has to be at the level of the most vulnerable with great sensitivity to climate change. The adaptation of fisheries sector to climate change should be fundamentally addressed at two levels: (i) improving the resilience of fish populations by adopting effective fisheries management measures; and (ii) improving the resilience of fishing communities by adopting effective livelihood and life protection measures.

HOW DOES THE ADAPTATION OCCUR?

The adaptations can occur by (i) improving fisheries management, (ii) increasing resilience of communities and (iii) adapting long-term plans (Figure 1).

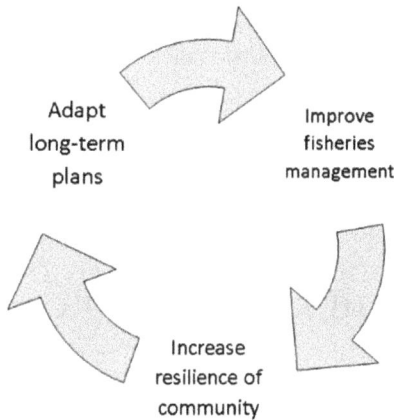

Adapt long-term plans

Improve fisheries management

Increase resilience of community

Fig. 1: Components of Climate Change Adaptation Plan for Fisheries

IMPROVING FISHERIES MANAGEMENT

Fishing and climate change are strongly interrelated pressures on fish production and must be addressed jointly. Reducing fishing mortality in the majority of fisheries, which are currently fully exploited or overexploited, is the principal means of reducing the impacts of climate change. Reduction of fishing effort (i) maximizes sustainable yields, (ii) helps adaptation of fish stocks and marine ecosystems to climate impacts, and (iii) reduces greenhouse gas emission by fishing boats (Brander, 2007). Hence, some of the most effective actions which we can take to tackle climate impacts are to deal with the old familiar problems such as overfishing and adopt Code of Conduct for Responsible Fisheries and Integrated Ecosystem-based Fisheries Management (FAO, 2007). To realize the full potential of marine fisheries, efforts should be directed towards fishing effort management; fleet-size optimization; mainstreaming biodiversity conservation in production processes; species-specific and area-specific management plans.

FAO's code of conduct for responsible fisheries (CCRF or Code) is today the most significant of the non-binding agreements in the global fisheries sector. It is global in scope and directed towards members and non-members of the FAO, fishing entities, organisations of all kinds, fishers, people engaged in the processing and marketing of fish and fishery products—in short, everyone concerned with management and development of fisheries. The Ecosystem Approach to Fisheries Management offers a practical and effective means to manage fisheries more holistically.

The global community has recognized the importance of small-scale fisheries as a principal contributor to poverty alleviation and food security and has agreed to the voluntary guidelines on sustainable small-scale fisheries (VG-SSF). The main objectives of the VG-SSF Guidelines are expected to be achieved through the promotion of a human rights-based approach in the context of food security and poverty eradication, by empowering small-scale fishing communities. The Government should make all efforts to implement the provisions of the VG-SSF keeping in view the complexities and divisions within the small-scale sector, particularly those involved in sustenance fishing.

As part of India's international commitments on climate change, the concept of green fisheries by reducing green house gases (GHG) emissions from fishing and fishing related activities also need to be encouraged. Marine Fishing Regulation Act is the most important vehicle for regulating the fisheries in the state. Climate change adaptation of fisheries should be mainstreamed into the Act.

INCREASING THE RESILIENCE OF COASTAL FISHING COMMUNITIES

Knowledge Management for Climate Change and Marine Fisheries

Knowledge management will be an approach to foster quick and easy availability of information on key attributes of the fisheries sector, such as resource abundance and distribution; real-time resource maps; productivity assessments; real-time potential fishing zone (PFZ) advisories; and weather forecasts for the benefit of fishers. The use of information technology (IT) and space technology have to be put to optimum use for harnessing the benefits in support of the fisher community.

Awareness Building

Effort is required to raise awareness of the impact, vulnerability, adaptation and mitigation related to climate change among the fishing communities and other stakeholders.

Monitoring, Control and Surveillance

As disaster risks due to increased intensity and frequency of storms are likely to increase, the existing mechanisms in place for a sound and effective MCS regime for marine fisheries sector need strengthening. While monitoring of fish catch and effort and control of fishing through registration and licensing is in place, MCS activities need to be further strengthened through greater engagement of the Department of Fisheries, Coastal Marine Police and the Coast Guard. Strengthening and improvements in MCS should be carried out in a phased manner, by using conventional means (use of log books, movement tokens, colour coding of fishing boats, biometric cards to fishers for their identity) and also space technologies and IT tools (e.g. Vessel Monitoring System and Automatic Identification System). Tracking the fishing boats by establishing high frequency (HF) ground stations and HF sets on board the fishing boats.

The marine fisheries sector is characterized by a range of fishing boats varying in design, use of material, size, engines and gear and area of operation. The legislations relating to registration, certification, sea-safety and manning norms of fishing vessels are required to be updated to cater to the needs of fishery sector and also to meet the international standards and norms prescribed by concerned agencies such as FAO, IMO, ILO etc.

Marine fishing is one of the riskiest professions in the world. The government should ensure that safety-at-sea measures are adequately strengthened and implemented. Such measures *inter alia* will include provisions of lifesaving appliances and communication equipment on the boat and adequate skills and capacity development of fishers and other concerned stakeholders.

In order to remedy this situation, the Government will consider providing public finance to fishermen with liberal terms and conditions. In this direction, the role of the National Bank of Agriculture and Rural Development (NABARD) and National Fisheries Development Board (NFDB) assumes significance and has to be considered in meeting the needs of the fishers.

Capacity Building at all Levels

The government should encourage focused studies on climate change impacts on fish stocks and fishing communities, besides implementation of adaptation options in a time bound manner. The government should initiate steps for training, capacity building as well as up-gradation of technological skills of traditional fishers in moving from artisanal fishing to more economic and efficient means of fishing.

Adapting Long-Term Plans

In the context of climate change, the primary challenge to the fisheries sector will be to ensure food supply, enhance nutritional security, improve livelihood and economic output, and ensure ecosystem safety. These objectives call for identifying and addressing the concerns arising out of climate change; evolve adaptive mechanisms and implement action across all stakeholders on a long-term. In response to shifting fish population and species, the fishing sector may have to respond with the right types of craft and gear combinations, on-board processing equipments etc. Governments should consider establishing Weather Watch Groups and decision support systems on a regional basis. Allocating research funds to analyze the impacts and establishing institutional mechanisms to enable the sector are also important. The relevance of active stakeholder participation and collaboration to exchange information and ideas is being felt now as never before.

A SWOT analysis can help understanding the key issues and identifying the opportunities for adaptation on a long term. From the available information, the strengths, weaknesses, opportunities and threats in the fisheries sector with reference to climate change are discernible (Table 1).

Table 1: SWOT Analysis for Adaptation of Fisheries Sector

Strengths	*Weaknesses*
• Resilience capacity of many tropical fishes • Fishers knowledge on environment • Entrepreneurship of fishers	• Inadequate knowledge on climate change • Resistance to adaptation advisories • Weak implementation of existing fisheries management policies
Opportunities	*Threats*
• Building awareness at all levels • Strengthening research to address immediate and long-term issues • Mainstreaming climate change adaptation into fisheries policies • Strict implementation of existing fisheries policies	• Overfishing, habitat degradation and pollution • Trade-related issues

CONCLUSION

For the fisheries sector, climate change notwithstanding, there are several issues to be addressed. Strategies to promote sustainability and improve the supplies should be in place before the threat of climate change assumes greater proportion. While the fisheries sector cannot do much to mitigate climate change, it could contribute to reduce the impact by following effective adaptation measures.

REFERENCES

Brander, K.M. (2007). Global fish production and climate change. *Proceedings of National Academy of Sciences of the USA,* 104: 19709–19714.

FAO (2007). Building adaptive capacity to climate change. Policies to sustain livelihoods, and fishers. New Directions in Fisheries—A Series of Policy Briefs on Development Issues, Food and Agriculture Organisation, 8:16 pp.

IPCC (1996). Summary for Policy Makers: Scientific-Technical Analysis of Impacts, Adaptations and Mitigation of Climate Change. Second Assessment Report of Intergovernmental Panel on Climate Change, p. 18.

Islam, M.M., Sallu, S., Hubacek, K. and Paavola, J. (2014). Limits and barriers to adaptation to climate variability and change in Bangladeshi coastal fishing communities. *Marine Policy*, 43: 208–216.

Smit, B., Burton, I., Klein, R.J.T. and Wandel, J. (2000). An anatomy of adaptation to climate change and variability. *Climatic Change,* 45: 223–251.

Vivekanandan. E. (2011). Climate change and Indian marine fisheries. Policy Brief 3, *CMFRI Special Publication,* 105: p. 97.

Adaptive Capabilities and Resilience of Indigenous Animal Genetic Resources to Climate Change

Arjava Sharma* and Sonika Ahlawat

ICAR-National Bureau of Animal Genetic Resource, Karnal 132 001, Haryana, India
*E-mail: arjava@yahoo.com

ABSTRACT: *Climate change is the most challenging environmental problem ever faced by the global community. Agriculture and livestock farming which ensure livelihoods of rural poor communities are considered to be the most climate-sensitive sectors. Most of the national growth achieved in agricultural sector is because of impressive improvement in livestock sector. Fortunately, India has a vast repertoire of indigenous animal genetic resources (AnGR) that act as a cushion to farming community during the period of distress caused by natural calamities. In addition to 160 recognized breeds of different livestock species, thousands of non-descript populations have been recognized worldwide for harboring unique attributes like heat tolerance, disease resistance, hardiness, ability to survive under harsh climatic conditions and sustenance on low quality roughages. Last few decades have witnessed crossbreeding of indigenous animals with exotic germplasm to meet the demand of animal products for the ever increasing human population in India. However, this has led to spread of homogenized, high performance breeds that produce well in controlled conditions only, with a concomitant decline in overall genetic diversity. The international community has acknowledged the need for maintenance of substantial AnGR diversity as a global insurance measure against inevitably changing climate scenario. There is a need to identify and strengthen local breeds with the ability to adapt to unforeseeable circumstances like climatic stress and feed scarcity. Grading up of non-descript indigenous breeds with defined superior breeds seems to be a viable alternative for overall genetic improvement. Both in-situ and ex-situ conservation strategies need to be adopted to conserve important adaptive genes and genetic traits. Finally, superiority of our native germplasm needs to be scientifically established using latest genomics approaches to trigger creative dialogues with the stakeholders.*

Keywords: Climate Resilient Livestock, Livestock Biodiversity, Adaptation, Thermo-Tolerance.

Climate change has inarguably been recognized as the most challenging environmental problem ever confronting the global community. The Inter-governmental panel on climate change (IPCC, 2007) predicts that by 2100 the increase in global surface temperature may be between 1.8 and 4.0°C posing a threat to 20–30% of plant and animal species. Evidence from International Fund for Agricultural Development (IFAD) is overwhelmingly convincing that the poorest and most vulnerable people will be most affected by the negative impacts of climate change. Rural poor communities in the developing countries rely greatly for their survival on agriculture and livestock keeping but unfortunately these two sectors are amongst the most climate-sensitive ones. Threat of climate change to the sustainability of livestock systems has been acknowledged globally. India's National communication to the United Nations Framework Convention on Climate Change (UNFCCC) reported that a rise in temperature by 2–4°C will negatively impact milk production by more than 15 million tons by 2050 in comparison to the current levels of production. Therefore, it is high time that measures for adaptation to, and mitigation of the detrimental effects of climatic extremes are given due importance (Sejian *et al.*, 2015).

In adopting the 'Global Plan of Action', the international community acknow-ledged that animal genetic diversity is critical not only for food security and rural development but also to combat ill effects of changing climatic conditions. In this regard, India is fortunate to be one of the largest mega biodiversity centers of the world. Our vast livestock wealth is distributed over a range of geographi-cal, ecological and climatic regions. As per Livestock Census 2012, the farm animal genetic resources in India are represented by cattle (37.28%), buffalo (21.23%), camel (37.28%), goat (26.04%), sheep (2.71%), pig (2.01%) and others such as yak, mithun, poultry (0.37%). ICAR–National Bureau of Animal Genetic Resources has recognized 160 breeds of different livestock species: cattle (40), buffalo (13), goat (26), sheep (42), pig (6), camel (9), donkey (1), horse and ponies (6) and chicken (17). The acknowledged breeds constitute only 20–25% of the total population and the rest are classified as non-descript. Interestingly, thousands of unrecognized or non-descript populations also contribute significantly to India's livestock production. For the year 2013–14, contribution of livestock sector to total GDP was 3.9% and to the agricultural GDP was 24.8%. The share of livestock in agricultural GDP has witnessed consistent increase from 15% in 1981–82 to 25% in 2013–14.

The indigenous germplasm has been acknowledged worldwide for possessing unique attributes such as heat tolerance, disease resistance, hardiness, ability to survive under harsh climatic conditions and sustenance on low quality roughages. Cattle breeds such as Sahiwal, Tharparkar, Hariana, Rathi, Gir, Nagori are credited for their ability to tolerate extremes of climate. Kankrejin addition to its adaptability to high temperature also has powerful draught capacity and resistance to tick borne diseases. Similarly, there are unique buffalo breeds known for milk production (Murrah); very high milk fat (Bhadawari); adaptability to brakish-water (Chilika); drought tolerance and nocturnal grazing (Banni buffaloes of Kutch-Gujarat) and excellent draft power (Swamp buffaloes of north-east India). There exist remarkable diversity and distinctive features in sheep genetic resources as well. For instance, adaptation to low temperate and low oxygen environment (Gaddi), survival under desert conditions (Malpura, Chokla, Marwari) and high prolificacy and adaptation to mangrove ecology (Garole). Many breeds of goat are also recognized for their adaptation to different ecologies with high temperature, high humidity or both. Andaman goat is acclimatized to saline conditions, Changthangi to high altitude, low temperate conditions and Jakhrana to semi-arid conditions.

Globally, human population is expected to increase from 6.2 billion today to 9.2 billion by 2020 and the demand for livestock products is likely to increase by two folds during the first half of this century, as a result of the growing human population and its growing affluence (FAO, 2009). India has also been confronted with a daunting challenge to meet the animal products demand of an expected population of 1.62 billion in 2050. To achieve quicker genetic improvement in terms of milk production, upgrading/large scale crossbreeding program of indigenous cattle with exotic germplasm (Holstein Friesian, Jersey or Brown Swiss) has been resorted to in the last few decades and consequently, today the crossbred cattle constitute 20.81% of the total cattle population of the country (Livestock Census, 2012). However, it has resulted in spread of homogenized, high performance breeds that produce well in controlled conditions only, with a simultaneous decline in overall genetic diversity. In the era of climate change, the policy of crossbreeding has received enough criticism because substantial differences in thermal tolerance exist between *Bos indicus* and *Bos taurus*. Unfortunately, the high-output breeds originating from temperate regions that contribute to the bulk of milk production today are not well adapted to heat stress. The basic reason for better tolerance in zebu breeds could be their emergence by natural selection through generations enabling them to adapt to stressful environment. The differences in physiological response of animals have been attributed to (Olson *et al.*, 2003, Hansen, 2004):

- Higher sweating rates because of large sized and higher density of sweat glands in tropically adapted *Bos indicus* cattle than *Bos taurus*.
- Inefficient transfer of metabolic heat to the skin due to greater tissue resistance in *Bos taurus*.
- Greater accumulation of heat at the skin because of greater resistance of the hair coat in *Bos taurus*.
- Poor utilization of low quality roughages in crossbreds compared to their indigenous counterparts.

An editorial in, *The Hindu* discussed the gross failure of the crossbreeding policy (Sainath, 2011). According to this editorial, in just three of the six crisis districts of Vidarbha region of Maharashtra where crossbreds of exotic cattle such as Jersey and Holstein were provided to the poor Adivasis at 75% subsidy under the Prime Minister's relief package, 28% were either dead, sold or could not be verified during the package period of 2006–07 to 2009–10. The main reasons for these results were poor adaptability of crossbreds to higher temperatures, frequent health ailments and higher consumption of fodder and water by crossbreds which the poor farmers could not afford. Another story published in the same newspaper also highlighted similar plight of livestock keepers in Kerala who were reported to prefer indigenous cattle like Vechur over the exotic ones. In addition to this, crossbred animals exhibit reduction in hybrid vigour due to which their productive years are reduced significantly (Indian Express, 2009).

The current policy of indiscriminate breeding of indigenous breeds with breeds from temperate regions, has been criticized by many researchers. Despite adaptive characters like heat tolerance, disease resistance and ability to survive under adverse conditions inherently harbored by indigenous breeds, they are rarely covered by structured breeding programmes. So, it is apparent that the policy of indiscriminate crossbreeding should be reviewed and this practice should be stopped particularly in drier areas where there is poor availability of feed, fodder and water. Rather than providing subsidy on crossbred cattle, it is a better alternative to provide subsidy on defined indigenous dairy breeds like Gir, Tharparkar, Sahiwal, Kankrej, etc. The focus now should be shifted to improving milk yields of indigenous cattle without resorting to crossbreeding. The scope for such genetic upgradation is evident from buffaloes, in which milk yields have gradually risen through the use of genetic material from proven bulls of superior indigenous breeds like Murrah and Nili-Ravi.

It has been accepted that climate change will have an impact on livestock per-formance (exotic, indigenous as well as crossbreds) and according to most estimates, the impact will be unfavourable. The changing climate scenario will adversely affect the livestock directly and indirectly. Direct consequences include impaired physiological (rectal temperature, respiration rate, dry matter intake etc.) as well as production parameters (milk production, meat production) (Davison *et al.*, 1996). Indirect consequences include reduced water availability, shortage of feed and fodder, reduced biodiversity and increase in the incidence of vector-borne diseases (Thornton *et al.*, 2007). In an attempt to maintain body temperature within physiological limits, heat-stressed animals initiate compensatory and adaptive mechanisms to re-establish homeothermy and homeostasis. Consequently, there is a reduction in productive potential of these animals (Hansen, 2004). Diseases and food scarcity further aggravate the situation. If this problem is not meticulously handled, rural communities banking on their livestock assets could collapse into chronic poverty because of loss of livelihoods.

Another area that deserves discussion is contribution of livestock to climate change. Besides being insurance to global food security, livestock also contributes to climate change in the form of green house gas (GHG) emissions (FAO, 2006). Livestock production accounts for approximately 18% of the global GHGs emissions including methane (CH_4) from enteric fermentation and manure management, nitrous oxide (N_2O) from animal manure, and carbon-dioxide (CO_2) from land use change caused by demand for feedgrains, grazing land and agricultural energy use. Intriguingly, methane emission from indigenous cattle (Sahiwal) is less than that of crossbreds (Karan Fries) maintained on the same feeding schedule, suggesting that the zebu cattle contributes less towards global warming. It has now become necessary to figure out what adaptation and mitigation strategies could reduce the impact of climate change on livestock and vice versa to ensure sustainability of this enterprise. In both cases, the immense animal genetic diversity appears to be a global insurance measure against unforeseeable circumstances in future.

STRATEGIES TO COMBAT CLIMATE CHANGE

The effects of climate change will be apparent in both developed and developing countries, but countries like India will be affected most because of lack of resources, knowledge, extension services and research technology development. To dilute such consequences, it is time we analyse the challenges and possibilities rationally and develop suitable policies to face the problem of climate change. Some of the steps that can be taken are:

- Characterization and evaluation of animal genetic resources (AnGR).
- Identifying and strengthening local breeds with the ability to adapt to local climatic stress and feed sources.
- Support traditional extensive livestock production systems rearing local breeds that are relatively resilient to environmental changes rather than relying on crossbreds that are actually high-input-output breeds needing specific husbandry practices and diet to express their full genetic potential and be economically viable.
- Develop national inventories including relevant spatial information and assess future breed distribution.
- Develop breeding indices for selection which include traits associated with thermal tolerance, low quality feed and disease resistance with more emphasis on genotype by environment interactions.
- Grading up of non-descript indigenous breeds with defined indigenous dairy breeds to maintain adaptive characteristics of both breeds along with concomitant increase in milk production.
- Develop livestock breeding policies that ensure *in situ* conservation by involving pastoralists/livestock keepers. The pastoralist communities are the main 'livestock gene keepers' which contribute to the resilience of food supply systems in addition to being the custodians of traditional knowledge.
- *Ex situ, in vitro* conservation could help conserve critical adaptive genes and genetic traits.
- Design physiological experiments that study the effects of multiple environmental stresses (heat, humidity, nutrition) on livestock rather than concentrating on one stress at a time to reach a logical interpretation.
- Research to mitigate the methane emission from the animals needs to be undertaken to find a viable solution to this problem. Metagenomics and

nutrigenomics studies to alter the rumen metabolism for reduced methane production could be planned.

- Technologies for production and conservation of fodder to ensure supply of animal feed during periods of scarcity and reduce malnutrition and mortality in herds.
- Generate systematic information on the impact assessment of climate change on livestock production.
- Monitor threats to breeds which may be caused by climate change or other pressures (e.g. geneflow), and develop some predictive warning systems.
- Generate species/breed specific information on climate change impacts.
- Capacity building of livestock keepers for increasing awareness of global climate changes.

National Innovations on Climate Resilient Agriculture (NICRA), a network project of the Indian Council of Agricultural Research (ICAR), has been launched with the aim to enhance resilience of Indian agriculture to climate change and climate vulnerability through strategic research and technology demonstration. There is a need for more such projects to make the effort worthwhile.

PROSPECTS IN THE GENOMICS ERA

There is no denying the fact that in order to meet the global demand for animal products, maintenance of substantial AnGR diversity is a pre-requisite under the inevitably changing climate scenario. The wide repertoire of livestock diversity that India possesses seems to be a saving grace to ensure the sustainability of Indian livestock production system. Research efforts to unravel the thermo-regulatory mechanism of resistance to heat stress in indigenous animals and vulnerability in crossbred animals are the need of the hour. The unique gene pool of our AnGR with superior adaptive traits presents an opportunity for mining specific allele(s). Superiority of our native germplasm needs to be scientifically acknowledged to trigger creative dialogues with the stakeholders. With rapid advancements in technologies, it is now possible to analyse functional genomic regions with potential associations with adaptation at physiological and cellular levels (Qian *et al.*, 2013). It would thus become possible to identify high-performance local-breeds and optimize their potential for multiple functions, rather than concentrate on a single trait. Some of the approaches can be:

- *SNP identification and genotyping:* The SNPs associated with a trait of interest (disease resistance, thermo-tolerance) can facilitate screening for presence of desirable alleles and ultimately pave way for introgression studies.
- *Phenotype-genotype association studies:* Integration of phenomics and genomics to systematically analyse different adaptive traits in livestock.
- *Genome-wide association study:* It has the potential to expedite both pure and crossbreeding programmes for adaptation, assuming that the phenotypes are available.
- *Metagenomics/nutrigenomics:* Strategies to mitigate methane emissions can be developed by adopting metagenomics and nutrigenomics studies for

manipulating methane producing microbes or identifying feed additives that can interrupt the methanogenesis process.

- *Transcriptomics and proteomics:* These approaches can be useful to delineate differentially expressed genes and proteins under heat stress and thermo-neutrality.

Table 1: Pathways and Genes that have been Identified as Potential Candidate Genes for Thermoregulation in Genomic Studies

Pathway/Function	Gene(s)	Reference
Cellular response to stress	STAC, WRNIP1, MLH1, RIPK1, SMC6, GEM1	Howard *et al.*, 2014
Response to heat	STAC	,,
Gap junction	TUBB2A, TUBB2B	,,
Cellular response to stress	CCNG, TNRC6A	,,
Apoptosis	FGD3, G2E3, RASA1, CSTB, DAPK1, MLH1, RIPK1, SERPINB9, HMGB1	,,
Ion transport	CACNG3, CLCN4, PRKCB, TRPC5, KCNS3, SLC22A23, TRPC4	,,
Thyroid hormone regulation	DIO2	,,
Body weight and feed intake	NBEA	,,
Heat shock protein response	HSPH1, TRAP1	,,
Respiration	ITGA9	,,
Calcium ion and protein binding	NCAD	Dikmen *et al.*, 2012
Protein ubiquitination	RFWD12, KBTBD2, CEP170, PLD5	,,
Thyroid hormone regulation	SLCO1C1	,,
Insulin signaling	PDE3A	,,
RNA metabolism	LSM5, SNORD14, SNORA19, U1, SCARNA3	,,
Transaminase activity	GOT1	,,
Apoptosis, cell signaling	FGF4	Hayes *et al.*, 2009

Not much work has been done on the direct impacts of climate change on heat stress in animals, particularly in the tropics and subtropics. Although there have been few studies dealing with thermo-regulation in India targeting few candidate genes at a time such as HSP1, HSP10, HSP40, HSP60, HSP70 and HSP90 (Kishore *et al.*, 2014; Kumar *et al.*, 2015), no comprehensive genome-wide study has been performed so far. There is a lack of appropriate physiological models that relate climate to animal physiology. Thermo-tolerance is a quantitative trait that is influenced by many regions of the genome and a handful of genomics studies have identified few pathways and genes that may have a role to play in governing the response to thermal stress. However, the results of these studies deserve further verification.

Identification of genetic merit of animals by latest biotechnology tools in order to select superior ones seems to be a possible solution to combat future adversities. Although it might seem challenging right now, the opportunities that this approach presents add new zeal and vigour in the scientific community.

REFERENCES

Davison, T., McGowan, M., Mayer, D., Young, B., Jonsson, N., Hall, A., Matschoss, A., Goodwin, P., Goughan, J. and Lake, M. (1996). Managing hot cows in Australia. Queensland Department of Primary Industry, 58.

Dikmen, S., Cole, J.B., Null, D.J. and Hansen, P.J. (2012). Heritability of rectal temperature and genetic correlations with production and reproduction traits in dairy cattle. *Journal of Dairy Science* 95: 3401–3405.

FAO (2009). The state of food and agriculture, Rome, Italy http://www.fao.org/docrep/012/i0680e/i0680e.pdf.

Hansen, P.J. (2004). Physiological and cellular adaptations of zebu cattle to thermal stress. *Animal Reproduction Science,* 82–83: 349–360.

Hayes, B.J., Bowman, P.J., Chamberlain, A.J., Savin, K., van Tassell, C.P., Sonstegard, T.S. and Goddard, M.E. (2009). A validated genome-wide association study to breed cattle adapted to an environment altered by climate change. *PLOS ONE,* 4(8): e6676.

Howard, J.T., Kachman, S.D., Snelling, W.M., Pollak, E.J., Ciobanu. D.C., Kuehn, L.A. and Spangler, M.L. (2014). Beef cattle body temperature during climatic stress: A genome-wide association study. *International Journal of Biometeorology,* 58: 1665–1672.

Indian Express (2009). http://www.indianexpress.com/news/crossbreeding-leaves-cows-alien-to-indianheat-reproduction-takes-hit/528006.

Kishore, A., Sodhi, M., Kumari, P., Mohanty, A.K., Sadana, D.K., Kapila, N., Khate, K., Shandilya, U., Kataria, R.S. and Mukesh, M. (2014). Peripheral blood mononuclear cells: a potential cellular system to understand differential heat shock response across native cattle (*Bos-indicus*), exotic cattle (*Bos-taurus*) and riverine buffaloes (*Bubalus bubalis*) of India. *Cell Stress Chaperones,* 19(5): 613–21.

Kumar, A., Ashraf, S., Goud, T.S., Grewal, A., Singh, S.V., Yadav, B.R. and Upadhyay, R.C. (2015). Expression profiling of major heat shock protein genes during different seasons in cattle (*Bos indicus*) and buffalo (*Bubalus bubalis*) under tropical climatic condition. *Journal of Thermal Biology,* 51: 55–64.

Olson, T.A., Lucena, C., Chase Jr., C.C. and Hammond, A.C. (2003). Evidence of a major gene influencing hair length and heat tolerance in *Bos taurus* cattle. *Journal of Animal Science,* 81: 80–90.

Qian, W., Deng, L., Lu, D. and Xu, S. (2013). Genome-wide landscapes of human local adaptation in Asia. *PLoS ONE,* 8: e54224.

Sainath, P. (2011). Cowed down by the Prime Minister. *The Hindu,* 13 May.

Sejian, V., Bhatta, R., Soren, N.M., Malik, P.K., Ravindra, J.P., Prasad, C.S. and Lal, R. (2015). Introduction to concepts of climate change impact on livestock and its adaptation and mitigation. In: Climate change Impact on livestock: adaptation and mitigation. Sejian, V., Gaughan, J., Baumgard, L., Prasad, C.S. (Eds), Springer-VerlagGMbH Publisher, New Delhi, India, pp. 1–26.

Thornton, P.K., Herrero, M., Freeman, A., OkeyoMwai Ed Rege, Jones, P. and McDermott, J. (2007). Vulnerability, Climate change and Livestock – Research Opportunities and Challenges for Poverty Alleviation. (http://www.icrisat.org/journal/SpecialProject/sp7.pdf.).

Mitigating Climate Change

Emission and Mitigation of Greenhouse Gases in Indian Agriculture

H. Pathak

ICAR-National Rice Research Institute, Cuttack, Odisha 753 006, India
E-mail: him_ensc@iari.res.in

ABSTRACT: *Agriculture, crucial for ensuring food, nutritional and livelihood security of India, is exposed to the stresses arising from climatic variability and climate change. Agriculture sector is also a major contributor to the enhanced greenhouse effect with the emissions of carbon dioxide, methane and nitrous oxide from agricultural soils and livestock. During 1970–2010, GHG emission from Indian agriculture has increased by about 75%. The increasing use of fertilizers and other agri-inputs, and the rising population of livestock are the major drivers for this increase in GHGs emission. The relative contribution of Indian agriculture to the total GHGs emission from all the sectors of the country, however, has decreased from 33% in 1970 to 18% in 2010. Mitigation of GHGs emission from agriculture can be achieved by sequestering C and reducing the emissions of methane and nitrous oxide through changes in land-use management and enhancing input-use efficiency. A win-win solution would be to develop such mitigation strategies that help in climate change adaptation and promote sustainable agricultural development.*

Keywords: Global Warming, GHGs, Emission, Mitigation, Adaptation.

Global warming, caused by the increase in concentration of GHGs in the atmosphere, has emerged as the most prominent environmental issue all over the world. These GHGs *viz.* Carbon Dioxide (CO_2), Methane (CH_4) and Nitrous Oxide (N_2O), trap the outgoing infrared radiations from the earth's surface and thus raise the atmospheric temperature. The Inter-Governmental Panel on Climate Change (IPCC), in its Fifth Assessment Report, has reiterated that warming of the climatic system is unequivocal. The anthropogenic influence on the climatic system is evident from the increasing concentrations of GHGs in the atmosphere and the positive radiative forcing. As a result, the temperature of atmosphere and ocean is going up, snow and ice are melting fast, and sea level is rising. This global climate change will have considerable impact on the crop, soil, livestock and fishery.

Agriculture engages almost two-third of the workforce in gainful employment and accounts for a significant share in India's gross domestic product. Several industries depend on agricultural production for their requirement of raw materials. Due to its close linkages with other economic sectors, agricultural growth has a multiplier effect on the entire economy of the country. The agricultural sector is believed to contribute to the greenhouse effect and the ensuring climate change is likely to have adverse impact on this sector. Various agricultural activities such as land clearing, cultivation of crops, irrigation, animal husbandry, fisheries and aquaculture have a significant impact on the emission of GHGs and consequently on climate change (IPCC, 2014). An in-depth understanding of trends in emission of GHGs, their drivers, and the relation between the two, is essential for comprehending the need for mitigation and adaptation. The objectives of this paper are to evaluate the emission of GHGs from Indian agriculture, analyze the drivers and trends of GHG emission and assess the potential of various mitigation options.

AGRICULTURE AS A SOURCE OF GREENHOUSE GASES

Agriculture contributes to greenhouse effect primarily through the emission and consumption of GHGs such as CH_4, N_2O and CO_2. CH_4 is produced in soil during microbial decomposition of organic matter under anaerobic conditions. Rice fields submerged underwater are the potential sources of CH_4 production. Continuous submergence, higher organic C content and use of organic manure in puddled soil enhance CH_4 emission. Burning of crop residues also contributes to the global methane budget. The enteric fermentation in ruminants is another major source of CH_4 emission.

Nitrous oxide is produced in soils through the processes of nitrification and denitrification. Nitrification is the aerobic microbial oxidation of ammonium to nitrate, and denitrification is the anaerobic microbial reduction of nitrate to nitrogen gas (N_2). Nitrous oxide is a gaseous intermediate in the reaction sequence of denitrification and a by-product of nitrification that leaks from microbial cells into the soil and ultimately into the atmosphere. One of the main controlling factors in this reaction is the availability of inorganic N in soil through additions of synthetic or organic fertilizers, manure, crop residues, sewage sludge or mineralization of N in soil organic matter following drainage/management of organic soils and cultivation/land-use change on mineral soils.

The main source of carbon dioxide in agriculture is tillage, which triggers emission of this gas through biological decomposition of soil organic matter. Tillage breaks the soil aggregates, increases the oxygen supply and exposes the surface area of organic material promoting the decomposition of organic matter. Use of fuel for various agricultural operations and burning of crop residues are the other sources of carbon dioxide emission. An off-site source is the manufacturing of farm implements, fertilizers and pesticides.

EMISSION OF GREENHOUSE GASES FROM INDIAN AGRICULTURE

In 2010, the global emission of GHGs was about 50 Billion ton (Bt) CO_2 eq. in which contribution of India was of about 2.34 Bt CO_2 eq. i.e., about 5% of the total emission. Agriculture contributed globally over 11% of the total GHGs emission. The share of Indian agriculture was about 7% of global emission. The updated inventory for the year 2010 showed that energy sector in India contributes the highest amount GHGs (65%) followed by agriculture (18%) and industry (16%) (Figure 1). Indian agricultural sector, including crop and animal husbandry, emitted 418 Mt of CO_2 eq. Enteric fermentation i.e., emission from ruminant animals contributed the highest (56%) amount of the emission from this sector, followed by agricultural soil (23%) and rice fields (18%) (Figure 1). Livestock manure management contributed 1% of the emissions and 2% was attributed to the burning of crop residues in field.

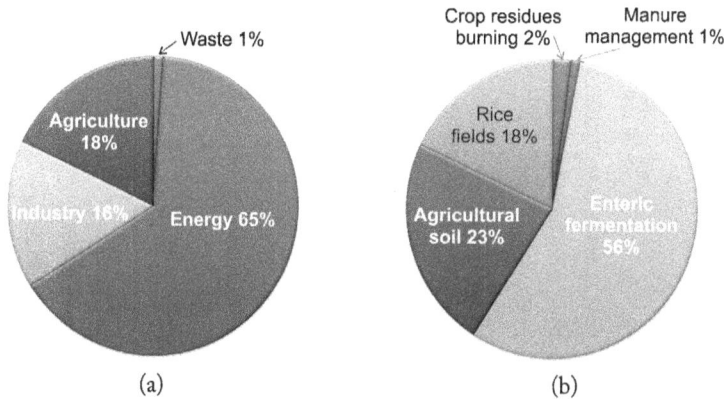

Fig. 1: Emission of Greenhouse Gases from (a) Various Sectors of Indian Economy and (b) Sub-Sectors of Agriculture

Source: Pathak (2015)

Several estimates of CH_4 emission from Indian rice fields have been made and the emission estimate has been rationalized from the earlier estimate of 37.5 Mt to 3.4 Mt (Figure 2). The highest emission was from the irrigated continuously flooded rice (34%), followed by rainfed flood-prone rice and irrigated single aeration (18%) (Bhatia *et al.,* 2013). Rainfed drought-prone, deep water and irrigated multiple aerations rice ecosystems contributed 16%, 8% and 6% of CH_4, respectively (Pathak, 2015). Fertilizer was the largest source contributing 77% to the total nitrous oxide emission. Emission of CH_4 and N_2O and global warming potential varies spatially at the national and state levels (Pathak, 2015). Emission of CH_4 ranged from 4 to 190 kg per hectare in different states. Higher CH_4 emission values observed in some of the states in southern and eastern India, are due to the continuous submergence of the fields under water and a higher soil organic C content. Emission of N_2O from the various states of India ranges from 0.18 to 9.11 kg ha^{-1}. Emissions of N$_2$O were higher from southern (Andhra Pradesh, Karnataka, Pondicherry) and northern (Punjab, Haryana) states of the country. The high emission of N$_2$O was due to large area under rice, and more use of N fertilizers. In northern India, rice is generally grown under intermittent drying and wetting conditions and considerable amounts of nitrous oxide emission could occur because of alternate wetting and drying of rice fields resulting in repetition of alternate nitrification and denitrification processes. The eastern and southern parts of the country have shown a higher global warning potential (GWP), mainly because of higher emissions of CH$_4$ and N$_2$O due to large area under rice cultivation. The highest GWP was observed in the states lying in the Indo-Gangetic Plains (IGP) as this is the major rice-wheat growing belt of the country with higher consumption of nitrogenous fertilizer and large area under rice cultivation.

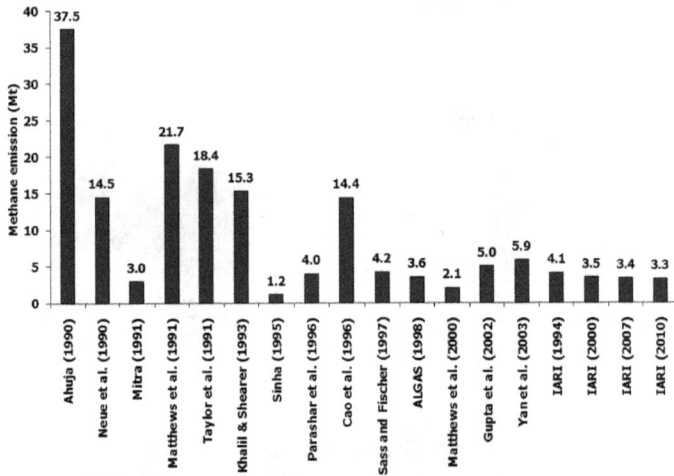

Fig. 2: Estimates of Methane Emission from Indian Rice Fields over the Years. IARI 1994, 2000, 2007 and 2010 are the Values Estimated by Indian Agricultural Research Institute, New Delhi for the Respective Years.

Adapted from Pathak (2015)

MITIGATING GREENHOUSE GASES EMISSION FROM AGRICULTURE

Mitigation in agriculture is possible through changes in management practices and technologies, replacing fossil-fuel based products with biomass energy, bio-plastics, and bio-fuel and sequestering C in soil. Emissions can be reduced by more efficient management of carbon and nitrogen in agro-ecosystems. Carbon can also be sequestered from the atmosphere and stored in soils or in vegetation. Crops and residues from agricultural lands can be used as a source of fuel to displace fossil fuel combustion, either directly or after conversion to fuels such as ethanol or diesel.

Agriculture is unique as it offers opportunities for mitigation from supply-side (management of land and livestock) and also from demand-side (C sequestration). Agriculture has the potential to mitigate GHGs cost-effectively through the adoption of changes in agricultural technologies and management practices. The IPCC (2014) has identified the following mitigation options in agriculture: (1) Plant management: Improved variety, rotation, cropping system; (2) Nutrient management: Type of fertilizer, inhibitor; (3) Tillage/residue management: Reduced tillage, residue retention; (4) Water management: Improved water application, drainage; (5) Land use change: Agro-forestry, bio-energy crops and (6) Biochar application.

Changes in animal agriculture that would reduce GHG emissions include raising fewer ruminants and shifting from ruminants to poultry, improving feeding and manure management practices and increasing efficiency in feed and livestock production. Other mitigation options include increased feeding efficiency, through improved forages, as well as dietary additives that suppress methanogenesis (IPCC, 2007; Smith *et al.*, 2008).

Table 1: Potential and Constraints of Greenhouse Gas Mitigation Options in Indian Agriculture

Technology	Mitigation (%)*	Yield (%)*	Requirements
Methane from rice field			
Intermittent drying	25–30	90–100	Assured irrigation
Direct-seeded rice	30–50	90–100	Sowing machine, herbicide
System of rice intensification	25–30	95–110	More labour, assured irrigation
Short duration variety	15–20	95–115	–
Nitrous oxide from soil			
Demand-driven N use	10–15	105–110	Knowledge intensive, tool needed
Nitrification inhibitor	10–15	105–110	Additional cost
Carbon sequestration in soil			
Conservation agriculture	5–10	95–110	Knowledge intensive, tool needed
Integrated nutrient management	5–10	95–110	Manure availability, additional cost

*Compared to the conventional practices.
Adapted from Pathak (2015).

CLIMATE CHANGE ADAPTATION OPTIONS AND THEIR MITIGATION CO-BENEFITS

Mitigation and adaptation are often viewed as separate activities. However, in agriculture, the adaptation measures can also generate significant mitigation effects, making them a highly worthwhile investment. It is unequivocal that concerted efforts are required for mitigation and adaptation to reduce the vulnerability of Indian agriculture to the adverse impacts of climate change and making it more resilient. However, mitigation options take a longer time to affect the climate change, whereas adaptation options are required almost immediately. The adaptive capacity of the poor farmers is often limited because of their subsistence agriculture and low formal education. Therefore, simple, economically-viable and culturally-acceptable adaptation strategies have to be developed and implemented. Fortunately, the adaptation strategies have several mitigation co-benefits (Pathak *et al.*, 2010). The co-mitigation benefits of such strategies, however, may be highly location-specific.

CONCLUSIONS

Agricultural practices are the major sources of GHG emission. But, agriculture can also mitigate GHG emissions through the reduction in N_2O and CH_4 emissions, as well as through carbon sequestration, particularly in the developing world. For Indian agricultural production systems to be viable in the future, there is a need to identify soil management systems that are climate change compatible, where soil organic C is enhanced or at least maintained and GHGs emission is reduced. It would require increased research and development efforts on mitigation and adaptation, capacity building, development activities and changes in land-use management. A win-win solution would be to develop such mitigation strategies that may help sustainable agricultural development such as increasing soil organic C content. Policies and incentives should be evolved that would encourage the farmers to adopt mitigation options, improve soil health, use water and energy more efficiently.

REFERENCES

Bhatia, A., Jain, N. and Pathak, H. (2013). Methane and nitrous oxide emissions from Indian rice paddies, agricultural soils and crop residue burning. *Greenhouse Gas Sci Technol, 3*, 196–211.

IPCC (2014). Climate Change: Mitigation of Climate Change. Contribution of Working Group III to the Fifth Assessment Report of the Intergovernmental Panel on Climate Change [Edenhofer, O., R. Pichs-Madruga, Y. Sokona, E. Farahani, S. Kadner, K. Seyboth, A. Adler, I. Baum, S. Brunner, P. Eickemeier, B. Kriemann, J. Savolainen, S. Schlömer, C. von Stechow, T. Zwickel and J.C. Minx (eds.)]. Cambridge University Press, Cambridge, United Kingdom and New York, NY, USA.

Pathak, H. (2010). Mitigating greenhouse gas and nitrogen loss with improved fertilizer management in rice: quantification and economic assessment. *Nutr Cycl Agroecosyst, 87*, 443–454.

Pathak, H. (2015). Greenhouse gas emission from Indian agriculture: Trends, drivers and mitigation strategies. *Proc Indian Natn Sci Acad, 81*(5): 1133–1149.

Production of Methane by the Livestock and its Mitigation Techniques

D.N. Kamra

ICAR, A.N. Division, IVRI, Izatnagar 243 122, India
E-mail: dnkamra@rediffmail.com

ABSTRACT: *Methane is one of the important green house gases and its production in India is 105 Tg/year, out of which about 60% is produced by enteric fermentation of ruminants, 92% of which is produced only by cattle and buffalo and 8% is generated by all other herbivorous animals like sheep, goat, yak, mithun, horse, elephants etc. The mitigation techniques of methane include plant secondary metabolites (tannins, saponins, essential oils and alkaloids containing plants), inorganic terminal electron acceptors (nitrates and sulphates), chemical composition of feed (concentrate/roughage ratio) and chemical and microbial interventions. These techniques individually or in combination, might lead to 30–40% methane inhibition.*

Keywords: Ruminants, Green House Gases, Methane Inhibition, Plant Secondary Compounds, Inorganic Compounds.

The ruminant feed is composed of ligno-cellulosic crop residues which the animal cannot digest by itself and therefore, depends upon the microbiome present in the gastro-intestinal tract of the animal, which converts the non-utilizable form of energy (lignocellulose) into a utilizable form of energy (volatile fatty acids) and microbial protein. During the course of feed degradation, hydrogen is generated in the form of reduced cofactors (NADH and NADPH) which are oxidized by reduction of carbon dioxide to methane. In the rumen of an adult cow/buffalo more than 200 liters of methane is produced every day. Several laboratories in India have estimated methane emitted by livestock and their figures also vary greatly, depending upon the method used for calculation (Kamra *et al.*, 2012; Patra, 2014). Swamy and Bhattacharya (2006) compared methane emission by Indian livestock. Recent analysis by the Department of Animal Husbandry, Dairying and Fisheries, Ministry of Agriculture, Govt. of India, indicate that the country is responsible for production of methane to the tune of 14.55 Tg/year (13.27 Tg from enteric fermentation and 1.28 Tg from livestock waste management), out of which cattle (6.73 Tg/year) and buffalo (6.56 Tg/year) collectively are responsible for 91.3% of total methane emitted by the livestock in India while the rest 8.7% is emitted by goats, sheep, yak and mithun.

In vitro screening of feed additives/their mixtures was done by *in vitro* gas production test (Menke and Steingass, 1988) using wheat straw and concentrate mixture in equal proportion by taking inoculum from buffaloes fed on the same feed. *In vivo* experiment was conducted in fistulated animals to study the biochemical and microbiological changes by real time PCR and diversity by metatranscriptomics.

CHEMICAL COMPOSITION OF FEED

There are many feed factors which influence methane production in the rumen. Methane production increases as the proportion of easily fermentable feed increases and lignocellulosic feed decreases. Methane production can be lowered by manipulating the feeding strategies of livestock. Improved nutritional quality of feed is one such method which can be used to inhibit methanogenesis. Animals grazing the lowest quality pasture will produce the highest amount of methane per unit product (Kirschgessner *et al.*, 1995). The treatment of paddy straw with urea resulted in a significant improvement in digestibility of nutrients and a significant depression in methane generation by the animals (Sahoo *et al.*, 2000). Even the different roughage sources like paddy straw, sugarcane bagasse and wheat straw caused different levels of methane production. Paddy straw supported more methane production than the other two roughage sources tested. Similarly lucerne caused less methane production than maize fodder (Kamra, 2003/2004). The feeding of high grain diet can be another option to reduce methanogenesis, but this option cannot be used practically in Indian conditions as coarse grain production is not increasing as per their requirement and whatever is available is used for the feeding of mono-gastric animals, especially poultry. Therefore feeding of grain to the ruminants cannot be recommended to the farmers.

PLANT SECONDARY METABOLITES

Plant secondary metabolites are compounds that provide protection to the plants against predators, pathogens, invaders etc. Several thousands of such metabolites have been identified. Majority of these compounds fall in the category of lignins, tannins, saponins, terpenoids/volatile essential oils, alkaloids etc. These plant secondary metabolites have anti-microbial activity, but their mechanism of action and inhibition of microbial growth is very specific and therefore these are active against a specific group of microbes. Their specificity against microbial groups can be used for selective manipulation of rumen fermentation. Therefore, there is a possibility that plant secondary compounds might act as a selective inhibitor of methanogens and can be used as a feed additive for the manipulation of rumen fermentation. The role of tannins, saponins and essential oils have been proved in inhibition of methanogens or the process of methanogenesis in the rumen (Agarwal *et al.*, 2009; Jayanegara *et al.*, 2009). In a project in Europe, a large number of plants have been screened *in vitro* for their potential to inhibit methanogenesis by the rumen microbes at the rate of 10% of substrate incubated in 120 ml serum bottles. A total of 35 plants inhibited methane more than 15% and six (*Carduus pycnocephalus* L., *Populus tremula* L., *Prunus avium* L., *Quercus robur* L., *Rheum nobile* Hook. F., Thoms and *Salix caprea* L.) more than 25% in comparison to their controls (Bodas *et al.*, 2008). Another major project has been funded by FAO/IAEA Joint Division, which ran for six years (2003-2009) at 8 international locations on using rumen molecular biological techniques to search for methods to inhibit methane emission by the livestock. The Indian participant of the project (IVRI, Izatnagar) worked on plant secondary compounds to determine their potential as a rumen modifier to reduce methane emission (Kamra *et al.*, 2006, 2008, Patra *et al.*, 2006).

Plants rich in secondary metabolites (saponins, tannins, essential oils, etc.) have antimicrobial activity which can be exploited for selective inhibition of methano-genesis. We have screened a large number of plant extracts for their potential to inhibit methanogenesis and ciliate protozoa in an *in vitro* gas production test using buffalo rumen liquor as the inoculum. A large number of extracts of ethanol, methanol and water were tested, 20 inhibited in alcohol and one water extract inhibited methanogenes. Methane inhibition was accompanied by a drastic fall in the number of methanogens as determined by real time PCR. Plants that appeared to have some potential as feed additives to control methanogenesis by the ruminants are: (i) seed pulp of *Sapindusmukorossi* (rich in saponins) and *Terminalia chebula* (rich in tannins); (ii) leaves of *Populus deltoides*, *Mangifera indica* and *Psidium guajava* (rich in tannins and essential oils); and (iii) flower buds of *Syzygium aromaticum* and bulb of *Allium sativum* (rich in essential oils). Some of the plants reported in literature exhibiting antimethanogenic activity include *Equisetum arvense*, *Lotus corniculatus*, *Rheum palmatum*, *Salvia officinalis*, *Sapindus saponaria*, *Uncaria gambir* and *Yucca schidigera* (Kamra *et al.*, 2008).

ALTERNATE ELECTRON ACCEPTORS

Fumaric Acid

Fumaric acid, a precursor of propionic acid, acts as an alternate sink for consumption of hydrogen generated during fermentation of feed in the rumen. When used as a feed additive, fumaric acid reduces methane emission, but the reports so far indicate that the results are not consistent and may vary from one laboratory to the other. But it does act as an alternate hydrogen sink in the rumen, as its inclusion in the diet increased total VFA concentration, increased propionate proportion and decreased acetate: propionate ratio (Beauchemin and McGinn, 2006), but the levels required to inhibit methanogenesis to a significant extent might cause a drop in pH which might affect feed fermentation adversely. Wallace *et al.* (2006) reported that by encapsulating fumaric acid in a shell of hydrogenated vegetable oil prevented a fall in pH, but retained the ability of inhibiting methanogenesis.

Nitrate

Nitrate and nitrite reducing bacteria in the rumen of goat are: *Selenomonas ruminantium, Veillonella parvula* and *Wollinella succinogenes*. Not all the types of these bacteria are able to reduce NO_3/NO_2, but only a small fraction of the total number of these three genera has been found to be able of reducing NO_3/NO_2 to ammonia. The reduction of nitrate to ammonia is accomplished through nitrite as an intermediate. But the reduction of nitrate is 2–3 times faster than that of nitrite, resulting in accumulation of nitrite in the rumen. Therefore, reduction of nitrite in the rumen should be enhanced to avoid intoxication of other rumen microbes. Nitrite combines with hemoglobin to convert it into methemoglobin which is a poor oxygen carrier. This leads to nitrite toxicity, but once the nitrite is converted to ammonia, it can be incorporated in amino acids and is available to animals as protein.

Nitrate inhibits methanogenesis significantly in sheep at the rate of 1.3 g $NaNO_3/kg\ W^{0.75}$ of the animals (Sar *et al.*, 2005), but it was accompanied with higher levels of plasma nitrite and blood methemoglobin. These negative effects of nitrate feeding could be reduced or completely removed by feeding *Escherichia coli* culture at the rate of 150 ml (2×10^{10} cells/ml) which had a capability to reduce nitrite.

Adaptation of nitrate and nitrite reducing bacteria by inoculating NRBB 57 resulted in reduction of methane production as gradual adaptation of animals to nitrate occurs (Figure 1) (Sakthivel *et al.*, 2010).

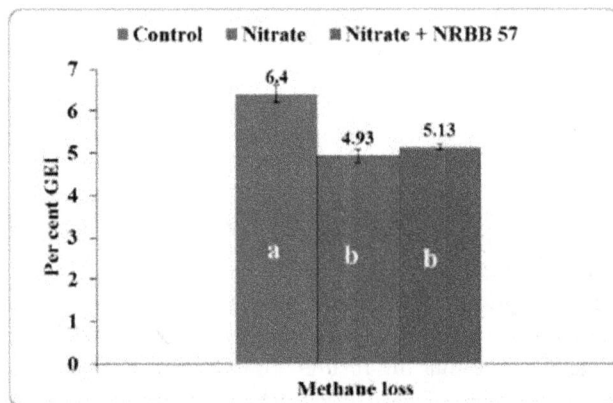

Fig. 1:   Effect of feeding Nitrate and Live Culture of NRBB 57 on Methane Energy Loss

MICROBIAL INTERVENTIONS

The microbial intervention includes suppression of methanogenic archaea, protozoa inhibition, reducing hydrogen generation or finding out an alternate hydrogen sink, where hydrogen generated during fermentation can be diverted to some more useful pathway where it can be brought back into the energy cycle through reductive acetogenesis.

The acetogens convert exogenous carbondioxide and molecular hydrogen (evolved during fermentation) into acetate. Thus it can serve as an alternate hydrogen sink in the rumen microbial eco-system. In comparison to methanogenesis, reductive acetogenesis has an added advantage that its end product (acetate) is an energy source for the animal, which is synthesized from energy less components (like carbon dioxide and hydrogen). One of the major reasons for the failure of conversion of carbon dioxide to acetate might be due to poor affinity of the reaction with molecular hydrogen as it requires higher hydrogen threshold level (1.26 *vs* 0.067 mmole/l) as compared to that required for methanogenesis. Efforts are needed to explore the biochemical pathways which determine the direction of hydrogen transfer either for methane generation or for acetate synthesis. If all the reducing power generated during rumen fermentation is diverted towards reductive acetogenesis, 5–10% of the gross energy intake by the animals can be saved and brought back into the energy cycle of the animals, which otherwise would have been lost in methane generation and released by the animals into the atmosphere.

The results of experiments conducted in our laboratory indicate that as ciliate protozoa remain attached with methanogens, elimination of protozoa results in inhibition of methanogenes, but it is never complete inhibition as all the ciliate protozoa are not attached.

METAGENOMICS

The methanogenic archaea are represented in the rumen by *Methanobrevibacter, Methanomicrobium, Methanosarcina, Methanomicrococcus, Methanosphaera, Methanobacterium* and *Methanoculleus* which are responsible for synthesis of methane in the rumen. Majority of the methanogens present in the rumen are hydrogen utilizers. The hydrogen generated in the rumen due to fermentation of feed is converted to methane by reduction of carbon dioxide and in this process 5–10% of gross energy intake of the animal is lost. If the activity of methanogens is limited and hydrogen is diverted to an alternate sink, the feed conversion efficiency can be improved and better livestock productivity can be attained with limited feed resources. *In vitro* methane production (ml/g DDM) was inhibited by 35.43 per cent without affecting in vitro true digestibility (IVTD) by using rumen liquor from additive fed animals as inocula as compared to control.

FUTURE RESEARCH NEEDS

- Search for newer feed additives or their mixtures which might inhibit methanogenesis without affecting the utilization of other nutrients.
- To identify feed additives which do not affect the quality of livestock products.

REFERENCES

Agarwal, N., Chandra, S., Kumar, R., Chaudhary, L.C. and Kamra, D.N. (2009). Effect of peppermint (*Mentha piperita*) oil on *in vitro* methanogenesis and fermentation of feed with buffalo rumen liquor. *Anim. Feed Sci. Technol.*, 148(2–4): 321–327.

Beauchemin, K.A. and McGinn, S.M. (2006). Methane emissions from beef cattle: effects of fumaric acid, essential oil and canola oil. *J. Anim. Sci,* 84: 1489–1496.

Bodas, R., L'opez, S., Fern'andez, M., Garc'ia-Gonz'alez, R., Rodr'iguez, A.B., Wallace, R.J. and Gonz'alez, J.S. (2008). *In vitro* screening of the potential of numerous plant species as antimethanogenic feed additives for ruminants. *Anim. Feed Sci. Technol.*, 145: 245–258.

Jayanegara, A., Togtokhbayar, N., Makkar, H.P.S. and Becker, K. (2009). Tannins determined by various methods as predictors of methane production reduction potential of plants by an *in vitro* rumen fermentation system. *Anim. Feed Sci. Technol.*, 150: 230–237.

Kamra, D.N. (2003/2004). Annual Report. Studies on interactions of fibre degrading microbes and methanogenic bacteria of rumen for reducing methanogenesis. IAEA/FAO Project, IVRI, Izatnagar, India.

Kamra, D.N., Agarwal, N. and Chaudhary, L.C. (2006). Inhibition of ruminal methanogenesis by tropical plants containing secondary compounds. *International Congress Series*, 1293: 156–163.

Kamra, D.N., Agarwal, N. and Chaudhary, L.C. (2012). NAIP project, ICAR, New Delhi.

Kamra, D.N., Patra, A.K., Chatterjee, P.N., Kumar, R., Agarwal, N. and Chaudhary, L.C. (2008). Effect of plant extracts on methanogenesis and microbial profile of the rumen of buffalo: a brief overview. *Australian Journal of Experimental Agriculture*, 48, 175–178

Kirchgessner, M., Windisch, W. and Muller H.L. (1995). Nutritional factors affecting methane production by ruminants. In: Ruminant Physiology: Digestion, Metabolism, Growth and Reproduction, (eds. Eng Aelhardt, W.V.; Leonhard-Marek, S.; Breves, G.; Giesecke, D.) Ferdinand Enke Verlag, Stuttgart, 333–348.

Menke, K.H. and Steingass, H. (1988). Estimation of the energetic feed value obtained from chemical analysis and in vitro gas production using rumen fluid. *Anim. Res. Dev.* 28, 7–55.

Patra, A.K. (2014). Trends and projected estimates of greenhouse gas emissions from Indian livestock in comparisons with greenhouse gas emissions from world and developing countries. Asian Australas. J. Anim. 27: 592–599.

Patra, A.K., Kamra, D.N. and Agarwal, N. (2006). Effect of plant extracts on *in vitro* methanogenesis enzyme activities and fermentation of feed in the rumen liquor of buffalo. *Anim. Feed Sci. Technol.*, 128: 276–291.

Sahoo, B., Sarawat M.L., Haque, N. and Khan, M.Y. (2000). Energy balance and methane production in sheep fed chemically treated wheat straw. *Small Ruminant Res.*, 35: 13–19.

Sakthivel, P.C., Kamra, D.N., Agarwal, N. and Chaudhary, L.C. (2010). Effect of sodium nitrate and nitrate reducing bacteria on *in vitro* methane production and feed fermentation with buffalo rumen liquor. Proc. Fourth Greenhouse Gases and Animal Agriculture Conference, Banff, Canada, p. 178.

Sar, C., Mwenya, B., Pen, B., Takaura, K., Morikawa, R., Tsujimoto, A., Kuwaki, K., Isogai, N., Shinzato, I., Asakura, Y., Toride, Y. and Takahashi, J. (2005). Effect of ruminal administration of *Escherichia coli* wild type or a genetically modified with enhanced nitrite reductase activity on methane emission and nitrate toxicity in nitrate infused sheep. *Brit. J. Nutr.,* 94: 691–697.

Swamy, M. and Bhattacharya, S. (2006). Budgeting anthropogenic green house gas emission from Indian livestock using country specific emission coefficients. *Current Science* 91: 1340–1353.

Wallace, R.J., Wood, T.A., Rowe, A., Price, J., Yanez, D.R., Williams, S.P. and Newbold, C.J. (2006). Encapsulated fumaric acid as a means of decreasing ruminal methane emissions. *International Congress Series.,* 1293: 148–151.

Greenhouse Gases Emission and Mitigation from Brackish Water Aquaculture Systems

M. Muralidhar*, M. Vasanth, V. Chitra, D. Thulasi, S. Kathyayani and R. Jannathulla

ICAR-Central Institute of Brackish Water Aquaculture,
75 Santhome High Road, R.A. Puram, Chennai 600 028, India
*E-mail: muralichintu@rediffmail.com, muralichintu@ciba.res.in

ABSTRACT: *Climate change is the greatest environmental threat of our time. Rapidly developing aquaculture sector is an anthropogenic activity, the contribution of which to global warming is little understood and estimation of Greenhouse Gases (GHGs) emission from the aquaculture ponds is a key practice in predicting the impact of aquaculture on global warming. A comprehensive methodology was developed for sampling through a fabricated floating chamber and simultaneous analysis of greenhouse gases (GHGs), carbon-di-oxide (CO$_2$), methane (CH$_4$) and nitrous oxide (N$_2$O) from the aquaculture ponds. GHGs emissions were quantified from shrimp and finfish species farming systems varying in stocking densities in different geographical locations of Andhra Pradesh and Tamil Nadu. GHGs emissions were high during summer and positively correlated with the stocking density. Environmental and management factors affecting the emission of GHGs and mitigation options to minimise the emission were also investigated.*

Keywords: Floating Chamber, Shrimp, Finish, Mitigation Aquaculture Ponds.

Greenhouse gases (GHGs), in one form or the other driven by anthropogenic activities are the root cause of the climate change (Brook *et al.,* 1996). Aquaculture production is playing an increasing role in meeting the demand for fish and other fishery products. Brackish water aquaculture, a rapidly growing sector in India is more or less synonymous with shrimp farming. Indian seafood exports reached an all-time high to the tune of 5.52 billion USD in 2014-15 and farmed shrimps alone account for 51% of the total earnings (www.mpeda. gov.in). Aquaculture, like agriculture is an important man made source of GHGs concentration, the contribution of which has not been estimated and not included in any of the GHGs inventory. The stocking density of aquatic species, large amount of feed required for their growth, types of chemical inputs for their survival, water logged condition, plankton density and animal biomass clearly indicate that aquaculture is a major anthropogenic activity. Continuous measurement of GHGs directly from the culture ponds provides meaningful information on gases emission trends and help to combat climate change.

The global N_2O-N emission from aquaculture, based on the indirect calculations will increase from 9.30×10^{10} g in 2009 to 3.83×10^{11} g which could account for 5.72% of anthropogenic N_2O-N emission by 2030 if the aquaculture industry continues to grow at the present annual growth rate of 7.10%. In case of the GHG emissions measurement from agricultural fields, closed chamber technique is frequently used (Smith *et al.*, 2000; Mer and Roger, 2000). Floating chambers are commonly used to measure GHGs emission rates in wetlands (Purvaja *et. al.*, 2004; Zhang *et. al.*, 2005) and estuaries (Borges *et al.*, 2004). A free floating chamber (Figure 1a) was fabricated with acrylic material, to collect GHGs flux from the aquaculture ponds (Vasanth *et al.*, 2016).

GHGs SAMPLING AND ANALYSIS

GHGs were collected once in a month from triplicate ponds of shrimp and finfish culture ponds using the floating chamber. Air sampling pump (SKC universal, PCXR8 model), tedlar gas sampling bags, thermometer and silicon tubing were used along with floating chamber to collect GHGs (Figure 1b). The chamber was allowed to float freely in the pond and the samples were collected from the dyke of the pond at a height of 0.5 m from the surface of the water at different time intervals in tedlar bags by shifting the valves of three way stopcock. GHG samples are stable upto 3 days from the date of sampling when stored at $10 \pm 2°C$ in tedlar bags (Vasanth *et al.*, 2016). GHGs were estimated simultaneously with greenhouse gases analyser (Agilent 7890A with G1888 network headspace sampler) and the emission fluxes were calculated.

GLASS JARS FABRICATION

In order to study the effect of environmental and management factors on GHGs emission and to study the mitigation options, experimental microcosm conditions were created in the fabricated glass jars (Figure 1c) simulating the pond conditions.

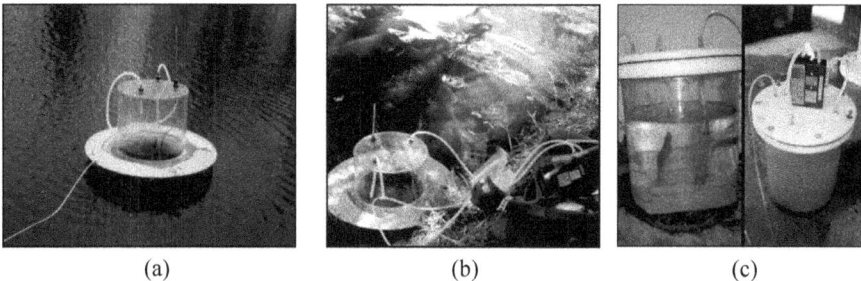

(a)   (b)   (c)

Fig. 1: GHGs Sampling from Aquaculture Ponds and Yard Experiments
a. Floating Chamber b. Collection of Gases Using Air Sampling Pump
c. Fabricated Glass Jar and Sampling from Microcosm Experiments.

QUANTIFICATION OF GREENHOUSE GASES FROM AQUACULTURE FARMING SYSTEMS VARYING IN SPECIES CULTURED, INTENSIFICATION AND SEASON

GHGs Emission were quantified from shrimp (Pacific White shrimp, Penaeus *vannamei* and Tiger shrimp, *P. monodon)* and finfish (Milk fish, *Chanos chanos;* Sea bass, *Lates calcarifer)* farming systems varying in stocking densities during summer and winter crops in different geographical locations of AP and TN from 2012 to 2016. Emission of all the three GHGs increased significantly ($P \leq 0.01$) with the days of culture and increase in stocking density ($r = 0.7, 0.81$ and 0.65 for N_2O, CH_4 and CO_2 respectively in shrimp farming systems). The correlation coefficient of 0.675 ($P \leq 0.05$) was registered between stocking density and overall GHGs emission in terms of CO_2 equivalent, a representative value for global warming. N_2O increased with the increase in dissolved inorganic N concentration in pond water. High GHGs emission in grow-out ponds of Sea bass compared to nursery ponds indicated the influence of biomass size on GHGs emission. In general, small lakes and ponds represent a net source of CO_2 to the atmosphere (Sobek *et al.,* 2005), and they are now recognized as important contributors to regional and global climate (Cole *et al.,* 2007). Brackish water culture ponds remain water logged and have a drying period of 3 months and are potential contributors of GHGs emission similar to wetlands and agricultural lands.

GHGs emission exhibited diurnal fluctuation in culture ponds (Figure 2). The GHGs emissions were high during the summer season compared to winter due to prevailing high temperatures and significantly reduced during rainy days. Microbial biomass carbon (MBC) in soil generally serves as an index for increased atmospheric CO_2 was significantly correlated with GHGs and soil organic carbon (OC). Diversity of methanogenic bacteria, responsible for CH_4 emission using Chao-1 richness estimator and Shannon index showed higher diversity and richness for methanomicrobiales (MMB) and methanococcales (MCC), in high saline culture ponds, whereas methanobacteriales (MBT) and methanosarcinales (MSL) were abundant in low saline culture ponds.

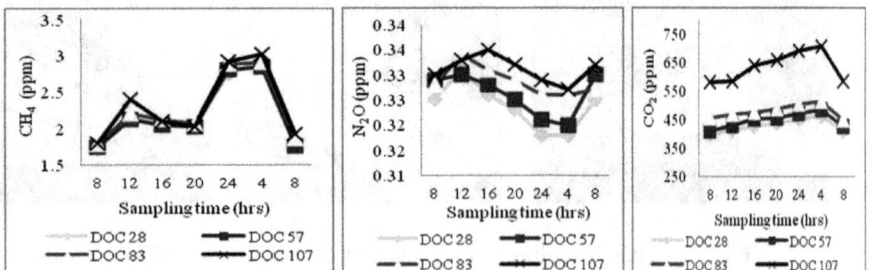

Fig. 2: Diurnal Variation in GHGs Emission

ENVIRONMENTAL AND MANAGEMENT FACTORS AFFECTING THE EMISSION OF GHGs

Several factors are responsible for the variation in the GHGs emission. Independent yard experiments in fabricated glass jars with varying factors such as i) different salinities, ii) redox potential (E_h) at varying salinity, iii) feeds varying in percent protein at two stocking densities, and iv) Fish Meal (FM) replaced feeds with plant protein sources (D1-FM 25%; D2-FM 20%; D3-FM 15%; D4-FM 10%; D5-FM 5%; D6-FM 0%) showed an increase in GHGs emission with the days in all the experiments. Inverse relation between salinity and methane, positive correlation between the feed protein percent and N_2O, and maximum CO_2 concentration in tanks fed with D1 & D6 and, least and maximum N_2O concentration in D1 & D6 respectively, were observed.

CONCLUSION

Methodology developed for the collection and simultaneous quantification of GHGs (CO_2, CH_4 and N_2O) emission from the aquaculture systems is useful to track the emission trends during the culture, quantify the contribution of aquaculture to global warming and develop mitigation strategies.

ACKNOWLEDGEMENT

The authors are thankful to Director of the Institute for providing the facilities and funding from National Innovations in Climate Resilient Agriculture (NICRA) project of Indian Council of Agriculture Research (ICAR).

REFERENCES

Borges, A.V., Vanderborght, J.P., Schiettecatte, L.F., Gazeau, F., Ferron-Smith, S., Delille, B. and Frankignoulle, M. (2004). Variability of gas transfer velocity of CO_2 in a macrotidal estuary (the Scheldt). *Estuaries,* 27: 593–603.

Brook, E.J., Sowers, T. and Orchardo, J. (1996). Rapid variation in atmospheric methane concentration during past 110000 years. *Science,* 273: 1087–1990. '

Cole, J.J., Prairie, Y.T., Carcao. N.F., Mc DSowell, W.H., Tranvik, L.J., Striegl, R.G., Duarte, C.M., Kortelainen, P., Downing, J.A. and Melack, J. (2007). Plumbing the global carbon cycle: Integrating inland waters into the terrestrial carbon budget. *Ecosystems,* 10: 171–184.

Hu, Z., Lee, J.W., Chandran, K., Kim, S. and Khanal, S.K. (2012). Nitrous Oxide (N_2O) emission from aquaculture: A review. *Environmental science & technology,* 46(12): 6470–6480.

Mer, J.L. and Roger, P. (2000). Production, oxidation, emission and consumption of methane by soils: A review. *European Journal of Soil Biology,* 37: 25–50

Purvaja, R., Ramesh, R. and Frenzel, P. (2004). Plant mediated methane emission from an Indian mangrove; *Global change biology,* 10(11): 1825–1834.

Smith, K.A., Dobbie, K.E., Ball, B.C., Bakken, L.R., Sitaula, B.K., Hansen, S., Brumme, R., Borken, W., Christensen, S., Prieme, A., Fowler, D., Macdonald, J.A., Skiba, U., Klemedtsson, L., Kasimir-Klemedtsson, A., Degorska, A. and Orlanski, P. (2000). Oxidation of atmospheric methane in Northern European soils, comparison with other ecosystems, and uncertainties in the global terrestrial sink. *Global Change Biology*, 6: 791–803

Sobek, S.L., Tranvik, J. and Cole, J.J. (2005). Temperature independence of carbon dioxide super saturation in global lakes. *Glob. Biogeochem. Cycles* 19: GB 2003, doi: 10.1029/2004GB002264.

Vasanth, M., Muralidhar, M., Saraswathy, R., Nagavel, A., Syama Dayal, J., Jayanthi, M., Lalitha, N., Kumararaja, P. and Vijayan, K.K. (2016). Methodological approach for the collection and simultaneous estimation of greenhouse gases emission from aquaculture ponds. *Environ Monit. Assess.* 188: 671. doi:10. 1007/s10661-016-5646-z.

Zhang, J., Song, C. and Yang, W. (2005). Cold season CH_4, CO_2 and N_2O fluxes from freshwater marshes in northeast China. *Chemosphere*, 59: 1703–1705.

Greenhouse Gas Exchange from Nordic Agriculture and Forestry: Perspectives on Climate Change Mitigation and Bioeconomy

Narasinha J. Shurpali* and Jukka Pumpanen

Biogeochemistry Research Group, Department of Environmental and Biological Sciences, University of Eastern Finland, Yliopistoranta, Kuopio, Finland
*E-mail: narasinha.shurpali@uef.fi

ABSTRACT: *The increase in the greenhouse gas concentration in the atmosphere leads to global warming and changes in climate systems. According to the future climate estimates projected by the Finnish Meteorological Institute, the average temperature in Finland could rise by 4–6°C and the average precipitation would grow by 15–25% by 2028. Extreme weather events, such as storms, droughts and heavy rains, are likely to increase. Owing to Finland being geographically located close to the arctic region, it is listed among the fastest warming biomes on the earth. Understanding what implications such changes will have on the country's agriculture, forestry and natural resources such as soil, vegetation, water and environment require data and models that help explain atmosphere-biosphere interactions under present climatic conditions. With this in view, the government of Finland has drawn a comprehensive strategy for mitigation and adaptation to climate change encouraging universities and research institutions across the country to generate high quality climate change research. We present here some case studies from past projects and describe a few ongoing projects that highlight the contribution of the Biogeochemistry Research Group at the University of Eastern Finland to the national climate strategy.*

Keywords: Eddy Covariance Technique, Finland, Peatlands and Drainage, Biochar, Carbon And Nitrogen Biogeochemistry.

The Nordic region, with its abundance of natural resources and advanced industry is relying more than ever on an economy built around bio-based resources (McKormick and Kautto, 2013). It is envisaged that by discovering and leveraging its strengths together, in a cross sectoral initiative, a positive impact can be made and thus a sustainable future for the Nordic region is ensured. Climate change, however, poses a serious threat to the region's ability to achieve these futuristic goals because, this region is among those regions of the world that are most vulnerable to the changing climate (Pedersen *et al.*, 2016; Easterling *et al.*, 2016). Agricultural and forestry associated land use options generate a substantial share of the total global greenhouse gas (GHG) emissions.

Actions to reduce net GHG emissions in the agriculture, forestry and other land use (AFOLU) sector provide valuable opportunities to build on and increase synergies with activities related to sustainable intensification, improved farm efficiency, climate change adaptation, food security and rural development. Therefore, with a broader goal of contributing to opportunities for mitigating climate change through reducing GHG emissions to the atmosphere, our research group at the University of Eastern Finland has been involved in several past and on-going research projects funded by the Finnish national and EU funding agencies. The objective of this paper is to highlight key research findings from the past projects with the aim of understanding the impact of climatic variability on the carbon and nitrogen biogeochemistry of investigated ecosystems and to introduce our on-going research work on the topic.

GHG EMISSIONS FROM A PERENNIAL BIOENERGY CROP

Finland is covered by nearly 11 M ha of peatlands. These peatlands store vast amounts of carbon. In their natural state, they have been shown to be sinks for atmospheric CO_2 (Turunen *et al.*, 2002) and sources of methane on a long-term basis. However, when they are drained for purposes such as forestry, agriculture or peat extraction for energy production, the thickness of the aerobic soil layer increases. This leads to an increase in the rate of organic matter decomposition. As a result, peatlands which are in general a sink for carbon are turned into atmospheric carbon sources after drainage (Minkkinen *et al.,* 2002). Drained organic soils have been found to be persistent sources of atmospheric carbon. Despite being cultivated with agricultural crops such as barley, wheat or potato or grasses, the organic soils in the boreal region have been reported to be emitting large amounts of CO_2 into the atmosphere (Lohila *et al.*, 2004). Drainage of peatlands for peat extraction for energy is a common practice in Fenno-Scandinavian region, Canada and Russia. These peatland areas abandoned after peat extraction are a threat to the environment (Waddington *et al.*, 2002). Significant CO_2 emissions have been reported from afforested organic soils (Mäkirauta *et al.*, 2007), although there are hardly any studies worldwide on the fate of carbon at the entire ecosystem scale in these soil types. Organic soils have been included among the areas with high risk of significant soil carbon losses and hence have been banned from biomass production for bioenergy (OECD, 2007). Therefore, finding an environmentally sound after-use option for peat extraction areas is an important land use issue in Finland.

With this in mind we hypothesized that cultivating a perennial bioenergy crop on a peat extracted site (a drained organic soil) would change the site from carbon source to a sink. To test this hypothesis, we performed GHG measurements from 2004–2011 on such a site in eastern Finland (Figure 1). The site was cultivated with a perennial bioenergy crop (Reed Canary Grass (RCG), *Phalaris arundinaceae*, L.). Net ecosystem CO_2 exchange was measured using eddy covariance technique and nitrous oxide and methane emissions were measured using the static chamber technique (Shurpali, *et al.*, 2009, 2010, 2013).

Fig. 1: An Eddy Covariance Tower Installed during 2004–11 on a Drained Organic Site (in eastern Finland) Cultivated with a Perennial Bioenergy Crop (Reed Canary Grass)

Based on several years of continuous CO_2 balance measurements, we found that during wet years, the RCG cultivation system acts as a distinct sink. During dry years, however, the ecosystem acts as a weaker sink bordering the sink–source line. We also found that the net carbon balance after accounting for the harvested biomass was still a net carbon sink during wet years, implying that a part of the captured carbon is left over in the ecosystem. Thus, the RCG cultivation was more than a 'carbon-neutral' system. According to the 2007 IPCC report, regions in the highl atitudes have been projected to receive over 20% higher annual precipitation and therefore, a higher frequency of wet years under the changed climate. Therefore, the capacity of the RCG cultivation to act as a carbon sink would increase in future climatic conditions. The results from this study have positive implications for the use of RCG as a bioenergy crop on organic soils. The results based on the net CO_2 balance alone providea strong evidence to suggest that the cultivation of RCG on such problematic soils is a promising land useoption. The annual NEE of the peat soil growing RCG during the entire study period ranged from -9 to -211 g $Cm^{-2}yr^{-1}$ (the negative sign indicates the amount of CO_2 taken up by the ecosystem). These results highlight that a perennial bioenergy crop such as reed canary grass can be cultivated successfully on an organic soil to mitigate CO_2 loss.

Based on encouraging results from the study on the drained organic soil, we performed a similar study on a mineral soil in eastern Finland (Figure 2). The objective here was to assess the GHG balance on a different soil type so that such information would help us in developing better bioenergy systems on various soil types at the Finnish national scale. We found that the total ecosystem respiration at the mineral site increased with soil temperature, green area index and gross primary productivity. Annual NEE was -262 and -256 gCm^{-2} in 2010 and 2011 respectively (Saara *et al.*, 2016). Throughout the study period from July 2009 until the end of 2011, cumulative NEE of the RCG crop was -575 gCm^{-2}. Carbon balance and its regulatory factors were compared to the published results the site drained organic soil cultivated with RCG in the same climatic region. On this mineral soil site, the RCG had higher capacity to take up CO_2 from the atmosphere than on the drained organic site.

Fig. 2: Eddy Covariance Tower Installed during 2009–13 at a Mineral Soil Study Site Cultivated with a Perennial Bioenergy Crop (Reed Canary Grass) in Eastern Finland

INDO-NORDEN, A BIOECONOMY RESEARCH CONSORTIUM

The project, INDO-NORDEN, is funded through the Science and Technology call of the INNO INDIGO Partnership Program (IPP) on Biobased Energy. The project is scheduled to begin in April 2017. The proposed project aims to address both sub-topics of the call, Biofuels and From Waste to Energy with research partners from Finland (with Dr. Narasinha Shurpali as the coordinator of the project), India and Estonia. The EU and India share common objectives in enhancing energy security, promoting energy efficiency and energy safety, and the pursuit of sustainable development of clean and renewable energy source (Figure 3). The main objective of INDO-NORDEN is to investigate, evaluate and develop efficient processes and land use practices of transforming forest and agricultural biomass, agricultural residues and farm waste into clean fuels (solid, liquid or gas), by thermochemical or biochemical conversions.

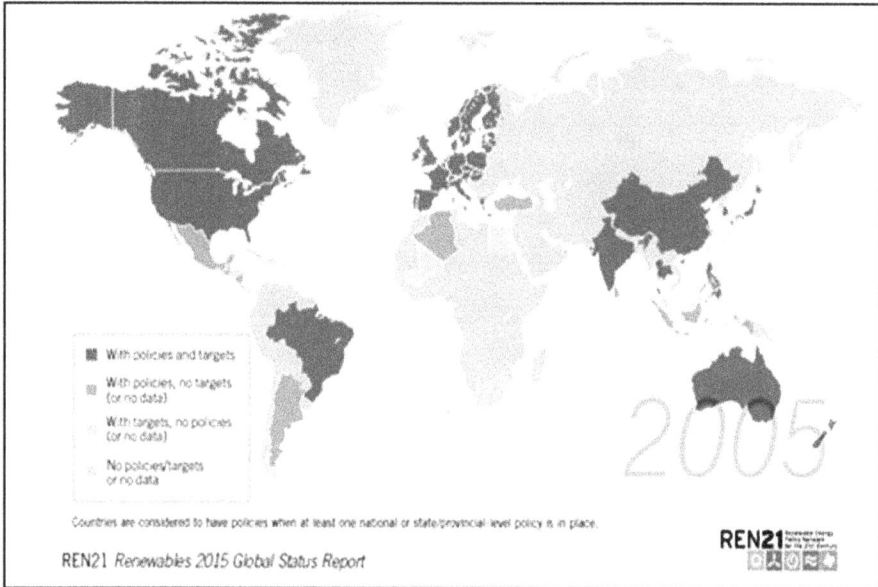

Fig. 3: The EU and India are among the Countries in the World
with Definite Renewable Energy Policies and Targets

With the work package structure (Figure 4) adopted in INDO-NORDEN, all forms of bio-based solutions (solid, liquid or gas) will be evaluated. Forestry and agriculture are the major bioenergy sectors in Finland. Intensive forest harvesting techniques are being used in Finland to enhance the share of bioenergy in the total energy consumption in the future. However, there are no clear indications how environmentally safe are these intensive forestry practices in Finland. We address this issue through field studies addressing the climate impacts on the ecosystem carbon balance and detailed life cycle assessment. For meeting Finland's bioenergy needs, the role of agriculture is expected to grow in the years to come. Here, we follow a holistic field experimental approach addressing several major issues relevant to Nordic agriculture under changing climatic conditions— soil nutrient management, recycling of nutrients, farm and agricultural waste management, biogas production potentials, greenhouse gas inventorying and entire production chain analysis. There is a considerable potential for process integration in the biofuel sector. This project plans to develop biofuel production processes adopted in Estonia and India with a major aim of enhancing biofuel yields. Additionally, the effects of biomass raw material on ash characteristics and behavior as well as on the fine particle and gas emissions in biomass-fired combustion plants will be evaluated. Thus, the project goes an extra mile in addressing both technological and environmental effects of bioenergy production with combustion processes. Finally, with a voluntary participation of companies with excellent track record in biogas production and CHP technology in participating countries, the project aims to bridge the gap between science, technology and industries.

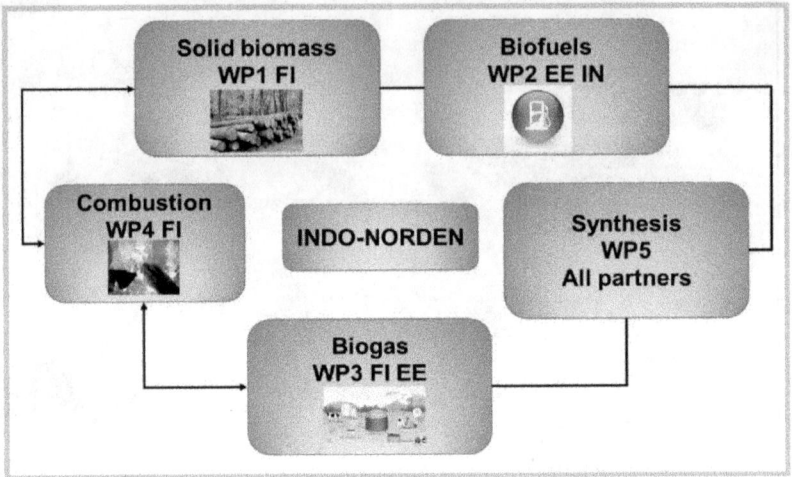

Fig. 4: The Work Package (WP) Structure of Indo-Norden

BIOCHAR AS A TOOL FOR IMPROVING SOIL QUALITY UNDER NORDIC CONDITIONS

Biochar (the charcoal produced from biomass by heating the biomass in conditions with low oxygen concentration) is used for improving the soil properties in agriculture, because it has positive effects on soil microbiology, water holding capacity (You-Ok-Youn *et al.*, 2013) and cation exchange capacity due to its porous structure and high surface area. It also increases soil pH leading to liming effect, which is often desired in acidic soils (Lehmann, 2007). However, the consequences of large-scale applications of fresh carbon compounds and their effects on the native soil organic matter pool are not yet understood. The long-term consequences of charcoal on the carbon and nitrogen cycle are largely unknown. The addition of large amounts of charcoal also affects soil pH, cation exchange capacity and water holding capacity and thus evidently alters the carbon and nitrogen turnover rate of the remaining SOM, which may ultimately affect the carbon storage capacity of the ecosystem resulting in feedback mechanism to carbon uptake and eventually atmospheric CO_2 concentration.

BIOCHAR, a research project awarded to Dr. Jukka Pumpanen by a Finnish national agency, aims to study the effect of charcoal (industrially produced biochar) on the decomposition of SOM in boreal forests. In this study, the changes in the size and quality of soil carbon and nitrogen pools because of large-scale charcoal addition to soil will be studied. The information obtained from this study can be used to assess the impacts of charcoal addition on soil carbon and nitrogen stocks and on the GPP of the forest and to the productivity of the forest in terms of wood biomass production. It will provide new and quantitative information on the effect of charcoal usage on soil carbon stocks and nutrient cycle of forest stands. This information is urgently needed for estimating the consequences of large-scale biochar utilization on soil CO_2 emissions and carbon stocks.

REFERENCES

Easterling, W., Aggarwal, P., Batima, P., Brander, K., Erda, L., Howden, M., Kirilenko, A., Morton, J., Soussana, J.-F., Schmidhuber, J. and Tubiello, F. (2007) in Climate Change 2007: *Impacts, Adaptation and Vulnerability*, eds Parry, M.L., Canziani, O.F., Palutikof, J.P., van der Linden, P.J., Hanson, C.E. (Cambridge Univ Press, Cambridge, UK), pp. 273–313.

Lehmann J. (2007). Bio-energy in the black. *Frontiers in Ecology and Environment* 5(7): 381–387.

Lohila, A., Aurela, M., Tuovinen, J.P., Laurila, T. (2004). Annual CO_2 exchange of a peat field growing spring barley or perennial forage grass. *Journal of Geophysical Research* D: Atmospheres, 109, D18116, doi: 10.1029/2004JD004715.

Mäkiranta, P., Hytönen, J. and Aro, L. *et al.* (2007). Soil greenhouse gas emissions from afforested organic soil croplands and cutaway peatlands. *Boreal Environment Research,* 12, 159–175.

McKormick, K. and Kautto, N. (2013): Bioeconomy in Europe. An overview. Sustainability 2013, 5.

Minkkinen, K., Korhonen, R., Savolainen, T. and Laine, J. (2002). Carbon balance and radiative forcing of Finnish peatlands 1900–2100 the impact of forestry drainage. *Global Change Biology*, 8, 785–799.

OECD (2007). Biofuels: Is the Cure Worse than the Disease? OECD Sustainable Development Studies. OECD Publications, Paris.

Pedersen, M.W. *et al.* (2016): Trends in marine climate change research in the Nordic region since the first IPCC report. *Climatic Change,* 134(1–2): 147–161.

Saara E. Lind; Narasinha J. Shurpali; Olli Peltola; Ivan Mammarella; Niina Hyvönen; Marja Maljanen; Mari Räty; Perttu Virkajärvi and Pertti J. Martikainen. (2016). Carbon dioxide exchange of a perennial bioenergy crop cultivation on a mineral soil. *Biogeosciences*, 13, 1255–1268.

Shurpali, N.J., Biasi, C., Jokinen, S., Hyvönen, N. and Martikainen, P.J. (2013). Linking water vapor and CO_2 exchange from a perennial bioenergy crop on a drained organic soil in eastern Finland. *Agricultural and Forest Meteorology* 168: 47–58 http://dx.doi.org/10.1016/j.agrformet.2012.08.006.

Shurpali, N.J., Hyvönen, N.P., Huttunen, J.T. *et al.* (2009), Perennial grass for bioenergy on a boreal organic soil – carbon sink or source? *Global Change Biology Bioenergy*, 1, 35–50.

Shurpali, N.J., Strandman, H., Kilpeläinen, A., Huttunen, J., Hyvönen, N., Biasi, C., Kellomäki, S. and Martikainen, P.J. (2010). Atmospheric impact of bioenergy based on perennial crop (reed canary grass, Phalarisarundinaceae, L.) cultivation on a drained boreal organic soil. *GCB Bioenergy* 2: 130–138.

Turunen, J., Tomppo, E., Tolonen, K. and Reinikainen, A. (2002). Estimating carbon accumulation rates of undrained mires in Finland – application to boreal and subarctic regions. *Holocene*, 12, 69–80.

Waddington, J.M., Warner, K.D. and Kennedy G.W., 2002. Cutover peatlands: A persistent source of atmospheric CO_2. *Global Biogeochem. Cycles*, 16(1).

Yu Ok-Youn; Raichle B. and Sink S. (2013). Impact of biochar on the water holding capacity of loamy sand soil. *International Journal of Energy and Environmental Engineering* doi: 10.1186/2251-6832-4-44.

Energy Conservation in Agriculture

K.K. Singh*, Anil K. Dubey and R.C. Singh

ICAR-Central Institute of Agricultural Engineering, Bhopal,
Madhya Pradesh 462 038, India
*E-mail: kk.singh@icar.gov.in

ABSTRACT: *India is the fastest-growing economy in the world. Energy sector is the most dominant source for greenhouse gas (GHG) emission with a global emission contribution of more than 75%. Global climate change has made immense pressure on Indian agriculture to produce more food with low production cost and at reduced environmental pollution. Climate change mitigation requires the use of new and energy efficient technologies which include usage of renewable energies and adoption of good management practices. In the country, about 5.2 million tractors and 25.45 million diesel or electricity operated pumps are in use. The diesel consumption in agriculture sector accounts for 14% in which tractors, pumps and other machineries consume 7.6, 3.3, 3.1%, respectively. The pumping systems used in field are inefficient and consume 50% more electricity and diesel irrespective of their actual consumption. The high energy intensive operations need to curtail excessive energy consumption in production agriculture for conserving energy and reducing the production cost to make farming profitable. A large variation is observed in energy requirement with varying production practices adopted in different regions. The specific energy requirement in production of wheat, paddy and maize crops are 5.10, 5.48 and 4.65 MJ/kg respectively. The reduction in specific energy consumption by 1.0 MJ/kg adopting improved machinery and optimized production practices, without compromising the productivity level, the country would be able to save energy in food grain production equivalent to 10 million tonnes of coal per year.*

Keywords: Climate Change, Energy Balance, Greenhouse Gas (GHG), Conservation Machinery, Energy Intensive.

Major challenge in agriculture during 21st century is to meet the food requirement for the growing population with limited per capita land availability without environmental degradation. In order to meet these growing demands, improved agronomical practices such as intensive tillage, optimized use of fertilizers, improved crop protection practices and crop residue management are being adopted. These practices are highly productive but are energy intensive, hence have contributed to a ten-fold increase in the global energy budget since the start of the 20th century and increase in anthropogenic emissions of GHG. The increase in energy inputs and GHG emissions in agriculture is mostly due to higher fossil fuel combustion during various farm operations. The growing concern on climate change has focused to reduce GHG emissions in agriculture. Saving in various inputs and increase in energy use efficiency in crop production is mostly needed as energy inputs (direct/indirect) and CO_2 emissions are directly related to each other. Further, increase in energy use efficiency and conservation of natural resources also offer opportunities for mitigation of climate change. Climate change mitigation involves efforts to reduce the emission of greenhouse gases with the use of new and energy efficient technologies, usage of renewable energies and adoption of good management practices. The shift in use of efficient technology, combined with the growth of renewable energy sources shows low carbon foot print with less emissions. In the current context of energy conservation and growing environmental concerns, conservation agriculture (CA) practices with reduced or zero tillage are essential. It would be able to reduce environmental pollution by reducing fossil fuel consumption which in turn reduces energy input and CO_2 emission with reduction in cost of cultivation. In recent years, minimum soil disturbance, residue retention and crop rotation have emerged as important management strategies to mitigate the GHG emission while maintaining crop productivity.

ENERGY SCENARIO IN AGRICULTURE

The increased mechanization in agriculture, crop production and rural agro-processing emerged as the major consumers of commercial energy. The overall mechanization level in India is only 40–45% even though 90% of the total farm power is contributed by mechanical and electrical power sources. The farm power availability has increased from 0.3 kW/ha (1971–72) to 2.02 kW/ha (2014–15). The demand of high speed diesel (HSD) has been increasing at the rate of 1.2% in agriculture sector to energize the various farm equipment and machinery. The increase in energy consumption is due to increase in farm mechanization, annual addition of tractors, and engines. In India, about 25.24 million pump sets energized with annual addition of 0.25 to 0.5 million pump sets. The electricity consumption has increased to two times in 2014–15 as compared to consumption level of the year 2000–01. Electricity consumption in agriculture sector has been increasing mainly because of greater irrigation demand for new crop varieties and subsidized electricity to this sector. Moreover, due importance is not given to proper selection, installation, operation, and maintenance of pumping sets, as a result of which they do not operate at the desired level of efficiency, leading to huge wastage of energy. The various study conducted by petroleum conservation research association (PCRA) has shown potential of energy savings in agriculture pumps by 25–30% (about 28 billion units).

ENERGY CONSERVATION

Over the last decade, India has been actively taking initiatives to promote energy efficiency in different sectors. It started with the enactment of the Energy

Conservation Act in 2001 and the establishment of the Bureau of Energy Efficiency (BEE) in 2002 to implement the provisions under the Act. Immediately after its formation, the BEE prepared an action plan for energy efficiency for the wider dissemination and implementation of standards set by the Bureau. The action plan gave a thrust to energy efficiency in the industrial sector, the setting of standards and labelling of appliances, agricultural and municipal demand-side management, energy efficiency in commercial buildings, capacity building of energy managers and auditors, energy performance codes and manuals preparation among others. Since its establishment, BEE has taken several initiatives across different sectors, based on the Energy Conservation Act and the action plan for energy efficiency. These energy efficiency schemes have been designed in such a way that their implementation would promote and improve energy efficiency for all categories of consumer of Indian electricity.

Though the initiatives taken by BEE are commendable, there is a paradox in the implementation of these initiatives: implementation is lower where the energy savings potential is higher. According to some estimates, the potential for energy savings is highest in the agricultural sector followed by the domestic (household) sector. The agricultural sector has a potential to save 27.79 billion kWh electricity, which is 30% of the sectoral consumption and more than 36% of total energy savings potential (NPC, 2009). According to this estimate, the agricultural sector offers the highest collective return in terms of energy savings and should be the priority area for energy efficiency initiatives. However, implementation by BEE shows a contradictory trend, whereby the industrial sector is being emphasised. Most of the BEE activities revolve around the industrial sector, while the agricultural sector is completely neglected though the potential for energy savings is highest in the agricultural sector. A possible explanation could be low incentive at the individual level for energy efficiency in the agricultural sector, although the collective incentive is high. Implementation of energy efficiency measures is higher when the individual incentive is high. The benefit accrued from energy efficiency for the individual owner of a pumpset in the agricultural sector is quite minimal, compared to that of an industrialist. This makes it difficult to motivate the pumpset owners to opt for energy efficiency and make the upfront investment. Also, the number of industrial and large commercial consumers is few, making it easier to target them. By contrast, the agricultural consumers are large in number and dispersed, making it difficult for BEE to reach them. BEE does not have the institutional capability to reach each and every consumer, but must work at this in coordination with other agencies. Finally, the technocratic orientation of the Indian electricity sector (Harrison and Swain, 2010) has led to an over-emphasis on technology based solutions for energy efficiency, even though promoting energy efficiency in the agricultural sector requires governance innovations along with technology.

ENERGY BALANCE OF SELECTED CROPS

The energy input-output of wheat, paddy, soybean and sugarcane crops in different growing conditions were analysed to highlight the energy intensive operation. This analysis indicates efforts needed for efficient utilization of various input resources and reduction in energy to reduce the GHG emissions. The energy consumption for wheat production in Madhya Pradesh is 8496 MJ/ha in which contribution of direct energy is 55% while indirect energy share is 45% of the total energy. In case of Punjab, total energy input for wheat production is 19364 MJ/ha. Seed bed preparation, irrigation and harvesting/threshing direct energy together contributes 47.57% of the total energy used while indirect energy share is 52.42% of the total energy. Total energy input for soybean production is 6280 MJ/ha comprising 47% direct energy and 53% indirect energy. The reduction in input energy would lead to the reduction in overall specific energy. Energy requirements for paddy crop in different states vary widely. The total input energy use in paddy cultivation is highest in Punjab (32892 MJ/ha) followed by Tamil Nadu (20726 MJ/ha), Uttar Pradesh (12482 MJ/ha), Odisha (11118 MJ/ha), West Bengal (8665 MJ/ha) and Madhya Pradesh (7510 MJ/ha). Fertiliser provides higher energy in Uttar Pradesh (about 52% of total input energy) while use of electricity is maximum in Tamil Nadu and Punjab as 33 and 31% of total energy input respectively. Operation-wise energy analysis indicates that irrigation, tillage and harvesting/threshing are major energy consuming operations. The operations such as irrigation, tillage and harvesting together consume more than 80 and 94% of the total operational energy in Maharashtra and Tamil Nadu, respectively. The total operational energy ranges from 6417 MJ/ha (Maharashtra) to 40851 MJ/ha (Tamil Nadu) while productivity ranges from 41133 kg/ha (Maharashtra) to 141120 (Tamil Nadu). Operation-wise energy consumption pattern of Tamil Nadu shows that farmers use more energy in irrigation operation to get the higher level of yield (Singh and Mittal 1992).

GHG EMISSIONS FROM AGRICULTURE SECTOR

Various agricultural activities such as land levelling, cultivation of crops, irrigation, animal husbandry, fisheries and aquaculture have a significant impact on the GHG emission (IPCC, 2014). The emissions from Indian agriculture have increased by 75% during 1970 to 2010. The increased use of fertilizers and other agricultural inputs along with rise in livestock population are the major sources to increase the GHGs emissions. The global emission of GHGs is about 50 billion tonnes of CO_2 equivalent in which contribution of India is about 2.34 billion tonnes *i.e.,* about 5% of the total emission which include contribution from agriculture sector over 11% of the total global GHGs emission. Indian agricultural sector including crop and animal husbandry emits 418 Mt of CO_2 equivalent. The details of emissions through different sources in agriculture are shown in Table 1.

Emission from ruminant animals (enteric fermentation) contributes the highest (56%) followed by agricultural soil (23%) and paddy fields (18%). Burning of crop residues in field contributes 2% and manure management contributes 1% of the emission. The annual growth of agricultural non-CO_2 emissions was 0.9% during 1970 to 2010. The paddy cultivation produces (9–11%) more CH_4 emissions followed by biomass burning (6–12%) and manure management (7–8%)

with average annual growth rates of 0.70% during 1961 and 2010 (FAO, 2013). Emissions from fertilizers grow at an average rate of 3.9% per year from 1961 to 2010 with absolute values more than nine-folds.

Table 1: Greenhouse Gas Emissions from Indian Agriculture in 2010

Source	Methane	Nitrous Oxide	GWP (CO_2 eq.)
Enteric fermentation	10.90	–	228.90
Manure management	0.13	0.08	27.50
Rice cultivation	3.40	–	85.00
Agricultural soil	–	0.26	77.80
Crop residue burning	0.30	0.01	9.60
Total	14.73	0.35	417.80

Source: Pathak (2015).

MEASURES FOR REDUCTION IN ENERGY USE IN AGRICULTURE

The use of various renewable and non-renewable sources of energy and technologies for agricultural operations like solar pumps, diesel engines, resource conservation machinery and practices are very effective to reduce the energy intensity in production and post-production agriculture. Deployment of new mitigation practices for agriculture and livestock systems and fertilizer applications will be essential to prevent an increase in energy use and emissions from agriculture after 2030.

- *Proper matching implements for the power source:* Tractor-implement operating efficiency depends considerably on how well the tractor and implement are matched. When ideally matched, there is less power loss, improved operating efficiency, reduced operating cost, and optimum utilization of capital on fixed costs. To improve operating efficiency, both units must be selected such that almost all of the power generated by the tractor is fully utilized under most operating conditions.

- *Conservation agriculture machinery:* Energy use for seed bed preparation, sowing/planting, and fertiliser application can be reduced by adopting energy efficient equipment such as laser land leveller, no till drill, happy seeder, rotary disc slit till drill, etc. Advances in weed control methods and farm machinery allow many crops to be grown with minimal tillage (reduced tillage) or without tillage (no-till). Since soil disturbance tends to stimulate soil carbon losses and soil erosion, the minimum or zero-tillage is increasingly being adopted, thus reduces the use of energy and often increasing carbon storage in soils. Conservation agricultural practices store carbon in the soil through retention of vegetative matter (crop residue). Since most conservation tillage practices reduce the number of trips across a field needed to grow and gather a crop, total energy required to grow a crop is reduced.

- *Equipment to reduce input of chemical fertilizers:* Variable rate fertilizer application with real-time sensing is the key for fertilizer management. Sensor based precision equipment, like sensor based variable rate fertilizer applicator and on-the-go variable rate urea application system integrated with spectral reflectance based sensor (Green seeker) have been developed at CIAE, Bhopal. An estimated 8–15% saving in urea is achieved with use of NDVI based variable rate fertilizer applicator in wheat and rice crops.

- *Renewable energy-operated machines for farm operations:* Major conventional power sources for agricultural machinery are fossil fuel-operated engines, which are major sources of GHGs emission. However, renewable energy sources such as solar photovoltaic (PV) have a lot of potential of being utilized to power irrigation pumps and small agricultural machineries for carrying out lighter farm operations. Solar PV-operated machinery can be used where electricity is neither available nor reliable. Machinery will be lighter in weight and can be easily transported from one location to another. PV panel can also be used for charging the battery to be used for powering the working elements. A considerable research and extension activities for the implements/machines discussed above are required for successful mechanization with improved energy efficiency.

- *Reduce energy use in irrigation:* Efficient irrigation measure can enhance carbon storage in soil through enhanced crop yields, reduced energy input and residue returns. Reduction in energy consumption for irrigation may be accomplished through use of energy efficient pump-sets and water-frugal farming methods. To improve the efficiency of irrigation pump sets, a number of technical measures are available. These include use of foot valves that have low-flow resistance; replacing undersized pipes and reducing number of elbows and other fittings that cause frictional losses; using high-efficiency pumps; selecting pumps better matched to the required lift characteristics; using rigid PVC pipes for suction and delivery; operating pumps at the recommended RPM; selecting prime mover for the pump (i.e., electric motor or diesel engine) matched to the load; selecting an efficient diesel engine or motor for the application; scheduling and performing recommended maintenance of the pump and the prime mover; and ensuring efficient transmission of mechanical power from the prime mover to the pump.

- *Real-time soil-moisture based sprinkler irrigation system:* A decision support system (DSS) based automatic controller for real-time irrigation scheduling has been installed at ICAR-CIAE, Bhopal for irrigation scheduling with sprinkler irrigation system. The capacitance-based soil moisture sensors (MP 406) are installed at 15 cm and 30 cm depths in irrigated wheat crop field. Soil-moisture data are collected using RM-Controller, radio telemetry and soil-moisture sensors in the experimental field. The real-time soil-moisture data recorded by the sensors are transmitted effectively to a PC through transmitter and receiver using the mobile phone network. It has been observed that variation of soil-moisture from 23% (lower threshold) to 34%

(upper threshold) has given maximum yield (5.1 to 5.5 t/ha) in the field (Singh *et al.*, 2012).

- **Drip irrigation in rice-wheat system:** Drip irrigation system was introduced in rice-wheat system as an innovative water saving device at the Project Directorate for Farming Systems Research (PDFSR), Modipuram. Drip irrigation in wheat @ 4 l/h, 40 cm spacing and 2.33 hours per irrigation in 5 times with 80, 120 and 160 kg N/ha at various critical growth stages and for rice 4 l/h and 30 irrigations at 3 days interval were applied. Highest mean grain yield (5.31 t/ha) of wheat has been recorded under drip irrigation with 120 kg N/ha. The water use efficiency is the highest (2.66 kg grain/m^3 water) under drip irrigation with 160 kg N/ha. Highest mean grain yield (5.45 l/ha), agronomic use efficiency of N (24.14 kg grain/kg N) and uptake efficiency of N (32.05 kg grain/kg N uptake) of rice has been recorded under flood irrigation with 120 kg N/ha. The water-use efficiency is highest as 0.79 kg grain/m^3 water, with 160 kg N/ha (Tiwari *et al.,* 2014).

- **Residue Management:** A large portion of the crop residues, about 140 million tonne, is burned in field primarily to clear the field from straw and stubble after the harvest of the preceding crop. Burning of agricultural residues, emits a significant source of chemically and radioactively important trace gases and aerosols such as CH_4, CO, N_2O, NO_X and other hydrocarbons to the atmosphere. It also emits large amount of particulates that are composed of wide variety of organic and inorganic species which causes severe environmental pollution. Under crop residues management practices, paddy crop residue incorporation increases the wheat yield (13%), net returns (15%) and B:C ratio (3%); but decreases energy output: input ratio (2%) and increases specific energy (2%) and specific cost (10%) compared to straw removal treatment. Zero till drilling has resulted maximum moisture content at all the growth stages of crop and also has maximum bulk density, cone index and mean weight diameter of aggregates than any other method of sowing under crop residue conditions. No and slit till drill also increases organic carbon (OC) by 13 and 9% and mean weight diameter of aggregates (MWD) by 49 and 20% after six crop cycles whereas conventional system reduces OC and MWD by 4 and 11% respectively (Singh and Sharma, 2005).

- **Strategies for saving electricity and HSD in agriculture:** The strategies for conservation to save electricity and HSD consumption in India for agriculture are given in Tables 2 and 3.

Besides these strategies most of the farms should be brought under micro-irrigation (drip and sprinkler based). A study indicates that only 3 million ha is under micro-irrigation (drip and sprinkler based) of the total 69.5 million ha of the potential (Rajput and Patel, 2010). Similarly, field studies undertaken in Punjab suggest that up to 40% reduction in input power for 5 hp electric pumps is possible through efficient pumping and stable electricity supply (TUV SUD, 2010). Ministry of Water Resources estimates that irrigation water-use efficiencies

Table 2: Strategies for Electricity Conservation in Agriculture

Energy Efficient Irrigation	*Possible Saving, %*	*Saving, GWh*			
		2020–21	*2030–31*	*2040–41*	*2050–51*
Design improvement in pumps	15–20	@2.5% 3972	@5% 9694	@7.5% 17166	@10% 26039
Matching drive, foot value, bends, pipes & fittings	8–15	@2.5% 3972	@5% 9694	@7.5% 17166	@10% 26039
Total saving in electricity GWh (% of total electricity consumption)		7944 (1.2)	19388 (2.3)	34332 (3.4)	52078 (5.0)
Solar pumping system (Additional % of total)	5–50%	@5% 7944 (1.2)	@10% 19388 (2.3)	@20% 45777 (5.0)	@30% 78118 (7.5)

Source: Singh and Jena (2014).

Table 3: Strategies for HSD Conservation in Agriculture

Energy Efficient Mechanization	*Possible Saving, %*	*Saving, Thousand Metric Tonne*			
		2020–21	*2030–31*	*2040–41*	*2050–51*
Matching machinery at rated capacity of prime movers	20–30	@ 5% 532	@10% 1294	@ 15% 2283	@20% 3502
Conservation agriculture	20–40	@5% 532	@10% 1294	@15% 2283	@20 3502
Diesel engines (BIS engine pump set) sfc = 200–250 g/bhp/h	8–15	@ 2.5% 266	@5% 647	@7.5 1370	@10 1751
Design improvement in pumps	15–20	@ 2.5% 266	@5% 647	@7.5 1370	@10 1751
Total saving, thousand metric tonne (% of total HSD consumption)		1596 (1.8)	3882 (3.27)	7306 (5.0)	10506 (6.0)

Source: Singh and Jena (2014).

vary between 35–40% for surface water and around 65–75% for ground water (Ministry of Agriculture, 2013). By 2050, about 22% of the geographic area and 17% of the population are estimated to be under absolute water scarcity. The per capita availability of water, which is 1704 cubic meters in 2010, is expected to fall to 1235 cm in 2050 (Ministry of Agriculture, 2013). In this context, energy and water-use efficiency can be identified as major game-changers for agriculture sector.

REFERENCES

BEE (2002). Agricultural Demand Side Management Programme. Bureau of Energy Efficiency, New Delhi.

FAO (2013). Tackling Climate Change Through Livestock: A Global Assessment of Emissions And Mitigation Opportunities. E-ISBN 978-92-5-107921-8 (PDF), FAO, Rome.

Harrison T. and Swain A.K. (2010). Manoeuvres for a Low Carbon State: India's Local Politics of Climate Change. Contemporary South Asia Seminar Series, Oxford Department of International Development, QEH, Oxford.

IPCC (2014). Climate Change: Mitigation of climate change. Contribution of working group III to the Fifth assessment report of the intergovernmental panel on climate change (Edenhofer, O., R. Pichs-Madruga, Y. Sokona, E. Farahani, S. Kadner, K. Seyboth, A. Adler, I. Baum, S. Brunner, P. Eickemeier, B. Kriemann, J. Savolainen, S. Schlömer, C. Von Stechow, T. Zwickel and J.C. Minx, Eds.). Cambridge University Press, Cambridge, United Kingdom and New York, NY, USA.

Ministry of Agriculture (2013). State of Indian Agriculture 2012–13. Department of Agriculture and Cooperation, pp. 29–36.

NPC (2009). State-wise Electricity Consumption & Conservation Potential in India. National Productivity Council & Bureau of Energy Efficiency, New Delhi.

Pathak, H. (2015). Greenhouse gas emission from Indian agriculture: Trends, drivers and mitigation strategies. Indian Agricultural Research Institute, New Delhi. *Proc. Indian National Sci. Acad.*, 81 (5), 1133–1149.

Rajput, T. and Patel, N. (2012). Micro Irrigation in India—Present Status and Future Scope. India Water Week 2012 – Water, Energy and Food Security: Call for Solutions. New Delhi.

Singh, C.D., Singh, R.C., Chandra, M.P. and Singh Ramadhar (2012). Telemetry assisted real time sprinkler irrigation scheduling based on soil moisture sensor in vertisol. *Journal of Crop Improvement*, Special Issue ISBN No. 0256-0933, pp. 99–100.

Singh, R.C. and Jena, P.C. (2014). Electricity and HSD Consumption in Indian Agriculture: Present and Future" paper presented in a interaction meet on National Consultation on Energy in Agriculture held at NASC, Complex New Delhi, during Nov. 15–16, 2014.

Singh, K.K. and Sharma, S.K. (2005). Conservation tillage and crop residue management in rice-wheat cropping system. In "Conservation Agriculture Status and Prospects" (Eds. Abrol IP, RK Gupta and RK Malik), Centre for Advancement of Sustainable Agriculture, New Delhi, p. 242.

Singh, S. and Mittal, J.P., 1992. Energy in production agriculture, Mittal Publications, New Delhi.

TUV SUD (2010). Pilot Agricultural Demand Side Management (Ag-DSM) Project at Muktsar & TaranTaran, Punjab-Detailed Project Report.

Tiwari, P.S., Tewari, V.K., Pal, S.S. and Singh, V.P. (2014). Final Report of NAIP sub-project on "Precision Farming Technologies base on Microprocessor and Decision Support Systems for Enhancing input Application Efficiency in Production Agriculture" Central Institute of Agricultural Engineering, Bhopal, pp. 1–115.

Global Warming Potential of Major Fishing Systems of India: A Life Cycle Assessment

Leela Edwin*, Renju Ravi and P.H. Dhiju Das

Central Institute of Fisheries Technology, Cochin, Kerala 682 029, India
*E-mail: leelaedwin@gmail.com

ABSTRACT: *Life Cycle Assessment (LCA) is a technique to assess environmental impacts associated with all stages of a product's life from cradle to grave. Here we attempt to highlight the utility of this technique for assessing the possible impact of different fishing systems on global warming. Primary data collected on the design and construction of fishing vessel, gear and other accessories used in trawl and ring seine fisheries were collected from fishermen and other stakeholders and used to relate to carbon dioxide equivalents. In both trawl and ring seine fishing systems the major contributor to the GWP was fuel followed by materials used for construction of fishing vessel and gear.*

Keywords: Fishing Systems, LCA, Global Warming Potential, Trawl, Ring Seine.

Marine fisheries contribute to nutritional security, livelihood and income generation to a large population in India. Fish production in India has shown an increasing trend during the last few decades. Globally, India ranked second in the world total fish production and the total marine fish landings production of India for the year 2015 is 3.40 million tonnes (CMFRI, 2016) of the nine maritime states and two union territories, Gujarat topped in landings with 7.22 lakh tonnes (21.2% share of the total), followed by Tamil Nadu with 7.09 lakh tonnes (20.9%) and Kerala with 4.82 lakh tonnes (14.2%). The following three categories of vessels operate in Indian marine fisheries, namely, mechanized (large boats with inboard engine), motorized (smaller boats with outboard engine) and non-motorized (traditional boats). There are about 1,99,141 fishing vessels in the sector, of which nearly 72,749 are mechanised vessels (36.5%), 73,410 are motorised (36.9%) and the rest 52,982 non-motorised (26.6%) (DADF, 2012).

The sector currently faces several sustainability issues such as overexploitation, pollution and habitat degradation. In recent years, concerns have been extended to environmental issues, and climate change in particular, has been recognized as one of the critical issues in fisheries. One of the characteristics of fishing is its dependence on fossil fuels and the resultant emission of Greenhouse Gases (GHGs). Fishing is considered as the most energy intensive food production method in the world. While the use of fossil fuels has increased the availability of fish to fisheries, the dependence of the fishing sector on fossil fuels raises concerns related to climate change, ocean acidification and economic vulnerability. When partial pressure of CO_2 increases in the atmosphere, miscibility in sea water increases, and the water absorbs more CO_2.

Using fuel consumption data has been recognized as a proxy to estimate CO_2 emission. For the estimation of CO_2 emission, diesel consumption was converted into CO_2 by considering the standard conversion factor that 1 litre of diesel produces 2.63 kg CO_2 (www.eia.doe/gov). All the mechanized and motorized fishing vessels in India use diesel or petrol for propulsion.

Life cycle assessment (LCA) is a technique to assess environmental impacts associated with all the stages of a product's life from cradle to grave. This technique is useful impact assessment methodology and its suitability for quantifying the impact associated with fisheries are proven (Pelletier *et al.,* 2007). Life cycle assessment (LCA) and carbon footprint (CF) studies will be useful for selecting energy efficient green fishing systems and for delineating approaches for fuel conservation in fishing operations. The aim of the present study was to assess the global warming potential (GWP) of trawlers and ring seiners operated from Kerala state.

Primary data collection on details on design and construction of fishing vessels, gears, operations, engines and other relevant information were collected from boat yard operators, net makers, fishermen and other stakeholders using structured questionnaire. The harbours and fish landing centres of Kerala were selected for the study, which was conducted during 2013–2015. Trawlers which forms 58% of the fishing vessels of the state were studied. Ring seiners which contribute to the major portion (51%) of fish landing of Kerala was also studied. The global warming potential was calculated in Carbon Dioxide Equivalents (CO2-Eq.). The residence time of the gases in the atmosphere is calculated for a period of 100 years and is customary.

The above fishing systems were selected for LCA analysis to be conducted on each of the systems separately. A total of 15 trawlers in the range 10.66 m to 27.43 m were selected of which six were wooden trawlers and nine were steel trawlers. A comparison has been attempted among two types of ring seine fishery *viz.,* mechanized and motorized ring seines operated in the same geographical space and time, to determine their environmental burdens. The analysis also included all activities pertaining to vessel construction, onboard equipment like purse winch, purse line reel, fishing gear construction, fuel production and consumption and all activities related to the manufacture of fishing accessories. The system boundary (Figure 1) defines which processes will be included in, or excluded from, the system and describes the processes and their relationships. LCA analysis for individual fishing unit (vessel and gear) and its operation was conducted using a cradle to gate approach. So in this study system, boundary has been limited to the point at which the catch reaches the harbour. Gabi 6 LCA software was used for analysis of global warming potential (GWP) the data. The data for each unit process can be classified as energy inputs, raw material inputs, ancillary inputs, other physical inputs, products, co-products, wastes, etc.

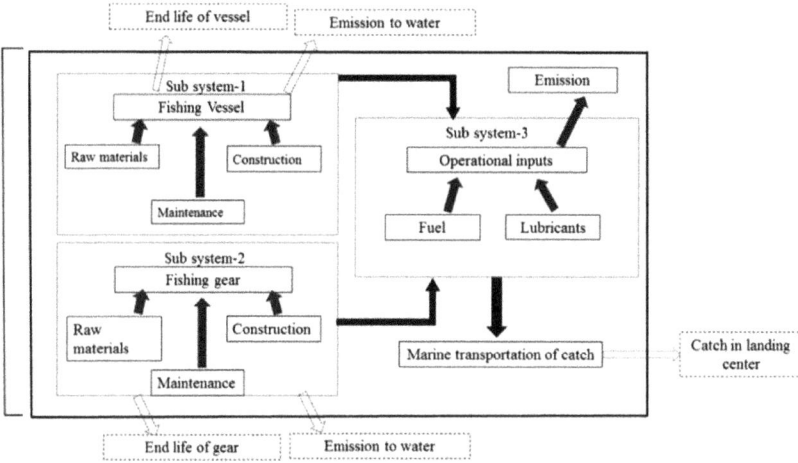

Fig. 1: Block Diagram of the System Studied

TRAWLERS

A comparison of global warming potential from different categories of trawlers is given in the Figure 2. Global warming potential ranged from 2165 to 4328 kg

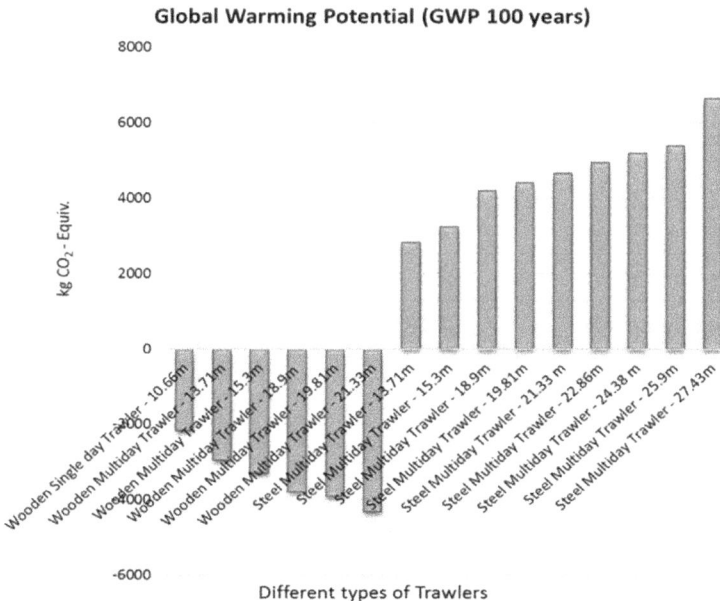

Fig. 2: Global Warming Potential from Different Categories of Trawlers

CO_2 Eq. in wooden trawlers and from 2824 to 6648 kg CO_2 Eq. in steel trawlers depending on the size. The GWP was higher in very large steel trawler due to inorganic emission to air especially carbon dioxide. In the case of 27.43 m very large multi day trawler, the GWP was highest. The GWP had a negative value for renewable resources i.e., wood used for construction, wooden otter board, marine plywood and cotton.

Among the materials used for construction of a 39.6 m trawl net GWP was maximum for iron sinker (64.6%) followed by high density polyethylene (HDPE) webbing (17.0%), polypropylene (PP) rope (10.3%), HDPE float (5.0%) and lead sinker (3.1%). The Life cycle inventory of trawl nets is given in Table 1.

Table 1: Life Cycle Inventory of Trawl Nets

	HDPE Webbing (kg)	PP Rope (kg)	Iron Sinkers (kg)	Lead Sinkers (kg)	Floats (kg)
39.6 m fish trawl	10.61	5.21	20	10	3.48
53.8 m fish trawl	90.00	7.59	40	20	28.00
72.0 m fish trawl	68.49	13.36	50	–	19.72
76.5 m fish trawl	67.75	13.91	30	20	23.86
81.0 m fish trawl	69.71	11.01	40	20	27.14
85.6 m fish trawl	224.86	19.91	45	25	32.14
33.4 m cephalopod trawl	75.55	5.26	30	10	19.72
45.6 m cephalopod trawl	81.75	8.95	–	60	28.00
54.0 m cephalopod trawl	83.60	9.82	30	20	19.43
57.6 m cephalopod trawl	36.68	7.84	–	45	16.95
34.2 m shrimp trawl	18.51	3.11	20	20	4.19
39.6 m shrimp trawl	10.61	5.21	20	10	3.48
40.0 m shrimp trawl	27.87	9.73	30	10	7.74
51.0 m shrimp trawl	21.00	12.67	25	8	13.27
58.0 m shrimp trawl	636.87	10.91	30	20	11.87

The fuel consumption is the major factor to GWP in both single day and multi day trawler operations and hence offers scope for impact reduction through operational fuel savings. The GWP was incrementally higher for multi day trawler operation corresponding to increase in size of trawlers. These observations are in agreement with earlier study conducted by Boopendranath (2000a, b, 2008, 2012), Ziegler *et al.* (2003), Thrane (2004), Hospido and Tyedmers (2005), Tyedmers *et al.* (2005).

RING SEINE

In case of fishing vessels in the ring seine sector, steel contributed 82.80% to GWP, FRP fishing vessel contributed 75.65% to GWP and in wooden fishing vessel GWP contribution shown negative value (–64.52%). Among motorized fishing vessels in the sector, FRP contributed a GWP of 50.53% whereas the contribution of wood is negligible and its GWP showed negative value (–21.26%). The ring seine fishing gear webbing made of polyamide contributed 77.94% to GWP. The GWP was 24% higher in the case of motorized vessel when compared to mechanized vessel as kerosene was the fuel used in outboard motors. The Global Warming Potential of fossil fuel used in both types of ring seine fishing vessel Figures 3 and 4.

LCA of Mechanised Ring Seine Unit
[kg CO₂-Equiv.]

$$\text{LCA of Mechanised Ring Seine Unit [kg CO}_2\text{-Equiv.]}$$

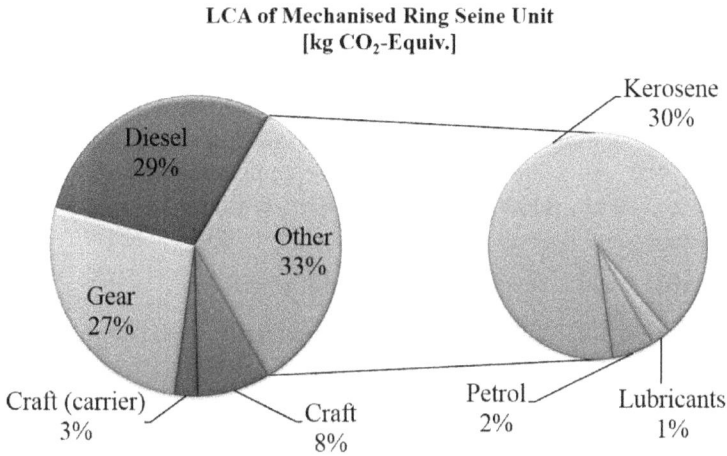

Kerosene 30%

Diesel 29%

Other 33%

Gear 27%

Craft (carrier) 3%

Craft 8%

Petrol 2%

Lubricants 1%

Fig. 3: Percentage Contribution of a Mechanized Ring Seine Fishing System in Operation to Carbon Emission

LCA of Motorised Ring Seine Unit
[kg CO₂-Equiv.]

$$\text{LCA of Motorised Ring Seine Unit [kg CO}_2\text{-Equiv.]}$$

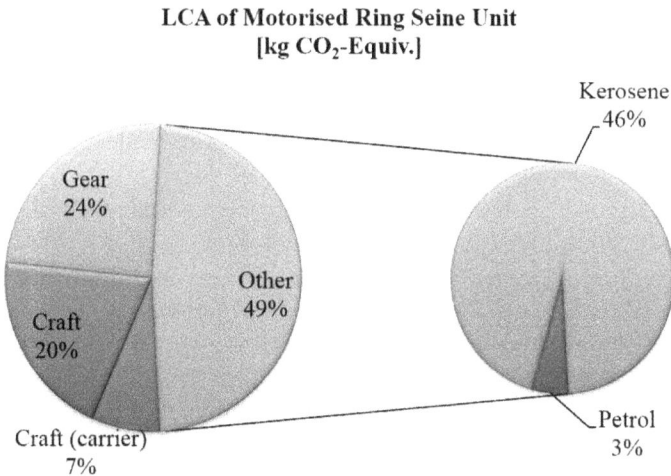

Kerosene 46%

Gear 24%

Other 49%

Craft 20%

Craft (carrier) 7%

Petrol 3%

Fig. 4: Percentage Contribution of a Motorized Ring Seine Fishing System in Operation to Carbon Emission

The higher environmental impact of the motorized fleets are mainly due to the operational issues like the intensive use of kerosene as fuel by inefficient outboard engines, whereas mechanized ring seiners are propelled by inboard engines run by diesel. A similar study conducted in Norwegian fleet shows that 90 kg fuel was consumed per tonne of mackerel landed (Ramos *et al.*, 2011). Another in Galicia on the horse mackerel fishery reports that 176 kg of fuel has been used for the production of one tonne of fish. Studies on Spanish tuna fishery shows 420 kg of fuel usage per tonne of production (Vázquez-Rowe *et al.*, 2010). To our knowledge there are no environmental impact studies based on CO_2 emission for any type of fishing method prevalent in the country.

Through this study some important interventions can be proposed for the improved efficiency of this fishery like fishing vessel hull optimization, reduction of engine rpm, periodic maintenance of hull and replacement of two stroke petrol engine to inboard diesel engine is recommended in order to reduce environmental impact related to fishing operation and knowledge about pelagic fish spatial distribution reduce the environmental impact by reduced shoal searching time. Use of high durability alternative webbing materials and appropriate use of lead sinkers will increase the life of gear and reduce the Global Warming Potential (GWP).

REFERENCES

Boopendranath, M.R. (2000a). Climate change impacts and fishing practices, Paper presented in Workshop on Impact of Climatic Change in Fisheries, 15th December 2008, ICAR, New Delhi.

Boopendranath, M.R. (2000b). Studies on Energy Requirement and Conservation of Selected Fish Harvesting Systems, Ph.D. Thesis, Cochin University of Science and Technology, Cochin, India.

Boopendranath, M.R. and Hameed, M.S. (2012). Energy analysis of the ring seine operations, off Cochin, Kerala. *Fishery Technology,* 49: 141–146.

CMFRI (2016). Annual Report 2015–2016. Central Marine Fisheries Research Institute, Cochin, p. 294.

DADF (2012). Handbook on Fisheries Statistics (2011), Department of Animal Husbandry, Dairying and Fisheries, Ministry of Agriculture, Govt. of India, New Delhi, October, 2012, p. 91.

Hospido, A. and Tyedmers, P. (2005). Life cycle environmental impacts of Spanish tuna fisheries. *Fisheries Research*, 76: 174–186.

Pelletier, N.L., Ayer, N.W., Tyedmers, P.H., Kruse, S.A., Flysjo, A., Robillard, G., Ziegler, F., Scolz, A.J. and Sonesson, U. (2007). Impact categories for life cycle assessment research of seafood production systems: reviews and prospectus. *Int. J. Life Cycle Assess.,* 12(6): 414–421.

Ramos, S., Vázquez-Rowe, I., Artetxe, I., Moreira, M.T., Feijoo, G. and Zufia J. (2011). Environmental assessment of the Atlantic mackerel (*Scomber scombrus*) season in the Basque Country. Increasing the timeline delimitation in fishery LCA studies. *Int. J. Life Cycle Assess.*, 16: 599–610.

Thrane, M. (2004). Energy consumption in the Danish fishery. *J. Ind. Ecol.*, 8: 223–239.

Tyedmers, P., Watson, R. and Pauly, D. (2005). Fuelling global fishing fleets. *AMBIO*, 34(8): 619–622.

Vázquez-Rowe I., Moreira M.T. and Feijoo G. (2010). Life cycle assessment of horse mackerel fisheries in Galicia (NW Spain): Comparative analysis of two major fishing methods. *Fisheries Research*, 106: 517–527.

Ziegler, F., Nilsson, P., Mattsson, B. and Walther, Y. (2003). Life cycle assessment of frozen cod fillets including fishery-specific environmental impacts. *Int. J. Life Cycle Assess.*, 8(1): 39–47.

Agri-Voltaic Systems of Solar Farming for Mitigating Climate Change in India

J.S. Samra

National Rainfed Agricultural Authority, GOI, New Delhi 110 012, India
E-mail: jssamra2001@yahoo.com

ABSTRACT: *Energy management is a primary concern of green house gasses liberation, global warming and mitigation of climate change. At present 61% of 308 GW of electricity is being generated from coal and energy demand is going to double (735 GW) of the existing installed capacity by 2047. It has been pledged by GOI to increase land demanding renewable energy from 13% at present to 40% by 2030. Photosynthetic process of harnessing bio-fuels or biomass converts hardly 3–4% of solar radiation into energy as compared to about 25% in photovoltaic process. There are technologies which are claiming even 40% conversion efficiency. Wind turbines are converting 40% of wind velocity into electricity. Co-generation of solar and wind energy has a potential of almost doubling the per ha electricity generation. Solar farming is yielding gross returns of ₹ 25 lakhs per ha per annum and land use for agriculture is going to be overtaken by wind and photo-voltaic farming. More than 20,000 farmers of Punjab have already offered their land for solar farming because of its very attractive economic returns. Inter cropping in between and below the wind turbines and solar arrays (agri-voltaic farming system) improved overall land productivity by about 30–40% especially of shade tolerant crops like lettuce and durum wheat in mediterranean region. This kind of agri-voltaic systems are also realizing environmental externalities by reducing wind erosion, soil and air temperature, evapo-transpiration, hail storms risks and conflicts about common lands with local communities. It requires further R&D for optimizing land use planning, leasing, contracting and sharing in various agro-ecologies and socio-economic situations.*

Keywords: Renewable Energy, Photosynthetic, Photo-Voltaic, Efficiency, Land Use Competition.

Energy is a vital engine of development and its per capita consumption is growing at a very fast rate in India. Historically wood was replaced by coal as a source of energy and subsequently other renewable sources were also considered to reduce carbon and global warming footprints. Ethanol (cereals), biodiesel (oilseeds) and other biomass production systems compete with food and inelastic land resources. Photovoltaic route of electricity generation from universally available solar radiation is presently 7–8 times more efficient as compared to photosynthetic (biomass production) route (Dupraz *et al.*, 2011).

Intercropping in between solar array and below photovoltaic panels has been experimented by Dupraz *et al.* (2011); Morrou *et al.* (2013 a, b & c), Cosson *et al.* (2014) and Harshavardhan, 2016 to optimise land use (Photo 2). Calvert and Malbee (2015) mapped land use conflicts among renewable sources for Ontario, Canada at the existing marketing rates. In good quality Class IV land photovoltaic capturing of solar radiation was 4–48 times more competitive than photosynthetic (biomass) systems under various situations. In poor quality Class VI land photovoltaic conversion of solar radiation into energy is still better than biomass production in the range of 13–72 times. Cost of photovoltaic generation (green energy) has fallen drastically due to better technologies and is equivalent to coal with unlimited potentials of its further cost reduction.

According to estimates of NITI Aayog (2015) and International Energy Agency (2015) energy demand in India is expected to reach 735 GW in 2047 which will be 2.3 times of the present installed capacity of 308 GW (upto Nov., 2016). At present about 61% of power is generated from coal and management of green house gases is a major concern of mitigation, adaptations and vulnerability of global warming (Table 1).

Table 1: Power Installed Capacity of India (MW) as on 30th Nov., 2016

Sl. No.	*Source*	*Installed Capacity (MW)*	*Percentage*
1.	Coal	1,87,803	61.0
2.	Hydro	43,133	14.0
3.	Wind	28,083	9.1
4.	Gas	25,282	8.2
5.	Solar	8,513	2.7
6.	Nuclear	5,780	1.9
7.	Biomass	4,882	1.6
8.	Small hydro	4,323	1.4
	Total	3,07,799 or 3.07 GW	

Significant climate change, higher frequency and intensity of devastating extreme weather events during 60 years (1951–2010) have been analysed by Rathore *et al.* (2013). Recently held COP 21 in Paris have also elaborated high priority to cut down the liberation of green house gases globally. Generation of renewable energy with very low carbon footprints is being advocated throughout the world (Harshavardan *et al.*, 2016; Nouhebel, S., 2005; Dupraz *et al.*, 2011). India has also pledged to increase its renewable energy generation from 13% at present to 40% by 2030 (30% by solar and wind power resources alone). Energy potentials of more than 10,000 GW of solar (more than 3 times of the present installed capacity) and 2000 GW of wind have been estimated by NITI Aayog (2015) and both can also be deployed by co-generation (Figure 1). However, land requirements @ two ha per MW for solar and 2.5 ha per MW for wind power will pose a serious competition for agriculture, food security and livelihood of rural communities (Calvert and Malbee, 2015). It has already been targeted to generate 1,00,000 MW of solar, 60,000 MW of wind, 10,000 MW of biomass and 5,000 MW small hydro power by 2022 in India (Table 2).

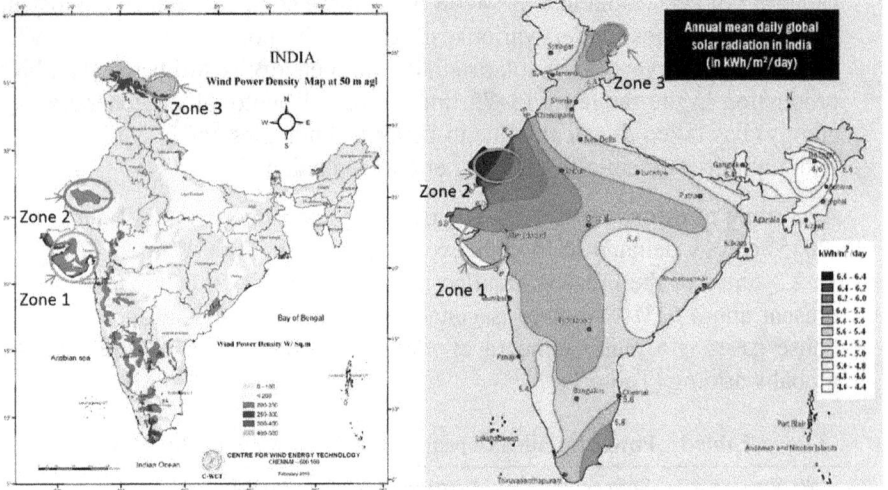

Fig. 1: Wind and Solar Energy Potentials of India

Table 2: Renewable Energy Target for 2022 of India

Sl. No.	Energy Sources	Power Generation (MW)
1.	Photovoltaic (Solar) generation	1,00,000
2.	Wind power	60,000
3.	Biomass	10,000
4.	Small hydro	5,000
	Total	1,75,000

Common or government wastelands are generally being spared for solar and wind parks, local communities are deprived of their traditional usufruct rights and various kinds of socio-economic conflicts are undesirable for the communal harmony.

Cost competitiveness and investment outlook of low carbon solar and wind energy is improving dramatically in India and even very productive and fertile lands are being leased by the private investors. Most of the government and common wastelands are generally in remote areas, in fragmented pieces, encroached or may be far away from the grid and are not acceptable to the private investors. A company has already leased land for 100 MW solar plant from 220 farmers in Bathinda, Punjab, @ ₹ 1,35,850 her ha per annum for 30 year period with annual escalation in the land rent. One Ha of land with investment of 2.5 crores can generate electricity worth ₹ 45 lakhs per annum @ ₹ 5 per unit. After accounting for interest of ₹ 20 lakhs @ 8% on 2.5 crores the per ha return of ₹ 25 lakhs per ha is better than agriculture land use only. Therefore, competition to food security is inevitable on economic considerations. It is therefore necessary to plan and promote hybrid system by converging or integrating generation of solar, wind power, inter-cropping and realization of some environmental externalities like reducing wind erosion, soil and air temperature and evapotranspiration and better soil moisture regimes in desert ecologies.

Cultivation of bio-fuels converts hardly 3 to 4% of solar radiations into energy by photosynthesis as compared to 25 to 30% conversion by photo-voltaic route and 50 to 60% per ha under co-generation of solar and wind. Some companies are claiming more than 40% efficiency of solar cells in their latest technologies. Wind turbines also convert about 40% of wind velocity into electricity. Therefore, harnessing of solar radiation and wind velocity into energy and intercropping for food production will be the best strategy to multiply overall productivity of limited land resources of India.

Agri-Voltaic System: This system is very similar to the agro-forestry systems of optimizing land productivity and diversifying of risks. Height of wind turbines has been raised from 50 m to 80 m or even 120 m due to technological development, solar panels can be installed above ground under wind towers and inter crops can also be raised on the ground. Photovoltaic panels and wind turbines are installed in arrays, rows and columns with intercropping in between and it becomes a three tiered system (Figure 2). This system reduces wind velocity, kinetic energy of rain drops, soil erosion, formation and shifting of sand dunes. Photovoltaic panels also protect intercrop against hail storms.

Fig. 2: Three Tiered Agri-Voltaic System of Wind Turbines, Photo-Voltaic Panels and Inter-Crops on the Ground

The same grid, service roads and other infrastructure can be shared by wind and solar energy, overall output increases and cost of generation decreases. Wind energy is generated mostly during night and solar power during day time and this convergence will improve grid supply also. Washing of the solar panel is essential and the same water resources can also be used for raising intercrops, forages and farmers can also be deployed for washing of panel, watch and ward.

Overall land productivity was increased by 30% as compared to solar farming alone or cropping alone. Land Equivalent Ratio (LER), varied from 1.2 to 1.5 in Agro-forestry system and still higher i.e. 1.3 to 1.6 in agri-voltaic system. Agri-voltaic system are being considered as controlled Environmental Agriculture of high productivity, better quality and environmentally benign preposition (Dupraz *et al.,* 2011; Harshavardhan *et al.,* 2016; Poncet *et al.,* 2012). As per European Parliament Directive 2009/28/EC France alone is obliged to dedicate more than 3 million hectare (10% of its agriculture) and Europe 25 million ha to bio-fuel stock production. Whereas photo-voltaic system is 150 times better than bio-diesel root of carbon foot prints (Podewils, 2007). Photovoltaic system (PV) are being integrated into green houses investments also. PV system have high Energy Return i.e. they generate 8 to 18 times more than the energy required in their manufacturing, installation and dismantling.

Shading, Micro-Climatic and Ecological Implications: Shading of the ground by photovoltaic panels, its shifting throughout the day and seasonal differences, consumtion of wind velocity by wind turbines are the major issues of managing agri-voltaic systems. Duproz *et al.* (2011) reported 43 to 71% reduction in photo-synthetic radiation (PAR) at different solar panel densities in Mediterranean region of Montpeller, France. This led to an average reduction of 29% of dry matter and 19% in durum wheat grain yield. This indicated relatively high photosynthetic efficiency of durum wheat grain as compared to total dry matter under shade. Cossou *et al.* (2014) reported 23% reduction in tomato productivity due to shading by photovoltaic panels. Morrou *et al.* (2013a), analysed 10–30% reduction in evapo-transpiration of lettuce when light availability was 50 to 70% of normal (open) and yield could be almost maintained due to increase in leaf size under partial shading conditions. Air, soil and crop canopy temperature and moisture regimes was also changed. Transparent PV panels could also be designed to minimize shading.

Land Equivalent Ratio (LER): This ratio depends upon the market prices of different components of agro-forestry and agri-voltaic system. Dupraz *et al.,* (2016) evaluated poplar-cereal agro-forestry system over 15 years and compared with agri-voltaic system at the same location. LER values ranged from 1.2 to 1.5 in agro-forestry and still higher (1.3 to 1.6) in agri-voltaic system. This indicated 30 to 60% overall higher productivity of land as compared to separate system each of photo-voltaic and crop production. Similar gains have also been reported in photo-voltaic green houses mainly due to high price of electricity generated as compared to food or bio-fuel crops. In agro-forestry system there is below ground competition of roots of trees and intercrops whereas such competition does not exist in agri-voltaic system.

REFERENCES

Calvert, K. and Mabee, W. (2015). More solar farms or more bio-energy crops? Mapping and assessing potential land-use conflicts among renewable energy technologies in eastern Ontario, Canada. *Applied Geography,* 56: 209–221.

Cosson, M., Murgia, L., Ledda, L., Deligios, P.A., Sirigu, A., Chessa, F. and Pazzona, A. (2014). Solar radiation distribution inside a green house with south oriented photo-voltaic roof and effect on crop productivity. *Applied Energy,* 133: 89–100.

Dupraz, C. Talbot, G., Marrou, H., Roux, S., Liagre, F., Ferard, Y. and Nogier, A. (2016). To mix or not to mix: evidences for the unexpected high productivity of new complex agri-voltaic and agro-forestry systems. www.dupraz@supagro.inra.fr.

Dupraz, C., Marrou, H., Talbot, G., Dufaour, L., Nogier, A. and Ferard, Y. (2011). Combining solar, photo-voltaic panels and food crops for optimizing landuse: Towards new agri voltaic schemes, *Renewable Energy:* 36: 2725–2732.

Harashavardhan, D. and Pearce, J.M. (2016). The potential of agri-voltaic systems. *Renewable and Sustainable Energy Reviews:* 54: 299–308.

International Energy Agency (2015). India Energy Outlook. www.worldenergyou tlook. org/india.

Morrou, H., Dufour, L. and Wery, J. (2013a). How does a shelter of solar panels influence water flow in a soil crop system? *European J. Agron.,* 50: 38–51.

Morrou, H., Guilione, L., Dufour, L., Dupraz, C. and Wery, J. (2013b). *Agriculture and Forest Meteorology,* 177: 117–132.

Morrou, H., Werfy, J., Dufour, L. and Dupraz (2013c). Productivity and radiation use efficiency of lettuce grown in partial shade of photo-voltaic panels. *European J. Agron.,* 44: 54–68.

NITI Aayog (2015). Report on India's Renewable Electricity Roadmap 2030: Towards Accelerated Renewable Electricity Deployment. www.niti.gov.in.

Nouhebal, S. (2005). Renewable energy and food supply. Will there be enough land? *Renewable and sustainable Energy reviews,* 9: 191–201.

Podewils, C. (2007). Organized wastefulness. *Photon International,* 04: 106–113.

Poncet, C., Muller, M.M., Brun, R. and Fatnassi, H. (2012). Photo-voltaic green houses, non-sense or a real opportunity for the Greenhouse systems? *Proc. XXVIIIth IHC-IS on Greenhouse 2010 and soilless cultivation (Ed) N. Castilla. Acta Hort,* 927, ISHS 2012.

Rathore, L.S., Attri, S.D. and Jaswal, A.K. (2013). State level climate change trends in India. Publ. Indian Meteorological Department, Lodi Road, New Delhi-110003, p. 147.

Carbon Sequestration is Not a Panacea for Climate Change

James Jacob

Rubber Research Institute of India, Rubber Board, Kottayam,
Kerala 686 009, India
E-mail: james@rubberboard.org.in

ABSTRACT: *Adaptation and mitigation strategies are urgently needed to address the issue of rising atmospheric CO_2 concentration. Both are socially and economically costly for poor and developing countries. Climate change mitigation aims at stabilizing and lowering atmospheric CO_2 concentration by reducing emission and increasing sequestration. The present analysis shows that CO_2 sequestration cannot be a panacea for climate change woes. At best, increasing this can slow down the rate of global warming, but this is not a permanent solution. Results of the first and the longest ongoing project on real time monitoring of eco-system level CO_2 flux measurements in India in a natural rubber plantation in Kerala are also discussed. Results show that almost 8% of the country's CO_2 emissions from the road transport sector is offset by CO_2 sequestration by natural rubber plantations. Almost 1.96% of the current rate of buildup of CO_2 concentration in the atmosphere is offset by CO_2 sequestered by rubber plantations globally. A market mechanism for incentivizing CO_2 sequestration activities will be beneficial.*

Keywords: Adaptation, Carbon Flux, Eddy Covariance, Natural Rubber, Synthetic Rubber.

Human activities releasing more amounts of greenhouse gases (GHGs) than what the planet can sequester result in their net accumulation in the atmosphere. It is estimated that close to 33 to 50% of the annual global emissions gets added to the atmosphere every year (Parikh and Parikh, 2016; Schimel *et al.*, 2001). It is also likely that a growing fraction of future emissions will be locked up in the atmosphere, because the geological and biological processes that remove CO_2 from the atmosphere have a finite capacity and this is reduced by climate warming (Reichstein *et al.*, 2013). Although there has been a small reduction recently (Keenan *et al.*, 2016), in general atmospheric CO_2 concentration has been increasing at an increasing rate aggravating global warming.

Adaptation and mitigation strategies are urgently required to address the growing threat from climate change. Adaptation can be financially, socially and politically very costly for the poor and developing countries whose historic contribution to climate change is insignificant (Jacob, 2005; Jakob *et al.*, 2014). The rich developed countries that are largely responsible for climate change have better technological and financial resources to adapt to climate change. Mitigating climate change by stabilising and reducing atmospheric CO_2 concentration is the permanent solution for climate change. Given the large dependence of the economy in poor and developing countries on cheap fossil fuels, reducing CO_2 emission by curtailing fossil fuel consumption amounts to hurting their developmental imperatives. Thus, neither in adaptation, nor in mitigation there is equity between the poor and the rich countries (Jacob, 2005).

The present study discusses how far mitigation through increasing sequestration be sufficient to stabilise and reduce atmospheric CO_2 concentration to a level that will not dangerously interfere with climate, as envisaged in the UNFCCC. It also discusses results from India's longest on-going study on real time monitoring of landscape level CO_2 sequestration in a natural rubber plantation at the Rubber Research Institute of India in Kerala using the *state-of-the-art* Eddy Covariance technique.

CLIMATE CHANGE MITIGATION

Climate change mitigation aims at stabilising and reducing atmospheric CO_2 concentration and this can be achieved by (i) reducing emissions and (ii) increasing sequestration. Reducing emissions without impacting economic growth calls for major shifts in the technology for production and use of carbon-neutral renewable energy, improving energy use-efficiency of equipment and automobiles, improving efficiency of primary production (including, agricultural and industrial production, storage, distribution and marketing), altering consumer behaviour (e.g. increased use of mass transit, consumer preference for renewables and products with small carbon footprint), preventing deforestation etc., (Edenhofer *et al.,* 2014). These emission reduction steps may be difficult to implement in poor and developing countries for social, political, economic and technological reasons (Jacob, 2015) and therefore the required scale to make any tangible impact on global carbon cycle may not be obtained.

MITIGATION THROUGH CO_2 SEQUESTRATION

Atmospheric CO_2 sequestration can occur through physical, geo-chemical and biological processes and the most abundant process is photosynthesis by terrestrial vegetation and marine ecosystems which amounts to 1.4 ± 0.7 and 1.7 ± 0.5 Pg C per year, respectively (IPCC, 2001). Marine ecosystems sequester and stock more carbon permanently than terrestrial ecosystems. Among the terrestrial sinks, forests and grasslands are a much bigger and more permanent sink than agricultural ecosystems. Globally, soils store far more carbon in a more permanent state than vegetation. Avoiding deforestation and increasing reforestation and afforestation are obvious mitigation options to reduce CO_2 emission and increase its sequestration.

Oceans sequester large quantities of CO_2 both through marine photosynthesis and physical absorption (IPCC, 2005). Both these processes will increase as atmospheric CO_2 concentration rises (CO_2 fertilization effect) which will also increase ocean acidification. Rising global temperatures will reduce solubility of CO_2 in water resulting in a reduction in the potential benefit of rising atmospheric CO_2 concentration on marine photosynthesis and physical dissolving of CO_2 in ocean waters. Increasing ocean acidification can lead to loss of stored carbon from coral reefs (Albright *et al.*, 2016). Similarly, as temperatures rise, carbon stored in arctic tundra, peat soils of rainforests and soil carbon stored almost everywhere else on the planet will be lost to increased microbial and physical oxidation (IPCC, 2013). Thanks to climate change, increasing incidences of wildfire as a result of severe drought and high temperature will also release more CO_2 into the atmosphere in future. Thus CO_2 sequestration by oceans and terrestrial ecosystems cannot be a panacea for rising concentration of CO_2 in the atmosphere.

Increasing carbon sequestration is just one of the numerous benefits of afforestation, preventing deforestation and forest degradation, improving agricultural practices, soil conservation *etc*. However, carbon sequestration through all these biological processes will be extremely inadequate to mitigate climate change as the amounts of CO_2 released through human activities are far more than the biological capacity of the planet to sequester it (Edenhofer *et al.*, 2014). More than one planet will be required to plant trees to sequester all the CO_2 that man is emitting today. At the present rate of global emissions, increasing CO_2 sequestration by planting more trees is a slow, inefficient and temporary solution to climate change. At best, planting trees may slow down the rate of global warming. Therefore, CO_2 sequestration through tree planting alone cannot mitigate climate change. Nevertheless, this should be promoted for the numerous co-benefits that tree planting will provide.

In theory, the physical process of carbon capture and storage (in geological formations etc.) from point sources that produce huge quantities of CO_2 (e.g., large fossil fuel power plants) has the potential to prevent CO_2 from getting emitted into atmosphere in large amounts (IPCC, 2005). Chemical scrubbing, deep sea injection and physical sucking of CO_2 from the atmosphere are also other concepts to reduce atmospheric CO_2 concentration. However, these technologies are yet to be perfected and their logistic and financial costs can be prohibitively high.

In the past, there was a school of thought among conservation groups as well as fossil fuel companies that planting more trees will be an adequate measure for combating climate change and some still follow this line. It is true that had it not been for the forests and plantations of the world, the rate of build-up of CO_2 in the atmosphere would have been much higher and faster than what it is today. Now the mantra is improving energy use efficiency.

There are number of good reasons to plant trees or improve energy use efficiency, but for combating climate change, these simply will not suffice. More green cover also can mean more absorption of solar heat energy and warming of earth's atmosphere. Improving fuel use efficiency is good, but when more people start using fossil fuel based equipment and automobiles as their purchasing capacity increases, the advantage of efficiency vis-à-vis total emissions is lost. Conscious efforts at reducing emissions in real terms will be the key to mitigating climate change and not just increasing CO_2 sequestration or improving fuel use efficiency. As things are what they are today, there will be little reduction in emissions unless the world reduces its fossil fuel consumption in absolute terms which is the single largest source of CO_2 emission (IPCC, 2013).

CO_2 SEQUESTRATION BY NATURAL RUBBER PLANTATIONS

Net CO_2 flux (which is the net balance between photosynthesis, respiration and microbial decomposition all occurring at the ecosystem level) from a natural rubber plantation was continuously measured by Eddy Covariance technique at the Central Experimental Station of Rubber Research Institute of India in Kerala (9° 26′N and 76° 48′E). This is the first such study in any ecosystem in India. This was started in March 2009 when the rubber plants were four years old. The system was commissioned on an 18 m tall flux tower and the sensors were fixed on the tower at 4 m above the canopy and these were moved up as the trees grew. Eddy Covariance method is a state-of-the-art micro-meteorological method through which the fluxes of CO_2 and water vapour and three-dimensional wind velocities are measured on real time basis (Baldocchi, 2003) and net ecosystem exchange (NEE) of CO_2 was calculated (Massman and Lee, 2002; Webb *et al.,* 1980) for each day since March 2009 and the study is still continuing.

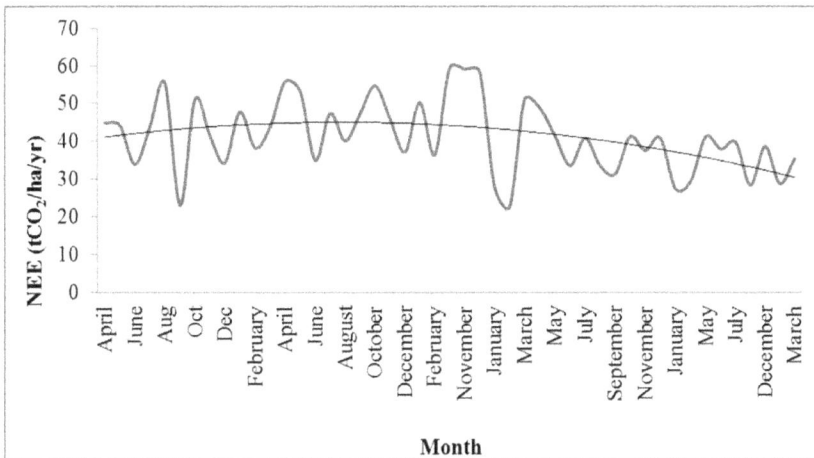

Fig. 1: Daily Net Ecosystem Exchange (NEE) of a Natural Rubber Plantation between March 2009 and December 2015 as Measured by Eddy Covariance Method

Over the six years of the study period since March 2009, there was a gradual increase followed by a slow decrease in NEE (Figure 1). On an average, the NEE during this growth period was about 40 ton CO_2/ha/yr. By extrapolating the trend for a 25 year plantation cycle, the mean NEE works out to roughly 30 ton CO_2/ha/yr. At the end of this period, there are around 350 to 375 mature trees in one hectare with a mean above-ground biomass of about one ton or so. Assuming that 45% of the biomass is carbon, the above mean NEE of 30 ton CO_2/ha/yr should produce a rubber tree weighing about 1.2 ton at the end of 25 years which is very close to what is normally realized from the field (roughly one ton per tree). The difference in biomass noticed could be attributed to root biomass, removal of self-thinning branches or fallen trees from the field as firewood etc. which are not accounted for here.

Road transport sector in the country releases about 260 million ton of CO_2 each year (Ramachandra and Swetmala, 2009). Taking a mean CO_2 sequestration rate of 30 ton CO_2/ha/yr, the 0.7 million ha of natural rubber plantations in India fix a total of 21 million ton CO_2 every year which is roughly 8% of the country's emission from the road transport sector. Globally, there is more than 10 million ha of rubber plantations which could be fixing more than 300 million ton of CO_2 every year which offsets the current rate of buildup of CO_2 in the atmosphere by around 1.96%.

Mitigation strategies may work well if a life style or practice that has a higher carbon footprint is taxed and the opposite is incentivised (Edenhofer *et al.,* 2014). For example levy a carbon tax on vegetarian food or use of own cars where public transit is available etc. Natural rubber produced by a tree has a carbon footprint that is far less than synthetic rubbers manufactured from petroleum bases. Synthetic rubber is a clear case that should attract a carbon tax. Conservation farming, zero tilling, improving fertilizer efficiency through practices like precision farming, organic farming, cover cropping *etc.,* help conserve soil carbon. A market mechanism for incentivising low carbon activities and products can boost CO_2 sequestration programs.

REFERENCES

Albright, R., Caldeira, L., Hosfelt, J., Kwiatkowski, L., Maclaren, J.K., Mason, B.M., Nebuchina, Y., Ninokawa, A., Pongratz, J., Ricke, K.L., Rivlin, T., Schneider, K., Sesboüé, M., Shamberger, K., Silverman, J., Wolfe, K., Zhu, K. and Caldeira, K. (2016). Reversal of ocean acidification enhances net coral reef calcification. *Nature,* 531: 362–365.

Baldocchi, D.D. (2003). Assessing ecosystem carbon balance. Problems and prospects of the eddy covariance technique. *Global Change Biology,* 9: 476–492.

Edenhofer, O., Pichs-Madruga, R. Sokona, Y., and others (2014). The IPCC special report "Climate Change 2014: Mitigation of Climate Change", UN reports, IPCC Working Group III Contributions to AR5, April 2014.

Fermann, G. (ed.), (1997). International politics of climate change: key issues and critical actors. Oslo: Scandinavian University Press.

IPCC (2013). Climate Change 2013: The Physical Science Basis. Contribution of Working Group I to the Fifth Assessment Report of the Intergovernmental Panel on Climate Change [Stocker, T.F., Qin, D., Plattner, G.-K., Tignor, M., Allen, S.K., Boschung, J., Nauels, A., Xia, Y., Bex, V. and Midgley. P.M. (eds.)]. Cambridge University Press, Cambridge, United Kingdom and New York, NY, USA.

IPCC (2005). IPCC Special Report on Carbon Dioxide Capture and Storage. Prepared by Working Group III of the Intergovernmental Panel on Climate Change [Metz, B., Davidson, O., de Coninck, H.C., Loos, M. and Meyer, L.A. (eds.)]. Cambridge University Press, Cambridge, United Kingdom and New York, USA.

IPCC (2001). Climate Change 2001: The Scientific Basis. Contribution of Working Group I to the Third Assessment Report of the Intergovernmental Panel on Climate Change [Houghton, J.T., Ding, Y., Griggs, D.J., Noguer, M., van der Linden, P.J., Dai, X., Maskell, K. and Johnson, C.A. (eds.)]. Cambridge University Press, Cambridge, United Kingdom and New York.

Jacob, J. (2005). The science, politics and economics of global climate change: Implications for the carbon sink projects. *Current Science*, 89(3): 464–474.

Jakob, M., Steckel, J.C., Klasen, S., Lay, J., Grunewald, N., Martínez-Zarzoso, I., Renner, S. and Edenhofer, O. (2014). Feasible mitigation actions in developing countries. *Nature Climate Change*, 4.

Keenan, T.F., Prentice, C., Canadel, J.G., Williams, C.A., Wang, H., Raupach, M. and Collatz, J. (2016). Recent pause in the growth rate of atmospheric CO_2 due to enhanced terrestrial carbon uptake. *Nature Communications,* 7: 13428.

Massman, W.J. and Lee, X. (2002). Eddy covariance flux corrections and uncertainties in long term studies of carbon and energy exchange. *Agriculture and Forest Meteorology,* 113: 121–144.

Parikh, K.S. and Parikh, J.R. (2016). Paris Agreement-Differentiation without Historical Responsibility. *Economic & Political Weekly,* LI 15: 21–25.

Roger, S.S. (2001). Sequestration of carbon by soil. *Soil Science,* 166(11): 858–871.

Ramachandra, T.V. and Swetmala (2009). Emissions from India's transport sector: State-wise synthesis. *Atmospheric Environment,* 43: 5510–5517.

Reichstein, M. and others (2013). Climate extremes and the carbon cycle. *Nature,* 500, 287–295.

Schimel, D.S. *et al.* (2001). Recent patterns and mechanisms of carbon exchange by terrestrial ecosystems. *Nature,* 414: 169–172.

Webb, E.K., Pearman, G.I. and Leuning, R. (1980). Corrections of flux measurements for density effects due to heat and water vapour transfer. *Quarterly Journal of Royal Meteorological Society*, 106: 86–100.

Spatial Biomass Estimation Using Remote Sensing Data and Advancements in Reduction of Estimation Uncertainty over Indian Forests

C.S. Jha*, R. Suraj Reddy, Rakesh Faroroda, G. Rajashekar and T.R. Kiran Chand

Forestry and Ecology Group, National Remote Sensing Centre,
Hyderabad 500 037, India
*E-mail: chandra.s.jha@gmail.com

ABSTRACT: *Forests contain about 80% of the global terrestrial above-ground carbon stocks and play a crucial role in the global carbon cycle by sequestrating the carbon emissions, produced due to anthropogenic activities. Tropical deforestation accounts for the one fifth of the total anthropogenic carbon dioxide (CO_2) emissions. Accurate estimation and understanding of biomass and its dynamics is therefore essential for terrestrial carbon accounting and climate change modelling. Forest above ground biomass (hereinafter referred as "AGB" or "biomass") provides the vital information about the total carbon in the ecosystem and also forest health. Thus, assessment of biomass plays an important role not only in conservation and sustainable forest management and enhancement of carbon stocks but also for analysing long term changes in Carbon (C) stocks (Houghton, 2005) and examining the dynamics of forest distribution.*

National level forest inventories provide the most extensive and accurate field observations of forest biomass. Although field measurements are a pre-requisite to generate any biomass database, they are expensive, time-taking and is often difficult in forests in terms of approach, terrain and accessibility. Field measurements in combination with remote sensing techniques provides maps of spatial biomass and its dynamics over wide geographical extent from regional to national scale (Saatchi *et al.*, 2007). Several methods exists to estimate spatial biomass from using different remotely sensed data based on the level-of-accuracy and the objective of the study. This paper describes various methods and their uncertainty over Indian forests.

STRATIFY AND MULTIPLY APPROACH

Stratify and multiply approach uses a simple but an effective technique to provide biomass estimate at regional and national scale. This method uses intense field data from national forest inventory (~6000 points across the country with plots of 0.1 ha) and secondary remote sensing data (forest density map) to estimate spatial biomass. Forest cover density maps were intersected with 5 km mesh and estimates of forest area, forest carbon density for each Agro-ecological sub region and forest carbon pools were linked to the 5 km grid coverage of India. At a 5 km level, this method had estimated a total forest carbon of > 4300 Tg C for entire Indian forests for 2014. This approach is a viable alternate to have quick estimate of entire area and understanding spatial C dynamics at a large scale but is often highly uncertain (~35%) due to limited description of forest using single parameter such as forest density.

SPECTRAL MODELLING APPROACH

In spectral modelling, canopy reflectance measured in different spectral ranges by passive optical and radar backscatter are related to biomass to establish a model and thereby provide a spatial biomass maps at regional level. However, these models hugely rely on image viewing geometry and acquisition season (dry or green) based on forest type. Since spectral modelling methods deal with relating pixel based reflectance values to biomass, a careful distribution of field plots should be ensured over different vegetation types and crown density. Jha *et al.* (2015), showed that the spectral models constructed for temperate (Sikkim), tropical (Tamil Nadu) and deciduous (Madhya Pradesh) forests of India using high temporal MODIS Normalised Difference Vegetation Index (NDVI) perform well ($R^2 > 0.7$) but are differed by choice of date in the satellite imagery. Spectral models constructed for temperate forest systems performed well using December NDVI whereas tropical and deciduous systems performed well using February NDVI. Further, the models constructed often depend on goodness of the fit and thus vary from a variety of regressions (power to exponential). Moreover, the models saturate with biomass of about 150–200 t ha^{-1}.

Radar backscatter acquired using synthetic aperture radars (SAR) had provided insights to estimate spatial biomass not only due to its capability to provide useful information during persistent cloud conditions over tropical forests but also since it gives structural information of the forest features which is more relevant to biomass. However, at low radar frequencies (L-band or P-band), higher the penetration and the backscatter relates significantly to biomass. A recent study estimates spatial biomass map of central Indian deciduous forests of Madhya Pradesh (M.P.) state, India, using advanced land observing satellite–phased array type L-band synthetic aperture radar (ALOS-PALSAR) L-band data of year 2010 and field based AGB estimates and empirical models (Thumaty *et al.*,

2015). A total of 415 sampling plots (0.1 ha) were used to establish an empirical model between HV backscatter and field biomass ($R^2 = 0.51$; Fig. **1**). The estimated model error is about 19.32 t ha^{-1} (i.e. ~32% relative to mean biomass value). It is also observed that the relationship saturates above 200–250 t ha^{-1}.

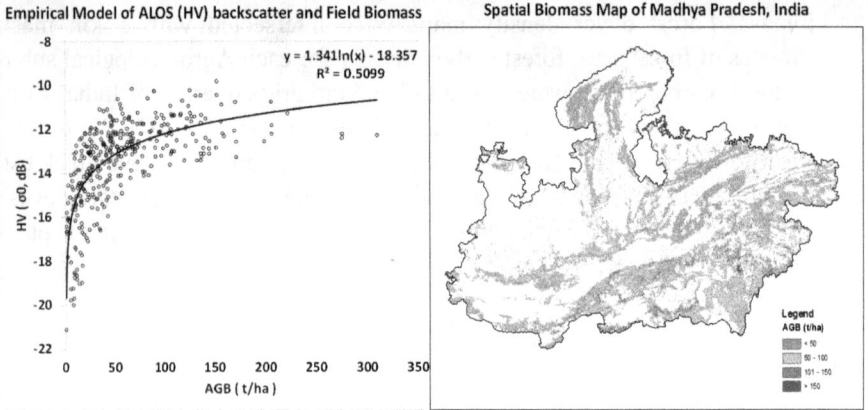

Fig. 1: Spatial Biomass Estimation using ALOS L-Band Data over Madhya Pradesh, India

Although, studies have shown that radar backscatter saturates at higher levels in comparison with optical data, the estimates are still highly uncertain. The reduction in uncertainty of biomass estimates is indeed highly necessary for carbon accounting over national scale. Since field inventories are efficient ways of assessing biomass, uncertainty in field biomass estimation would propagate with establishing empirical models with satellite data. Primarily it has to be ensured to select a species specific volume equation (or allometric equation) to assess field biomass to minimize the uncertainty in biomass estimation. Further, field plot size would also contribute to uncertainty in field biomass, small plot size gives higher standard error since rare large tree contributes a large fraction of the overall plot biomass.

REDUCTION IN FIELD BIOMASS UNCERTAINTY

Tree level errors average out in large plots (1 ha, *i.e.* 100 × 100 m), and thus would be more representative of the forest area. Several large plots (> 40) are established in different forest types over India, and it is observed that the coefficient of variation decreases with increase in plot size, which indicates higher sampling error with the smaller plots. Further, it is observed that the coefficient of variation stabilised with plot sizes > 0.5 ha (70 × 70 m), indicating that it is advisable to establish large permanent plots with a minimum of 0.5 ha to reduce uncertainty in field biomass (Fig. **2**).

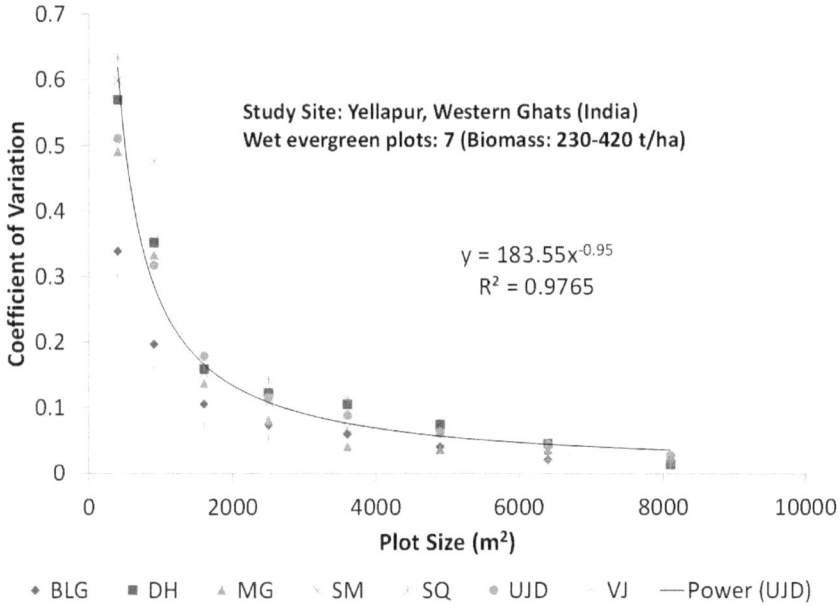

Study Site: Yellapur, Western Ghats (India)
Wet evergreen plots: 7 (Biomass: 230-420 t/ha)

$y = 183.55x^{-0.95}$
$R^2 = 0.9765$

♦ BLG ■ DH ▲ MG ⬠ SM ⬩ SQ ● UJD ⬦ VJ —Power (UJD)

IRS CARTOSAT BASED SPATIAL BIOMASS

In addition to the large plots and efforts towards reduction in spatial biomass estimation, it is necessary to use information other than reflectance of pixel and biomass. It is found that the use of texture (information from neighbouring pixels) would help to estimate the biomass with a non-saturating relationship even in the high biomass range of about 500–600 t ha^{-1}. High spatial resolution data, i.e. a spatial resolution of about < 5 m, are now widely available from various satellites (*viz.*, Ikonos, Quickbird, Spot-5 and Cartosat), helps us to use geometrical/ textural characteristics rather than just considering the individual pixels.

In the high biomass forests of Uppangala, Western Ghats (maximum biomass of 632 t ha^{-1}), texture information derived from IRS Cartosat data was used to estimate spatial biomass map of the study area (Reddy *et al.* 2016). Fourier transform based textural ordination (FOTO) technique, which involves deriving radial spectrum information via 2D fast fourier transform (FFT) and ordination through principal component analysis (PCA) was used for characterizing the textural properties of forest canopies. Plot level estimated biomass from 15 (1 ha) plots was used to relate with texture derived information from high resolution satellite datasets (viz., IKONOS and Cartosat-1). The results show that the FOTO method using stereo Cartosat (A&F) images at 2.5 m resolution are able to perform well in characterizing high biomass values since the texture-biomass relationship is only subjected to 18% relative error to that of 15% in case of IKONOS (Fig. **3**) and could aid in reduction of uncertainty in biomass estimation at a large landscape levels.

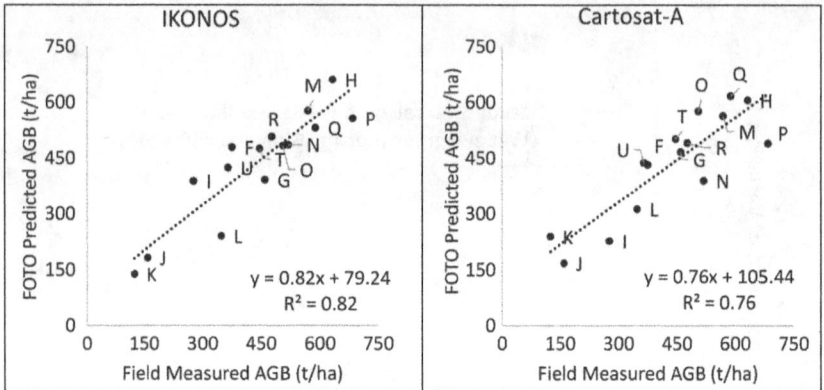

Fig. 3: Predicted vs. Observed AGB Estimation Using IKONOS
and Cartosat Images over Uppangala, Western Ghats, India

LIDAR BASED BIOMASS

Image based techniques, from optical or microwave sensors can usually describe only the horizontal distribution of biomass and not its vertical components, a complex issue in multi-layered forests. The emergence of LiDAR in the late 1990s has provided new insights in quantifying vegetation distribution in both vertical and horizontal directions and aided in estimation of biomass at landscape level. In Uppangala (Western Ghats), LiDAR derived forest stand structure parameters were related to field biomass from large plots to establish an empirical model. The results suggested that LiDAR derived H50 metric (median of the top-of-canopy height, derived from the canopy height model) was the best predictor of field biomass. A logarithmic regression model led to a relative RMSE error of 17.9% with about 23 large (1 ha) plots (Fig. **4**). The study identifies the importance of the vertical structure information in reducing uncertainty in estimation of spatial biomass.

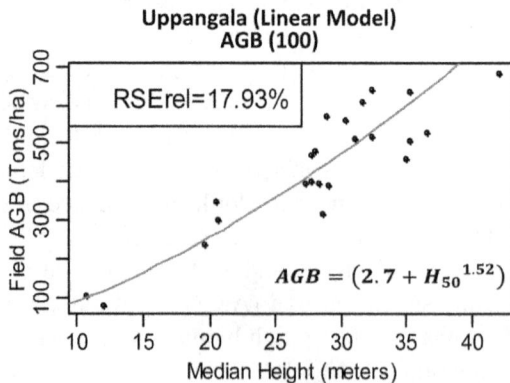

Fig. 4: Empirical Model between LiDAR Derived Median Height
and AGB over Uppangala, Western Ghats, India

CONCLUSION

Spatial biomass estimation using remote sensing is possible by integrating field measurements to satellite derived vegetation components and has the advantage of spatio-temporal coverage over large areas. However, the uncertainty of estimation depends on the type of data used and the established empirical model. In order to reduce uncertainty of estimation not only it is necessary to reduce uncertainty in field measurements but also have to improve the methods of estimation and introduce vertical structure information. Though large plots have proven to reduce the uncertainty in field estimations, establishment of such network over different forest types over large areas is a stupendous task. Also, large plots in combination with advanced methods to relate texture to biomass have also shown significant prominence to reduce uncertainty in spatial biomass estimation. Further, the advancements of technology, LiDAR, have also shown the importance of vertical structure in estimation of spatial biomass. However, considering the cost of operations of LiDAR's, it might be difficult to obtain data over large areas. Thus further methods to explore the possibility of how 2D information from satellite imagery could be fused with LiDAR to provide enhanced assessment of biomass at national level.

REFERENCES

Baccini, A., Laporte, N., Goetz, S.J., Sun, M. and Dong, H. (2008). A first map of tropical Africa's above-ground biomass derived from satellite imagery. *Environmental Research Letters*, 3(4), p. 45011.

Jha, C.S., Fararoda, R., Gopalakrishnan, R. and Dadhwal, V.K. (2015). Spatial Distribution of Biomass in Indian Forests Using Spectral Modelling. Geospatial information systems for multi-scale forest biomass assessment and monitoring in the Hindu Kush Himalayan region. Kathmandu: ICIMOD.

Houghton, R.A. (2005). Aboveground forest biomass and the global carbon balance. *Global Change Biology*, 11(6), pp. 945–958.

Houghton, R.A., Butman, D., Bunn, A.G., Krankina, O.N., Schlesinger, P. and Stone, T.A. (2007). Mapping Russian forest biomass with data from satellites and forest inventories. *Environmental Research Letters*, 2(4), p. 45032.

Reddy, R.S., Gopalakrishnan, R., Jha, C.S. and Couteron, P. (2016). Estimation of Above Ground Biomass Using Texture Metrics Derived from IRS Cartosat-1 Panchromatic Data in Evergreen Forests of Western Ghats, India. *Journal of the Indian Society of Remote Sensing*.

Saatchi, S.S., Houghton, R.A., DosSantos Alvala, R.C., Soares, J.V. and Yu, Y. (2007). Distribution of aboveground live biomass in the Amazon basin. *Global Change Biology*, 13(4), pp. 816–837.

Tang, G., Beckage, B., Smith, B. and Miller, P.A. (2010). Estimating potential forest NPP, biomass and their climatic sensitivity in New England using a dynamic ecosystem model. *Ecosphere*, 1(6), pp. 1–20.

Thumaty, K.C., Fararoda, R., Middinti, S., Gopalakrishnan, R., Jha, C.S. and Dadhwal, V.K. (2015). Estimation of Aboveground Biomass for Central Indian Deciduous Forests Using ALOS PALSAR L-Band Data. *Journal of the Indian Society of Remote Sensing*, 44: 31–39.

Mitigating Climate Change: Plant Disease Scenario and Management Strategies

Harender Raj Gautam and I.M. Sharma

Dr. Y.S. Parmar University of Horticulture and Forestry, Nauni, Solan 173 230, Himachal Pradesh, India
E-mail: hrg_mpp@yahoo.com

ABSTRACT: *Climate change is the currently the biggest threat to mankind. The economic loss of climate change is projected to be US $ 1.2 trillion further its impacts are likely to cause 0.4 million deaths a year worldwide. Climate change is also affecting agriculture due to an increase in temperature of 0.85°C (from 0.65 to 1.06°C) between 1880 and 2012 and increased CO_2 concentration from 280 ppm (pre-industrial value) to 401 ppm in 2015. Such changes are projected to have a drastic effect on the growth and cultivation of the different crops on the earth. Simultaneously, these changes will also affect the reproduction, spread and severity of many plant pathogens, thus posing a threat to our food security. Climate change is also putting stem rust resistance due to Sr31 under threat of Ug99 race of stem rust caused by Puccinia graminis f. sp. tritici. Elevated temperature and CO_2 concentration are also posing higher threat perception of late blight (Phytophthora infestans) disease of potato and important diseases of rice, namely blast (Pyricularia oryzae) and sheath blight (Rhizoctonia solani). Changing disease scenario due to climate change has highlighted the need for future studies on models which can predict the severity of important pathogens of major crops under field conditions. Simultaneously, disease management strategies need reorientation under the changing conditions with amalgamation of new strategies for sustainable food production.*

Keywords: Climate Change, Disease Threat, Food Security, Plant Pathogens.

Climate change is the biggest threat of the present century and it is already contributing to the deaths of nearly 400,000 people a year and costing the world more than US $ 1.2 trillion, thus wiping 1.6 per cent annually from global GDP. Latest report of Inter-Governmental Panel on Climate Change has reported an increase of 0.85°C [0.65 to 1.06°C] in temperature between 1880 and 2012 (IPCC, 2014). Further, the CO_2 concentration has increased from 280 ppm (pre-industrial value) to 401 ppm in 2015. The possible changes in temperature, precipitation and CO_2 concentration are expected to significantly impact crop growth. However, with successful adaptation and adequate irrigation, global agricultural production could be increased due to the doubling of CO_2 fertilization effect such that the overall impact of climate change on worldwide food production is low to moderate (Gautam, 2009a and Gautam, 2009b). Wheat is one of our important food grains and climate change is expected to affect wheat crop production both, during the processes of plant growth and development. Climate change is also expected to critically influence the occurrence and severity of plant diseases. Climate change models for Central Europe apparently suggest an increase in winter precipitation and a decrease in summer rains leading to higher drought frequencies. Changes in climatic conditions will affect the development of plant diseases depending on the region and the crop considered (Bregaglio *et al.*, 2013; Mikkelsen *et al.*, 2015). Climate change may increase the impact of pests by allowing their establishment in areas where they could previously not establish. Changes in temperature can result in changes in geographic ranges and facilitate overwintering. Some species could therefore extend their geographic range towards the pole and to higher altitudes (IPCC, 2014; Svobodová *et al.*, 2014). In general, studies carried out on a wide range of regions and crops worldwide indicate negative impacts of climate change on crop yields are more common than positive impacts (IPCC, 2014).

EFFECT OF INCREASED CO_2 CONCENTRATIONS ON PATHOGENS

The concentration of CO_2 in the atmosphere reached 379 ppm in 2005, which exceeds the natural range of values of the past 650,000 years (IPCC, 2007). An increase in CO_2 levels may encourage the production of plant biomass; however, productivity is regulated by water and nutrient availability, competition against weeds and damage by pests and diseases. Consequently, a high concentration of carbohydrates in the host tissue promotes the development of biotrophic fungi such as rust (Chakraborty *et al.*, 2002). Thus, an increase in biomass can modify the microclimate and affect the risk of infection. In general, increased plant density will tend to increase leaf surface wetness duration to regulate temperature thus leading to increased likelihood of infection by foliar pathogens. Overall, the effects of elevated CO_2 concentration on plant disease can be positive or negative, although in a majority of the examples reviewed by Chakraborty *et al.* (2000), disease severity increased.

THE EFFECTS OF AN INCREASE IN TEMPERATURE AND ULTRAVIOLET RADIATION ON PATHOGENS

Due to changes in temperature and precipitation regimes, climate change may alter the growth stage, development rate and pathogenicity of infectious agents, and the physiology and resistance of the host plant (Chakraborty and Datta, 2003). A change in temperature could directly affect the spread of infectious diseases and survival of the pathogen between seasons. Ultraviolet radiation plays an important role in natural regulation of diseases. Evidence suggests that sunlight affects pathogens due to the accumulation of phytoalexins or protective pigments in host tissue. A change in temperature may favour the development of different inactive pathogens, which could induce an epidemic. Increase in temperatures with sufficient soil moisture may increase evapotranspiration resulting in humid microclimate in crop and may lead to incidence of diseases favoured under these conditions (Mina and Sinha, 2008). Temperature is one of the most important factors affecting the occurrence of bacterial diseases such as *Ralstonia solanacearum, Acidovorax avenae* and *Burkholderia glumea*. Similarly, the incidence of vector-borne diseases is expected to be altered since climate can substantially influence the development and distribution of vectors. Changes may result in altered geographical distribution, increased overwintering, changes in population growth rates, increases in the number of generations, extension of the development season, changes in crop-pest synchrony of phenology, changes in inter-specific interactions and increased risk of invasion by migrant pests. Furthermore, increase in temperature could determine the distribution of areas favorable for overwintering (Garrett *et al.*, 2006) or even more lethal zones where the insect cannot survive.

EFFECT OF CLIMATE CHANGE ON PLANT DISEASES

The climate influences the incidence as well as temporal and spatial distribution of plant diseases (Gautam *et al.*, 2013). The main factors that control growth and development of diseases are temperature, light and humidity and water. Changes in rainfall patterns and temperature can induce severe epidemics in plants because the changes may tend to favoursome types of pathogens over others. Moreover, if these changes cause unfavorable condition for pathogens, diseases could be reduced or may not occur. In the presence of susceptible hosts, pathogens with short life cycles, high reproduction rates and effective dispersion mechanisms respond quickly to climate change, resulting in faster adaptation to climatic conditions (Coakley *et al.*, 1999). Harvell *et al.* (2002) demonstrated that warm winters with high night temperatures facilitate the survival of pathogens, accelerate life cycles of vectors and fungi, and increase sporulation and aerial fungal infection. Climate change will also modify host physiology and resistance, and alter the stages and rates of the development of pathogens. Temperature and moisture govern the rate of reproduction of many pathogens (Caffarra *et al.*, 2012). The longer growing seasons resulting from global warming will extend the amount of time available for pathogen reproduction and dissemination. Climate change may also influence the sexual reproduction of pathogens (Legler *et al.*, 2012), thereby increasing the evolutionary potential of individual populations. In wheat belt of India in Punjab, while changes in temperature and humidity will reduce the importance of yellow rust (*P. striiformis*) and Karnal bunt (*T. indica*); the importance of leaf rust, foliar blights, *Fusarium* head blight and stem rust may increase in the future, particularly in the absence of resistance in wheat cultivars

(Kaur *et al.*, 2008). On the other hand, the importance of leaf rust, foliar blights, *Fusarium* head blight and stem rust may increase in Punjab in the future. In plant diseases, there are some pathogens of major crops which have a huge potential to cause losses in those crops. In wheat, *Sr31* stem rust resistance has been effective in cultivars for over 30 years, which can be overcome by new races of *Puccinia graminis* f. sp. *tritici* like Ug99 (Duveiller *et al.*, 2007). According to estimates, Ug99 race of the stem rust can result in up to 10 per cent yield loss in Asia alone, amounting to US $ 1–2 billion per year. Similarly, banana wilt caused by *Xanthomonas* will severly affect the food security of 70 million people in Uganda. In potato, economic production is often impossible without the application of pesticides. Coffee rust epidemics caused by the fungus, *Hemileia vastatrix* is another indicator of climate change with effect on number of countries including Colombia, Central America, Mexico, Peru and Ecuador with higher intensities of the disease between 2008 and 2013 (Avelino *et al.*, 2015).

Late blight of potato caused by *P. infestans*, is considered to be the most economically important disease of potato worldwide. The disease can destroy a potato crop within a few weeks. Estimates of losses to late blight in developing countries vary between US $ 3 and US $ 10 billion each year, and about US $ 750 million is spent on pesticides alone. In the temperate Indian hills which occupy about 20 per cent of the acreage, a severe epiphytotic (epidemic) of late blight recurs every year resulting in 40–85 per cent yield loss. The disease now appears earlier in the northern part (November) and later in the eastern part (February) and within a wider temperature range, i.e. 14–27.5°C than at 10–25°C recorded in earlier years (Luck *et al.*, 2012). Pesticide usage in potato may increase if changing crop physiology interferes with the uptake and translocation of pesticides or changes in other climatic factors (e.g. more frequent rainfall, washing away residues of contact pesticides) may force more frequent applications. Faster crop development at increased temperature could also increase the need for application of pesticides. Fungicide and bactericide efficacy may change with increased CO_2, moisture and temperature (Schepers, 1996). Systemic fungicides could be affected negatively by physiological changes that slowdown uptake rates, such as smaller stomatal opening or thicker epicuticular waxes in crop plants grown under higher temperatures. The same fungicides could be affected positively by increased plant metabolic rates that could increase fungicide uptake.

Climate change is affecting apple cultivation in Himachal Pradesh and diseases also have an important role in causing the yield reduction in the crop (Gautam *et al.*, 2014). Effect of climate change has also been observed on different diseases in Himachal Pradesh. In apple the incidence of cankers is increasing due to scanty rains during the rainy season and severe summers. Similarly, incidence of apple scab (*Venturiaina equalis*) has been reduced due to lower rainfall in winter and also in March-April. Early rise in seasonal temperature is resulting in greater severity of powdery mildew (*Oidium mangiferae*) in mango. Wide difference in winters in day and night temperature is resulting in higher severity of die-back (*Botryodiplodia theobromae*) of mango. In stone fruits, due to rise in temperature and low humidity in winters, there is low incidence of fungal leaf curl caused by *Taphrina deformans*. Incidence of rust and powdery mildews has also increased as these diseases are more severe when night temperature is higher. Consequently, these diseases are now appearing on the French bean also. Incidence of fungal and bacterial wilts in vegetable crops is usually observed in low elevation areas but due to rise in temperature, these diseases are now observed in the mid-hills too.

CONCLUSION

Effects of climate change are visible on agricultural production and there is a need to combine climate change, crop growth and crop disease models to predict impacts of climate change on crop diseases. Disease risk analyses based on host-pathogen interactions should be performed and research on host response and adaptation should be carried out to understand how an imminent change in the climate could affect plant diseases. With changing climate, there is need to change disease management strategy. Based on different climate change prediction models, forewarning of the diseases can be issued since disease forecasting models based on weather exist for many important plant diseases. Currently, there are two favoured methods to predict changes in geographic ranges and average severity of a disease. Out of the two, more widely used method assumes that climate and host currently limit the pathogen and then an attempt is made to match the current geographic range with suitable climatic measures. Based on this methodology changes in disease severity and geographical range in relation to future climate are predicted (Desprez-Loustau *et al.,* 2007; Steffek *et al.,* 2007). If we hypothesize that climatic factors limit the abundance of a pathogen and understand quantitatively their relation to weather, then it may be appropriate to use a weather-based prediction system to predict future abundance (Semenov, 2007). Micro-climate advisory (Leaf Wetness Index Model) has been found effective against leaf spot disease of groundnut which helped to reduce the pesticide and fungicide usage without affecting the yield (Koshy *et al.,* 2014). With our past experience in disease forecasting, there is a need for dynamic model which incorporates the recorded data of each crop season for a particular pest to suitably revise itself and thus remains stable, relevant enough to continue providing accurate forecasting (Kumar *et al.,* 2016). Thus, climate change will affect the timing of measures taken by farmers to effectively manage disease as well as the feasibility of particular cropping systems in particular regions.

REFERENCES

Avelino, J., Cristancho, M., Georgiou, S., Imbach, P., Aguilar, L., Bornemann, G., Läderach, P., Anzueto, F., Hruska, A.J. and Morales, C. (2015). The coffee rust crises in Colombia and Central America (2008–2013): impacts, plausible causes and proposed solutions. *Food Security, 7*(2), 303–321.

Bregaglio, S., Donatelli, M. and Confalonieri, R. (2013). Fungal infections of rice, wheat, and grape in Europe in 2030–2050. *Agron Sustain Dev.,* 33, 767–776.

Caffarra, A., Rinaldi, M., Eccel, E., Rossi, V. and Pertot., I. (2012). Modeling the impact of climate change on the interaction between grapevine and its pests and pathogens: European grapevine moth and powdery mildew. *Agric. Ecosyst. Environ.,* 148, 89–101.

Chakraborty, S. and Datta, S. (2003). How will plant pathogens adapt to host plant resistance at elevated CO_2 under a changing climate? *New Phytol.,* 159, 733–742.

Chakraborty, S., Tiedermann, A.V. and Teng, P.S. (2000). Climate change: potential impact on plant diseases. *Environmental Pollution,* 108: 317–326.

Chakraborty, S., Murray, G. and White, N. (2002). Potential impact of climate change on plant diseases of economic significance to Australia. *Australas. Plant. Pathol.,* 27, 15–35.

Coakley, S.M., Scherm, H. and Chakraborty, S. (1999). Climate Change and Plant Disease. *Annu. Rev. Phytopathol.,* 37, 399–426.

Desprez-Loustau M.L., Robin C., Reynaud G., Deque M., Badeau V., Piou D., Husson C. and Marcais, B. (2007). Simulating the effects of a climate change scenario on the geographical range and activity of forest pathogenic fungi. *Canadian Journal of Plant Pathology,* 29, 101–120.

Duveiller, E., Singh, R.P. and Nicol, J.M. (2007). The challenges of maintaining wheat productivity: pests, diseases, and potential epidemics. *Euphytica,* 157, 417–430.

Garrett, K.A., Dendy, S.P., Frank, E.E., Rouse, M.N. and Travers, S.E. (2006). Climate change effects on plant disease: genomes to ecosystems. *Annu. Rev. Phytopathol.,* 44, 489–509.

FAO (2016). Climate change and food security: risks and responses, Rome 98 P.

Gautam, H.R., 2009a. Challenges of climate change and bio-energy to world food security. *Open Learning (July–December),* 49–51.

Gautam, H.R. (2009b). Effect of climate change on rural India. *Kurukshetra,* 57(9), 3–5.

Gautam, H.R., Bhardwaj, M.L. and Kumar, R. (2013). Climate Change and its impact on plant diseases. *Current Science,* 105(12), 1685–1691.

Gautam, H.R., Sharma, I.M. and Kumar, R. (2014). Climate change affecting apple cultivation in Himachal Pradesh. *Current Science,* 106(4), 498–499.

Harvell, H.C., Mitchell, C.E., Ward, J.R., Altizer, S., Dobson, A.P., Ostfeld, R.S., Samuel, M.D. (2002). Climate warming and disease risks for terrestrial and marine biota. *Science,* 296, 2158–2162.

IPCC (2007). The Fourth IPCC Assessment Report, Cambridge University Press, Cambridge, UK.

IPCC (2014). Climate change 2014: synthesis report. Contribution of Working Groups I, II and III to the Fifth Assessment Report of the Intergovernmental Panel on Climate Change [Pachauri RK., Meyer LA. (eds.)]. IPCC, Geneva, Switzerland, pp. 151.

Kaur, S., Dhaliwal, L. and Kaur, P. (2008). Impact of climate change on wheat disease scenario in Punjab. *J. Res.* 45 (3 & 4), 161–170.

Koshy, S.S., Nagaraju, Y., Pallil, S., Prasad, Y.G. and Pola, N. (2014). Wireless Sensor Network based Forewarning Models for Pests and Diseases in Agriculture – A Case Study on Groundnut. *International Journal of Advancements in Research & Technology,* 3(1), 74–82.

Kumar, A., Bhattacharya, B.K., Kumar, V., Jain, A.K., Mishra, A.K. and Chattopadhyay, C. (2016). Epidemiology and forecasting of insect-pests and diseases for value-added agro-advisory. *MAUSAM,* 67(1), 267–276.

Legler, S.E., Caffi, T. and Rossi, V. (2012). A nonlinear model for temperature dependent development of *Erysiphe necator* chasmothecia on grapevine leaves. *Plant Pathol.,* 61, 96–105.

Luck, J., Spackman, M., Freeman, A., Tre̦bicki, P., Griffiths, W., Finlay, K. and Chakraborty, S. (2011). Climate change and diseases of food crops. *Plant Pathology,* 60, 113–121.

Luck, J., Asaduzzaman, M., Banerjee, S., Bhattacharya, I., Coughlan, K., Debnath, G.C., De Boer, D., Dutta, S., Forbes, G., Griffiths, W., Hossain, D., Huda, S., Jagannathan, R., Khan, S., O'Leary, G., Miah, G., Saha, A. and Spooner-Hart, R. (2012). The effects of climate change on pests and diseases of major food crops in the Asia-Pacific region. Final Report for APN (Asia-Pacific Network for Global Change Research) Project, p. 73.

Mikkelsen, B.L., Jørgensen, R.B., Lyngkjær, M.F. (2015). Complex interplay of future climate levels of CO_2, ozone and temperature on susceptibility to fungal diseases in barley. *Plant Pathol.,* 64, 319–327.

Mina, U. and Sinha, P. (2008). Effects of Climate Change on Plant Pathogens. *Environ. News.,* 14(4), 6–10.

Schepers, H.T.A.M. (1996). Effect of rain on efficacy of fungicides on potato against *Phytophthora infestans. Potato Res.,* 39, 541–550.

Semenov M.A., 2007. Simulation of extreme weather events by a stochastic weather generator. *Climate Research,* 35, 203–212.

Sharma, I.M., 2012. Changing disease scenario in apple orchards: Perspective, challenges and management strategies. In: Proceedings of the National Symposium on Blending conventional and modern plant pathology for sustainable agriculture, Indian Institute of Horticultural Research, Bangaluru, 2012.

Svobodová, E., Trnka, M., Dubrovský, M., Semerádová, D., Eitzinger, J., Stěpánek, P. and Zalud, Z., 2014. Determination of areas with the most significant shift in persistence of pests in Europe under climate change. *Pest Manag. Sci.,* 70(5), 708–715.

Steffek R., Reisenzein H., Zeisner N. (2007). Analysis of the pest risk from Grapevine flavescencedoréephytoplasma to Austrian viticulture. *EPPO Bulletin,* 37, 191–203.

Weather based Forewarning Mustard Aphid Model for Weather based Agro-Advisories

N.V.K. Chakravarty

Indian Agricultural Research Institute, New Delhi 110 012, India
E-mail: nvkchak@gmail.com

ABSTRACT: *Among the factors responsible for avoidable yield losses in rapeseed- mustard, the aphid, Lipaphis erysimi, is most important causing nearly 50 per cent loss. A weather based forewarning model was developed based on heat accumulation in terms of growing degree days limitations and exceptions of the model are discussed including emerging issues in Agro-advisories.*

Keywords: Growing Degree Days, Aphids, Temperature, Rainfall and Fog.

Rapeseed-mustard (*Brassica* spp.) is a major group of oilseed crops being grown in 53 countries across six continents. India is the second largest cultivator after China. In India it is cultivated on 6.70 million hectares with production of 7.96 mt and productivity of 1188 kg/ha (2013–14). The time of sowing, the most essential non-monetary input for obtaining higher production, decides the environmental conditions the crop is likely to get exposed during its different growth stages, but seeding time can be so adjusted that the various physiological stages of the crop can coincide with specific (most suitable) temperature during crop growth cycle and also to escape the aphid attack. However, in northern parts of our country the sowing date depends on the preceding crop's harvest and hence cannot be sown on the optimum date at some places.

This crop needs special attention because there is still considerable gap between yield potential and harvest. The avoidable yield losses due to mustard aphid (*Lipaphis erysimi,* Kaltenbach) alone are reported to be between 20 and 50 per cent and in extreme conditions, it might exceed further. Weather plays an important role in deciding the crop yield in presence of insect pests and diseases. Probably due to inadequate information on the quantitative relationship between weather parameters and the aphid dynamics, appropriate or satisfactory models could not be developed to forewarn the mustard aphid incidence and severity well in time with a sufficient lead time. Though there had been some sincere attempts in this direction, it was felt that a farmer-friendly forewarning model was the need of the hour. Daily variation of temperatures was found to be highly varying and hence an attempt was made to use the Degree Day concept to develop such a simple forewarning model.

Weather based Agro-advisories have been proved to be of immense value to the needy farmers and appreciable progress was made over the last more than two decades in this direction. It becomes absolutely necessary to provide highly accurate weather forecast along with Agro-advisory services to the farming community to minimize the weather-related or weather-influenced losses and maximize the production through early fore-warning systems regarding pests/ diseases alerts, their remedial actions etc. Developing simple forewarning thumb rules for pests would be of immense value to the farmer. These thumb rules need to be integrated with the decision support system so that they become user-friendly and come handy in saving the imminent losses from the pest attack.

Field experiments were carried out in the farm of Indian Agricultural Research Institute for several years starting from 2000–01 *rabi* season on various cultivars of mustard to understand the crop-pest-weather dynamics. Three varieties were continuously grown for four *rabi* seasons at weekly intervals starting from 2000–01 to create different thermal environment and this was continued till 2005–06. Data generated earlier on the crop as a part of the in-house research project was also used. Daily weather data recorded in the Agro-met Observatory adjoining the experimental site was collected and analysed. Degree day accumulations were computed with 5°C as base temperature. Intensive data was generated on the pest population in terms of first appearance, their multiplication and dispersal (Chakravarty and Gautam, 2002; 2004).

BASIS FOR DEVELOPMENT OF FOREWARNING MODEL

In Delhi, the aphid generally appears during the first/second week of January, reaches a peak by third week of February and disappears from the crop by mid-March. Hence, initially daily temperature data (both maximum and minimum) were analyzed for about fifteen years from 1st October to 20th March every year. When this data was plotted with an aim to relate the variations to the aphid build up and peak population, no meaningful relation could be found. Hence, it was thought that probably heat accumulation, in terms of growing degree days (GDD) could give some stable relation. So, GDD were computed. Interestingly, the variation in the heat units was found to be relatively less. Moreover, it was observed that the variation among years was relatively appreciable from 1st January to 25th January in all these years analyzed (Figure 1). Further detailed analysis formed the basis of the development of the forewarning model. The heat accumulations from 1st January to 25th January during different crop seasons were corroborated with the peak aphid population reached during the corresponding seasons following Sastry (1996).

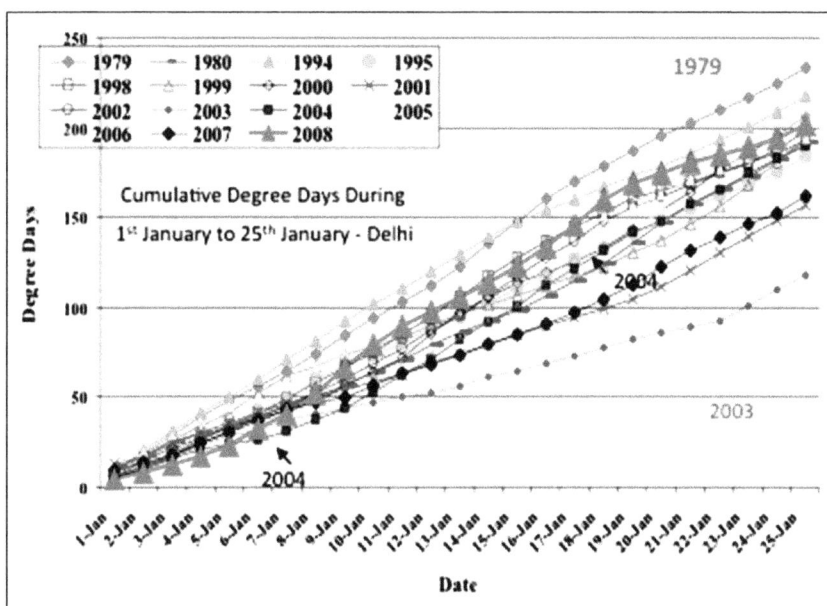

Fig. 1: Growing Degree Days from 1st to 25th January in Different Years (IARI)

Based on intense field studies, as the degree day concept was observed to be reliable and stable in forewarning the aphid infestation levels, a hypothesis was developed (Chakravarty and Gautam, 2002; 2004) as "Aphid population may be more in a year when the degree day accumulation from 1st January to 25th January is slower and vice versa". This hypothesis was widely used., adopted and validated by many researchers in the mustard growing areas in the country and observed to be fairly accurate in forecasting the aphid peak levels much in advance, giving a lead time of about 20 days for the farmers to take preventive steps.

It was felt that the hypothesis relating the degree-days with the peak aphid population be used as a simple thumb rule and might serve as an indicator to predict the levels of aphid populations well in time so that control measures could be suggested which if implemented would go a long way in increasing the mustard crop production as well as productivity, leading to sustainability in this crop yield. Moreover, being simple and involving only temperature as the input, this thumb rule could go into decision support system. The forewarning model was validated using IARI data for three consecutive years 2014, 2015 and 2016 (Figure 2). In the first two years, the GDD were 185 and 172D indicating that the peak aphid population would be moderate to low while in the third year 2016 the GDD were 208 indicating low infestation and were found to be valid. The peak aphid population in the first two years was observed to be moderate while in 2016 it was low. Thus this simple temperature based powerful thumb rule may go a long way in forewarn the farmers about the aphid infestation and take preventive/protective measures besides saving unnecessary sprayings.

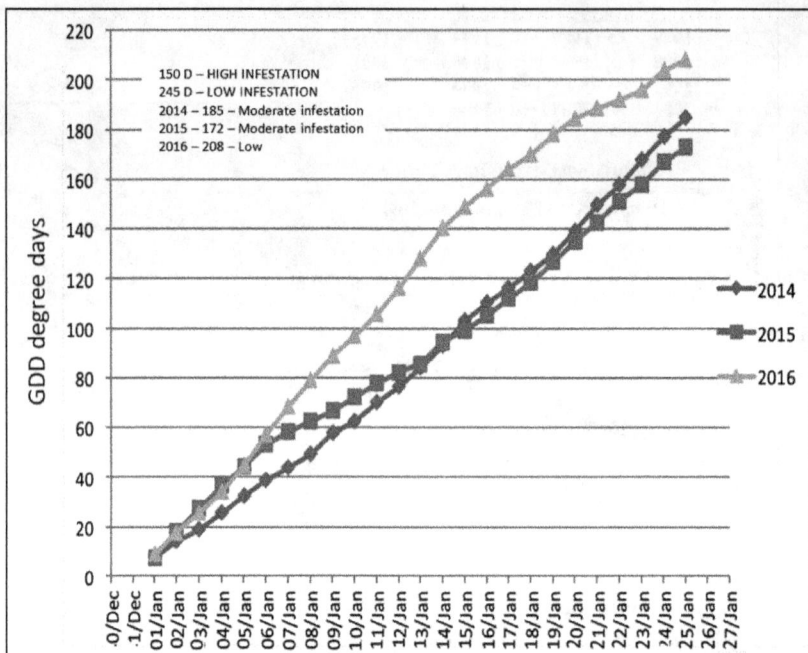

Fig. 2: Growing Degree Days and Peak Aphid Infestation in Three Years (IARI)

LIMITATIONS/EXCEPTIONS OF THE VALIDATION

Rainfall: Rainfall has a physical effect on the aphid in the sense that it would dislodge them to the base of the plant and does not have any physiological influence. Hence, when a forewarning is made using this concept, care must be taken to see whether there was any rainfall before the peak aphid infestation or at the time of forewarning. Though some reports were seen wherein rainfall was correlated with the aphid multiplication, our study established the fact that it had only physical impact and it would be misleading to include rainfall in the correlation rainfall studies or developing forewarning model. As a case study, when two years (2007 and 2008) degree day accumulations from 1st to 25th January were plotted (Figure 3), it was observed that as per the hypothesis, the aphid population would have been higher in 2007 as compared to 2008. But, actual aphid data revealed that in 2007 when the GDD accumulation was low, the population was also less. When the weather data was further scrutinized, it was observed that on 11th February, a rainfall of 30.6 mm was received resulting in dislodging all the aphids to the bottom.

Fog: Just as rainfall, dense fog also dislodges the aphid population down to the bottom. It was observed that in the crop season 2004–05, the aphid infestation was severe in the mustard crop and dense fog persisted on 12th night & 13th Dec. morning dislodged the aphids out of the shoot (Figure 4). In such cases, caution must be taken to apply this forewarning thumb rule.

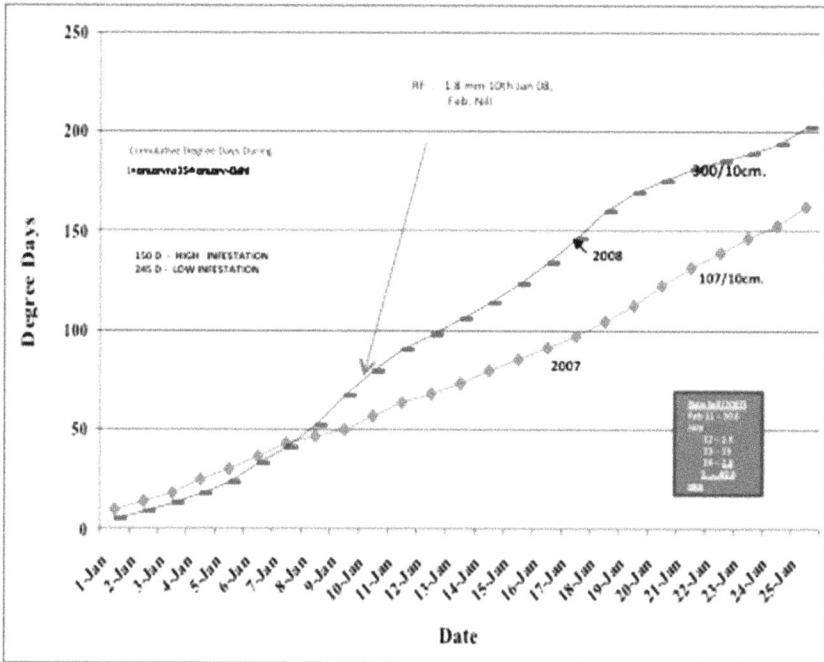

Fig. 3: Growing Degree Days in Two Years along with Peak Aphid Infestation

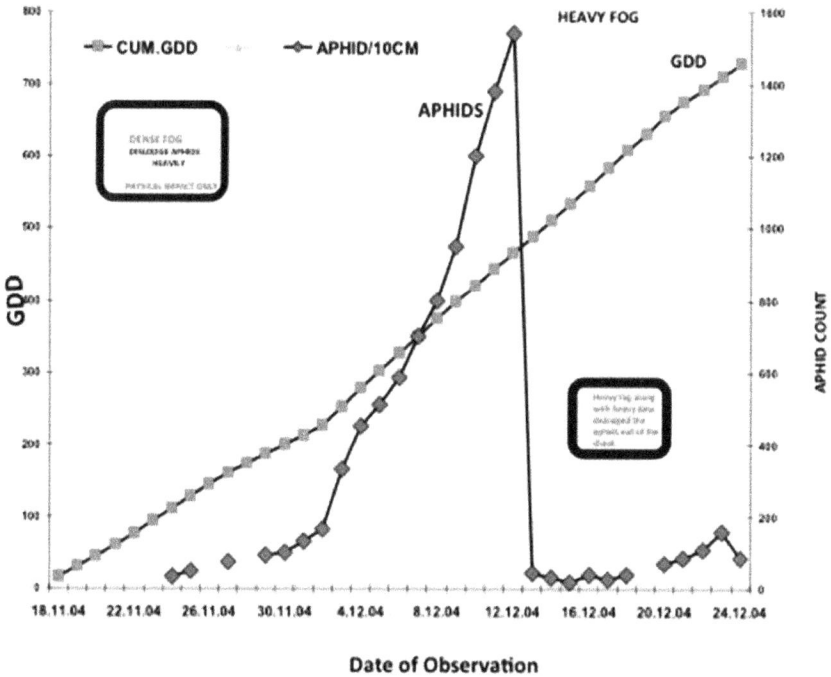

Fig. 4: Example of Fog Dislodging Aphids to the Base of the Plant

Maximum Temperature Impact on Aphid Multiplication

Many studies focused on the mean temperatures during a week or during a certain crop growth period. While several authors (Chattopadhyay *et al.,* 2005; etc.) predicted aphid occurrence using the preceding weather conditions, Prasanta *et al.* (2004) observed that GDD of preceding three days were important to explain the dynamics of aphid population. Earlier, Prasad and Chakravarty (2000) worked out a model to estimate the aphid population using previous week's aphid population and weather conditions as inputs. But, one interesting point was brought out from our intense observations carried out in the field was that for aphid multiplication, day (maximum) temperature above 20°C would be congenial irrespective of the minimum temperature. It was observed that whatever may be the minimum temperatures, so long as the day maximum temperatures were below 20°C, the aphid multiplication was inhibited and on the other hand, once the maximum temperatures crossed that value, the multiplication started fast. Hence, it can be said firmly that the day maximum temperatures play key role in their multiplication.

WEATHER BASED AGRO-ADVISORIES

It is possible for the farmers to modify his decision if he has advance information on the weather conditions along with proper guidelines to take tactical decisions based on weather forecast; which ultimately reduce the associated risk of his decision for sustainable agriculture. Looking to these important aspects, Department of Science and Technology (DST) started through National Center for Medium Range Weather Forecasting (NCMRWF), Agro-Meteorological Field Units (AMFUs) in the country during 1991 at different centers with the objective of formulating and disseminating weather based Agro-advisories to the farmers of the surrounding villages of each unit. Till the year 2009, about 130 such units established in 83 agro climatic zones of India. The medium range weather forecast being given by IMD for following four days (twice a week) is being corroborated with the current crop status and the Agro-advisories for different crops would be formulated which are disseminated to the farmers through various means and also through web.

Being simple and only temperature based, the mustard aphid forewarning thumb rule discussed above formed the integral part of the Agro-advisories and the decision support system in some parts of the country. It was observed that whenever weather based Agro-met advisory is adopted by the farmers, there would be profound reduction in rice crop production risks. Parvinder Maini and Rathore (2011) reported that farmers who adopted medium range weather based Agro-met advisories did accrue a net physical benefit of 10 to 15 per cent higher yield for the agricultural crops tested and also there was reduction in the cost of cultivation by 2 to 5 per cent as compared to the non adoption.

To conclude, probably, it is the holistic approach integrating soil, weather and physiological aspects of crops with remote sensing/GIS techniques that may address these emerging issues in the next two decades which stand a great challenge to the Agro-meteorologists. As an attempt to meet the future challenges, one step forward could be introduction of some of the weather based agribusiness topics in the curriculum of Agro-meteorlogy. Food supplies ultimately depend upon the skill with which farmers can exploit the potential of good weather and minimize the impact of bad weather. Recent developments in instrumentation, data management systems, climate prediction, crop modelling, dissemination of agro-meteorological information provide Agro-meteorologists with the necessary tools to help the farmers improve such skills. The future for agricultural meteorological services appears bright, and could contribute substantially to promoting sustainable agriculture and alleviating poverty.

ACKNOWLEDGEMENTS

The contributions by Dr. Ananta Vashisth (Principal Scientist, Agricultural Physics, IARI, New Delhi) and Ms. Mehnaz Tharranum (SRF, GKMS, IMD, New Delhi) are acknowledged.

REFERENCES

Chakravarty, N.V.K and Gautam, R.D. (2002). Forewarning mustard aphid. NATP project report. Division of Agricultural Physics, IARI, New Delhi, p. 63.

Chakravarty, N.V.K. and Gautam, R.D. (2004). Degree-day based forewarning system for mustard aphid. *J. Agro-meteorology,* 6(2): 215–222.

Chattopadhyay, C.; Agrawal, R.; Kumar, A.; Singh, Y.P.; Roy, S.K.; Khan, S.A.; Bhar, L.M.; Chakravarty, N.V.K.; Srivastava, A.; Patel, B.S.; Srivastava, B.; Singh C.P. and Mehta, S.C. (2005). Forewarning of Lipaphis erysimi on oilseed Brassica in India–A case study. *Crop Prot.,* 24, 1042–1053.

Parvinder Maini and Rathore, L.S. (2011). Economic impact assessment of the Agro-meteorological Advisory Service of India. *Current Sci.,* 101(10): 1296–1310.

Prasad, S.K. and Chakravarty, N.V.K. (2000). A model to predict the aphid population on rapeseed crop. Proc. Nat. Workshop on dynamic crop simulation modelling for Agrometorological Advisory Services (Eds: Singh S.V., Horels, R.A., Saseendran, S.A. and Singh, K.K.). National Centre for Medium Range Weather Forecasting, New Delhi, pp. 315–320.

Prasanta, N., Chakravarty, N.V.K., Srivastava A.K. and Bhagawati, G. (2004). A Forewarning model for mustard aphid on real time basis. *J. Agril. Phys.,* 4, 44–50.

Sastry, P.S.N. (1996). Project completion report ESS/71/012/91 (Unpublished Dept. Science & Technology, New Delhi Project Report) p. 28.

Adaptation to Climate Change: Management Strategies for Pests and Diseases

R.K. Singh[1] and S.V.S. Malik[2]

[1]ICAR-Indian Veterinary Research Institute, Izatnagar 243 122, U.P., India
[2]Division of VPH, ICAR-IVRI, Izatnagar 243 122, India
[1]E-mail: directoriviri@gmail.com

ABSTRACT: *Climate change across the globe is not only affecting the plants but also the humans, animals, particularly those malnourished besides causing expansion in the geographic range and incidence of vector-borne, zoonotic, and food- and water-borne diseases as well as in the prevalence of the diseases associated with air pollutants and aero-allergens. Emerging and re-emerging trends are being driven by climate change in case of diseases like malaria, chikungunya, dengue, Japanese encephalitis, leptospirosis, influenza, plague, anthrax and other food-borne and water-borne illnesses. The impact of climate-related diseases is most palpable in low socio-economic groups owing to their least adaptive and resilience capacities. Therefore, effective and pragmatic mitigation strategies including robust surveillance equipped with early warning systems need to be put in place for prompt prevention and control of these diseases for ensuring long-term health and economic benefits.*

Keywords: Vector-Borne Diseases, Zoonoses, Food-Borne Disease, Water-Borne, Control and Prevention.

Climate change refers to long-term statistical shifts in the weather, including changes in the average weather condition or in the distribution of weather conditions around the average (i.e. extreme weather events). These alterations in weather patterns can affect natural and physical systems. The direct effects of climate change include frequent/increased heat stress, drought, floods and storms, with the indirect effects of threatening population health through adverse changes in air pollution, the spread of disease vectors, food insecurity and under-nutrition, displacement, and mental ill health. These changes cause expansion in geographic range and incidence of vector- food- and water-borne diseases including zoonoses; and the prevalence of diseases associated with air pollutants and aeroallergens. Such changes affect the malnourished most and impact the survival, reproduction, or distribution of disease pathogens and hosts, as well as the availability and various means of their transmission routes (Wu *et al.,* 2014). A warming and unstable climate is playing an ever-increasing role in driving the global emergence, resurgence and redistribution of infectious diseases affecting man and animals (Singh and Malik, 2014). Many of the most common infectious diseases, and particularly those transmitted by insects/vectors, are highly sensitive to climate variation (Tian *et al.*, 2015). An overview of inter-relationship between climate change and health is illustrated in Figure 1.

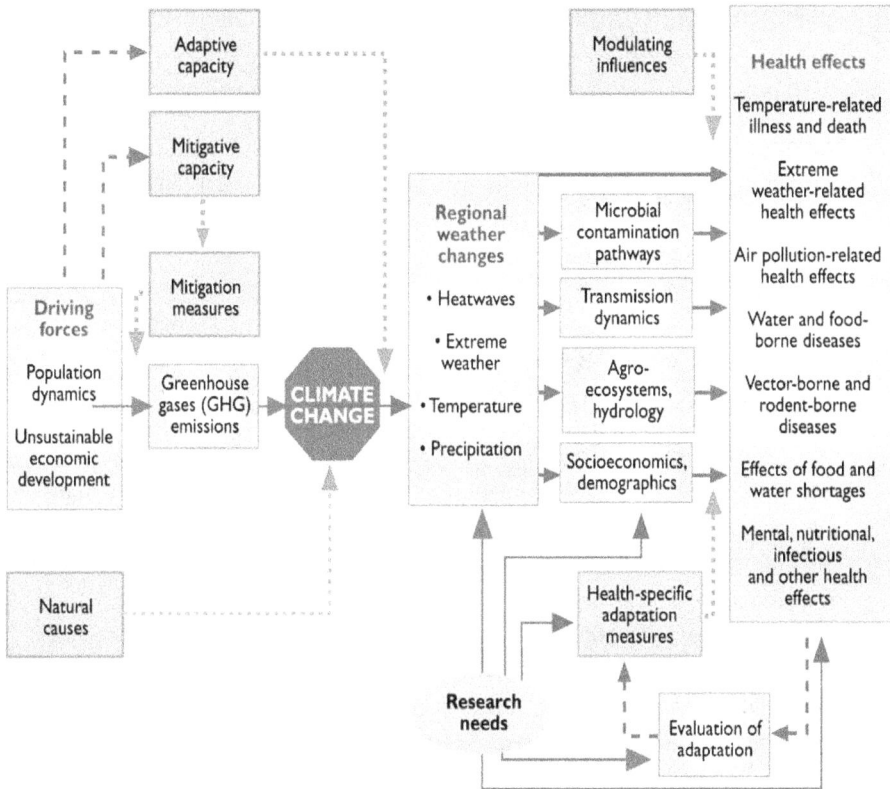

Fig. 1: Climate change and health: pathways from driving forces, through exposures to potential health impacts (Adapted from: Climate change and human health: risks and responses; WHO, 2003).

CLIMATE CHANGE AND HUMAN HEALTH: AN OVERVIEW

According to the recent estimates of World Health Organization (WHO), around 12.6 million deaths a year are associated with the environmental pollution. Of these, an estimated 6.5 million deaths (11.6% of all global deaths) are associated with air pollution, from household and outdoor sources (WHO, 2016). The mean global temperature is expected to increase by another 1.8 to 5.8°C by the end of this century. Moreover, mountain glaciers are contracting, ocean water has become more acidic, and extreme weather events occur more often. These unpredictable climate patterns are expected to have a substantial effect on the burden of infectious diseases that are transmitted by vectors and through contaminated water. Recent emergence of vector-borne diseases like dengue in Europe (Bouzid *et al.*, 2014) has shown that climate change can have drastic adverse effect on human health including the mental health (Willox *et al.*, 2015).

The three essential components for most infectious diseases include, an agent (or pathogen), a host (or vector) and transmission environment (Epstein, 2001).

Effect on Pathogens

Climate change can affect not only the quantity but also the geographic and seasonal distributions of pathogens. The impact of climate change on pathogens can be:

- *Direct:* by influencing the survival, reproduction, and life cycle of pathogens, or
- *Indirect:* by influencing the habitat, environment, or competitors of pathogens.

Effect of Various Climatic Parameters on Disease Transmissions

Effect of temperature on life cycle of pathogens: Temperature variations may affect disease patterns by influencing the life cycle of pathogens in many ways.

- A pathogen needs a certain temperature range to survive and develop, e.g. the two thresholds, i.e., the maximum temperature of 22–23°C required for the mosquito development and the minimum temperature of 25–26°C for the Japanese Encephalitis Virus (JEV) transmission, play key roles in the ecology of JEV (Mellor and Leake, 2000; Tian *et al.*, 2015).
- Rising temperature can influence the reproduction and extrinsic incubation period (EIP) of pathogens, e.g. due to temperature variation, the EIP for *P. falciparum* has been observed to reduce from 26 days at 20°C to 13 days at 25°C (Bunyavanich *et al.*, 2003).
- Rise in the average temperature of water bodies and food environment may provide an agreeable environment for microorganism reproduction cycles and algal blooms. In a major food-borne disease–Salmonellosis, the repro-duction of the bacteria increases as temperature rises in that range between 7°C and 37°C (IWGCCH, 2010).
- Rising temperature may limit the proliferation of a pathogen through favoring its competitors. For example, *Campylobacter* spp. causing food-borne disease was found to be more concentrated at low temperature and during winter as it is believed that warmer temperature supports other bacteria to out-compete *Campylobacter* spp. (Jones, 2001).

Effect of shifts in precipitations: Climate change may cause shifts in precipita-tion, which affects with the increase of fecal pathogens as heavy rain may stir up sediments in water, leading to the accumulation of fecal microorganisms (Jofre *et al.*, 2010).

Effect of humidity: The pathogens of air-borne infectious disease such as influenza tend to be responsive to humidity condition, e.g. humidity and temperature were found to affect influenza virus the dissemination of water-borne pathogens. Rainfall plays an important role in the development of various water-borne disease pathogens. Rainy season is found to be correlated transmission and survival (Xu *et al.,* 2014).

Effect of wind: Wind is a key factor affecting the pathogens of air-borne diseases. A positive correlation has been found between the dust particle association/attachment and the virus survival/transportation (Chen *et al.,* 2010). It has been reported that the presence of desert dust in the atmosphere during Asian dust storms is associated with increased concentration of cultivable bacteria, cultivable fungi, and fungal spores (Griffin, 2007). The viruses of infectious diseases can be transported across ocean by dust particles, which may facilitate the transmission of viruses between distant hosts (Griffin, 2007).

Climate Change and Vectors/Hosts

The spatial pattern and population changes of insect vectors are closely associated with the changes of climate. Temperature affects the spatial as well as temporal distribution of disease vectors. As the mean global temperature continues to rise, the vectors in low-latitude regions may find new habitats in mid- or high-latitude regions and in the areas of high altitude, leading to their geographical expansion of diseases. Studies have found that many of the vector-borne human infectious diseases, including malaria, African trypanosomiasis, Lyme disease, tick-borne encephalitis, yellow fever, plague, rift valley fever, Japanese encephalitis, dengue and most recently Zika disease have distributed to a wider range (Singh *et al.,* 2016; Wu *et al.,* 2016). Most of these diseases have extended into the higher latitude areas following the habitat expansion of their vectors. Many of the vector-borne diseases are found to be positively correlated with the rainfall conditions. The larval development of mosquito vectors accelerates with increased rainfall and rising temperature (Gage *et al.,* 2008). Many disease vectors and hosts tend to respond strongly to change in humidity. Generally, low humidity coupled with high temperature forms the unfavorable condition for ticks and fleas (e.g. grasslands or forestlands), hence, limiting the spread of the related infectious diseases (Gage and Kosoy, 2005). Wind exerts dual effects on disease vectors/hosts. Strong wind can reduce the biting opportunities for mosquitoes, but can help in extending their flight range (Reid, 2000).

Climate Change and Disease Transmission

Depending on the route, disease transmission can be direct or indirect. Many studies have proved that climate change may affect disease transmission, either directly by influencing the viability or survivability of pathogens or indirectly by changing their transmission routes. Temperature change alone, or together with other climatic variables change such as rainfall, wind and dust storms may alter the transmission routes of infectious diseases. Climate change can not only affect the transmission of infectious diseases by altering the contact patterns between host, pathogen and vectors but also can harm or affect the human immunity and susceptibility to disease (Xu *et al.,* 2016). Global warming is also influencing the water availability, this creates the scarcity of clean water, which may lead to more diarrheal diseases worldwide (Lloyd *et al.,* 2007). Overall, the climate change causes ecosystem degradation and instability, which possibly brings negative effect on agricultural productivity that leads to malnutrition and starvation, increased population displacement, resource conflict and break down of public health measures.

MANAGEMENT STRATEGIES

Climate change is increasingly being recognized as the biggest global health threat and opportunity (Watts *et al.*, 2015). Therefore, understanding its present effects and the likely future impact on population health and survival is of vital scientific and public health importance in local, national, and international settings. Climate change also affects many natural and social systems and processes that are highly essential for life (Watts *et al.,* 2015), as it disrupts the earth's life-support systems that underpin the world's capacity to supply adequate food and water, as well as the eco-physical buffering against natural disasters.

Five Key Strategic Elements of the Integrated Vector Management (IVM) Framework (WHO, 2009)

Integrated vector management (IVM) is an adaptive, evidence-based vector management that draws on vector control measures from both within and outside the health sector. A successful IVM program has been proposed as an ideal approach for tackling environmental issues. Five key strategic elements of the IVM framework are:

1. *Advocacy, social mobilization and legislation*
   - Promotion and embedding of IVM principles in designing policies.
   - Establishment or strengthening of regulatory and legislative controls for public health.
   - Empowerment of communities.
2. *Collaboration within the health sector and with other sectors:*
   - Collaboration within and between public and private sectors.
   - Strengthening 'one-health' approach.
3. *Capacity building:* Provision of evidence-based essential material infrastructure, financial resources and human resources at local and national level to manage IVM strategies.
4. *Evidence-based decision-making:* Adaptation of strategies and interventions to local ecology, epidemiology and resources, guided by operational research.

5. *Integrated Approach:* Ensure rational use of available resources by addressing several diseases, integrating various vector-control methods with other disease-control methods.

Framework of indicators for monitoring critical areas of climate change (Watts *et al.,* 2015).

Health Impacts

Need to assess an updated review of evidence on the health impacts of climate change. Various priority areas are:

- Climate-sensitive dynamic of infectious diseases
- Food insecurity
- Climate-related population migration
- Air pollution and allergic conditions
- Ecosystem degradation.

Actions to reduce greenhouse gas emissions that improve public health

Need an International progress and compliance with:
- A strong and equitable international agreement.
- Low-carbon and climate-resilient technology innovation and investment.
- Climate governance (finance, decision making, coordination, legislation).

Adaptation, resilience, and climate-smart health systems

- Poverty reduction and reductions in inequities.
- Vulnerability and exposure reduction in high-risk populations.
- Food security in poor countries.
- Communication of climate risks and community engagement for local solutions.
- Development of climate resilient, low-carbon health systems.

In the fast changing global climate scenario, we need to be pro-active, supportive and sincere in tackling the climate change related burgeoning risks and challenges as an emergency. Therefore, all the health and allied agencies need to collaborate to develop and improve the existing research and monitoring programs to better understand the gravity of the threat to human and animal health so as to act within a much larger and more comprehensive framework.

REFERENCES

Bouzid, M., Colón-González, F.J., Lung, T., Lake, I.R. and Hunter, P.R. (2014). Climate change and the emergence of vector-borne diseases in Europe: case study of dengue fever. *BMC Public Health* 14, 781.

Bunyavanich, S., Landrigan, C.P., McMichael, A.J. and Epstein, P.R. (2003). The impact of climate change on child health. *Ambul. Pediatr,* 3, 44–52.

Chen, P.S., Tsai, F.T., Lin, C.K., Yang, C.Y., Chan, C.C., Young, C.Y. and Lee, C.H. (2010). Ambient influenza and avian influenza virus during dust storm days and background days. *Environ. Health Perspect*, 118, 1211–1216.

Epstein, P.R. (2001). Climate change and emerging infectious diseases. Microbes Infect. 3,

Gage, K.L., Burkot, T.R., Eisen, R.J. and Hayes, E.B., 2008. Climate and vectorborne diseases. *Am. J. Prev. Med.* 35, 436–450.

Gage, K.L. and Kosoy, M.Y. (2005). Natural history of plague: perspectives from more than a century of research. *Annu. Rev. Entomol.,* 50, 505–528.

Griffin, D.W. (2007). Atmospheric movement of microorganisms in clouds of desert dust and implications for human health. *Clin. Microbiol. Rev.,* 20, 459–477.

IWGCCH (2010). A human health perspective on climate change. In: Tart, K.T. (Ed.), A Report Outlining the Research Needs on the Human Health Effects of Climate Change: Environmental Health Perspectives and the National Institute of Environmental Health Sciences.

Jofre, J., Blanch, A.R. and Lucena, F. (2010). Water-borne infectious disease outbreaks associated with water scarcity and rainfall events. In: Sabater, S. and Barcelo, D. (Eds.), *Water Scarcity in the Mediterranean: Perspectives under Global Change.* Springer.

Jones, K. (2001). Campylobacters in water, sewage and the environment. *J. Appl. Microbiol.* 90, 68S–79S.

Lloyd, S.J., Kovats, R.S. and Armstrong, B.G. (2007). Global diarrhoea morbidity, weather and climate. *Clim. Res.* 34, 119–127.

Mellor, P.S. and Leake, C.J. (2000). Climatic and geographic influences on arboviral infections and vectors. *Rev. Sci. Tech.,* 19, 41–54.

Reid, C. (2000). Implications of Climate Change on Malaria in Karnataka, India. Brown University.

Singh, R.K., Dhama, K., Malik, Y.S., Ramakrishnan, M.A., Karthik, K., Tiwari, R., Saurabh, S., Sachan, S. and Joshi, S.K. (2016). Zika Virus–Emergence, evolution, pathology, diagnosis and control: current global scenario and future perspectives–A comprehensive review. *Veterinary Quarterly.* pp. 1–43.

Singh, R.K. and Malik, S.V.S. (2014). Climate Change and Animal Health. In: XXVIII Annual Convention of IAVMI and International conference on "Challenges and Opportunities in Animal Health at the face of globalization and climate change", 30th Oct.–1st Nov., DUVASU, Mathura (UP), pp. 10.

Tian, H.Y., Zhou, S., Dong, L., Van Boeckel, T.P., Cui, Y.J., Wu, Y.R., Cazelles, B., Huang, S.Q., Yang, R.F., Grenfell, B.T. and Xu, B. (2015). Avian influenza H5N1 viral and bird migration networks in Asia. *Proc. Natl. Acad. Sci. U.S.A.,* 112, 172–177.

Watts, N., Adger, W.N., Agnolucci, P., Blackstock, J., Byass, P., Cai, W., Chaytor, S., Colbourn, T., Collins, M., Cooper, A. and Cox, P.M. (2015). Health and climate change: Policy responses to protect public health. *The Lancet, 386*(10006), pp. 1861–1914.

WHO (2003). Climate change and human health: risks and responses: summary.

WHO, 2009. http://apps.who.int/iris/bitstream/10665/70054/1/WHO_HTM_NTD_VEM_2009.1_eng.pdf

WHO (2012). Guidance on Policy-Making for Integrated Vector Management, WHO.

WHO Public Health, Environmental and Social Determinants of Health (PHE) e-News. (2016). http://www.who.int/phe/news/nov-2016/en

Willox, A.C., Stephenson, E., Allen, J., Bourque, F., Drossos, A., Elgarøy, S., Kral, M.J., Mauro, I., Moses, J. and Pearce, T. (2015). Examining relationships between climate change and mental health in the Circum-polar North. *Reg. Environ. Chang.*, 15, 169–182.

Wu, X.X., Tian, H.Y., Zhou, S., Chen, L.F. and Xu, B. (2014). Impact of global change on transmission of human infectious diseases. *Sci. China Earth Sci.* 57, 189–203.

Wu, X., Lu, Y., Zhou, S., Chen, L. and Xu, B. (2016). Impact of climate change on human infectious diseases: Empirical evidence and human adaptation. *Environment International*, 86, pp. 14–23.

Climate Change and its Influence on Freshwater Fish Diseases

P.K. Sahoo* and Anirban Paul

Fish Health Management Division, ICAR-Central Institute of Freshwater Aquaculture, Kausalyaganga, Bhubaneswar 751 002, India
*E-mail: pksahoo@hotmail.com

ABSTRACT: The global climate change is the most threatening challenge for all living creatures. With increasing temperature, most animals are facing problems related to temperature adjustment and an increase in infectious diseases. Fishes are going to face rough water condition in the form of water temperature and other related parameters. The rise in temperature will lead to increase in physiological stress in the fishes as there will be changes in the physicochemical parameters of waters. This will make them more susceptible to pathogenic organisms. With the changing climate, there is every chance of change in the pathogenicity pattern of different pathogens, that may favour the host or the pathogen. It is accepted that the adverse effect of climate change will be more on the temperate fishes than on tropical fishes. However, in the tropical freshwater fishes there are chances of increasing susceptibility to parasites, bacteria and fungi as these pathogens will complete their life cycle faster in warmer waters. There may be decrease in the susceptibility of viruses as virus needs specific temperature for their replication. There is every chance that the opportunistic pathogens will take advantage of this climate change and infect the fishes and there may be emergence of new pathogens in the aquaculture system. This paper deals with examples of freshwater fish disease emergence in different countries with respect to climate change and also discusses about the Indian aquaculture scenario and probable climate change effects.

Keywords: Temperature, Infectious Diseases, Pathogenicity Pattern, Stress, Tropical Fishes.

Climate change and global warming are the most threatening challenges ever imagined intoday's world. Current climate pattern is changing rapidly, and mean temperatures and frequency of extreme weather events are likely to increase. Mean global surface air temperature is projected to increase by 1.4° to 5.8°C by 2100 relative to 1990, with the magnitude of the increase varying both spatially and temporally. Animal kingdom as a whole is likely to be affected by this climate change. Infectious diseases are the destructive forces which can threaten biodiversity by catalysing population declines and accelerate species extinction. Infectious diseases are responsible for the recent decline of Australian and Central American frogs (Daszak *et al.,* 1999; Daszak *et al.,* 2000), Hawaiian forest birds, and African wild dogs (Daszak and Cunningham, 1999). Extinctions of invertebrates associated with infectious diseases include the Polynesian tree snail (Cunningham and Daszak, 1998) and a marine limpet (Harvell *et al.,* 1999). In the fisheries and aquaculture, aquatic environment plays a very crucial role as fishes inhabit in the water itself. Fishes are poikilothermic animals thus their physiology is directly affected by the ambient temperature and hence, changes in the water temperature can cause stress in the fish body which may lead to the outbreak of different diseases through compromising host immune system. Because of the change in the climate or water temperature there may be change in the pathogenicity pattern of different fish pathogens. Some of the pathogens may become more pathogenic or some may become less or there may be change in the host range of the pathogens or change in the geographic distribution of the host and the pathogens. The multiplication/replication rate of parasites and other pathogens are greatly affected by temperature. At higher temperatures, the generation time of bacteria, fungi and parasites with direct lifecycles is shorter, on the other hand, each virus has its own optimal temperature range for replication. So in this paper the probable effect of climate change on different freshwater fishes and their disease occurrences is discussed.

CONSEQUENCES OF CLIMATE CHANGE

Increase in Water Temperature and its Effect on Fishes

Fishes cannot regulate their body temperature, with the changing water temperature their body temperature also changes accordingly. Most of the fishes are having a suitable temperature range in which they remain comfortable by regulating their body metabolic activity, however, due to extreme changes in the water temperature disruption in body biochemical reaction rates occur. Therefore, increasing global water temperatures can affect individual fish by modifying physiological functions such as thermal tolerance, growth, metabolism, food consumption, reproductive ability and the aptitude to maintain internal homeostasis against a variable external environment (Fry, 1971). Different fishes are having different thermal tolerance limit, above or below of that particular range of temperature they will fall under stress. Any kind of stress leads to release of an array of catecholamines or cortisols, prolactin in freshwater fish which in turn increases the permeability of the surface epithelia leading to hydro-mineral imbalance, damage to the immune-competent organs and increase in susceptibility to pathogens.

Decreased Dissolve Oxygen

Biologically available oxygen (DO) in water is much less in concentration than in air (Moyle and Joseph, 2004). Dissolved oxygen concentrations of 5 mg/ l or more are favourable for most aquatic organisms (Stickney, 2000), and concentrations below 2–3 mg/l are considered hypoxic (Kalff, 2002). Availability of oxygen in the water is inversely related to the water temperature, *i.e.* with increasing temperature the water will have less amount of DO. The metabolic rates of most cold-blooded aquatic organisms increase with increasing temperature thus, increase in global water temperature acts dually in fish by decreasing the DO supply and increasing the Biological Oxygen Demand (BOD) (Kalff, 2002; Ficke *et al.*, 2007). With the changing climate we can expect extremes of both cold and warm temperatures. Hence, both the cases will have adverse effects on the fishes.

Change in Toxicity of Pollutants and their Effect in Fishes

With the changing global average temperature there may be change in the solubility of the different pollutants and chemicals, and may increase or decrease in there toxicity level. The temperature-dependent accumulation of toxicants in the fish body has many adverse effects and economic impacts. With the increasing temperature, there is every chance of increase in the uptake of pollutants by the fish tissues. For example, accumulation of endogenously produced ammonia and its metabolites is a serious problem in aquaculture systems. A combination of high temperature, low dissolved oxygen in culture water, and sub lethal concentration of ammonia have proved to cause gill necrosis in common carp (Jeney *et al.*,1992).

Eutrophication

Eutrophication is a phenomenon which is likely to increase with global climate change. Eutrophication may not affect that extent to our farm aquaculture system but it will definitely play a crucial role in the lake and reservoir ecosystems. Most cases of eutrophication result from the inflow of excess nutrients from urban and agricultural runoff and from sewage discharge (Lammens, 1990; Karabin *et al.*, 1997; Nicholls, 1998). However, increase in temperature can also accelerate the productivity of a water body by increasing algal growth, bacterial metabolism, and nutrient cycling rates (Klapper, 1991). A scientific study showed that a 2–3°C increase in temperature could cause a 300–500% increase in shoot biomass of the aquatic macrophyte *Elodea canadensis* (Kankaala *et al.*, 2002). But the exact relation between the temperature increase and the waterbody biomass is very complex, and a very comprehensive study is required to understand the effects.

Disease Scenario Related to Climate Change

The disease scenario of the freshwater system may alter to a great extent with the changing environment. The scenario may be very much different between the cold water and the warm water fishes. It is expected that the climate change will more severely affect the temperate fishes than the tropical fishes. The effects of climate change on micro- and macro-organisms causing disease will be superimposed onto the effects of other forms of stressors in the environment, fish food organisms, contaminants, species introductions and habitat loss. With the increasing temperature there will be change in the generation time of the bacteria, parasites and fungi. Thus the pathogenicity of these will also change, since fishes will already be in stress, due to change in water temperature. At the cellular level there will be delay in repairing proteins which is an important component for maintenance of metabolism. These proteins are vulnerable to heat damage, as temperatures increase, more protein repairs become necessary. Parasite infection and transmission depends on host condition, water quality, temperature and presence of intermediate hosts (Macrogliese, 2001). It is expected that parasites having direct life cycle will become more threatening as they will complete their life cycle fast and infect the fishes. The fungal and the bacterial diseases are expected to become more threatening as with increasing temperature the fish will become stressful with immunocompromised conditions and the pathogen life cycle will be shortened. Temperature also affects release of eggs or larvae by adult parasitic worms, embryonic development and hatching, longevity of free-living stages, infectivity to intermediate hosts, development in these hosts, infectivity to definitive hosts, maturation time, and the life span of adults (especially in fish hosts). Further, temperature plays a key role in host feeding and behaviour, host range and ecology, and host resistance in fish (Figure 1).

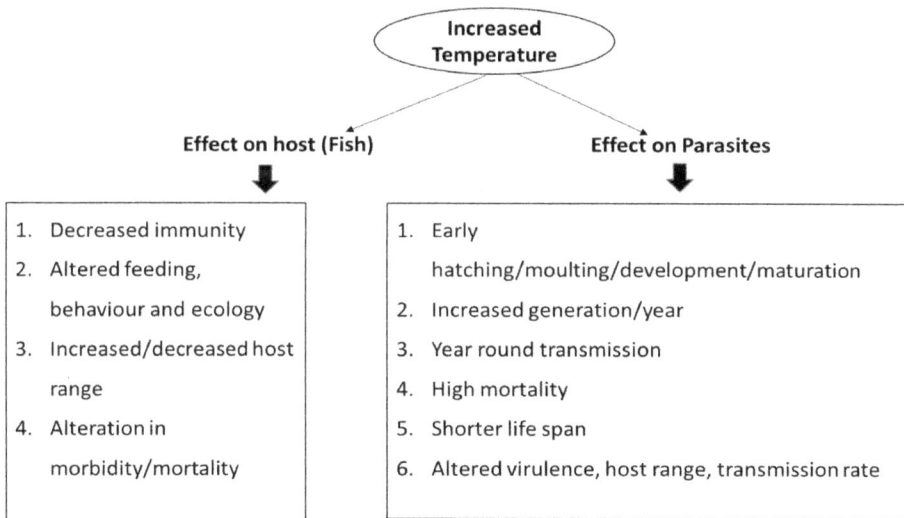

Fig. 1: Effect of Temperature Rise on Host and Parasites

Most of the viruses have their own optimum temperature range of growth, so with increasing temperature there may be less threat of viral infection in the fishes, but there is every chance of emergence of new viruses, particularly due to mutation in the genome, which can infect fishes. CyHV2, a Cyprinid herpes virus 2 disease leads to high mortality in juveniles at temperature ranging from 15 to 25°C, with the fish exhibiting lethargy, anorexia, pale gills, and swollen and necrotic spleen and kidney tissues in goldfish (Goodwin *et al.*, 2006; Jeffery *et al.*, 2007; Sahoo *et al.*, 2016). At higher temperature, the virus seems to be less effective. However, this dormant virus with increase in temperate may cause

Table 1: Climate Change and Probable Disease Prediction
for Freshwater Aquaculture Sector of India

Water Temperature		Pathogen Type	Effect on Pathogen/Parasite/Host
26-28°C	Parasite	*Argulus* Sp.	• Increased Fecundity, Prevalence, Year-Round Prevalence • Increased Parasitic Load and Secondary Infection • Decreased Fish Production • Increased Fish Production
		Ichthyophthirius Sp.	• Decreased Parasitic Load
		Trematodes, Cestodes and Myxozoans	• Increased Prevalence
	Bacteria	Opportunistic Pathogens	• Increased Intensity of Infection
	Virus	Koi Herpes Virus, Spring Viraemia of Carp, Cyprinid Herpesvirus-2, or any Other Viral Disease Outbreak	• Decreased Outbreaks • Increased Latency
	Fungus	*Aphanomyces, Saprolegnia*	• Decreased Prevalence
20-25°C	Parasites	*Ichthyophthirius* and Other Ectoparasites	• Increased prevalence
	Bacteria	Coldwater bacterial diseases i.e. Coldwater vibriosis, Furunculosis	• Increased prevalence
	Virus	Coldwater viral diseases	• Increased prevalence
	Fungus	*Saprolegnia, Aphanomyces*	• Increased prevalence
11-15°C	Parasite	Ichthyophthirius	• Increased prevalence
	Bacteria	Bacterial kidney disease, Enteric red mouth, Furunculosis, Coldwater vibriosis, Rainbow trout fry syndrome	• Increased prevalence
	Virus	Viral hemorrhagic septicemia virus, Infectious hematopoietic necrosis virus, SVCV, KHV, Epizootic haematopoietic necrosis Virus	• Increased prevalence
	Fungus	*Saprolegnia*	• Increased prevalence

serious mortality in goldfish due to high load of secondary bacterial pathogens (Sahoo *et al.*, 2016). Many of the viruses may enter into latency state and flare up suddenly when opportunity arrises. The optimum range for the growth and other normal metabolic activity of warm water fishes is 26.7 to 29.4°C (McLarney, 1998). Increase in water temperature gives the opportunistic pathogens a good platform to cause diseases. Some examples of pathogenicity with respect to temperature rise inclued anumber of endemic diseases of salmonids *e.g.* enteric red mouth, furunculosis, proliferative kidney disease and white spot, which will become more prevalent and difficult to control as water temperatures increase (Marcos-Lo´pez *et al.*, 2010). It is also expected that warm water fish bacteria will also become more virulent with increased temperature. The risk of Viral Haemorrhagic Septicaemia Virus (VHSV), Infectious Hematopoietic Necrosis Virus (IHNV) and Spring Viraemia of Carp Virus (SVCV) reduces as infection generally only establishes when water temperatures are less than 14°C (for VHSV and IHNV) and 17°C (for SCVC) (Marcos-Lopez *et al.*, 2010).

The emergence of disease is mostly associated with changes in ecology of host or pathogens or both. The abundance and geographical range of pathogens may likely to increase. For example, the secondary invader or opportunistic pathogen *Aeromonas hydrophila* greatly increases its pathogenicity in host at higher temperature (Das Mahapatra *et al.*, 2008). The virulent strains of *Flavobacterium column* are causing columnar is infection in fish, adhere more strongly to gill tissue at higher temperature and in water with high organic load (Decostere *et al.*, 1999).

Similar is the case with the most abundant parasites in freshwater fish, *Argulus* sp. The hatching and incubation time will be shorter and the parasite will complete its life cycle in shorter time, thereby will increase its infectivity load at high temperature. Indirectly, the farmers would be more compelled to depend upon frequent medication in aquaculture systems, thereby polluting the environment further besides the residual issues of medications. The possibility of continuous year-round prolonged infection or transmission also cannot be ruled out.

CONCLUSION

Although it is speculative to conclude climate change associated environment modulations, a large array of pathogens would increase in their emergence, pathogenicity and mortality in aquatic organisms as well as human beings in the climate change scenario. Global warming has clearly led to parasite disease transmission with increased virulence. The indirect effects affecting the economy and livelihood would be much more due to superimposed effects of climate change and associated anthropogenic environmental changes. The pathogens which are more likely influenced by ambient temperature (*viz.* aeromoniasis, argulosis, koi herpes virus infection, spring viraemia of carp virus), would strongly influence disease scenario in freshwater aquaculture systems.

REFERENCES

Anagnostakis, S.L. (1987). Chestnut blight: the classical problem of an introduced pathogen. *Mycologia*, 79(1): 23–37.

Cunningham, A.A. and Daszak, P. (1998). Extinction of a species of land snail due to infection with a microsporidian parasite. *Conservation Biology*, 12(5): 1139–1141.

Daszak, P., Berger, L., Cunningham, A.A., Hyatt, A.D., Green, D.E. and Speare, R. (1999). Emerging infectious diseases and amphibian population declines. *Emerging Infectious Diseases,* 5(6): 735.

Daszak, P. and Cunningham, A.A. (1999). Extinction by infection. *Trends in Ecology & Evolution,* 14(7): 279.

Daszak, P., Cunningham, A.A. and Hyatt, A.D. (2000). Emerging infectious diseases of wildlife—threats to biodiversity and human health. *Science*, 287(5452): 443–449.

Das Mahapatra, K., Gjerde, B., Sahoo, P.K., Saha, J.N., Barat, A., Sahoo, M., Mohanty, B.R., Ødegård, J., Rye, M. and Salte, R. (2008). Genetic variations in survival of rohu carp (*Labeorohita*, Hamilton) after *Aeromonas hydrophila* infection in challenge tests. *Aquaculture*, 279: 29–34.

Decostere, A., Haesebrouck, E., Turnbull, J.F and Charlier, G. (1999). Influence of water quality and temperature on adhesion of high and low virulence *Flavobacterium columnare* strains to isolated gill arches. *Journal of Fish Diseases*, 22: 1–11.

Ficke, A.D., Myrick, C.A. and Hansen, L.J. (2007). Potential impacts of global climate change on freshwater fisheries. *Reviews in Fish Biology and Fisheries*, 17(4): 581–613.

Fry, F.E.J. (1971). 1 The effect of environmental factors on the physiology of fish. *Fish Physiology,* 6: 1–98.

Goodwin, A.E., Merry, G.E. and Sadler, J. (2006). Detection of the herpesviral hematopoietic necrosis disease agent (Cyprinid herpesvirus 2) in moribund and healthy goldfish: Validation of a quantitative PCR diagnostic method. *Diseases of Aquatic Organisms*, 69(2–3): 137–143.

Harvell, C.D., Kim, K., Burkholder, J.M., Colwell, R.R., Epstein, P.R., Grimes, D.J., Hofmann, E.E., Lipp, E.K., Osterhaus, A.D.M.E., Overstreet, R.M. and Porter, J.W. (1999). Emerging marine diseases—climate links and anthropogenic factors. *Science*, 285(5433): 1505–1510.

Jeffery, K.R., Bateman, K., Bayley, A., Feist, S.W., Hulland, J., Longshaw, C., Stone, D., Woolford, G. and Way, K. (2007). Isolation of a cyprinid herpes virus 2 from goldfish, *Carassius auratus* (L.), in the UK. *Journal of Fish Diseases*, 30(11): 649–656.

Jeney, G., Nemcsok, J., Jeney, Z.S. and Olah, J. (1992). Acute effect of sublethal ammonia concentrations on common carp (*Cyprinus carpio* L.). II. Effect of ammonia on blood plasma transaminases (GOT, GPT), G1DH enzyme activity, and ATP value. *Aquaculture*, 104(1–2): 149–156.

Kalff, J. (2002). Limnology: inland water ecosystems. Prentice–Hall. Upper Saddle River, New Jersey.

Kankaala, P., Ojala, A., Tulonen, T. and Arvola, L. (2002). Changes in nutrient retention capacity of boreal aquatic ecosystems under climate warming: A simulation study. *Hydrobiologia*, 469(1–3): 67–76.

Karabin, A., Ejsmont-Karabin, J. and Kornatowska, R. (1997). Eutrophication processes in a shallow, macrophyte dominated lake–factors influencing zooplankton structure and density in Lake Łuknajno (Poland). *Hydrobiologia*, 342: 401–409.

Klapper, H. (1991). Control of eutrophication in inland waters. Ellis Horwood Ltd.

Lammens, E.H. (1990). The relation of biotic and abiotic interactions to eutrophication in Tjeukemeer, The Netherlands. In Trophic Relationships in Inland Waters. Springer Netherlands. pp. 29–37.

Marcogliese, D.J. (2001). Implications of climate change for parasitism of animals in the aquatic environment. *Canadian Journal of Zoology*, 79(8): 1331–1352.

Marcos López, M., Gale, P., Oidtmann, B.C. and Peeler, E.J. (2010). Assessing the impact of climate change on disease emergence in freshwater fish in the United Kingdom. *Transboundary and Emerging Diseases*, 57(5): 293–304.

McLarney, W.O. (1998). Freshwater aquaculture: a handbook for small scale fish culture in North America. Hartley & Marks Publishers.

Moyle, P.B.C. and Joseph, J. (2004). Fishes: an introduction to ichthyology (No. 597: MOY).

Nicholls, K.H., 1998. El Nino, ice cover, and Great Lakes phosphorus: implications for climate warming. *Limnology and Oceanography*, 43(4): 715–719.

Sahoo, P.K., Swaminathan, T.R., Abraham, T.J., Kumar, R., Pattanayak, S., Mohapatra, A., Rath, S.S., Patra, A., Adikesavalu, H., Sood, N. and Pradhan, P.K. (2016). Detection of goldfish haematopoietic necrosis herpes virus (Cyprinid herpesvirus-2) with multi-drug resistant Aeromonas hydrophila infection in goldfish: First evidence of any viral disease outbreak in ornamental freshwater aquaculture farms in India. *Acta Tropica*, 161: 8–17.

Stickney, R.R. (2000). Encyclopedia of aquaculture. John Wiley and Sons.

Forewarning Systems and Pest Management Strategies

Y.G. Prasad

ICAR-Agricultural Technology Application Research Institute (ATARI),
Hyderabad 500 059, India
E-mail: ygprasad@gmail.com

ABSTRACT: Forewarning of pest attacks before they actually take place is desired in pest management programmes, so that measures can be planned with maximum efficiency. Models are useful ways of synthesizing the available information and knowledge on population dynamics of pests in agroecosystems and natural habitats. Forewarning systems including phenology models, pest simulation models, decision support systems and integration of pest and crop simulation models including the importance of agromet networks and strategies to combat climate change and variability are discussed here.

Keywords: Forewarning Pest Outbreak, Phenology Models, Simulation Models, Agromet Networks.

Pest associated crop losses are estimated between 10-30% of the total agricultural production and additonally involves costs in the form of pesticides applied for pest control. Knowledge and information is the key for correct pest management decisions. Integrated pest management (IPM) is a system that emphasizes appropriate decision-making, which depends heavily on intensive, accurate and timely information for field implementation by practi-tioners. Pest forecast is an important component of the broad IPM philosophy. Forewarning models provide lead time for managing impending pest attacks and thus minimize crop loss and optimize pest control leading to reduction in cost of cultivation and least disruption to the environment.

Pest populations display fluctuations in timing and intensity depending on location and season. Mostly, they tend to fluctuate over a mean level. This average population over time, when computed across several years, results from the sum of action of all positive and negative factors influencing pest populations. Unlike pests of host plants in undisturbed habitats such as forestry which exhibit natural cycles, pests of agroecosystems are exposed to rapidly changing environments due to changes in cropping systems and a host of management interventions. As a result, crop pests show a greater degree of instability in population levels. Hence, forewarning of agricultural pests is much more challenging.

Pests vary in their biology and in their response to their environment. Pests in colder climates in general have discrete generations and resting phases in their life cycles, while in the warmer climates, most species exhibit polymodal patterns of occurrences, with several generations in a year, resulting from continuous breeding opportunities and food availability. Seasonal temperatures and rainfall patterns constitute major factors that determine the distributions of organisms. Reliable forewarning is much more complex in tropics than in the temperate ecosystems.

In nature, pests are regulated by their natural enemies including parasitoids, predators and pathogens, which are in turn influenced by biophysical factors (Thomson *et al.*, 2010). Exclusive reliance on physical factors alone in forecast systems is bound to influence the accuracy of predictions. Therefore, a precise understanding of population dynamics can result from comprehensive ecological studies. However, despite our best efforts, gaps in pest ecological databases remain as a result of the complexity of interactions among the ecosystem components.

Worldwide, one important outcome of understanding population dynamics is to aim for a forecasting capability for appropriate management decisions. Successful forecasting techniques are those that are as simple as possible and that are based on knowledge of the biology and ecology of the pests concerned.

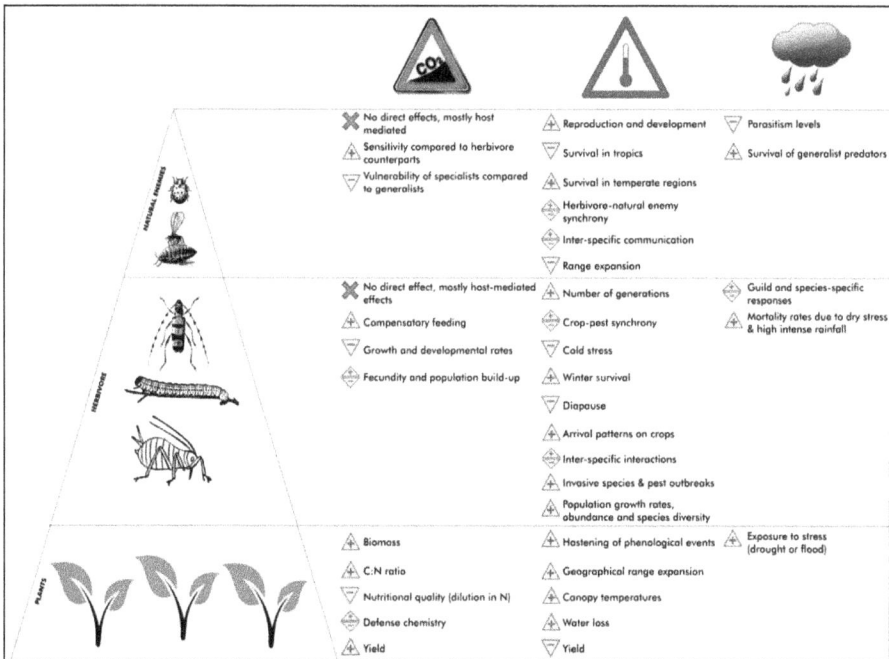

Fig. 1: Effects of climate change and variability at different trophic levels: Plant-Herbivore-Natural enemy; (+) = increase, (−) = decrease, (±) = increase or decrease, (×) = no effect

Climate change and variablility have further exacerbated the uncertainity in pest population dynamics across both tropical and temperate regions. The induced effect of climate change and variability on plant-herbivore-natural enemy interactions is complex and has diverse implications for pest management. Some of the effects attributed to individual climate change parameters (CO_2, temperature and rainfall) are given in Figure 1. However, interaction effects of these parameters are diverse, complex and unpredictable.

DEVELOPEMENT OF FOREWARNING SYSTEMS

Pest Monitoring

Monitoring of pests is fundmental in IPM programmes. Pests are monitored through a variety of monitoring tools such as pheromone traps, light traps, coloured sticky traps, pitfall traps and suction traps. The trap capture data serves several purposes: ecological studies, tracking insect migration, timing of pest arrivals into agroecosystems for initiating field scouting and sampling procedures, timing of pesticide applications, starting date or biofix for phenology models and prediction of later generations based on size of earlier generations. Monitoring through a network of sites is most useful for studying spatial distributions of pests, early detection of infestations and identification of hot-spot locations to initiate appropriate management interventions on a spatial scale. Monitoring at the regional level improves the reliability of population moni-toring for implementa-tion of appropriate area-wide IPM systems (Ayalew *et al.*, 2008).

Phenology Models

Insects are incapable of internal temperature regulation and hence their development depends on the temperature to which they are exposed. Studies of population dynamics often involve modelling growth as a function of temperature. The rate summation methodology has perhaps proved to be the most viable approach to such modelling (Stinner *et al.*, 1974). The most common development rate model, often called degree-day summation, assumes a linear relation-ship between development rate and temperature between lower and upper development thresholds (Allen, 1976). This method works well for optimum temperatures (Ikemoto, 2005). Phenology models help predict the time of events in an insect's development and are important analytical tools for predicting, evaluating and understanding the dynamics of pest populations in agroecosystems under a variety of environmental conditions.

Degree-day models (Higley *et al.*, 1986) have long been used as part of decision support systems to help growers predict spray timing or begin pest scouting (Welch *et al.*, 1978). Phenology models are also used as one component of risk analysis for predicting exotic pest establishment (Baker, 1991; Jarvis and Baker, 2001). A well-known example is the DYMEX modelling package (Yonow *et al.*,

2004). Some other modelling packages, such as CLIMEX, although not strictly phenology models, may also use some developmental requirements for risk assessment (Sutherst *et al.*, 1999). Another example is the web-based North Carolina State University APHIS Plant Pest Forecast (NAPPFAST) modelling system, which links daily climate and historical weather data with biological models to produce customized risk maps for phytosanitary risk assessments (Borchert and Magarey, 2005). Resources like the Crop Protection Compendium (CAB International, 2004) often summarize insect development, while the University of California Statewide IPM programme lists development data for insects on their website (http://www.ipm.ucdavis.edu/MODELS) for use in degree-day models. An Insect Development Database containing the developmental requirements for over 500 insect species has been created (Nietschke *et al.*, 2007). Insect Life Cycle Modeling (ILCYM) software, a generic open-source computer-aided tool, can be used to facilitate the development of phenology models and prediction of pest activity in specific agroecologies (Sporleder *et al.*, 2009).

Decision Support Systems

A decision support system (DSS) integrates a user-friendly front end to often-complex models, knowledge bases, expert sytems, and database technologies. DSSs have emerged as essential tools to bridge the gap between science-based technology and end-users who make day-to-day management decisions. An IPM-DSS provides user all necessary information on pest developmental models including sampling, decision making criteria and management options. Such systems are essential to enhance the adoption and spread of IPM in the country. The efforts made under the National Agricultural Innovation Project led to development of model based DSS (CROP PEST DSS) using data generated from laboratory and field studies for major pests of rice and cotton crops (Prasad *et al.*, 2014, Prasad and Prabhakar, 2012).

Pest Simulation Models

Simulation models based on mathematical descriptions of biological data as influenced by the environment are more easily applied across locations and environments. Computer programs or software to run these models facilitate the practical application of these models in understanding population dynamics and dissemination of pest forecasts for timely pest management decisions (Coulson and Saunders, 1987). Simulation approaches offer flexibility for testing, refinement and sensitivity analysis as well as field validation of developed models over a wide range of environmental conditions. Thorough descriptions of cropping systems being managed or studied are needed to explain the interactions among pests, plants and the environment (Colbach, 2010). Systems models or other prediction schemes can be used with appropriate biological, environmental, economic or other inputs to analyse the most effective management actions, based on acceptable control, sustainability and assessment of economic or other risks (Strand, 2000). Several such models have been developed and operationally used across the globe.

Coupled Crop and Pest Simulation Models

Crop system models can be used to generate information on the status of the crop, its pests and its environment under different scenarios, including different management options (Chander *et al.*, 2007). In practice, there are few examples of these models that include all the necessary components and can be used for practical decision making. However, a more practical approach has been the development of individual crop and pest components that can be analysed at the same time to give information that can improve decisions.

Establishment of Agromet Networks for Operational Pest Forecasting

Farmers are mainly interested in current disease and pest severity data, preferably for their localities to aid their decision making in crop protection. Pest monitoring data along with complementary monitoring of weather data is crucial to run pest forecast models and make available forecasts for operational use. Weather measurements under field conditions from several geo-referenced sites in the crop-cultivated regions additionally provides spatial information that can be used for generating pest forecast maps (Huang *et al.*, 2008).

Adaptation Strategies to Combat Climate Change and Variability

Global climate change and variability influence plant-herbivore-natural enemy interactions with implications for pest management. Many of the gaps in our understanding on trophic interactions can be attributed to experiments addressing individual trophic levels and that too the response studies are for individual climate change parameters. As more experiments address the complexities involved in plant-herbivore-natural enemy systems under the influence of a combination of climate change parameters (elevated CO_2 and temperature), the direction and magnitude of changes due to global climate change will be unravelled. IPM systems that emphasize and accord priority to host plant resistance, cultural, natural and biological control tactics would emerge as the best-bet, cost-effective and sustainable options to reduce pest infestations to acceptable levels in the tropics (Prasad and Bambawale, 2010).

CONCLUSIONS

Pest monitoring is the foundation for the issue of early warnings and the development and validation of pest forecast models and decision support systems, which are crucial for the design and implementation of successful IPM programmes. Models are useful ways of synthesizing the available information and knowledge on population dynamics of pests in agroecosystems and natural habitats. The development of long-term monitoring spatial data on crop–pest–weather relationships will narrow the gaps in knowledge required for reliable forecasts. In the tropics, agroecosystems are characterized by greater crop diversity in small parcels of land and dynamically changing weather. Available generic simulation models need to be tested with location-specific inputs for greater accuracy. In developing countries, there is a strong need to establish agro-meteorological networks for specific crop sectors with the major objective of pest forecasting through models and decision support systems.

REFERENCES

Allen, I.C., 1976. A modified sine wave method for calculating degree days. *Environmental Entomology,* 5, 388–396.

Ayalew, G., Sciarretta, A., Baumgartner, J., Ogol, C. and Lohr, B., 2008. Spatial distribution of diamondback moth, *Plutella xylostella* L. (Lepidoptera: Plutellidae), at the field and the regional level in Ethiopia. *International Journal of Pest Management,* 54, 31–38.

Baker, C.R.B., 1991. The validation and use of a life-cycle simulation model for risk assessment of insect pests. *EPPO Bulletin* 21, 615–622.

Birch, L.C., 1957. The role of weather in determining the distribution and abundance of animals. *Cold Spring Harbor Symposia On Quantitative Biology* 22, 203–18.

Borchert, D.M. and Magarey, R.D., 2005. A guide to the use of NAPPFAST. <http://www.nappfast.org/usermanual/nappfast-manual.pdf>.

CAB International, 2004. Crop Protection Compendium. CAB International, Wallingford, UK.

Chander, S., Kalra, N. and Aggarwal, P.K., 2007. Development and application of crop growth simulation modelling in pest management. *Outlook on Agriculture* 36, 63–70.

Colbach, N., 2010. Modelling cropping system effects on crop pest dynamics: How to compromise between process analysis and decision aid. *Plant Science* 179, 1–13.

Coulson, R.N. and Saunders, M.C., 1987. Computer-assisted decision making as applied to entomology. *Annual Review of Entomology* 32, 415–437.

Higley, L.G., Pedigo, L.P. and Ostlie, K.R., 1986. DEGDAY: A program for calculating degree-days, and assumptions behind the degree-day approach. *Environmental Entomology* 15, 999–1016.

Huang, Y., Lan, Y., Westbrook, J.K. and Hoffmann, W.C., 2008. Remote sensing and GIS applications for precision area-wide pest management: Implications for homeland security. *Geo Journal Library* 94, 241–255.

Ikemoto, T., 2005. Intrinsic optimum temperature for development of insects and mites. *Environmental Entomology* 34, 1377–1387.

Jarvis, C.H. and Baker, R.H.A., 2001. Risk assessment for nonindigenous pests, I. Mapping the outputs of phenology models to assess the likelihood of establishment. *Diversity Distribution* 7, 223–235.

Nietschke, B.S., Magarey, R.D., Borchert, D.M., Calvin, D.D. and Jones, E., 2007. A developmental database to support insect phenology models. *Crop Protection* 26, 1444–1448.

Prasad Y.G. and Bamabawale O.M., 2010. Effects of climate change on natural control of insect pests. *Indian Journal of Dryland Agricultural Research Development* 25(2): 1–12

Prasad Y.G. and Prabhakar M., 2012. Pest monitoring and forecasting. (in) Integrated Pest Management Principles and Practice (Abrol, D.B. and Uma Shankar, Eds.). p. 41–57, CAB International, UK.

Prasad Y.G., Prabhakar M., Katti G., Nagrare V.S., Vennila S. and Sujay Dutta, 2014. Development of decision support systems for insect pests of major rice and cotton based cropping systems (NAIP 4167), Central Research Institute for Dryland Agriculture (CRIDA), Hyderabad, p. 120.

Sporleder, M., Simon, R., Gonzales, J., Chavez, D., Juarez, H., De Mendiburu, F. and Krosche, J., 2009. ILCYM—Insect Life Cycle Modelling: a Software Package for Developing Temperature Based Insect Phenology Models with Applications for Regional and Global Risk Assessments and Mapping. International Potato Centre, Lima, Peru.

Stinner, R.E., Gutierrez, A.P. and Butler, G.D. Jr., 1974. An algorithm for temperature-dependent growth rate simulation. *Canadian Entomologist* 106, 519–524.

Strand, J.F., 2000. Some agrometeorological aspects of pest and disease management for the 21st century. *Agricultural and Forest Meteorology* 103, 73–82.

Sutherst, R.W., Maywald, G.F., Yonow, T. and Stevens, P.M., 1999. CLIMEX: Predicting the Effects of Climate on Plants and Animals. CSIRO Publishing, Collingwood, Australia.

Thomson, L.J., Macfadyen, S. and Hoffman, A.A., 2010. Predicting the effect of climate change on natural enemies of agricultural pests. *Biological Control* 52, 296–306.

Welch, S.M., Croft, B.A., Brunner, J.F. and Michels, M.F., 1978. PETE: An extension phenology modeling system for management of multi-species pest complex. *Environ. Entomol.*, 7: 482–494.

Yonow, T., Zalucki, M.P., Sutherst, R.W., Dominiak, B.C., Maywald, G.F., Maelzer, D.A. and Kriticos, D.J., 2004. Modelling the population dynamics of the Queensland fruit fly, *Bactrocera (Dacus) tryoni*: A cohort-based approach incorporating the effects of weather. *Ecological Modeling* 17, 39–30.

Farmers' Innovation
for Climate Change

Crucible of Crop Diversity: Forging Partnership with Farmer Breeders and Innovators for Higher Climate Resilience

Anil K. Gupta[1] and Anamika R. Dey[2]

[1]Honey Bee Network, Centre for Management in Agriculture, IIM Ahmedabad, Gujarat 380 015, India

[2]Gujarat Grassroots Innovation Augmentation Network, Ahmedabad, Gujarat 380 015, India

[1]E-mail: anilg@iima.ac.in

ABSTRACT: *Farmer breeders and innovators have rich insights to offer to agricultural scientists and plant breeders for developing climate resilient technologies. However, sufficient institutional arrangement is still lacking to blend their knowledge and institutional strength with formal R&D systems in agriculture. In this paper we take the case of developing climate resilient crop varieties by a) promoting in situ conservation of agrobiodiversity, b) blending farmers' selection criteria with scientists objectives, c) developing modified descriptors for crop germplasm characterization so as to incorporate new traits that help in resilience and food processing for overcoming market fluctuations, d) strengthening linkage among livestock, craft, tree and other agroecosystem components for better adaptation and e) creating new institutional platforms nationally and internationally to forge durable scientific partnership among farmers, workers and scientists. Examples are drawn from the Honey Bee Network's experience of 28 years and decadal plot-wise study of agrobiodiversity pursued for 30 years in the same villages of eastern Uttar Pradesh, India. Recent policy thrust on organic and sustainable agriculture has not yet triggered a corresponding thrust among the scientists for understanding science of farmers' sustainable innovations. This paper hopes to bridge this hiatus and forge a reciprocal, responsible, respectful open innovation partnership between informal and formal sector.*

Keywords: Agrobiodiversity, Grassroots Innovation, Climate Resilient Breeding, Conservation.

One of the important challenges that plant breeders face while selecting varieties for higher climate resilience, is to select for better adaptation to stress in a low external input environment. The widening gap between breeders from formal and informal sector has been long acknowledged (Gupta *et al.,* 1990), more so in the case of marginal farmers in marginal environment

(Cleveland and Soleri, 2002, Sillitoe *et al.,* 2002). This has expanded the scope for participatory plant breeding programmes. Much of the participatory plant breeding has focused on identifying farmers' selection criteria in a segregating or early stage breeding material developed by institutional breeders or later in the testing stage (Ashby *et al.,* 1989, Witcombe *et al.,* 2006, Morris and Bellon, 2004). In some cases, farmers have also been brought to the research station to make selections (Nkongolo, *et al.,* 2008). However, formal R&D systems have paid less attention on farmers' own innovations that identify unique traits of varieties, and other agrobiodiversity conservation practices for climate resilience (Gupta, 1992). Farmers' varieties are generally referred to a mix of farmer conserved land races and in some cases, varieties developed and conserved by community based selection. In this paper, we use the term 'farmer varieties' to denote varieties developed, diffused and conserved through farmer selection, propagation and lateral diffusion. Only in a few cases, scientists may have characterized these varieties besides testing their reliability and stability (Balakrishna *et al.,* 2016, Selvi *et al.,* 2013, Rao *et al.,* 2013, Tumwegamire, 2011). Rarely but surely, some of these varieties have been used in crossing programme for developing new varieties. There have not been many cases when contributions of farmer breeders have been celebrated in a befitting manner except through the awards given by the President of India through grassroots innovation campaigns organized by National Innovation Foundation (NIF). For the first time, the Honey Bee Network organized a full-fledged session atthe International Plant Physiology Conference, New Delhi in 1988. Similar sessions on farmers' innovations were organized at International Agronomy Conference, 1990 and International Soil and Water Conservation Conference, 1994. The Indian Council of Agricultural Research has also honoured several farmer breeders. Several such farmers were invited to the first International Agrobiodiversity conference 2016. Perhaps, time has come for every scientific association to schedule at least one session to a dedicated dialogue among farmer innovators, breeders, scientists, policy makers and other stakeholders. Earlier, a review of postgraduate thesis in five agricultural disciplines, including plant breeding had shown a grave disjunction between sustainability imperatives and research trends (Gupta *et al.,* 1990).

In this paper, we provide a framework of partnership between formal and informal sectors and providing incentives to encourage reciprocal, responsible and respectful open innovations (Gupta *et al.,* 2016, Gupta, 2016). We also illustrate the system ecology view to understand the role of farmers, particularly women farmers and workers in conservation, characterization and augmentation of *ex-situ* and *in-situ* by agrobiodiversity collection.

FORGING PARTNERSHIP THROUGH NEW PATHWAYS

(a) *Changing the framework of collection and characterization of germplasm in ex-situ gene banks*

The plant breeders have made collections from roadside farms, local and regional markets and the fields of specific farmers. Seldom do we record the details of the germplasm provider or GPS coordinates (Hijmans *et al.,* 1999). It is not easy to compare the same germplasm as they may have been modified 1) during successive growouts on the research stations (Zeuli *et al.,* 1995) or 2) by the farmers through cycles of climate fluctuations and in some cases, change. In future, this anomaly may be corrected. Some scholars have argued that there is more genetic erosion in gene banks than at the community level (Fowler and Mooney, 1990). Disregarding some exaggeration here, the point remains that in a pulverized irrigated farm, genes, for roots that penetrate hard soil and go deep in search of moisture, may not survive. Similarly, many genes that will express only in a specific agro-climatic stress may not be expressed in a controlled micro-environment at the research stations. The case for *in-situ* conservation can be rested on this ground alone, though there are many other reasons. Different incentive models can be worked upon to encourage farmers to participate in *in situ* conservation of land races. In a study at IIMA in 2003, ten incentive models were proposed based on a matrix of material vs. nonmaterial and individual vs. collective incentives (Gupta and Chandok, 2010)

(b) *Characterization of germplasm using local knowledge system of women and other farmers: storage, food processing quality, therapeutic uses, eco-indicator function, weed-crop association and associated uses, etc., (Gupta, 1993, 2003; Nazarea, 2006)*

Sometimes, farmer breeders discover traits that might be missed by the institutional breeders (Brush and Meng 1998; Winarto and Ardhianto, 2007). Excessive focus on yield rather than reduction in cost, wastage or other inputs may be partly responsible for this mismatch (Ceccarelli *et al.,* 2013). Farmers may select for traits suitable for the condition of the particular field in which he wants to sow the crop next time (*i.e.* ecological niche) (Richards 1985; Gladwin, 1979) or may also select for different uses and different markets (Jarvis, 2008 in Maxted *et al.,* 2008, Smale *et al.,* 2001). A farmer from Gujarat, Thakarshi bhai had showcased his groundnut variety viz., morla at the session on farmers' innovations at International Conference on Plant Physiology, Vigyan Bhawan, New Delhi, 1988. In the presence of eminent breeders like Dr. Norman Borlaug, Dr. Swaminathan and others, he presented two traits of his variety which were often ignored by the groundnut breeders: (i) absence of ridges on the pods and (ii) strong peg. The first one prevented or reduced the amount of soil attached to the pod. The digging of pod became easier. The stronger peg reduced the number of pods left in the soil. The breeders while evaluating this variety went into traditional evaluation mode and parameters, did appreciate its higher oil content but failed to appreciate the importance of these two traits for future breeding. It had many other desirable characteristics such as taste and good yield even at low fertility conditions (Gupta *et al.,* 1997, 2001; Gupta, 1999, 2007).

Richa variety of pigeon pea, developed by Raj Kumar Rathore, had bearing behavior of pods from top to the bottom of the plant and is perennial. It gave higher yield due to better partitioning efficiency. Jagdish Parekh developed cauliflower variety known for three crops a year, besides potential for very large flower. Santosh Pachar, a very enterprising woman farmer developed a carrot variety (Laxmangarh Selection) which had long slender root and did not fork (National Innovation Foundation, 2016). Surjeet Singh developed *Surjeet Basmati-1,* a salt tolerant, high yielding paddy variety selected from the variety Pusa 1460. It is foot-rot and sheath blight resistant. It grows well even in degraded sodic or saline soils. All of these farmers have been awarded by the NIF at the Presidential Grassroots Innovation award function. There are many more farmer breeders who have selected unique traits while selecting crop varieties which have potential for enriching the repertoire of institutional breeders. Some of these traits may be useful in developing climate resilient varieties which could be a very useful *ex ante* coping strategy to deal with climate fluctuations.

Ex post coping strategies like salvage technologies are employed for minimizing loss over different scales like time, space or alternate usage. For example, sometimes after the floods, crop is harvested prematurely and used as fodder. This helps to save time and use residual moisture for timely sowing of the next crop. Alternatively, in the floodplains the stalks of the surviving paddy plants are cut and sowed, just like sugarcane sets and new plants emerge from the rooting at nodes. The asexual propagation saves farmer's time as he has to sow a new nursery otherwise.

Concurrent coping strategies employed by women include using weeds and other uncultivated plants for food during the lean months to tide over the stress period. For dealing with droughts in the crop season, Harbhajan Singh from Hissar employed a simple practice of irrigating cotton in alternate rows to reduce water usage by half and also minimize use of pesticide (due to lesser pest attack) without losing productivity.

In the wake of climate fluctuations, it has become all the more important to understand both optimal and sub-optimal coping strategies so that room for maneuver can be explored. The farmers may not know the linkage of mono or multi genic controlled traits with other variables that have to be traded off. Breeders can explain that and build the capacity of farmer breeders to blend modern scientific knowledge with their intuitive selection index.

(c) *Livestock-crop germplasm interaction*

Suitability of fodder, peculiar nutritional supplement quality, animal grazing on the crop to overcome apical dominance and change the phenotype, crop-animal interaction (suitability of paddy varieties for fish cultivation in the same field), suitability for allelopathy particularly since stress increases such effects etc., may play an important role in shaping the community selection criteria.

Scientists have verified the morphological modification effect of letting sheep graze on a standing crop of chickpea. Once the sheep nibble on the young shoots, the apical dominance is overcome triggering lateral branching. Cytokinin concentration in lateral buds of chickpea was found to increase within six hours of topping stimulating lateral branches (Turnbull *et al.,* 1997). Khattak *et al.,* 2007 reported that sheep and goats were made to graze over chickpea crop at the seedling stage by the farmers in the irrigated areas of Dera Ismail Khan, Pakistan as was also practiced in Mahendragarh district, Haryana, noticed during our field work there in 1983.

With better capture of sun light and pruning advantage of grazing, the yield goes up. Similarly, slow release of nutrients from the manure of sheep and goat droppings collected through penning is more advantageous for certain varieties of crops. The fodder quality has often been ignored even in the rainfed regions where livelihood security by small farmers is achieved primarily through livestock (Gupta, 1984). In Syria, farmers allowed frequent grazing of immature barley during the winter months and scientists noted that the farmers attached similar value to the straw quality and grain yield. Such observation led to extensive research on developing dual purpose barley (Nordblom, 1983).

Seldom do breeders screen F1 or F2 or even F4 for livestock palatability and digestibility. In the event of delayed or no rain towards the maturity period of crops, premature harvest for fodder purposes is advised to grow next crop on time. In the wake of too much rain, similar action is preferred because vegetative growth picks up and the ripening gets delayed or reduced. Suitability of varieties for such climate fluctuations is yet to be properly assessed.

The use of weeds as fodder, medicinal and food purposes, requires allelopathic effects to be relooked at. It has been shown that with the increase in the temperature, these effects in some of the varieties gets enhanced. Depending upon what minerals and other nutrients are provided by these weeds, the importance of inhibitory effects of crop vis-à-vis competitive effect of weed can be traded off. Comprehensive look at these relationships from household nutritional perspective is still awaited. In a recent study based on decadal survey data for 30 years from the same villages, eastern Uttar Pradesh, India, it was noticed that the weed diversity and quantity was higher in the case of improved traditional varieties of paddy such as Sarju 52 vis-à-vis new hybrid varieties.

(d) *Cultural attributes of wild varieties and their functional role as a source of important climate resilience traits*

One of the wild relatives of rice conserved in eastern Uttar Pradesh, *Oryza rufipogon* is allowed to be harvested by anyone irrespective of who owns the land. Generally, the poor landless workers or marginal farmers collect these easily shattered rice grains. On a particular festival, these grains are offered by the community members to a deity as a part of sacred ritual. At that time the price of this grain goes up and poor people are able to earn a amount from their efforts of collecting it. This weed grows in swampy areas of plots. It has also been used for developing a CMS line for further breeding [It has been screened for higher climate resilience and also for high yield-potential (Marri *et al.,* 2005)]. Importance of cultural institutions in conserving weeds or wild rice needs to be appreciated. Similarly, a weed sometime grown as a crop viz., *colona* (sama) grows in paddy field and is not often removed. We were intrigued as to why would a weed be conserved apart from its value as a sacred food in cultural rituals. Some studies have shown that it is a host for several rice pests (Bharati *et al.,* 1990; Tanwar *et al.,* 2010). It acts as an antifeedant for brown plant hopper (Kim *et al.,* 2008).

(e) *Farmers' varieties as heuristics of frugal breeding for frugal cultivation, what do farmers bring on breeders' table*

Farmer breeders, especially marginal farmers in marginal environment, select plants and consequent varieties which work well in low external input environment given the limited material resources. Invariably, these have better partitioning efficiency. Study of the selection criteria used by the farmers' breeders, one can proactively breed varieties for such constrained input environment. The design of farm machinery and the interculture practices depends upon the nature of rooting system of crops. If weeding is too close to the base of the crop affecting the root system, it may affect the yield adversely. The design of machinery for harvesting vegetable crops like tomato influenced the selection of thick skinned tomato varieties to enable mechanical harvesting. Later, the transportation and packaging influenced the flatter tomato shape instead of oval or round (de la Pena, 2013). The history of such adjustments has been well documented in the classic study, 'Hard Time, Hard Tomato' (Hightower, 1973).

The case for caring to listen, learn and leverage the knowledge of farmer breeders is well made by numerous varieties that have been awarded by the Honey Bee Network and its partner organizations including SRISTI, GIAN and NIF. Studies in drought and flood prone areas have shown that with an increase in climate fluctuations, need for flexible and multi line varieties cannot be ignored. Farmer breeders obviously select varieties in constrained environment. That doesn't mean that these varieties do not respond to better input environment conditions. In some cases, they do. However, given the increasing agro ecological heterogeneity, a need for giving farmers a semi stitched cloth rather than a readymade garment (Kumar, personal communication, 1979) cannot be over-stressed, i.e., instead of recommending a final and fixed variety, farming communities may be encouraged to select from segregating population genes for resilience. Some scientists like Ashby have already tried this for improving productivity.

The feed and fodder need of livestock have to be factored in the breeding of crops. Similarly, given the high tree density on the small farm lands in rainfed dry regions (Gupta, 1984a) indicates the reliance on livestock as a criteria for selecting cropping system and the relevant varieties. The tree shade and leaf manure also play a role in modifying soil fertility, providing wind break which in turn affect the productivity of a variety. The role of the crop, livestock, trees and craft interaction in the selection of appropriate varieties and their future evolution also remains to be properly explored.

There is a need for setting up a taskforce of national and international inter-disciplinary scientists to identify the dynamic breeding goals for a diversified agro-ecosystems jointly with farmers breeders. The Honey Bee Network and SRISTI, GIAN and National Innovation Foundation will be very happy to help in the matter.

It is also necessary that *in situ* storage and processing of crops and livestock products are given due importance to increase community resilience with climate fluctuations. Breeding of crops amenable to such uses will need fresh characterization of germplasm. The descriptors of different crops need to be redefined so that new traits identified by farmers women and men are incorporated in the process of characterization.

REFERENCES

Ashby, J.A., Quiros, C.A. and Rivers, Y.M. (1989). Farmer participation in technology development: Work with crop varieties. In: Chambers, R., Pacey, A. and Thrupp, L.A. (Eds.). Farmer first. Farmer innovation and agricultural research, IT Publications Ltd., London. pp. 15–132. Accessed at http://www.future-agricultures.org/farmerfirst/files/FF115-122Ashby_etal.pdf

Balakrishna, B., Reddy, V.C. and Reddy, K.V.S. (2016). Distinctness, uniformity and stability (DUS) characterization on fiber quality traits of cotton (Gossypium spp.) germplasm. *Environment and Ecology*, 34(1A), 262–266.

Bharati, L.R., Om, H. and Kushwaha, K.S. (1990). Alternate plant hosts of rice leaffolder. International Rice Research Newsletter, 15(4), 21–22. Accessed at http://www.cabdirect.org/abstracts/19911152105.html;jsessionid=437B3F1F45B D7C1B77E9E15403BD97CF#

Brush, S.B. and Meng, E. (1998). Farmers' valuation and conservation of crop genetic resources. *Genetic resources and crop evolution*, 45(2), 139–150.

Ceccarelli, S., Galie, A. and Grando, S. (2013). Participatory breeding for climate change-related traits. In *Genomics and breeding for climate-resilient crops* (pp. 331–376). Springer Berlin Heidelberg. Accessed at https://www.researchgate.net/publication/285990133_Participatory_Breeding_for_Climate_Change-Related_Traits

Cleveland, D.A. and Soleri, D., (2002). Indigenous and scientific knowledge of plant breeding: Similarities, differences, and implications for collaboration. In: Sillitoe, P., Bicker, A. and Pottier, J. (ed.) Participating in Development: Approaches to Indigenous Knowledge. Routledge, London.

de la Peña, C. (2013). Good to Think with: Another Look at the Mechanized Tomato. *Food, Culture & Society*, 16(4), 603–631.

Fowler, C. and Mooney, P.R. (1990). *Shattering: food, politics, and the loss of genetic diversity.* University of Arizona Press.

Gladwin, C.H. (1979). Cognitive strategies and adoption decisions: A case study of nonadoption of an agronomic recommendation. *Economic development and cultural change*, 28(1), 155–173.

Gupta, Anil K. (1984), matching Farmers' Concerns with Breeders' objectives, CMA, IIM Ahmedabad mimeo.

Gupta, Anil K. (1984a), Small farmer Household Economy in semiarid regions, CMA, IIMA mimeo.

Gupta Anil, K., Patel, N.T. and Shah Rekha, N. (1990). Review of Post-Graduate Research In Agriculture (1973–1984): Are We Building Appropriate Skills For Tomorrow? IIM, Ahmedabad. W.P. no. 843. 1990, p. 13.

Gupta, A.K. (1992). *Sustainability through biodiversity: Designing crucible of culture, creativity and conscience.* Ahmedabad: Indian Institute of Management.

Gupta Anil, K. (1993), Can We Add Without Subtraction: Some Awkward Questions for Revitalizing Agricultural Research Strategies, in Proceedings of First Agricultural Science Congress, Ed. Prem Narain, New Delhi, National Academy of Agricultural Sciences, 1993, 122–129.

Gupta, A.K. (1999). Making Indian agriculture more knowledge intensive and competitive: The case of intellectual property rights. *Indian Journal of Agricultural Economics*, 54(3), 343.

Gupta, A.K. (2003). Conserving biodiversity and rewarding associated knowledge and innovation systems: Honey bee perspective. *Intellectual property: Trade, competition, and sustainable development*, 3, 373–402.

Gupta, A.K. (2007). Harnessing Community and Individual Knowledge in Plant Genetic Resources Management. *Search for New Genes*, 311.

Gupta, A.K. and Vikas, C. (2010). *Cradle of Creativity: Strategies for In-Situ Conservation of Agro Biodiversity* (No. WP2010-09-03). Indian Institute of Management Ahmedabad, Research and Publication Department.

Gupta, A.K., Patel, K.K., Chand, P.G., Pastakia, A.R., Suthar, J., Shukla, S. and Sinha, R. (1997). Participatory research: Will the koel hatch the crow's eggs? Paper presented in the International Seminar on Participatory Research and Gender Analysis for Technology Development, organised by CIAT, Colombia, 1996; In: New Frontiers in Participatory Research and Gender Analysis, 209–243.

Gupta, A.K.; Dey, Anamika, R.; Shinde, Chintan; Mahanta, Hiranmay; Patel, Chetan; Patel, Ramesh; Sahay, Nirmal; Sahu, Balram; Vivekanandan, P.; Verma, Sundaram; Ganesham, P.; Kumar, Vivek; Kumar, Vipin; Patel, Mahesh and Tole, Pooja, (2016), Theory of open inclusive innovation for reciprocal, responsive and respectful outcomes: Coping creatively with climatic and institutional risks, Journal of Open Innovation: Technology, Market, and Complexity, December 2016, 2:16, 2:16. doi:10.1186/s40852-016-0038-8.

Gupta, Anil K. (2016), Grassroots Innovations: Minds on the margin are not marginal minds, New Delhi: Penguin Randomhouse.

Gupta, A.K. and Chandak, V.S. (2010). Cradle of Creativity: Strategies for in-situ conservation of Agro Biodiversity. W.P. No. 2010-09-03 Indian Institute of Management Ahmedabad 380 015, India.

Hightower, Jim (1973), Hard Tomatoes, Hard Times; Cambridge, Mass.: Schenkman Publishing Company, 1973, xiii + 268.

Hijmans, R.J., Schreuder, M., De la Cruz, J. and Guarino, L. (1999). Using GIS to check co-ordinates of genebank accessions. *Genetic resources and crop evolution*, 46(3), 291–296.

Jarvis, A., Lane, A. and Hijmans, R.J. (2008). The effect of climate change on crop wild relatives. Agriculture, Ecosystem and Environment, 126: 13–23.

Khattak, G.S.S., Ashraf, M., Zamir, R. and Saeed, I. (2007). High yielding DESI chickpea (Cicer arietinum L.) variety "Nifa-2005". *Pakistan Journal of Botany*, 39(1), 93.

Kim, J.H., Lee, S.Y., Myung, S.C., Kim, Y.S., Kim, T.H. and Kim, M.K. (2008). Clinical significance of the leptin and leptin receptor expressions in prostate tissues. Asian J Androl, 10, 923–928.

Marri, P.R., Sarla, N., Reddy, L.V. and Siddiq, E.A. (2005). Identification and mapping of yield and yield related QTLs from an Indian accession of Oryza rufipogon. *BMC genetics*, 6(1), 33. Accessed at https://www.ncbi.nlm.nih.gov/pmc/articles/PMC1181812

Maxted, N., Guarino, L., Myer, L. and Chiwona, E.A. (2002). Towards a methodology for on farm conservation of plant genetic resources. Genetic Resources and Crop Evolution, 49: 31–46.

Morris, M.L. and Bellon, M.R. (2004). Participatory plant breeding research: Opportunities and challenges for the international crop improvement system. *Euphytica*, 136(1), 21–35.

National Innovation Foundation (2016). http://nif.org.in/innovation/improved variety-of-carrot/525

Nazarea, V.D. (2006). Local knowledge and memory in biodiversity conservation. *Annu. Rev. Anthropol.*, 35, 317–335. Accessed at http://www.annualreviews.org/doi/pdf/10.1146/ annurev.anthro.35.081705.123252

Nkongolo, K.K., Chinthu, K.K.L., Malusi, M. and Vokhiwa, Z. (2008). Participatory variety selection and characterization of Sorghum (Sorghum bicolor (L.) Moench) elite accessions from Malawian gene pool using farmer and breeder knowledge. *African Journal of Agricultural Research*, 3(4), 273–283.

Nordblom, T.L. (1983). Livestock-crop interactions. *The case of green stage grazing. ICARDA Disc. Paper*, (9). Accessed at http://pdf.usaid.gov/pdf_docs/PNAAN861.pdf

Richards, P. (1985). Indigenous Agricultural Revolution: Ecology and Food Production in West Africa. Methuen, London.

Rao, L.V., Shiva Prasad, G., Chiranjivi, M., Chaitanya, U. and Surendhar, R. (2013). DUS Characterization for Farmer varieties of rice. *IOSR Journal of Agriculture and Veterinary Science (IOSR-JAVS)*, 4(5), 35–43.

Selvi, D.T., Srimathi, P., Senthil, N. and Ganesan, K.N. (2013). Distinctness, uniformity and stability (DUS) characterization on phenological traits and assessing the diversity of inbreds in maize (Zea mays L.). *African Journal of Agricultural Research*, 8(48), 6086–6092.

Smale, M., Bellon, M.R. and Aguirre Gómez, J.A. (2001). Maize diversity, variety attributes, and farmers' choices in Southeastern Guanajuato, Mexico. *Economic development and cultural change*, 50(1), 201–225.

Sillitoe, P., Bicker, A. and Pottier, J. (2002). Participating in development: Approaches to indigenous knowledge. ASA Monographs; 39. London: Routledge.

Tanwar, R.K.; Anand Prakash, S.K; Panda, N.C; Swain, D.K; Garg, S.P; Singh, S. Sathyakumar and Bambawale, O.M. (2010). Rice swarming caterpillar (Spodoptera mauritia) and its management strategies Technical Bulletin 24 National Centre for Integrated Pest Management, New Delhi. Accessed at http://www. ncipm.org.in/NCIPMPDFs/ Publication/Swarming_caterpillar.pdf

Tumwegamire, S. (2011). Genetic variation, diversity and genotype by environment interactions of nutritional quality traits in East Aftican sweet potato. Ph.D. thesis, Makerere University, Kampala, Uganda, pp. 1–105.

Turnbull, C.G.N., Raymond, M.A.A., Dodd, I.C. and Morris, S.E. (1997). Rapid increases in cytokinin concentration in lateral buds of chickpea (Cicer arietinum L.) during release of apical dominance. *Planta*, 202 3: 271–276. doi:10.1007/s004250050128

Winarto, Y. and Ardhianto, I. (2007). Becoming Plant Breeders, Rediscovering Local Varieties: The Creativity of Farmers in Indramayu, Indonesia. *The HoneyBee Magazine*, 18, 1–11. Acccessed at http://www.sristi.org/cms/files/creativity_of_farmers.pdf

Witcombe, J.R., Gyawali, S., Sunwar, S., Sthapit, B.R. and Joshi, K.D. (2006). Participatory plant breeding is better described as highly client-oriented plant breeding. II. Optional farmer collaboration in the segregating generations. *Experimental agriculture*, 42(01), 79–90.

Zeuli, P.L., Sergio, L. and Perrino, P. (1995). Changes in the genetic structure of wheat germplasm accessions during seed rejuvenation. *Plant breeding*, 114(3), 193–198.

Farmers Innovation for Climate Change

A.V. Balasubramanian

Centre for Indian Knowledge Systems, Chennai 600 035, India
E-mail: info@ciks.org; ciksorg@gmail.com

ABSTRACT: *According to the IPCC the green house gas (GHG) emissions from agricultural fields is about 10 to 12% of the global emission. If external factors including the deforestation for agriculture and other related activities are taken into consideration the emissions would rise to 32%. The focus here is on approaches based on sustainable agriculture which have been part of traditional farmers' practices as well as innovations building or adapting these practices to the present day situation. Some of the specific activities under both these approaches for reducing CHGs in agricultural systems are discussed here.*

Keywords: Sustainable Agriculture, Mitigation, Adaptation, System of Rice Intensification (SRI), Millets.

Agriculture is amongst the three major causes for GHG emissions compromising 10–12% of global emission according to intergovernmental panel on climate change (IPCC). However, these figures do not take into account "external factors" such as—agriculture driven deforestation. It is estimated that if such factors were taken into account the number would be 32% of global emissions. The major GHG contributions by agriculture are methane and nitrous oxide (The Royal Society; 2010; Ranjan *et al.,* 2013). In terms of dealing with climate change there may be two approaches namely, mitigation and adaptation. They can be broadly described as follows.

- Mitigation is an attempt at a gradual reversal of the effects of climate change. The approach would be to decrease and sequester GHG being released into the atmosphere.
- Adaptation is to make adjustments to the changes brought forth by climate change.

However, a particular intervention may be contributing to both the approaches.

The main focus would be on approaches based on sustainable agriculture which have been part of traditional farmers' practices as well as innovations building on or adapting these practices to the present day situation. Some of the approaches through which mitigation can be brought about are:

- avoidance of chemical fertilizers and herbicides,

- building soil carbon and soil fertility,
- avoiding bare soil, appropriate tillage, combining perennial and annual crops,
- sustainable livestock management,
- optimal manure management,
- improved grassland management,
- System of rice intensification (SRI),
- local production and consumption.

Some of the approaches through which adaptation can take place are—preventing and reversing soil erosion and restoring degraded land, drought and flooding resilience and water use efficiency, resilient crops, agro-genetic biodiversity, diversification and local farmer knowledge.

MITIGATION AND ADAPTATION—ILLUSTRATIONS FROM SUSTAINABLE AGRICULTURE

System of Rice Intensification (SRI) is an example of mitigation and the cultivation of resilient crops such as millets. In the following section some of the elements that contribute to mitigation and adaptation are outlined.

SRI and Climate Change Mitigation

The use of nitrogen fertilizers can be substituted by bio-fertilizers such as—Azolla, vermicompost, and neem cake. Overall nitrous oxide emissions are completely eliminated. The use of bio-fertilizers increases the organic matter in the soil, which in turn translates to an increase in carbon level in the soil. Organic SRI actually sequesters carbon. Carbon emissions from transportation of chemical fertilizers are also eliminated. SRI avoids the constant flooding of the fields. This reduces the amount of methane emissions.

SRI—Adaptation Component

In fact, as mentioned earlier, some of the interventions may contribute to both mitigation as well as adaptation. In the case of SRI it also contributes to adaptation in two different ways listed below because of the economy in the usage of water. Firstly, the fields are not constantly flooded and so water requirements are lower. Secondly, by using organic fertilizers the organic matter of the soil increases. Such soil stores 20–30 percent of its weight in water. Paddy grown under SRI methods grows healthier and stronger due to the decrease in root competition as seedlings are spaced out. As plants become stronger they are able to withstand the problems caused by erratic rainfall. Overall, SRI alongside organic agricultural practices has a great of potential of contributing to both mitigation and adaptation (Senthil *et al.*, 2009; Adhikari *et al.*, 2010).

Millets and Climate Change Adaptation

The cultivation of millets helps in adapting to the following three factors arising from climate change with respect to agriculture.

Decreased Rainfall

It is predicted that a decrease in the amount of rainy days will be one of the stresses brought forth by climate change. India and particularly Tamil Nadu are expected to suffer from water stress. Water dependence for millets is significantly less than that of crops such as rice and sugar cane.

Increased Pests

The changing climate is expected to lead to an increase in the number of pest. When grown under traditional methods most millets do not require any pesticides. In fact, several traditional varieties are quite hardy and have natural resistance against pests of the crop. Some millets such as foxtail can actually be used as anti-pest agents for storing sensitive crops.

Soil Deterioration

As a result of all the different stresses brought forth by climate change, the soils are also expected to deteriorate. Different varieties of millets can adapt to various forms of deteriorated soil such as, acidic, saline, sandy and low fertility soils.

PROSPECTS FOR THE FUTURE

Currently, many of the above practices such as SRI and the cultivation of millets is taking place largely due to the efforts of farmers with very little or no support from the Government or the State agencies. In general, the support offered by the State for various sustainable agriculture initiatives is quite small in comparison with the support received for chemical agriculture practices. Hence, the prospects for the spread of these practices to larger areas really depends on the support that can come from the State.

The ninth report of the Committee on estimates (2015–16) of the Lok Sabha on the "National Project on Organic Farming" of the Ministry of Agriculture (Dept. of Agriculture and Cooperation) suggested a series of steps to strengthen and support sustainable agriculture. Some of the observations made in this report are significant and have a great bearing on our topic.

- Reduce fertilizer subsidy from about ₹ 80,000 crore (which is the current annual figure) by three-fourths.
- To formulate a National policy on organic farming to systematically tackle the "... adverse consequences of chemical-based farming".
- Establishing an Organic Finance and Development Corporation.
- It can raise the level of employment in agriculture by 30%.

It is clear that there can be a very substantial increase in the area under SRI and millet cultivation if the recommendations of the committee are implemented and suitable schemes are drawn up along these lines.

REFERENCES

Adhikari, P., Sen, D. and Uphoff, N., 2010. System of Rice Intensification as a Resource-conserving Methodology: Contributing to Food Security in an Era of Climate Change. State Agricultural Technologists Service Association. *Annual Technical Issue,* 14. 26–44.

IFOAM, 2009. Organic Agriculture a Guide to Climate Change and Food Security. p. 23.

The Royal Society, 2010. Climate Change: A Summary of the Science. p. 16.

Ranjan, M., Senapati, Bhagirathi, B. and Ranjan, M.S., 2013. Impact of Climate Change on Indian Agriculture and its Mitigating Priorities. *American Journal of Environmental Protection,* 1(4): 109–11.

Senthil, K., Suresh, V., Parimala, K. and Arumugasamy, S., 2009. System of Rice Intensification. Centre For Indian Knowledge Systems. p. 16.

Sindhu, J.S., 2011. Potential Impacts of Climate Change on Agriculture. *Indian Journal on Science and Technology,* 4(3): 348–353.

Revisiting Resource Conservation Technologies for Climate Smart Agriculture

Sreenath Dixit

ICAR-Agricultural Technology Application Research Institute, Hebbal, Bengaluru 560 024, India

E-mail: atari.bengaluru@icar.gov.in

ABSTRACT: *Farming everywhere is weather dependent. In India, nearly 50 per cent of the land under cultivation depends on rainfall for source of moisture. Although farmers have been dealing with the uncertainties of rain-fall over the past 5000 years, never before in the history of agriculture farmers were seen as vulnerable to climate change as they are in the present context. However, the current climate change scenario can be looked at as a situation with a high degree of climate variability. Accordingly, several resource conservation technologies (RCTs) that have a potential to impart resilience to farming were demonstrated under the project National Innovations in Climate Resilient Agriculture (NICRA) across 100 vulnerable districts of the country. The basic assumption was that adoption of resource conservation technologies help farmers adapt to the high degree of climate variability witnessed in the recent past. The demonstrations were implemented through farmer participatory strategy by selecting appropriate RCTs suitable to particular agro-ecological condition. The demonstrations were grouped into four modules viz., Natural Resource Management (NRM); Crop Production; Livestock and Fisheries; and Institutional Interventions. Although NRM module dealt with demonstration of resource conservation technologies, it had implications on all other modules. The base line survey conducted at the beginning of the project provided clarity to the project implementing Krishi Vigyan Kendra (KVKs) to work out a feasible action plan including appropriate RCTs. The community was involved in deciding the appropriateness and potential benefits of the selected RCTs; which included farm-based as well as community-based interventions. This paper provides an account of the process adopted to drive home the point among the farming community that RCTs can indeed help them to cope with climate variability.*

Keywords: Vulnerability, Climate Variability, Krishi Vigyan Kendras, Techno-Social Approach.

Global climate change is more likely to affect 'subsistence' or 'smallholder' farmers, predominantly in developing countries. Their vulnerability to climate change comes from the fact that they are mostly located in the tropics, and that they belong to various socioeconomic, demographic, and policy backgrounds which tend to limit their capacity to adapt to change. However, these impacts will be difficult to model or predict because of (*i*) the lack of standardized definitions of these sorts of farming systems, (*ii*) intrinsic characteristics of these systems, particularly their complexity, their location-specificity, and their integration of agricultural and non-agricultural livelihood strategies and (*iii*) their vulnerability to a range of climate-related and other stressors (Morton, 2007).

In response to the growing concern over the impact of changing climate on agriculture, Indian Council of Agricultural Research embarked on a nation-wide initiative at the end of XI Five Year Plan, called the National Initiative on Climate Resilient Agriculture (NICRA). The strategic and applied research required for a long-term response to climate change was taken up a consortium of several leading research institutes of the country. While the technology demonstration component of the project sought to engage with the farming community to provide some immediate answers to the problem by demonstrating available agricultural technologies by adopting a techno-social approach.

Adaptation in agriculture to climate change is important for impact and vulnerability assessment and for the development of climate change policy. A wide variety of adaptation options has been proposed as having the potential to reduce vulnerability of agricultural systems to risks related to climate change (Smit and Skinner, 2002).

The districts of the country were ranked based on their vulnerability to climate variability (CRIDA, 2010). And 100 most vulnerable districts across the country were chosen for implementing the technology demonstration. Krishi Vigyan Kendra (KVKs) of each concerned vulnerable district was designated as the project implementing agency. The KVKs identified a cluster of villages representing the vulnerability of the district to implement the project. The project was implemented in four modules *viz.* Natural resource management; Crop Production; Livestock and Fisheries, and Institutional Interventions. A baseline survey was conducted to collect information about the climate related challenges faced by the communities and their response to such challenges. Different types of climate vulnerabilities in the selected sites across different zones of the country are depicted in Figure 1.

The KVKs were engaged in a series of vision workshops that enabled to prepare appropriate action plans to address the vulnerabilities faced by the communities. Emphasis was placed on proposing resource conservation technologies, to help farmers adapt to the problems of climate variability.

Under the NRM module, several *in situ* and *ex situ* rainwater harvesting interventions were implemented in drought-prone areas, while in areas prone to floods, better drainage was provided to minimize damage due to water-logging. Communality participation was ensured in implementing all NRM interventions.

The harvested rainwater was used judiciously by employing water saving devices to achieve higher water productivity (Table 1). Land configuration like broad bed and furrows, conservation furrows, mulching, green manuring, plowing and sowing across the slope, de-silting of water ways, irrigation channels, percolation ponds, farm ponds, check dams, community water bodies were taken up with integrated approach (Figure 2). The choice of the intervention was based on the local situation and feasibility at the site or the project. The overall impact of the resource conservation technologies not only demonstrated their potential to provide the communities with much needed resilience, but also instilled confidence among them to face situations due to changing climate.

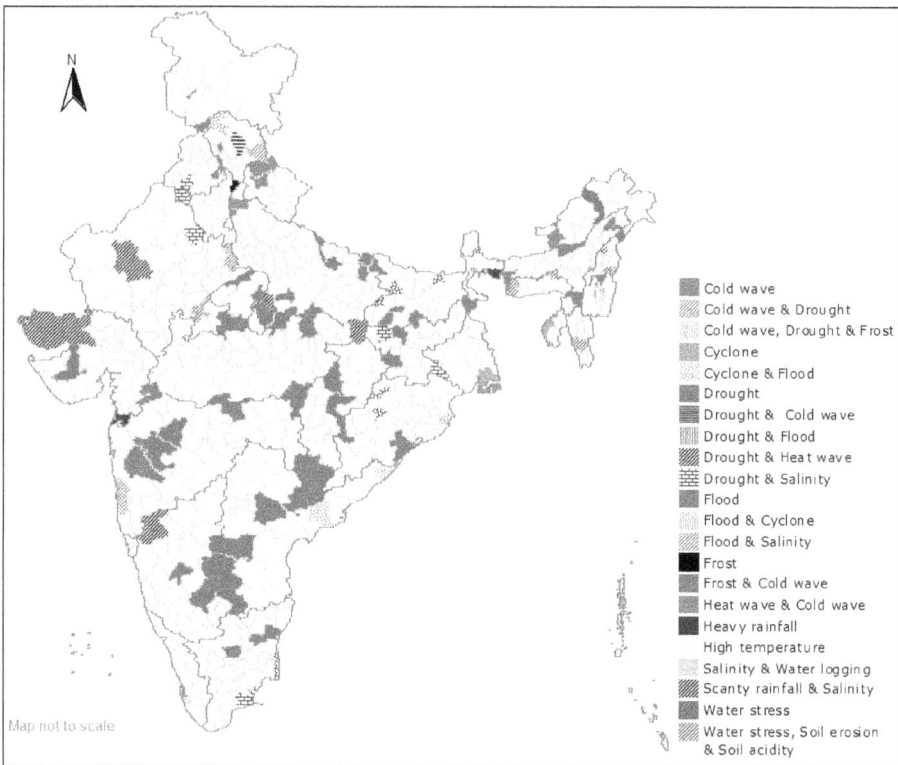

Legend:
- Cold wave
- Cold wave & Drought
- Cold wave, Drought & Frost
- Cyclone
- Cyclone & Flood
- Drought
- Drought & Cold wave
- Drought & Flood
- Drought & Heat wave
- Drought & Salinity
- Flood
- Flood & Cyclone
- Flood & Salinity
- Frost
- Frost & Cold wave
- Heat wave & Cold wave
- Heavy rainfall
- High temperature
- Salinity & Water logging
- Scanty rainfall & Salinity
- Water stress
- Water stress, Soil erosion & Soil acidity

Map not to scale

Fig. 1: Vulnerabilities Addressed Across the Sites by the Project

Table 1: Use of Harvested Rainwater for Improving Productivity: Aurangabad

Water Saving Method	Rainwater Harvested (m³)	Water Used (m³)	Irrigated Area (ha.)	Yield (q/ha)	Yield Increase (%)
Sprinkler irrigation from ponds	4600	2628.5	2.0	10.50	172.37

Fig. 2: Small Quantity of Harvested Rainwater by Ingenious Methods
Used Judiciously by Farmers

REFERENCES

CRIDA (2010). ICAR Network Project on Impact, Adaptation and Vulnerability of Indian Agriculture to Climate Change. p. 130.

Morton, J.F. (2007). The impact of climate change on smallholder and subsistence agriculture. *Proceedings of National Academy of Sciences*, 104 (50): 19680–19685.

Smit, B. and Skinner, M.W. (2002). Adaptation options in agriculture to climate change: A typology. *Mitigation and Adaptation Strategies for Global Change*, 7(1): 85–114.

Institutional Interventions Facilitating Climate Change Adaptation

A.K. Singh[1], Y.G. Prasad[2] and R. Roy Burman[3]

[1]Indian Council of Agricultural Research, New Delhi 110 012, India
[2]Agricultural Technology Application Research Institute, Santoshnagar,
Hyderabad 500 059, India
[3]Indian Agricultural Research Institute, New Delhi 110 012, India
[1]E-mail: aksicar@gmail.com

ABSTRACT: *Enhancing resilience of agriculture to climate risk is of paramount importance for protecting livelihoods of small and marginal farmers. The Indian Council of Agricultural Research (ICAR) launched the NICRA project in 2011 for enhancing resilience of Indian agriculture to climate change. Climate resilient technology demonstrations were undertaken by Krishi Vigyan Kendras (KVKs) in more than 100 vulnerable districts across the country. Districts were identified based on their exposure to recurrent climatic vulnerability such as drought, floods, cyclones, heat/cold stress etc. Bottom-up planning and participatory technology demonstration approach was adopted to enhance the coping ability of farmers to climate variability. Village Climate Risk Management Committee (VCRMC) was constituted in the selected villages comprising of villagers and empowering them to make decisions on appropriate interventions that are to be incorporated in local adaptation plans.*

Keywords: NICRA, VCRMC, Climate Smart Villages.

India accounts for about 2.3% of world's geographical area and 4.2% of water resources that support almost 18% of world's human population and 15% of the livestock. Agriculture remains the most important sector of Indian economy with 18% share in gross domestic product (GDP) at 2011–12 prices, 11% of exports and 53.3% share in total employment or workforce in 2013–14. Operational holding in the country has almost doubled from 71.01 million in 1970–71 to 138.35 million in 2010–11 (Agriculture Census, 2010–11). During the same period, the average size of operational holding has declined to 1.15 ha from 2.28 ha in 1970–71 (GOI, 2014). The small and marginal holdings taken together (below 2.0 ha) constituted 85.01% with operated area of 44.58% in 2010–11 against 83.29% with 41.14% of operated area in 2005–06.

Among the important cereals, India is the second largest producer globally of rice and wheat, and the top producer of pulses. India is also the largest producer of milk and second largest producer of groundnut, vegetables, fruits, sugarcane, and cotton. Long-term trends in household-level consumption patterns show that per capita direct consumption of food grains has been declining and of livestock products, fruits, and vegetables has been increasing for a fairly long time (Kumar *et al.*, 2007; Mittal, 2007; Chand, 2009). Total food grain demand is estimated to be 291 Mt by 2025 and 377 Mt by 2050, whereas the total production is estimated to be 292 Mt by 2025 and 385 Mt by 2050, which is 2.0% more than the demand. However, production deficits are projected for other cereals and pulses (33 and 3% in 2025 and 43 and 7% in 2050, respectively).

Agricultural production in India is closely linked to the performance of summer monsoon (June–September) which contributes about 75% of the annual precipitation. Apart from the inter-annual variability in summer monsoon rainfall, occurrence of many of the hydro-meteorological events is found to influence Indian agriculture at different spatial scales. Apart from the summer monsoon rainfall, India receives about 15% of annual precipitation during the winter months of December–March. This precipitation is very important for rabi (winter) crops.

In spite of the uncertainties about the precise magnitude of climate change at regional scales, an assessment of the possible impacts of changes in key climatic elements on agricultural resources is important for formulating response strategies (Rajeevan, 2013). Climate change projections made for India indicate an overall increase in temperature by 1 to 4°C and precipitation by 9 to 16% by 2050s (Krishna Kumar *et al.*, 2011). It is important to identify regions that are more vulnerable to climate change and variability in order to develop and target appropriate adaptation measures. Vulnerability refers to the propensity of the entity to face a climate shock and suffer loss in production and/or income from agriculture, though the latter is not always specified explicitly. The vulnerability assessment of 572 rural districts in India was carried out by Rama Rao *et al.* (2013). At present, many districts in states of Rajasthan, Gujarat, Maharashtra, and Karnataka and some districts in Andhra Pradesh, Uttar Pradesh, Bihar, Uttarakhand, and Jharkhand exhibit high and very high vulnerability. Most districts along the eastern and western coast and northeastern states are less vulnerable. By midcentury (2021–50), districts in Rajasthan, Gujarat, Madhya Pradesh, Karnataka, Maharashtra, Andhra Pradesh, Tamil Nadu, eastern Uttar Pradesh, and Bihar will exhibit very high and high vulnerability.

There are many evidences of the impacts of climate change and variability on Indian agriculture. Birthal *et al.* (2014) projected the effects of climate change on crop yields for three time scales (2035, 2065, and 2100) at minimum and maximum changes in temperature and rainfall. In general, the production of pulses will be affected more by climate change than other crops. By the year 2100, with a significant change in climate, the yield of chick pea and pigeon pea will be lower by around 25% vis-a-vis without climate change. The climate impacts on cereals will vary widely in rainy season as well as winter seasons. In the winter season, wheat yield will be less by about 22%, almost 3 times that of barley. Similarly, among rainy season cereals, rice will be affected more than maize and sorghum by the changing climate.

Rice yield will decline by over 15% with significant changes in climate as compared to loss of 7% in sorghum and of 4% in maize. By 2065, India's population is likely to cross 1.7 billion mark demanding more and diversified foods. With climate change, ensuring food security with more food production under limited resources will be a big challenge. It is, however, necessary for farmers and other stakeholders to adapt to climate change and reduce the losses. Simple adaptations, such as change in crop variety, planting dates, rainwater conservation, adoption of resilient intercropping systems, particularly in rainfed areas could help in reducing impacts of climate change. For example, losses in wheat production can be reduced from 4–5 to 1–2 Mt, if a large percentage of farmers could change to timely planting.

INITIATING CLIMATE CHANGE ADAPTATION

Farmers need to intelligently adapt to the changing climate in order to sustain crop yields and farm income. Enhancing resilience of agriculture to climate risk is of paramount importance for protecting livelihoods of small and marginal farmers. In the context of climate change and variability, farmers need to adapt quickly to enhance their resilience to increasing threats of climatic variability such as droughts, floods and other extreme climatic events. Adoption of climate resilient practices and technologies by farmers appears to be more a necessity than an option in the current scenario of increasing frequency in occurrence of climate aberrations.

The Indian Council of Agricultural Research (ICAR) launched the National Initiative on Climate Resilient Agriculture (NICRA) project in 2011 with the aim of enhancing resilience of Indian agriculture to climate change and climate variability. Climate resilient technology demonstrations were undertaken by Krishi Vigyan Kendras (KVKs) in 121 vulnerable districts across 28 states in the country. Districts were identified based on their exposure to recurrent climatic vulnerability such as drought, floods, cyclones, heat/cold stress etc.

INSTITUTIONAL INTERVENTION

Bottom-up planning and participatory technology demonstration approach was adopted to enhance the coping ability of farmers to climate variability. In each of the 121 climatically vulnerable districts, a representative village cluster was selected for field implementation. A village climate risk management committee (VCRMC) was established with the approval of *gram sabha*. An action plan was prepared to address the climate vulnerability faced including technology interventions covering agriculture and allied sectors such as horticulture, animal husbandry, poultry and fisheries. Accordingly, technology interventions fell into three modules (1) Natural resource management (2) Crop production systems (3) Livestock and fisheries production systems. A fourth module was the most crucial for successful project implementation which comprised of *institutional interventions* and support mechanisms such as establishment of custom hiring centre for farm implements to promote and facilitate adoption of climate resilient field interventions, seed and fodder bank, common interest groups for water, various commodities and marketing groups for climate literacy through establishment of manual weather stations and preparation of monsoon action plan.

Climate Smart Villages

Climate smart agriculture is an integrative approach to address the interlinked challenges of food security and climate change that explicitly aims for three objectives: (1) sustainably increasing agricultural productivity to support equitable increases in farm incomes, food security, and human development; (2) adapting and building resilience of agricultural and food security systems to climate change at multiple levels, and (3) reducing greenhouse gas (GHG) emissions from agriculture (including crops, livestock, and fisheries) to the extent possible (FAO, 2013). The concept of climate resilient villages (CRVs) consists of implementing these resilient practices at a scale to cover the entire village in a saturation mode depending on the resource endowments of the farmers with one or several interventions for imparting resilience to the production systems. Though it is planned to saturate the entire village gradually with resilient interventions, the number and kind of interventions implemented are largely determined by the resources available, vulnerability status of the village and involvement of communities (Ch. Srinivasa Rao *et al.*, 2016).

INNOVATIVE INSTITUTIONAL SETUP IN CRVS AND THEIR ROLES

Village Climate Risk Management Committees (VCRMC) have been constituted, each comprising of 12–20 villagers with nominated members as President, Secretary and Treasurer. A bank account is opened in the name of VCRMC and is operated by any two signatories. The committee fixes the charges for hiring of different implements and hiring rates are displayed prominently. Farmers' contributory share towards inputs like seeds, fertilizer, animals etc. is also deposited in the bank account. The revenue and expenditure details must be shared with the general body (Prasad *et al.*, 2015).

Custom Hiring Centers (CHCs) for farm implements were established in 121 NICRA villages which could successfully empower farmers to tide over the shortage of labour and improve efficiency of agricultural operations. A committee of farmers, nominated by the *gram sabha* manages the custom hiring centre. The rates for hiring the machines/implements are decided by the VCRMC, an innovative institutional mechanism put in place at the village level for management of the custom hiring centre for farm machinery. This committee also uses the revenue generated from hiring charges for repairs and maintenance of the implements and 25% of the revenue goes into the sustainability fund. There are 27 different types of farm machinery stocked in 121 CHCs, the most popular are rotavator, zero till drill, drum seeder, multi-crop planter, power weeder and chaff cutter. Each centre was established at a capital cost of Rs. 6.25 lakhs funded by the NICRA project (Prasad *et al.*, 2014).

Seed bank for access to improved crop cultivars: Village level seed production of short duration, drought and flood tolerant varieties was taken up by farmers and seed societies in several NICRA villages with the technical support of KVKs in rice, wheat, soybean, foxtail millet, greengram, pigeonpea, finger millet, chickpea, wheat, rapeseed and mustard. It has become a regular practice to source seed of drought tolerant and short duration cultivars from few NICRA villages as interested farmers and seed societies have taken up this as a livelihood activity. Examples include SIA3085 and Suryan of foxtail millet in Kurnool, ML-365 of finger millet in Tumkur, JS-93-05 of soybean in Satna & Datia, many paddy varieties in several districts in Bihar, Jharkhand, Odisha, Chhattisgarh and Valsad, Gujarat. VCRMCs facilitate seed bank activities in the village. Spread is through farmer to farmer sale as truthfully labeled seed (Prasad *et al.*, 2015 and Reddy *et al.*, 2015).

Fodder Bank: Availability of improved varieties of fodder seed for delayed planting situation is a major constraint for fodder production in NICRA villages. Examples include Yagantipalle village of Kurnool, where green and dry fodder shortage was acute. Fodder bank was established in the village under NICRA project in an area of 10 acres with high yield in hybrid Napier varieties viz. APBN-1, CO-3 and CO-4. Fodder stem cuttings were supplied to farmers from this fodder bank in the village. The fodder was shared by 40l and less farmers who had purchased crossbred cows. The fodder area in the village has increased from 16.18 acres in 2011 to 79.2 acres in 2016. The green fodder shortage was reduced from 86% to 36% within four years of NICRA project. In several NICRA villages, seeds of improved cultivars of fodder sorghum, maize, pearlmillet, berseem, lucerne and oats were produced for use in regular and contingency situations. About 528 farmers participated in this activity in 148 ha area (Prasad *et al.*, 2015).

LESSONS LEARNT

Adaptation to climate change and climate variability are context and location specific. Community awareness, participation and involvement in decision making was crucial for enhancing the coping ability of farmers. This was achieved through establishment of VCRMC empowering the villagers to make decisions on appropriate interventions that were to be incorporated in local adaptation plans. Many of climate smart practices and technologies required access to farm machinery and implements. Increased access of such equipment to small and marginal farmers was demonstrated through establishment of custom hiring centers. Since changes in rainfall and increase in temperature lead to shortening of growing season of many crops, adaptation to such contingency situations is possible only when access to short duration and stress tolerant varieties is facilitated. The immediate consequence of drought situation was fodder shortages for livestock. Establishment of seed and fodder bank at village level was helpful to the community for continued adaptation.

REFERENCES

Birthal, P.S., Khan, M.J., Negi, D.S. and Agarwal, S., 2014. Impact of climate change on yields of major food crops in India. Implications for food security. *Agric. Econ. Res. Rev.*27 (2), 145–155.

Chand, R., 2009. Demand for Food Grains during the 11th Plan and Towards 2020. Policy Brief No. 28. National Centre for Agricultural Economics and Policy Research, New Delhi, p. 1–4.

Ch. Srinivasa Rao; Gopinath, K.A.; Prasad, J.V.N.S.; Prasannakumar and Singh, A.K., 2016. Climate Resilient Villages for Sustainable Food Security in Tropical India: Concept, Process, Technologies, Institutions, and Impacts. Advances in Agronomy (Ed.) Donald L. Sparks, Volume 140, p. 101–206.

FAO, 2013. Climate Smart Agriculture Source Book. Food and Agriculture Organization of the United Nations, Rome, Italy.

GOI, 2014. Agriculture Census 2010–11. All India Report on Number and Area of Operational Holdings. Ministry of Agriculture, Government of India, p. 87.

Krishna Kumar, K., Kamala, K., Rajagopalan, B., Hoerling, M.P., Eischeid, J.K., Patwardhan, S.K., Srinivasan, G., Goswami, B.N. and Nemani, R., 2011. The once and future pulse of Indian monsoonal climate. *Climate Dynam.* 36(11), 2159–2170.

Kumar, P., Mruthyunjaya, D. and Dey, M.M., 2007. Long-term changes in food basket and nutrition in India. *Econ. Polit. Wkly.* 42(385), 3567–3572.

Mittal, S., 2007. What affect changes in cereal consumption? *Econ. Polit. Wkly.* February, 444–447.

Prasad, Y.G., Maheswari, M., Dixit, S., Srinivasarao Ch, Sikka, A.K., Venkateswarlu, B., Sudhakar, N., Prabhu Kumar, S., Singh, A.K., Gogoi, A.K., Singh, A.K., Singh, Y.V. and Mishra, A., 2014. Smart Practices and Technologies for Climate Resilient Agriculture. Central Research Institute for Dryland Agriculture (ICAR), Hyderabad. p. 76.

Prasad, Y.G., Srinivasa Rao, Ch., Prasad, J.V.N.S., Rao, K.V., Ramana, D.B.V., Gopinath, K.A., Srinivas, I., Reddy, B.S., Adake, R., Rao, V.U.M., Maheswari, M., Singh, A.K. and Sikka, A.K., 2015. Technology Demonstrations: Enhancing resilience and adaptive capacity of farmers to climate variability. National Innovations in Climate Resilient Agriculture (NICRA) Project, ICAR-Central Research Institute for Dryland Agriculture, Hyderabad. p. 109.

Rajeevan, M., 2013. Climate change and its impact on Indian agriculture. In: Shetty, P.K., Ayyappan, S. and Swaminathan, S. (Eds.), Climate Change and Sustainable Food Security. National Institute of Advanced Studies, India, pp. 1–12.

Rama Rao, C.A., Raju, B.M.K., Subba Rao, A.V.M., Rao, K.V., Rao, V.U.M., Ramachandran, K., Venkateswarlu, B. and Sikka, A.K., 2013. Atlas on Vulnerability of Indian Agriculture to Climate Change. Central Research Institute for Dryland Agriculture, Hyderabad p. 116.

Reddy, R.G., Sudhakar, N., Srinivasa Rao, Ch., Prasad, Y.G., Dattatri, K., Chari Appaji, Prasad, J.V. and Reddy, A.R., 2015. Compendium of Climate Resilient Technologies. Agricultural Technology Application Research Institute (Zone-V), CRIDA Campus, Santoshnagar, Hyderabad-500059, Telangana, India. p. 83.

Framework of Entrepreneurial Venture in Climate Smart Agriculture

Dipak De

Institute of Agricultural Sciences, Banaras Hindu University,
Varanasi 221 005, India
E-mail: dipakde1953@gmail.com

ABSTRACT: *Climate-Smart Agriculture (CSA) is an approach that helps the farming community to transform and reorient agricultural systems to ensure food security and support development effectively in a changing climate, where 'Job Creator' not 'Job Seeker' is the present day need of the country. Government of India has taken up important initiatives in this direction by launching 'Start-up India-Stand-up India'. Startup India-is a flagship initiative of Government of India for building a strong ecosystem of Entrepreneurship Development in context of Climate Smart Agriculture.*

The Framework of Entrepreneurial Venture proposed in this paper argues that there is an interactive dynamics for evolving entrepreneurial process among individuals to create venture. It has been explained enlisting four important Phases of venture creation namely i) Identification of Enterprise; ii) Initiation/Creation of enterprise; iii) Nurturing of enterprise and iv) Transition Phase by an individual known as Entrepreneur. The activities carried out by an entrepreneur will make them to be identified as (1) Mobile entrepreneurs (2) Managerial entrepreneurs (3) Innovative entrepreneurs (4) Empire Builders by creating Entrepreneurial Firms like 1. Economic Core Firms 2. Constrained Growth Firms 3. Ambitious and 4. Glamorous Firms.

Keywords: CSA (Climate Smart Agriculture), Entrepreneur, Entrepreneurship, Phases of Entrepreneurship Development, Typology of Entrepreneurs, Typology of Entrepreneurial Farm.

All round development of agriculture is possible only with effective exploitation of entrepreneurial behaviour skills as well as material resources. Our country has abundant material resources and human resources. So, we can identify individuals in all segments of the farming population, who have the requisite entrepreneurial behaviour skills. The changing climate scenario which has increased risk in the farming occupation has a flip side also. The farming community should make it an opportunity by mitigating the risk through adopting various entrepreneurial ventures in Climate Smart Agriculture. The entrepreneur is an economic man, who tries to maximise his profits by identification and adoption of innovations. However, the entrepreneurs are not simply innovators but they are the persons with a will to act to assure risk and bring about a change through organisation of human efforts. It plays a key role in economic development of the country. Importance of development of entrepreneurship as an ingredient of economic development has been recognised long time back (Himachalam, 1990).

FRAMEWORK OF ENTREPRENEURIAL VENTURE

Entrepreneurship process starts when the entrepreneur recognizes an opportunity in the environment. There are certain words viz. Entrepreneur, Entrepreneurship and Entrepreneurial Behaviour which requires elaboration. For a long time there was no equivalent term for 'Entrepreneur' in English language. Three words were commonly used to connote the sense the French term carried; adventurer, undertaker and projector; these were used interchangeably and lacked the precision and characteristics of a scientific person (Gopakumar, 1995). Hence the term entrepreneur did not find any prominence in the history of economic thought. Ramana (1999) defined entrepreneurs as those people who work for themselves. The word entrepreneur derived from French word "Entreprendre" meaning to "Undertake". The entrepreneur is thus a person who organizes and manages an activity/ organization, undertaking the risks for fulfilling some of his needs. His job involves the quality of boldness, courage, dynamism and risk taking in sufficient measure. The entrepreneur in this context is defined as one who could start new activity or a new enterprise which is a deviation from his traditional family occupation or profession. De (1986) stated that a farmer does not become an entrepreneur only by adopting a new agricultural technology but he becomes an entrepreneur only when he comes to be an operator of a farm business. A business involves rational decision on investment after assessing risk, other alternatives and possibilities or profit and loss. An entrepreneur is a dynamic agent of change or the catalyst who increasingly transforms physical, natural and human resources into corresponding production possibilities.

Entrepreneurship is that factor which urges an individual to take advantage of favourable situations by understanding innovative practices with a concern for excellence and assessment of self and the environment. Entrepreneurship is a way of life, a thought process, to bring any sustainable change; effort has to be broader based. Cole (1949) stated that entrepreneurship comprises any purposeful activity that initiates, maintains or develops a profit oriented business in interaction with the internal situation of the business or with the economic, political and social circumstances surrounding the business. Hegan (1998) stated that entrepreneurship is not only conceiving the idea behind a venture but also designing and maintaining the organization for carrying it out. According to Schumpter (1970) entrepreneurship is a function of group level pattern, a function of managerial skill and leadership, an organizational building function, a function of high achievement, input complementing and gap filling, a function of status withdrawal and a function of social, political and economic structure.

Singh (1986) stated that Entrepreneurial Behaviour was a function of an individual's personality characteristics and environmental factors, it was represented as $EB = f (PE)$ where EB = Entrepreneurial Behaviour, P = Personality characteristics and E = Environmental factors. Entrepreneurial Behaviour has been defined as package of personality characteristics and environmental factors related to dynamic agent of change for transforming physical, natural and human resources into corresponding production possibilities.

PHASES OF ENTREPRENEURIAL VENTURE

Framework for entrepreneurial venture has been developed in line with Ramachandran and Ray (1998). Entrepreneurial activities originate from individuals; the framework of entrepreneurial venture is described as under:

- **Phase–I:** *Identification of enterprise* - by an entrepreneur may come from unsatisfied need of people as well as unsatisfied personal need. This is reflected by the alertness of entrepreneur him/herself, his ability to identify opportunity, family and educational background, professional experience, formal and informal networks etc.
- **Phase–II:** *Initiation/Creation of Enterprise*—The entrepreneur needs to create an organization for transforming the concept into a marketable product by accumulating and combining physical and other resources. The ideas of entrepreneurs and environment interact to turn the ideas into reality. The new venture takes away large portion of time and attention of entrepreneur from the management of environment to the management of organization. Interaction with several components like bankers, regulatory agencies, experts and advisors, suppliers and so on entrepreneur alone has to constantly interact with them to have the control over the enterprise that he/she created.
- **Phase–III:** *Nurturing of Enterprise*—Organisation translates the business concept into marketable product and offers it to the customer. The entrepreneur gets feedback on market response in terms of profitability, sales etc. The feedback is measured in terms of efficiency, competitiveness, effectiveness, innovativeness, flexibility etc.
- **Phase–IV:** *Concluding/Transition Phase*—Entrepreneurial process does not end with achieving stability and reaches success. At this stage organizations require different managerial style as venture competition gradually builds up more and more.

TYPOLOGY OF ENTREPRENEUR

Ramachandran and Ray (1998) described the typology of entrepreneurs based on the outcome of entrepreneurial ventures. Empirical evidence suggests the existence of following four types of entrepreneurs.

1. *Mobile Entrepreneurs:* They are the persons who leave the ventures as soon as venture is created. They are the true Schumpeterian entrepreneurs as

according to Schumpeter the entrepreneurial activity ends as soon as the venture is created. Ted Nieren berg, founder of Dansk designs in the US. Mohan of Good knight and Vikram Sarabhai both from India come closer to this type.

2. *Managerial Entrepreneurs:* They are the persons who prefer to continue in the same venture by transforming themselves to fit the changing demands. As for example persons like Henry Ford of Ford Motors, Pierre S. Du Pont, and George Eastman of Kodak.

3. *Innovative Entrepreneurs:* They are the persons who create an organization and remain engaged in their pursuits of innovation and creation of novel product's and technology. Walt Disney, Ibuka and Akio Morita of Sony Corporation, Sochiro llinda of llinda motor Company and Bill Gates of Microsoft are some of the notable entrepreneurs belonging to this type.

4. *Empire Builders:* They are the persons engaged in creating chain of new ventures having an ownership. They have the qualities of vast vision, flair of innovation and managerial capability to build an empire for themselves. John D. Rockfeller of the US, J.N. Tata and Ghanshyam Birla both from India, Konosuke Matsushita of Japan and Chung Ju Yung of Hyundai and Kim Woo Chong of Daewoo both from South Korea are some of examples of this type.

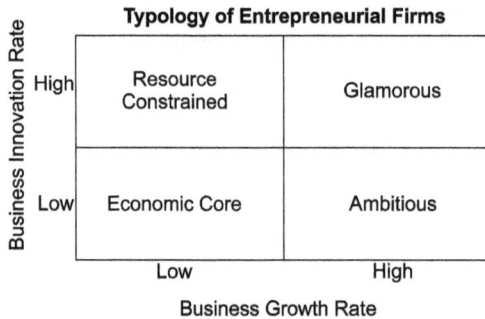

Typology of Entrepreneurial Firms

High	Resource Constrained	Glamorous
Low	Economic Core	Ambitious
	Low	High

Business Innovation Rate (vertical axis)

Business Growth Rate (horizontal axis)

TYPOLOGY OF ENTREPRENEURIAL FIRMS

This is based on rates of innovation and growth, developed by Kirchhoff (1994). He has developed a matrix having four classes of entrepreneurial firms, viz., 1. Economic core firms 2. Constrained growth firms 3. Ambitious firms and 4. Glamorous firms.

1. *Economic Core Firms:* These firms are low in innovation and are low growth firms. These firms are most common form of entrepreneurship. These firms primarily satisfy the owner's desire for independence. It fulfils a specific need in a small market and it does not obtain a significant growth. Initially it achieves a degree of growth. But once it reaches a size that can satisfy the owner's needs, growth stops.

2. *Constrained Growth Firms:* These firms have high rates of innovation, they lack adequate resources for growth, therefore it fails to grow. These firms

can be classified into two classes, i. Firms unable to acquire the needed resources and ii. Firms internally make decision to constrain growth. The owner of resource constrained firms is willing to pay the cost of resources but is unable to attract the capital needed to sustain high growth. Whereas the owner of an internally constrained firm makes decisions not to acquire the resources needed to support high growth. These firms are easy prey to better financed competitors.

3. *Ambitious:* These firms have high growth rate with limited number of innovations in a large market. A single successful product can sustain high growth for many years. The growth rate will eventually decline if new products are not introduced because markets are not stable.

4. *Glamorous:* These firms have high growth rate with high rates of innovation. These firms are technology branded firms which require regular updating. Kirchoff calls these firms glamorous firms because they often attract high media attention and receive local and national awards for their success.

CONCLUSION

The conceptual framework presented here identifies several key components of entrepreneurs. Entrepreneurship process starts when the entrepreneur recognizes an opportunity in the environment. Entrepreneurship Development process goes through i. identification of enterprise ii. Creation of enterprise iii. Nurturing of enterprise and iv. Concluding/transition phase of enterprise. The creation is 1. Economic Core Firms 2. Constrained Growth Firms 3. Ambitious and 4. Glamorous firms.

REFERENCES

Cole, A.H. (1949). Entrepreneurship and entrepreneurial history. In change and the entrepreneur. Prepared by the research centre in entrepreneurial history. Harvard University Press, Cambridge.

De, D. (1981). A *study of axiomatic approach to theory building in adoption of farm technology.* Ph.D. Thesis (Unpublished). Division of Agricultural Extension, IARI, New Delhi.

De, D. and Rao, M.S. (2001) Entrepreneurial behaviour of farmers: An Axiomatic theory. Ganga Kaveri Publishing House, D.35/77 Jangamwadi Math Varanasi.

De, Dipak (1986). Factors affecting entrepreneurial characteristics of farmers. *Indian Journal of Social Work*, Vol. XLVI, No. 4, pp. 541–546.

De, D. (1985). Status symbol and innovative entrepreneurship are predictors of farmers progressivism. *Journal of Extension System*, Vol. 1, No. 1, pp. 81–84.

Gopakumar, K. (1995). Entrepreneurship in economic thought: A thematic review. *Journal of Entrepreneurship*, 4(1), pp. 1–17.

Hagen, E.E. (1998). Theory of social change: *How economic growth begins*, Dorsey Press, Illinois, New York.

Himachalam, A. (1990). Developing entrepreneurial industrialist–An alternate approach. Productivity, Vol. 34(2), pp. 281–287.

Kirchhoff, Bruce A. (1994). Entrepreneurship and Dynamic Capitalism: The Economics of Business Firm Formation and Growth Praeger Publishers 240 pages.

Ramachandran, K. and Ray, Sougata (1998). A framework and entrepreneurial typology for developing a comprehensive theory of entrepreneurship. In Entrepreneurship and innovation: Models for development. Ed Kanungo R.N. Sage Publications, New Delhi. pp. 40–63.

Ramana, A.V. (1999). Entrepreneurship and economic development. Kurukshetra. Vol 48, No. 2, pp. 8–10 & 45.

Schumpter, J.A. (1970). The entrepreneur as innovator. Reading in management (Second edition) Mc. Grew Hill. New York.

Schumpter, J.A. (1961). *The theory of economic development: An enquiry into profits, capital, interest and the business cycle.* Cambridge, Harvard University Press. Cambridge.

Singh, P.N. (1986). Factors influencing entrepreneurship. Developing entrepreneurship for economic development. Vikas Publishing House Pvt. Ltd. New Delhi, pp. 46–92.

Farmer Innovation System: Rethinking the Way We Look at Farmer Innovations[1]

Bharat S. Sontakki* and S.P. Subash

National Academy for Agricultural Research Management, Rajendra Nagar, Hyderabad 500 030, India
*E-mail: bharatss@naarm.ernet.in

ABSTRACT: *Growing evidences, the world over, are pointing towards a new perspective on 'farmer innovation system' consisting of on-farm innovations and innovators. With strong roots in social learning and co-learning, this framework presents a holistic view on farmer innovations as manifestations of adaptability to local problems including those caused by climate change, besides the curiosity and opportunity dimensions. The present paper tries to understand what is innovation? what are its types? and who are those innovators? These questions are answered by an exploratory research carried out on profile of documented farmer innovations in India. Results indicate that most of the farm innovations are outcome followed by process, and innovators have diverse socio-economic profile. Research and policy support are imminent for mainstreaming 'informal farmer innovation system' for livelihood secure and climate resilient farming. The paper recommends facilitative role by formal institutions for an enabling ecosystem to promote farmer innovations for local adaptations.*

Keywords: Farmer Innovation, Farm Innovator, Farmer Innovation System.

Farmers have been innovators ever since they started domestication of animals and plants for human consumption. Farmer-led research and small-scale agricultural innovations are fundamental for sustainable farming solutions (MacMillan and Benton, 2014). But the critical role of farmers' innovation in addressing the emerging challenges like climate change and food system is not acknowledged (Smith and Bargdon, 2015).

Driven by creativity, necessity and opportunity, informal experimentation and innovation by farmers have always taken place. However, such knowledge and expertise of farmers are historically undervalued. Farm level innovations are often unaccounted, ignored and not available to scientists and academicians in formal sector (Beckford and Baker, 2007b).

---

[1] We would like to acknowledge the review document by Smith and Bargdon (2015), which had influenced the current article to a great extent. Views expressed in this article are authors and do not reflect views of the organization they are affiliated to.

Most of the documented farmer innovations remains as grey literature in project reports or documents (Wettasinha *et al.,* 2014). A few such efforts in India worth mentioning are, grass root innovations by Honey Bee Network (Gupta *et al.,* 2003), Farm Innovators (ICAR, 2010), Farmer Innovations (GoI, 2011) and Grassroot innovators (Sastry and Tara, 2014). Though these documents had highlighted the importance of on-farm innovations, efforts to understand the role and measure on-farm innovation are still in a nascent phase (Lapple *et al.,* 2015). Globally, few studies which had explored them at the farm level are Diederen *et al.* (2003), Karafillis and Papamgiotou (2011), Ariza *et al.* (2013), Wu and Zhang (2013), Lapple *et al.* (2015). These studies emphasize the need to promote on-farm innovation for empowerment of farmers (Lema and Kapange, 2006; Dolinska and d'Aquino, 2016).

Set against this backdrop, the paper delves on the role of farmer innovations in a larger perspective of livelihood security and empowerment of farming communities. In doing so, the paper examines ways to comprehensively understand 'farm-innovation', and 'farm-innovator' and argues for creating, nurturing and sustaining 'enabling eco-system' for promoting them in a value chain approach.

The literature on farmer innovation can be categorised into three paradigms: i) conventional agricultural innovation ii) agricultural innovation systems and iii) farmers' innovation system (see Table 1). These paradigms have evolved over time, but they had not replaced the older ones rather they coexist and are gradually being implemented (Coudel, 2013).

Table 1: Understanding Innovation in Agriculture

Features	*Conventional Agricultural Innovation*	*Agricultural Innovation System*	*Farmers Innovation System*
Focus	Innovation for farmers	Innovation with farmers	Innovation by farmers
Innovation type	Output	Process	Output & Process
Primary actors	Formal institutions and organisations	Formal institutions and organisations	Farmers
Role of formal sector	Innovate and facilitate technology transfer	Facilities research process and technology adoption	Provide resource and facilities
Role of farmers	Adopt new technologies	Participate in innovation process	Innovate and adapt
Type of Innovation	Modern varieties and farm management practices	Modern varieties, farm management practices and alternative ways of organising	Adaptation of modern varieties and practices, integration of knowledge system, on-farm experimentation
Major themes in literature	Investment in R&D, improving technology transfer	Investment in R&D and extension services, multiple stakeholder platforms, participatory research	Innovation as a social learning process, building social capital, roles of supporting actors

Source: Adopted from Smith and Bargdon (2015).

Conventionally, agricultural innovation is defined as invention, output or a concrete improvement (Berdegue, 2005). The conventional technology transfer model emphasizes on development of technologies by a formal institution and transferring them to farmers. This definition focuses on development of technologies by public or private institutions with intellectual property rights by professional researchers for the benefit of farmers (Rotman, 2013). The science and technology policies of most of the countries aim at investment in developing technologies and improving the diffusion rates (OECD, 2014). Roger's theory (Roger, 1962) on diffusion of innovation dominates this view which states how such innovations could benefit the society (Water-bayers *et al.,* 2009). These studies focus on how public R&D through technology change could lead to economic development. Another set of literature also argues on role of private firms in gradual alleviation of poverty by incentivising them and strengthening intellectual property (Rotman, 2013). There is also an increasing body of literature on development of technologies which are more relevant and applicable to end users through participatory plant breeding (Witcomb, 1996; Chambers *et al.,* 1989; Halewood, *et al.,* 2007). The level of participation varies with respect to the role and stage. In this model again role of farmer is passive (Sperling *et al.,* 2001). The criticism against the conventional understanding of innovation as linear technology diffusion was, as it doesn't reflect the complexity of the whole agricultural system (Spielman *et al.,* 2009; Coudel 2013; Smith *et al.,* 2014), less benefits to farmers (Wettasinha *et al.,* 2014), equity issues (Hall *et al.,* 2001) and other issues such as erosion of indigenous knowledge, biodiversity loss and degradation of natural resources (Water-Bayer *et al.,* 2009). These criticisms had drawn the discussion on agricultural innovation into a holistic view from being an outcome to a process of social transformation (Spielman *et al.,* 2009; Kraemer-Mbula and Wamae, 2010).

Agricultural Innovation Systems (AIS) perspective emerged out of complex systems theory and was widely promoted in works of Hall and Clark (1995), Engel (1997), Hall *et al.* (2001, 2003). The innovation system was considered as a social process embedded within the complex systems (Spielman *et al.,* 2009). AIS perspective views innovation as a process where innovation is done by formal institutions with farmers. Innovation system is defined as bringing together various actors (public, private, civil) for developing a product or process into economic use. The key difference from the conventional understanding of innovations was that it accounts not only technology development but also on institutional and organisational arrangements (Yang *et al.,* 2014). The scope of innovation also broadened in terms of access to markets, labour and technology (Adjei-Nsiah *et al.,* 2008; Ton *et al.,* 2014; Pamuk *et al.,* 2014). This literature focuses on engagement of farmers and other formal actors to improve the relevance and applicability of innovation (Smith and Bargdon, 2015). The consensus developed from this study was that farmers should have a greater influence on the innovation (Douthwaithe, 2002; Klerkx and Leeuwis, 2008; Klerkx *et al.,* 2006; Neef and Neubert, 2011; Poulton *et al.,* 2010; Ton *et al.,*

2015). However, the role of formal institutions in introduction and diffusion of innovation remain the same as conventional approach (Kraemer-Mbula and Wamae, 2010). This becomes challenging as the generic technologies develop by such institutions may not suit to the context specific problems (Smith *et al.*, 2014). Also the challenges in re-orientation of institutional perspectives and asymmetrical power relationship between the actors remains a biggest hurdle (Kraemer-Mbula and Wamae, 2010; Wettasinha *et al.*, 2014). This approach has so changed the direction of flow of knowledge and expertise.

Recently, there is an increasing body of literature emphasising on farmer's capacity to innovate (C2I). This 'farmer innovation' perspective recognises the capacity of farmers to innovate and formal institutions performs a supporting role (Water-Bayer *et al.*, 2007). The fundamental concept is that farmers are innovators themselves and have the capacity to innovate, experiment and adapt. This was coined in various studies as embedded innovation process (Van Rijn *et al.*, 2012), farmer-led innovation (Wettasinha *et al.*, 2014), informal innovation system (Lapple *et al.*, 2015), demand driven innovation (von Hippel 2005, 2007, Kramer-Mbula and Wamae, 2010) and grass root innovation system (Seyfang and Smith, 2007; Smith *et al.*, 2014; Sastry and Tara, 2014). Social learning and co-learning are underlying concepts to this perspective (Smith and Bargdon, 2015). These works emphasize upon collaborative partnership between formal institutions and farmers and redefined the role of such institutions as supportive and farmers as active and central to innovation. Though this approach is gaining recognition, it is rarely discussed in policy or scientific communities (Smith *et al.*, 2014). Such approach to innovation has a potential to improve the adaptive capacity of farmers especially in experimenting with new varieties or practices (Tittonell *et al.*, 2012) and bring institutional innovation (Cheetri *et al.*, 2010; Chhetri and Esterling, 2010) in the context of climate change.

The current work is an exploratory study carried out through literature review supported by desk research. To understand the typology of innovations, we reviewed various documented innovations (primarily drawn from ICAR 2010, SFAC 2012 and Kalpana Sastry and Tara (2014) and classified them based on the framework developed by Smith and Bargdon (2015). To understand the socio-economic characteristic of innovators we used data of innovators docu-ment of ICAR named 'Farm innovators 2010' (ICAR 2010). This document consists of 163 farm innovators in various categories from different states of the country. Their profile was analysed using summary statistics.

Farmer innovation types: Farmer innovations could be a tangible outcome (technology or techniques) or a process which can derive a tangible outcome (new ways of organising) or a mix of both (Smith and Bargdon, 2015). We classified the documented innovations in to these three categories (Table 2). The output could be technology or institutional change and the process could be application of traditional knowledge, management practices or adaptation of innovations (exogenous) to local contexts. The results show that most of the documented innovations are outcome, followed by process and a fewer in the third category. Wettasinha *et al.* (2014) also reported similar trend, whereby the innovations are mostly hard (product based) rather than soft (social-institutional based). Though we tried to look in to the innovations that are relevant to climate change, we could not find many in this compendium. The very rationale that most of the farmer innovations are manifestations of adaptation to locally relevant problems and issues including climate resilience is thus, evident from our observations.

Table 2: Types of Innovations

Source	Types of Innovations (%)			Total (number)
	Outcome	Process	Outcome and Process	
ICAR (2010)	66.91	24.46	8.63	139
SFAC (2012)*	65.00	23.00	12.00	100
Kalpana Sastry & Tara (2015)	46.15	23.08	7.69	13

Note: * SFAC (2012) document on Profile of Agricultural innovators in India has the innovations which belong to conventional paradigm of innovation (Institutional driven) also non-farmer innovations.

Farm innovators: To understand the profile of farm innovators, we analysed the socio-economic characteristics of the innovators documented in Farm Innovators (ICAR 2010). The data presented in Table 3 shows the profile diversity of farm

Table 3: Profile of Farm Innovators

Variable	Observa-tions	Mean	Standard Deviations	Minimum	Maximum
Age	137	48	12.55	21	92
Farming experience	134	26	13.71	5	70
Landholding*	139	7.74	14.73	0	100
Beneficiaries**	13	2232.08	4680.24	50	15000

Education	Frequency	Percentage
Illiterate	19	13.67
Primary	8	5.76
Middle	20	14.39
Secondary	45	32.37
Higher Secondary	11	7.91
Graduate and above	34	24.46
Others	2	1.44

Note: * Converstions 1 Bigha = 2.5 ha, 1 Kani = 0.167 ha. ** Only data of 14 innovators available and a farmer who reported 5 lakh beneficiaries were removed to moderate the results.

Source: Calculated by Authors based on ICAR (2010).

innovators in terms of age, education, farming experience and landholding. Limited data available on number of beneficiaries reached by farm-innovators makes it difficult to draw any generalizations. These results indicate that farm-innovators vary a great deal in terms of their age, farming experience, education and land-holding. The fact that necessity, opportunity and curiosity are the main drivers of farmer-innovations, thus, gains substantial ground from the trend of observed results. Similar observations were recorded by Wettasinha *et al.* (2014).

CONCLUSION AND WAY FORWARD

The literature reviewed and empirical analysis of farm innovations and farm innovators establishes the need to relook at innovation in a much larger perspective than is being done currently. Growing evidences suggest that social-learning and localized problem-solving are deeply embedded in the concept and practice of on-farm innovations by farmers. Farmers do have the capacity to innovate, in spite of being most vulnerable to biotic, abiotic and market shocks and food and nutritionally insecure. Nurturing this capacity calls for in-depth research and policy support to create and enabling ecosystem with both pull and push measures. Aligning public policy incentives with farmers' motivations to innovate will encourage the type of innovation that yields public benefit, will promote diversity, and contribute towards a climate-resilient farming system.

REFERENCES

Adjei-Nsiah, S., Kuyper, T.W., Leeuwis, C., Abekoe, M.K., Cobbinah, J., Sakyi-Dawson, O. and Giller, K.E., 2008. Farmers agronomic and social evaluation of productivity, yield and N2-fixation in different cowpea varieties and their subsequent residual N effects on a succeeding maize crop. *Nutrient Cycling in Agro ecosystems*, 80: 199–209.

Ariza, C., Rugeles, L., Saavedra, D. and Guaitero, B., 2013. Measuring innovation in agricultural firms: A methodological approach. *Electr J Kn Manage*, 11: 185–198.

Beckford, C., Barker, D. and Bailey, S., 2007. Adaptation, innovation and domestic food production in Jamaica: Some examples of survival strategies of small-scale farmers. *Singapore Journal of Tropical Geography*, 28: 273–286.

Chambers, R., Pacey, A. and Thrupp, L.A. (eds.), 1989. Farmer first: Farmer innovation and agricultural research. Intermediate Technology Publications London.

Chhetri, N.B., Chaudhary, P., Tiwari, P.R. and Yadaw, R.B., 2012. Institutional and technological innovation: Understanding agricultural adaptation to climate change in Nepal. *Applied Geography*, 33: 142–150.

Chhetri, N.B. and Easterling, W.E., 2010. Adapting to climate change: Retrospective analysis of climate technology interaction in the rice-based farming system of Nepal. *Annals of the Association of American Geographers*. 100: 1156–1176.

Coudel, E., (ed.) 2013. Renewing innovation systems in agriculture and food: How to go towards more sustainability? Wageningen Academic Publishers.

Diederen, P., van Meijl H., Wolters, A. and Bijak, K., 2003. Innovation adoption in agriculture: Innovators, early adopters and laggards. *Cah. Econ. Sociologie Rurales*, 67: 30–50.

Dolinskaa, A. and Aquinob, P., 2016. Farmers as agents in innovation systems. Empowering farmers for innovation through communities of practice, *Agricultural Systems*, 142: 122–130.

Douthwaite, B., 2002. Enabling innovation: a practical guide to understanding and fostering technological change, ZED Books: London.

Engel, P.G.H., 1997. The social organization of innovation: A focus on stakeholder interaction. Royal Tropical Institute, Amsterdam.

Gupta, A.K., Sinha, R., Koradia, D., Patel, R., Parmar, M., Rohit, P. and Vivekanandan, P., 2003, 'Mobilizing grassroots' technological innovations and traditional knowledge, values and institutions: Articulating social and ethical capital. *Futures*, 359: 975–987.

Hall, A. and Clark, N., 1995. Coping with change, complexity and diversity in agriculture—The case of rhizobium inoculants in Thailand. *World Development*, 239: 1601–1614.

Hall, A., Bockett, G., Taylor, S., Sivamohan, M.V.K. and Clark, N., 2001. Why research partnerships really matter: innovation theory, institutional arrangements and implications for developing new technologies for the poor. *World Development*, 29: 783–797.

Hall, A., Sulaiman, R., Clark, N. and Yoganand, B., 2003. From measuring impact to learning institutional lessons: an innovation systems perspective on improving the management of international agricultural research. *Agricultural Systems*, 78: 213–41.

ICAR, 2010. Farm Innovators 2010, Division of Agricultural Extension Indian Council of Agricultural Research', New Delhi Available from: http://wwwicarorgin/files/Farm-Innovators-2010pdf. Accessed on [17 December, 2015].

Sastry, K.R. and Tara, O.K., 2014. Rural innovation @ Grassroots-Mining the minds of masses. National Academy of Agricultural Research Management, Hyderabad 500 030, India.

Karafillis, C. and Papanagiotou, E., 2011. Innovation and total factor productivity in organic farming. *Applied Economics*, 43: 3075–3087.

Klerkx, L. and Leeuwis, C., 2008. Institutionalizing end-user demand steering in agricultural R&D: Farmer levy funding of R&D in The Netherlands. *Research Policy*, 373: 460–472.

Klerkx, L., De Grip, K. and Leeuwis, C., 2006. Hands off but strings attached: The contradictions of policy-induced demand-driven agricultural extension. *Agriculture and Human Values*, 232: 189–204.

Kraemer-Mbula, E. and Wamae, W., 2010. Innovation and the Development Agenda OECD/IDRC.

Läpple, D., Renwick, A. and Thorne, F., 2015. Measuring and understanding the drivers of agricultural innovation: Evidence from Ireland. *Food Policy*, 51: 1–8.

Lema, N.M. and Kapange, B.W., 2006. Farmers organisation and agricultural innovation in Tanzania. The sector policy for real farmer empowerment KIT. Case study available from http://wwwkitnl/sed/wpcontent/uploads/publications/912_Case%20Study%20Tanzaniapdf [26 December, 2015].

Mac Millan, T. and Benton, T.G., 2014. Engage farmers in research. *Nature*, 509: 25–27.

OECD, 2013. Agricultural Innovation Systems: A Framework for Analyzing the Role of the Government. OECD Publishing Available from http://dxdoiorg/ 101787/ 9789264200593-en.

Pamuk, H., Bulte, E. and Adekunle, A.A., 2014. Do decentralized innovation systems promote agricultural technology adoption? Experimental evidence from Africa. *Food Policy*, 44: 227–236.

Pingali, P.L., 2012. Green Revolution: Impacts, limits, and the path ahead. *Proceedings of the National Academy of Sciences*, 10931: 12302–12308.

Rogers, E.M., 1962. Diffusion of innovation New York: Free Press.

Rotman, D., 2013. "Why we will Need Genetically Modified Foods" MIT Technology Review Available from http://wwwtechnologyreviewcom/featuredstory/522596.

Seyfang, G. and Smith, A., 2007. Grassroots innovations for sustainable development: Towards a new research and policy agenda. *Environmental politics*, 164: 584–603.

SFAC (Small Farmers' Agribusiness Consortium), 2012. *KrishiSuthra*, Small Farmers' Agribusiness Consortium (SFAC), New Delhi. Available online on http://agritech. tnau.ac.in/farm_innovations/KRISHI_SUTRARA.pdf. Accessed on [17 December, 2015].

Shiferaw, B.A., Okello, J. and Reddy, R.V., 2009. Adoption and adaptation of natural resource management innovations in smallholder agriculture: Reflections on key lessons and best practices Environment. *Development and Sustainability*, 11: 601–619.

Smith, C. and Bargdon, S.H., 2015. Small-scale farmer innovation system: A review of current literature. Quaker United Nations Office, Geneva.

Smith, A., Fressoli, M. and Thomas, H., 2014. Grassroots innovation movements: Challenges and contributions. *Journal of Cleaner Production*, 63: 114–124.

Sperling, L., Ashby, J.A., Smith, M.E., Weltzien, E. and McGuire, S., 2001. A framework for analyzing participatory plant breeding approaches and results. *Euphytica*, 1223, pp. 439–450.

Spielman D.J., Ekboir J. and Davis K. 2009. 'The art and science of innovation systems inquiry: Applications to Sub-Saharan African agriculture' *Technology in Society*, vol. 314: 399–405.

Tittonell, P., Scopel, E., Andrieu, N., Posthumus, H., Mapfumo, P., Corbeels, M. and Mkomwa, S., 2012. Agroecology-based aggradation-conservation agriculture ABACO: Targeting innovations to combat soil degradation and food insecurity in semi-arid Africa. *Field Crops Research*, 132: 168–174.

Ton, G., Klerkx, L., Grip, K. and Rau, M., 2015. Innovation grants to smallholder farmers: Revisiting the key assumptions in the impact pathways. *Journal of Food Policy*, 51: 9–23.

Van Rijn, F., Bulte, E. and Adekunle, A., 2012. Social capital and agricultural innovation in Sub-Saharan Africa. *Agricultural Systems*, 108: 112–122.

Von Hippel, E., 2005. Democratizing innovation: The evolving phenomenon of user innovation. *Journal fürBetriebswirtschaft*, 551: 63–78.

Waters-Bayer, A., van Veldhuizen, L., Wongtschowski, M. and Wettasinha, C., 2009. Recognising and enhancing processes of local innovation, In Sanginga, PC (ed.). Innovation Africa: Enriching farmers' livelihoods. Earthscan, pp. 239–254.

Wettasinha, C., Waters-Bayer, A., van Veldhuizen, L., Quiroga, G. and Swaans, K., 2014. Study on impacts of farmer-led research supported by civil society organizations Penang, Malaysia: CGIAR Research Program on Aquatic Agricultural Systems Working Paper AAS, pp. 2014–40.

Wu, B. and Zhang, L., 2013. Farmer innovation diffusion via network building: A case of winter greenhouse diffusion in China. *Agriculture and Human Values*, 30: 641–651.

Yang, H., Klerkx, L. and Leeuwis, C., 2014. Functions and limitations of farmer cooperatives as innovation intermediaries: Findings from China. *Agricultural Systems*, 127: 115–125.

NICRA: Towards Adaptation and Mitigation of Climate Change
Thus Far and Way Forward

M. Prabhakar* and Ch. Srinivasa Rao

Central Research Institute for Dryland Agriculture, Santoshnagar, Hyderabad 500 050, India
*E-mail: mprabhakar@crida.in

ABSTRACT: NICRA is a flagship network programme launched by the Indian Council of Agricultural Research with the objective of developing and identifying technological and other options to enhance the resilience of Indian agriculture in crops, livestock and fisheries to climatic variability and climate change; to demonstrate the site specific technology packages on farmers' fields for adapting to current climate risks; and for capacity building in climate resilient agricultural research and its application. As a result, a large platform related to climate change research has been created in the country. There are some positive lessons and experiences emerging out of technology demonstration component, but there is still considerable need to continue this activity to identify and demonstrate technologies that help deal with climate change.

Keywords: Mitigation of Climate Change, International Panel on Climate Change, NICRA.

International Panel on Climate Change (IPCC) in its Fifth Assessment Report observed that 'Warming of climate system is unequivocal, and since the 1950s, many of the observed changes are unprecedented over decades to millennia. The atmosphere and oceans have warmed, the amount of snow and ice have diminished, and sea levels risen' (IPCC, 2014). As per the latest IPCC report, Earth is warmer by 0.75°C since pre-industrial time (1850). The years 1995–2006 had 12 warmest years since 1850. India accounts for 4.5% of the total global GHG emissions, and as per COP21 Paris Agreement India is committed to reduce Green House Gas emissions by 33–35% from the levels on 2005. This goal has to be achieved by 2030 (BUR, 2015). The impacts of climate change are global, but countries like India are highly vulnerable in view of its large population depending on agriculture. The predicted temperature rise for India is in the range of 0.5–1.2°C by 2020, 0.88–3.16°C by 2050 and 1.56–5.44°C by the year 2080, with different regions expected to experience differential change in the amount and distribution of rainfall. It is evident from the shifting weather patterns in recent years, causing increased rainfall variability (8 out of 15 years between 2000 and 2015 were drought years), heat and cold waves, hail storms, rising sea levels contaminating coastal freshwater reserves and increased frequency of flooding affecting many areas.

Agriculture sector contributes to about 17.4% of India's GDP. Studies have indicated significant negative impacts of climate change in India, predicted to reduce yields by 4.5 to 9.0 %, depending on magnitude and distribution of warming. A 4.5 to 9% negative impact on production implies a cost of climate change to be roughly up to 1.5% of GDP per year. Therefore, Govt. of India has accorded high priority on research and development to cope with climate change in general and agriculture in particular. The Prime Ministers National Action Plan on Climate change has identified Agriculture as one of the 8 National Missions.

To meet the challenges of sustaining domestic food production in the face of changing climate, to generate technologies towards adaptation and mitigation of climate change in agriculture to support and articulate the country's views at different global fora like UNFCCC, the Indian Council of Agricultural Research (ICAR) launched a flagship network project *'National Initiative on Climate Resilient Agriculture'* (NICRA) during XI Plan in February 2011, and in the current Plan period it is referred to as *'National Innovations in Climate Resilient Agriculture'* (NICRA).

OBJECTIVES AND COMPONENTS

The major objective of NICRA is to generate and identify technological and other options that help enhance the resilience of Indian agriculture in crops, livestock and fisheries to climatic variability and climate change; to demonstrate the site specific technology packages on farmers' fields for adapting to current climate risks; and to enhance the capacity (knowledge, skill and management) of scientists and other stakeholders in climate resilient agricultural research and its application. The project encompasses 4 major components, viz. Strategic Research (40 ICAR Institutes), Sponsored and Competitive Grants (18 + 34 Projects), Technology Demonstration (121 KVKs, 25 AICRPAM, 23 AICRPDA Centers, 7 core institutes of ICAR), Capacity Building and Knowledge Management.

Adaptation and mitigation are two complementary strategies for responding to climate change. Adaption is the process of adjustment to actual or expected climate change and its effects in order to either lessen or avoid harm or exploit beneficial opportunities. Mitigation is the process of reducing emissions or enhancing sinks of green house gases, so as to limit future climate change (IPCC, 2014). Considerable progress has been under the project towards adaptation and mitigation of climate change.

SALIENT ACHIEVEMENTS OF NICRA

Over the past five years considerable progress has been made under the NICRA project at several partner ICAR Research Institutes, State Agricultural Universities, KVKs & NGOs across the country towards climate resilient agriculture. One of the major achievements of the project is creation of *state of the art* infrastructure for climate change research such as High throughput plant phenomics facilities, Free air temperature enrichment (FATE), Carbon dioxide temperature gradient chamber (CTGC), Eddy covariance towers, automatic weather stations (AWS), Satellite data reception system, Rainout shelter facility, Animal calorimeter, CO_2 Environmental chambers, Custom designed animal sheds, Research shipping vessels, etc. for supporting strategic research towards understanding how plant, animal and fish species respond to changing climate. This effort is long term in nature and thus warrants commitment of resources. Following are some of the salient achievements (SrinivasaRao *et al.*, 2016):

- Identified the districts that are relatively more vulnerable to climate change.
- Standardization of the techniques for measurement of GHG emissions for different production systems, marine ecosystems and quantification of carbon sequestration potential of major agroforestry systems in the country.
- Extensive phenotyping of germplasm and breeding in rice, wheat, maize, pigeon pea and tomato to multiple abiotic stresses.
- Mapping unique traits for thermal tolerance in livestock, invention of heat care mixture for poultry, development of several technologies consisting of feed, breed and shelter management to cope with heat stress and diseases in livestock.
- Relationships were established between increase in Surface Sea Temperature (SST) and catch and spawning in major marine fish species.
- Simulation modeling was used for assessing climate change impacts at regional/national level.
- Real time pest dynamics in the changing climate was studied and developed weather based forewarning systems for major crop pests.
- Initiated efforts to identify and fine tune NRM technologies viz. biochar, conservation agriculture (CA) and emission reduction through efficient energy management that offer adaptation and mitigation benefits.

Technology demonstration component (TDC) is being implemented in farmer participatory mode in the climatically most vulnerable districts of the country through 121 Krishi Vigyan Kendras (KVKs) spread across the country in 28 States & one Union Territory. Location specific technologies developed by the national agricultural research system, which can impart resilience against climatic vulnerability, are being demonstrated in a representative village. The technologies demonstrated are broadly divided into four modules, natural resource management, crop production, livestock and fisheries and the creation of institutional structures for scaling up of interventions.

WAY FORWARD

NICRA is a unique project, which brings all sectors of agriculture viz. crops, horticulture, livestock, fisheries, NRM and extension scientists on one platform for addressing climate concerns. It is very important to sustain the efforts made in the past few years and take forward the project for some more years. Over the past five years, the *state of the art* infrastructure facilities have been established, standardized and put into function in core institutes of ICAR to undertake the climate change research. Manpower (Scientists, Research Associates, Research Fellows, Technical Officers, etc.) have been trained to handle and operate these facilities. However, some of these precious research facilities are yet to be utilized to full potential. In other words, a large platform related to climate change research has been created in the country. Crop improvement for multiple stresses takes several years of research and multi location testing. Efforts made under this project, in some cases resulted in development of varieties/hybrids ready for large-scale cultivation. Whereas, many are under different stages of development, which may require few more years to be released as variety/hybrid/breed. Simulation modeling to assess the impact of climate change at regional level is still at initial stage. Standardization of minimum data sets and compilation of data from different sources have shown good progress. In the next phase, these data sets will be used for modeling. Capacity building for this activity will be emphasized and a dedicated group will be formalized. Research, essentially long term in nature, should continue further to achieve the intended outputs and outcomes.

Though there are some positive lessons and experiences emerging out of technology demonstration component, there is still considerable need to continue this activity to identify and demonstrate technologies that help deal with climate change. In fact, the technologies found to be performing well are getting fed into programs such as NMSA. There is still a need to develop variety of adaptation options for different sub-sectors within agriculture, for different regions and for farmers with varying resource endowments. Such an effort is to be accompanied by identification of factors that help adopt technologies on a wider scale.

The commitment of the country for emission reduction requires to generate appropriate information and data on emissions as well as options that help reduce emissions. Techniques standardized so far under NICRA for estimation of GHG emissions from different management practices will be used for further reducing the carbon footprint of production systems in the country. Government of India has committed for the reduction of emission intensity of GDP by 32–35% by 2030 from 2005 levels. The outputs of NICRA project are contributing to several national project reports like, intended nationally determined contribution (INDC), biennial update report (BUR), nationally appropriate mitigation action (NAMAs), national mission on sustainable agriculture (NMSA) and several other Missions under National Action Plan on Climate Change. The system-wide impacts and responses to climate change need to be understood more comprehensively. The efforts in this direction, which have begun recently have to be taken through their logical course, for such an understanding is necessary to identify and prioritize various adaptation options.

REFERENCES

BUR, 2015. First Biennial Update Report to the United Nations Framework Convention on Climate Change. Ministry of Environment, Forest and Climate Change Government of India, 184 p.

IPCC, 2014. Climate Change 2014: Synthesis Report. Contribution of Working Groups I, II and III to the Fifth Assessment Report of the Intergovernmental Panel on Climate Change [Core Writing Team, R.K. Pachauri and L.A. Meyer (eds.)]. IPCC, Geneva, Switzerland, 151 p.

SrinivasaRao, Ch., Prabhakar, M., Maheswari, M., SrinivasaRao, M., Sharma, K.L., Srinivas, K., Prasad, J.V.N.S., Rama Rao, C.A., Vanaja, M., Ramana, D.B.V., Gopinath, K.A., SubbaRao, A.V.M., Rejani, R., Bhaskar, S., Sikka, A.K. and Alagusundaram, K., 2016. National Innovations in Climate Resilient Agriculture (NICRA), Research Highlights 2015-16. Central Research Institute for Dryland Agriculture, Hyderabad, 112 p.

Capacity Building for Climate Resilient Agriculture

Power of ICT to Connect Each Farm and Each Farmer—mKRISHI® CCA: A Case Study

Dinesh Kumar Singh

Innovation Lab, Tata Consultancy Services Ltd., Mumbai 400 601, India
E-mail: dineshkumar.singh@tcs.com

ABSTRACT: *Knowledge is key to decision making and timely advice can help plan better and manage the risks appropriately. India with ever growing population and shrinking land and natural resources, have a greater challenge to feed billions of people and also sustain its economy, which, still, largely depends on agriculture – directly or indirectly. There are more than 120 million cultivators and another 300 million persons involved in the allied sectors for their livelihood. Majority of these resides in the countryside with inadequate infrastructure growth. On the other hand India also boasts of a growing knowledge economy in excess of 100 billion USD per year and also well-established network of agricultural research, under the umbrella of Indian Council of Agricultural Research. Despite these, the research outputs don't reach to last mile, i.e. the farmers. Spread of the mobile telephony gives a hope. It has brought a revolution which is now being tried to usher the country in a "digital India". Digital can extend the knowledge from Universities to KVKs to Krish Sevak and to farmers, but should be done in a systematic manner and at a larger scale. Services like mKRISHI® CCA helps create an access channel to "knowledge pool" and "knowledge bank". But technology can only do so much, and unless a framework of "collaboration" – among all stakeholders are not established, this would also become just one of the research.*

Keywords: Digital Extension, KVK, Digital India, Collaborative Research.

Climate variability is a very complex phenomenon. It has its own mind and behavior based on multiple environmental parameters. But climate regulates or controls everything on this earth; hence it's important to understand how it behaves, what the variations over long and short terms are and how to cope with it (Karl *et al.*, 1995). Agriculture and climate is very tightly coupled. Climatic change in the rain pattern can have significant impact. Onset or departure of rain, rain intensity and distribution pattern is crucial for crop planning, sowing and crop management. Change in climate makes the old knowledge of growing food irrelevant and hence farmers need to adopt a different strategy (Crane *et al.*,

2011). They need to know what has changed, how much is the change and what is the impact. Scientist use various simulation models to predict the change in the climate and accordingly they study the impact. This helps in devising the "adaptation strategy" to combat the change in climate. The adaptation strategy includes change in crop varieties such as resistant varieties, water, disease and other management practices (Singer, 2011). It is important that such renewed knowledge is expanded in real-time to the rest of the stakeholders like agro input manufacturers, suppliers, farmers and the extension officers at Taluka and village level. Information and communication technology (ICT) can create a digital bridge to bring these stakeholders, together on a platform (Singh and Singh, 2012). Such digital information highway will help reduce the "time gap" in transferring the knowledge from "lab to land" and get timely feedback.

We need a multi-model knowledge extension mechanism to ensure the increased reach as well as the effective compliance.

Most popular modes of ICT based knowledge extensions are:
- Radio and TV based knowledge extension
- SMS based mobile advisory
- Online query resolution (like Kisan call center (KCC).

But each of these mechanisms has their own challenges, which are discussed below.

ADVISORY IS GENERIC IN NATURE

TV and Radio based programs are more generic in nature and have scope at the national, state or district level. Hence, though they provide a generic guidance and not exactly the actionable advice at the farm level.

SCALABILITY

KCC kind of services have regular working hours and number of associates per farmer is very less. Also, the advisory is based on the symptom provided by farmers and hence may not be completely accurate.

CHALLENGES OF SMS BASED ADVISORY

Most of the mobile phones don't support local language. Majority of Indian farmers do not understand English. Content length for every SMS is limited to 120 characters for English and only 50 characters or 7–10 words for local languages. If the content length is more, then the content is broken into multiple SMSs, which may be delivered out of sequence i.e. part 2 of SMS may reach first, and vice-versa. The content effectively loses the meaning. Also SMS delivery is not reliable – 20% SMS don't reach to end users for various reasons. Because of telemarketers, users get lot of SMS (SPAM) and in this process they miss important messages or don't bother to read each message.

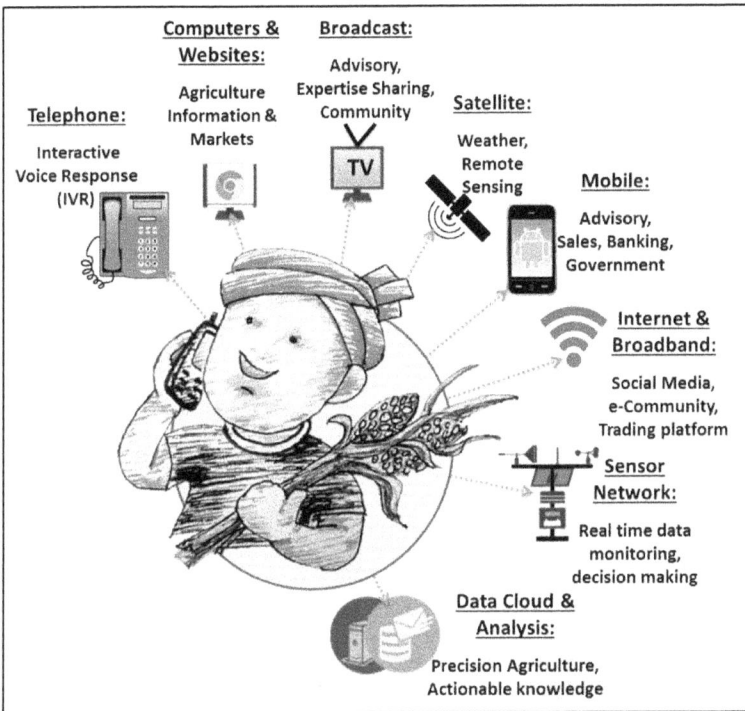

Fig. 1: Various Modes of ICT Based Digital Knowledge Extension

Besides these approaches public extension officers like Krishisevak and the private extension officers of the agro input companies too visit the villages and advice farmers on the actions to be taken. These extension officers visit only once in 15 days or so, and hence they are not able to provide the timely advice.

There are now multiple mobile apps for the advisory too. But with so many mobile apps available, there is confusion on which ones to use. Also, most of the apps are Android based, which is a popular mobile operating system. But in the rural areas the penetration of Android mobile phones is still low, coupled with poor data network. Only 40% of India is covered by data network, majority of which are Tier 1 and 2 cities.

Hence, we need to device a mechanism to unify all such extension services and offer an "Integrated Knowledge Extension services".

mKRISHI® CCA provides an integrated advisory service. It offers IVR based mKRISHI® Lite and feature phone Java OS and Android OS based mKRISHI® Regular services.

Human computer interaction (HCI) methodology has been applied to make the services easy to use by less literate person (Shinde *et al.*, 2014). Both the services give a 24 × 7 kind of environment enabling to raise their query anytime. The IVR services prompt for a user to record their queries. Mobile app helps user to select simple menu and icon to learn about the Best Management Practices, Frequently Asked Questions, and Alerts. Such small, knowledge capsules, helps resolve 60–70% of the queries. Farmers can ask any particular query by just recording an audio along with the images of the symptoms. This helps farmers ask specific question to seek a personalized advisory.

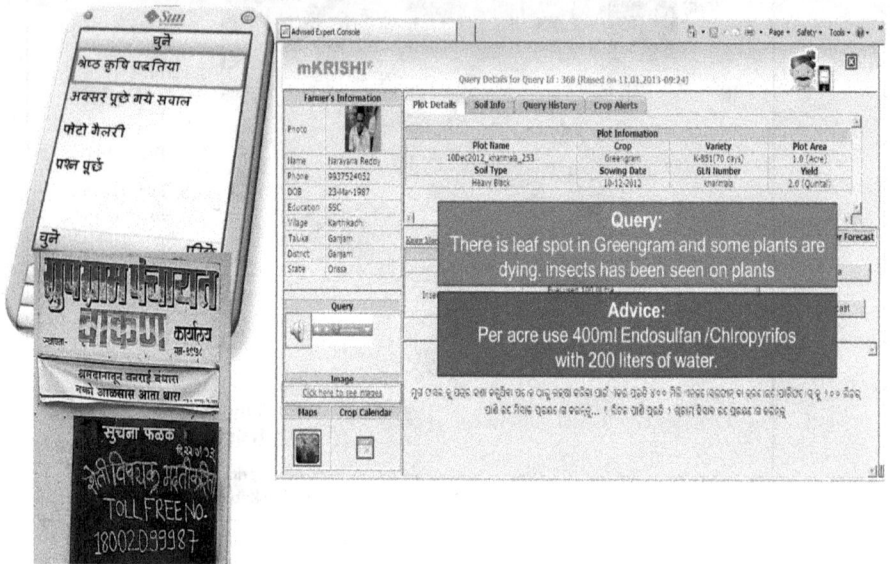

Fig. 2: mKRISHI® ICT Platform: Lite, Regular and Expert Console

The services also help farmers to record their farm operations, which helps increase the personalization. All transactions including queries are stored and shown together on an expert console to the agricultural experts, along with various other environmental information like the weather forecast for the given region for which the farmer has asked the questions, real-time sensor data, if any, available, etc. Hence, experts tend to get a "digital farm view" on his/her web console, enabling him to give a much better and improved advice.

The services were deployed in four climatically vulnerable districts of India viz., Dhar (MP) and Mewat (HR) that face frequent drought and Ganjam (OD) and Raigad (MH) which are prone to flash flood. The services helped connect more than 11,000 farmers across 500 villages. Two levels of agro experts resolved more than 23,000 queries. Besides these, more than 5 lakhs weather based alerts were sent, especially during the Phailin, Hudhud and Komen cyclones and more than three lakh mobile app based weather hits were recorded.

Our study has shown that we need to have a multi-model extension system, which helps offer similar knowledge through different media. The amount of information and personalization may vary in a broadcast medium, but it can be increased by data archiving and personalizing the farmer input. Such personalization helps to have improved impact.

REFERENCES

Crane, Todd A., Carla Roncoli and Gerrit Hoogenboom (2011). Adaptation to climate change and climate variability: The importance of understanding agriculture as performance. *NJAS-Wageningen Journal of Life Sciences,* 57, no. 3: 179–185.

Karl, Thomas R., Richard W. Knight and Neil Plummer (1995). Trends in high-frequency climate variability in the 20th-century. *Nature,* 377, no. 6546: 217–220.

Shinde, S., Divya P., Srinivasan, K., Dineshkumar Singh, Rahul Sharma and Preetam Mohnaty (2014). mKRISHI: Simplification of IVR Based Services for Rural Community. In Proceedings of the India HCI 2014 Conference on Human Computer Interaction, p. 154. ACM.

Singer, P. (2011). The future of animal farming: renewing the ancient contract. Edited by Marian Stamp Dawkins, and Roland Bonney. John Wiley & Sons.

Singh, R.K.P. and Singh, K.M. (2012). Climate Change, Agriculture and ICT: An Exploratory Analysis. ICT for agricultural development under changing climate: 17–28.

Human and Institutional Capacity for Climate Resilient Agriculture: Lessons from Bangladesh, Ghana, India and Vietnam

Suresh Chandra Babu* and Alex De Pinto

International Food Policy Research Institute (IFPRI), Washington D.C., USA

*E-mail: s.babu@cgiar.org

ABSTRACT: *The risks farmers face due to climate change have been increasing, particularly in developing countries. While governments and the international community have made a number of efforts to combat climate change issues, national level capacity continues to act as major constraint in making agriculture climate resilient. There is a need to build a comprehensive framework which helps in identification of capacity gaps and assist policy makers to develop complete solutions to combat climate change issues. This research presents a comprehensive framework for the assessment of national level human, institutional, and policy capacity for climate resilient agriculture. The framework is applied to four countries, namely, Bangladesh, Ghana, India, and Vietnam to determine capacity gaps and comparatively asses their status.*

Keywords: Climate Resilience, Climate Change, Agriculture, Human, Institutional, Policy, Capacity.

INTRODUCTION

Over the past few decades, the issue of climate change has become an important concern for policymakers. This is particularly true for the agricultural sector in developing countries, in which farmers are adversely affected by changing temperatures and weather conditions. In fact, the impact of climate change is most severe for communities that primarily rely on agriculture for their livelihood with important implications for food security. Hence, it is important for countries to reform and develop policies in agriculture and allied sectors in order to adapt to, and mitigate climate change while promoting economic growth and development. Significant attention should be given to building an agriculture production system that is resilient to changing climate regimes. While most of the emphasis has been given to enabling farmers to cope better with changing climate without compromising yield increases, less work has been done on the institutional and governance response to these challenges.

Weak and inadequate human, institutional, and policy capacities are a major constraint in building climate resilience in developing countries. New international agreements, such as, the Paris agreement and the related national intended contributions (NDCs), introduced at the 21st conference of the parties (COP21) and ratified later in 2016, record individual countries' plans to contribute to limiting global warming below 2°C relative to pre-industrial levels. NDCs are ambitious plans that provide countries with an opportunity to develop agendas for social and environmental change, with potential to create benefits across all economic sectors and a broad range of geographical scales. While some of the benefits can be realized in the short run, the long-term planning horizon of the NDCs requires policy and decision makers to design policies that are economically and politically sustainable. They require institutions and governance structures that are capable of devising and monitoring the sound implementation of policies that fully account for the effects of climate change. This study presents a framework for the assessment for policy, institutional, and human capacity for climate resilient agriculture.

METHODOLOGY

Building on existing literature, this analysis uses the conceptual framework shown in Figure 1 to derive its findings (Zurek *et al.*, 2014; Richerzhagen and Scholz, 2008). The figure has been adapted from Babu and Blom (2014) to elaborate how agricultural systems can transform to become more climate resilient. Drawing from their narrative on food system resilience, this analysis extends the idea to climate resilience, and moves beyond achieving food security to a larger goal of conserving the environment.

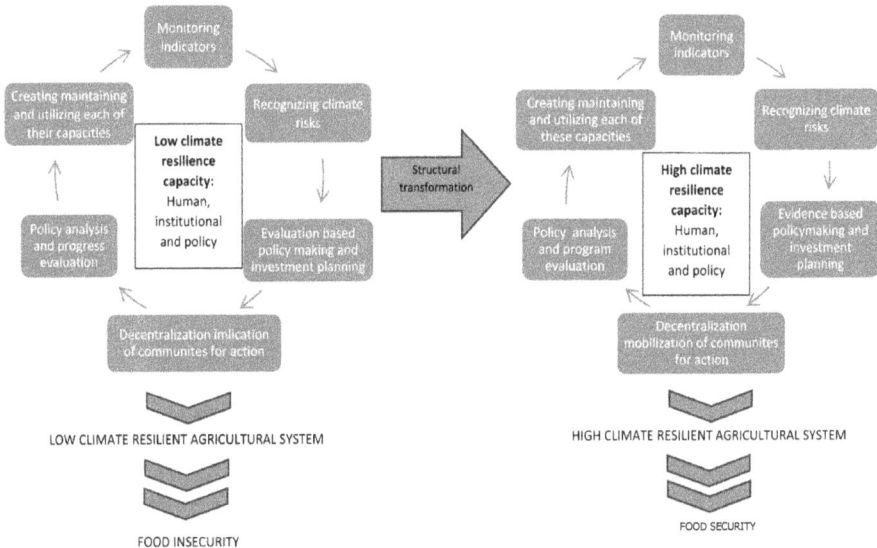

Fig. 1: Transformation of Climate Resilient Agricultural Systems

(*Source:* Adapted from Babu and Blom (2014))

The framework shows how climate resilience in agriculture is a multi-dimensional concept and requires a transformation of human, institutional, and policy capacity. With this framework in mind, the study uses a case study approach to reach its findings. The rationale for using a case study approach is the qualitative nature of the investigation. The case study approach is a preferred method when a researcher seeks to answer "how" and "why" questions with little control over events influencing the outcome (Yin, 2009).

The findings presented are a result of extensive field work with local country collaborators in Bangladesh, Ghana, India, and Vietnam. In each country the collaborators conducted a series of in depth interviews with policy-makers, researchers, and other stakeholders. Hence, rooted in rigorous conceptual framework, this study provides practical guidance for the policy makers, administrators and other stakeholders to incorporate climate resilient interventions in existing policies and programs.

PRELIMINARY INSIGHTS FROM CASE STUDIES

The four countries selected for this analysis have a few things in common. They are developing countries with major resource constraints and multiple urgent policy issues to be dealt with. Additionally, they are agriculture based economies, with a large section of farmers owning small or medium land holdings. Further, all the four countries are extremely vulnerable to climate change albeit for different reasons. Despite the similarities, these countries have different agricultural systems, cropping patterns, and policy processes, hence giving complete representation of the issues related to agricultural climate resilience in developing countries. The preliminary findings of the analysis are presented below.

Bangladesh

The idea of climate resilience in agriculture was first incorporated in the policies in Bangladesh in 2008, with the Climate Change Cell as the main inter-ministerial national committee dealing with these matters. Most policies are finalized by an extensive formal consultative process, involving all related ministries. Sometimes informal channels like the media are also used to receive feedback concerning policies. Apart from that, the Right to Information Act is the primary accountability mechanism, but it is hardly used for all policy issues. For research, BARC has the responsibility to coordinate research and foster inter-institute collaboration, monitor and review the research program of NARS institutes. However, the research communities generally interact on an issue on a need basis. There are no formal platforms to share multi-disciplinary research in the country. Bangladesh is also part of a number of climate change conventions and treaties. The latest government document charting out the plan to deal with climate change issues is the Perspective Plan 2010–2021, which is being implemented in two phases. It must be noted that over time, the focus of most policy documents related to climate change has shifted from infrastructural development, relief, and rehabilitation to include food security, sector-specific issues and nutrition.

Government institutions are currently the most prominent role in the climate resilience space. The Ministry of Environment and Forests and the Ministry of Agriculture are at the center of policy process. The Climate Change Cell was formed under the Ministry of Environment to deal with issues related to climate change in the country. This is the central organization for formulation and implementation of polices on climate change. Bangladesh also lacks in institutional capacity for data collection and analysis. Additionally, very few organizations (13 percent) play a key role in implementing programs related to climate resilience. As for research institutes, Bangladesh has some government and private research institutions that conduct research in climate resilience in agriculture. A key role of coordinating research is played by the Bangladesh Agriculture Research Council (BARC), which is the apex of the NARS. Organizations such as the Bangladesh Institute of Development Studies (BIDS) and the Bangladesh Centre for Advanced Studies (BCAS) conduct important research activities on climate change. Bangladesh also has a number of NGOs working on climate change. However, a mere five per cent of the organizations play a key in advocacy and lobbying. Bangladesh faces constraints in its institutional capacity because of the lack of funding channels and a well-articulated institutional vision and strategy for climate resilience in agriculture.

The responses indicate that, human capacity to understand and meet the challenges of increasing agricultural climate resilience in Bangladesh is very poor. Interviews with stakeholders from some of the key organizations engaged in climate related work reveal that about 47 per cent of respondents had a Master's degree. The percentage of PhDs is about 15 per cent in key institutions for climate resilience in Bangladesh, indicating that the capacity for high-level and specialized research is quite low. Additionally, there is a large skill gap in the individual capacity for research across institutions. Currently, about 52 per cent of individuals involved in climate related work help in the development of agricultural practices and technology (adaptation to climate change). Due to this, individuals have very little time to update their skills. Lastly, the linkages between researchers and policy makers is very weak. In fact, the survey indicates that only seven per cent of the respondents were members of any policy group at the national level. This low level of participation of respondents in national level policy, relative to other countries, indicates the need for the establishment of stronger linkages between research and policy.

Ghana

Over the last few years, Ghana has made strides in its policy responses to climate change. In 2010, the government constituted the National Committee on Climate

Change (NCCC) composed of different ministries, government technical agencies, and civil society groups, responsible for reviewing policies and programs to complement national priorities on climate change. In 2010, the government developed its action plan to combat climate change. Out of this document came the National Climate Change Adaptation Strategy (2010), the National Climate Change Policy Framework (2010) and finally, the National Climate Change Policy (2013) which represents Ghana's integrated response to climate change and provides a clearly defined pathway for dealing with the challenges of climate change within the country's current and future socio-economic context. By establishing the NCCC, the government has created a platform to facilitate discussions, disseminate research findings, implement and coordinate initiatives across different sectors and collaborate and share ideas concerning climate resilient agricultural practices. Despite these positive advances in policy capacity, there remain several gaps. The National Committee's work is thwarted by inadequate funding to support activities such as data collection, knowledge management, and monitoring and evaluation. Furthermore, funding is not specifically channeled to support specific areas. Additionally, there is no institutionalized process for engaging policy specialists with technical capacity to engage in the climate change policy process.

Institutional capacity in Ghana for climate resilience in Ghana is very low. Only 11 percent and four percent of the institutions selected are regularly involved in work related to climate change. Further there is very low level of involvement of institutions in conducting regular risk and vulnerability assessments (five percent only). Institutions lack the critical capacities needed to ensure that Ghana adopts climate resilient agricultural practices. In fact, only four percent of organizations are regularly involved in monitoring food security, agricultural production and markets, compared to five percent in climate change, weather and environmental factors. Finally, only 33 percent of organizations have the capacity for collective action, effective leadership, management and communication.

Human capacity in Ghana's climate change sphere consists of actors in the governmental, non-governmental and scientific and research community. Survey results show that 31 percent of individuals within these institutions have less than a high school certificate whereas nine per cent and less than one percent have master's degrees and PhDs respectively. This low level of higher education indicates a low capacity for high-level, specialized research. To compound this, 89 percent of individuals have no formal climate change training. With regards to individuals' capacity for policy work, only 23 percent of individuals are professionally involved in climate resilience for agriculture, specifically, working on adaptation of technologies and practices. The results from the survey are a clear indication that there is a human resource constraint in Ghana's climate change research and policy sphere.

India

Over the years India has made a number of efforts to improve its polices on climate resilience in agriculture. The first big step towards mitigation of climate change risks was with the government adopting the National Action Plan on Climate Change in 2008 (Government of India, 2008). The plan charts eight priority 'mission', one among which is the National Mission for Sustainable Agriculture. It clearly lays down goals till 2017, but does not articulate clearly how they will be achieved. India is also part of various international treaties on climate change.

With the government as the prominent actor in policy formulation and implementation, there are a number of institutions working on climate resilience in India. The Ministry of Agriculture is the primary agency which formulates and implements policies and programs related with the overall climate change goals are articulated by the Ministry of Environment, Forest and Climate Change. The Central Research Institute for Dry Land Agriculture (CRIDA) is the nodal agency for coordinating research and implementation of projects related to climate resilience in agriculture throughout the country. The Indian Council for Agricultural Research (ICAR) is the apex body for coordinating, guiding and managing agricultural research and education institutes, research centers and universities under it, spread throughout the country. Climate change is the running theme across all divisions in ICAR. In addition, there are a number of state level organizations working on building climate resilience in agriculture. However, there is a lot of state to state variation in the type of organizations depending on the state's political and fiscal commitment towards climate change. India also has a number prominent NGOs, primarily focusing on reducing knowledge gaps and implementing climate resilient programs and practices.

As compared to other countries India has a relatively higher level of human capacity. Data shows that about 30 per cent of the respondents have a Masters or a PhD degree. Further, only 27 per cent of the respondents were qualified with a specialization in agricultural sciences and about 42 per cent of the respondents had received formal training on climate change issues. In terms of experience, only 27 per cent of the respondents have prior experience (of 5–10 years) in climate change related work. The proportion of respondents involved as primary actors at local level climate resilience efforts is the highest at 18 per cent, followed by involvement at the sub-national (12 per cent) and national level (9 per cent). Overall the focus on climate resilience work is quite low with weak links between research and policy.

Vietnam

The big push in building policy capacity for climate resilient agriculture has come from donors and the international community in Vietnam. In the recent past, MARD implemented the Action Plan Framework for Adaptation to Climate

Change in the Agriculture and Rural Development Sector for 2008–2020. Vietnam is also a part of various international treaties on climate change. Although Vietnam has made significant progress, currently, the content and purpose of climate change regulation mostly focuses on disaster mitigation and preparedness. Currently, there is a lack of coordination between institutional efforts on climate resilience across the country. There is little focus on encouraging stakeholder participation particularly, of farmers and industrial groups. Additionally, there is very little inter-sectoral cooperation, and a severe lack of human and financial resources. This is combined with the lack of detailed regulation and unclear responsibility allocation among ministries leading to sub-optimal system level outcomes.

International organizations have played an important role in setting up the institutional capacity for climate resilience in Vietnam. The country receives grants, loans, and technical assistance from a number of multilateral organizations with UNDP, FAO, and USAID playing a key role in the country's agricultural sector. Currently, the Ministry of Natural Resources and Environment (MONRE) functions as the main implementer of climate change policy in the country. Additionally, the Ministry of Planning and Investment (MPI) collaborates with other ministries, sectors, in incorporating climate change into socio-economic development strategies, programs, schemes and plans. MPI works with the MONRE to craft regimes of monitoring and assessing the implementation of the Strategy. For climate resilience, the Ministry of Agricultural and Rural Development (MARD) is the most important organization. The Ministry applies the climate change goals to the agricultural sector. It has adopted measures to reduce greenhouse and CO_2 emission levels, along with work on agricultural extension activities for farmers to cope with climate change. Among the research institutes, the Centre for Agrarian System Research and Development (CASRAD), Institute of Water and Environment (IWE), Can Tho University, and Mekong Delta Institute play an important role in this area.

Vietnam has a major knowledge gap and lack of coordination between researchers and policy makers, relative to other countries. Only five percent of the respondents have a PhD, but more than half had taken at least one course in agriculture. Despite this, the participation of respondents in the national level policy process was seen to be the highest amongst all countries in this study (53 percent). The coordination between researchers, policymakers and NGOs was seen to be very weak.

SYNTHESIS: FINDINGS AND CONCLUSIONS

Effectiveness of efforts on agricultural climate resilience depends on the human, institutional, and policy capacity of countries. Further, a country's climate readiness, or its capacity to manage plan, implement, and monitor climate finance and activities related to climate change, is important to sustain its efforts.

It is important to note that, a country's ability to access climate finance will depend on their capacity to carry out actions and measuring results, and countries should be driven by the goal of increasing the efficiency of the food production system, its resilience and reassure foreign investors. Most countries lack the larger institutional technical capacities and infrastructure necessary for climate related programs at large scales. Building these capacities will require new information, new technologies, novel finance instruments, and new skill-sets and responsibilities in existing institutions. Many countries are building these capabilities now, but there is uncertainty about what is needed and what is feasible. The goal of the framework presented in this study is to help understand how existing systems can evolve to meet new needs or how existing protocols can be modified to include climate-related information and considerations. Developing a common understanding of the role of agriculture in a changing climate would be useful and would promote the alignment of objectives across different initiatives. This could be achieved through facilitated inter-ministerial dialogues, creation of inter-disciplinary communities of practice across relevant ministries, research institutes, planning units, farmer unions, joint planning exercises, and multi-stakeholder consultation. This research indicates that there are major capacity gaps across developing countries. Findings from the case studies offer some general insights:

1. Constrained financial resources for climate change issues is a common concern across countries. Due to this countries are not able to invest in quality human resources needed for climate resilient agriculture.

2. There is a need to focus on increasing the trust between government extension workers and farmers to ensure that climate resilient practices are adopted. This needs to be combined with a comprehensive incentive structure that induces behavioral change amongst farmers.

3. Lack of technical capacity on climate change is an issue across countries. Even in India, even though technical capacity is relatively higher, the linkages between research and policy is very weak.

4. Clear institutional roles and responsibilities need to be laid out to ensure that there is no overlap on work related to climate resilience in agriculture.

5. There is a need to initiate greater involvement of the private sector and NGOs to reduce the burden of climate resilience interventions on the government alone.

A national level process to develop the capacity to deal with the challenges of climate change could be structured in three components, modeled along the existing REDD+ readiness framework: governance, strategy and monitoring and verification (Zurek *et al.*, 2014). The progress towards developing this resilience framework could be monitored using multi-level indicator of the type indicated in the Table 1.

Hence, the study shows that there are major capacity gaps in developing countries. However, a systematic approach – as presented in this study – can be used to identify and address these gaps. The analysis presented here, helps to assess individual countries and compare it to its peers. This also gives policymakers the opportunity for learning from different contexts. The scope of this study is limited to providing a guide to policymakers to estimate the minimal threshold level of capacity required for climate resilience in agriculture. It identifies factors that affect capacity the most, and the ones that might have been left out, which need to be studied further. The findings presented here are tentative and provide immense scope for further research.

Table 1: Multi-Level Indicators to Monitor Development of a
Framework for Climate Resilient Agriculture

Indicator label	Definition (where appropriate, provide an example of measurement units used, such as $/kg)
Governance	A Ministry or inter-ministerial body is designated to manage and coordinate climate-related activities. Institutional roles in agencies and local or sectorial jurisdictions are defined.
Strategy	Existence in official documents of an agreed-upon vision and goals for the agricultural sector that balance food security, adaptation and mitigation. Climate resilient agricultural policies consistent with other development priorities have been identified. Information systems, analyses of socio-economic and biophysical suitability to CSA interventions are available. Areas of interventions have been prioritized.
Monitoring and verification	Protocols to update activity data with a regular frequency are in place. Baseline projections are in place for measuring GHG reductions in main production systems, sectors, or regions. National inventory uses country-specific emissions factors for main production systems GHG accounting framework and baselines are integrated with other national statistical and data collection or monitoring systems.

Increasing the capacity of policy-makers to better align policies across multiple policy areas and coordinate policy formulation horizontally across national government entities, and vertically from local to national levels, is essential to produce solutions that deliver across objectives. Improved consultation and coordination could lead to more coherent institutional support within and across the government as well as increase the efficient use of use of scarce resources by identifying areas of synergistic interests.

REFERENCES

Babu, Suresh Chandra and Blom, Sylvia (2014). Capacity development for resilient food systems: Issues, approaches and knowledge gaps. 2020 Conference Paper 6. May 17–19, Addis Ababa, Ethiopia. Washington, D.C.: International Food Policy Research Institute (IFPRI). http://ebrary.ifpri.org/cdm/ref/collection/p15738coll2/id/128151.

Richerzhagen, C. and Scholz, I., (2008). China's capacities for mitigating climate change. *World Development,* 36(2), 308–324.

Yin, R.K. (2013). Case study research: Design and methods. Sage publications.

Zurek, M., Streck, C., Roe, S. and Haupt, F. (2014). Climate readiness in smallholder agricultural systems: lessons learned from REDD+. CCAFS Working paper No. 75. CGAIR Research Program on Climate Change, Agriculture and Food Security (CCAFS), Copenhagen, Denmark.

Connected We Stand, Disconnected We Fall: A Perspective on Implementing ICTs in Crop Health Management

Y.B. Srinivasa

Tene Agricultural Solutions Pvt. Ltd., Yelahanka New Town,
Bengaluru 560 064, India
E-mail: yb@tene-ag.com

ABSTRACT: *A perspective for developing Information and Communication Technologies (ICTs) in agriculture, especially to tackle the issue of crop health is portrayed here. For a farmer, crop health is perhaps the most perplexing aspect in crop production. It makes crop production an uncertain business and leaves the farmer vulnerable to exploitation. This vulnerability comes from the lack of a functioning extension system in the nation. Unfortunately, modern ICTs in agriculture, especially webpages and apps, are either promoting self-diagnosis or remote-diagnosis, both of which have potentially dire consequences on the farming sector. Instead, a two-pronged approach, which not only leads to rural employment but also leads to digital empowerment of the rural sector, is proposed to handle crop health issues locally. Although issues shall be dealt locally, the digital system shall work to connect all the relevant stakeholders concerning crop health in the country. The focus of the system shall not only be the delivery of information to the rural sector, but also the delivery of real-time information regarding farm situations to the other stakeholders of agriculture. Real-time data on farm situations shall be of vital importance to tackle the challenge of the changing climate.*

Keywords: Agriculture, Climate Change, Crop Health Management, ICTs, Real-Time Data.

Agriculture is going through a challenging phase. It is ironical to note that the current situation has resulted when each stakeholder seems to aim at increasing the quality and quantum of agricultural produce. Be it the government, scientist, bank or the corporate, every stakeholder intends to drive the progress of agriculture. But, despite the intentions and the efforts, agricultural growth has stagnated; in fact there are reports of it having declined. Soils have lost their structure and texture; in many places the chemistry has deteriorated beyond the bounds of repair. Water resources have thinned down rapidly, forcing some farmers to flee from farming. Pest devastations of various orders from across the country are more frequently reported today; the latest of technologies, even genetically engineered crops, are on shaky grounds. Agriculture has become a capital-intensive activity, and with the random behaviour of markets, it has become a very risky business indeed, especially for the cash-strapped farmer. The threat of climate change is looming large; agriculture-intensive nations like India are painted in deep red by those who are estimating the impact of the changing climate. And, it is primarily the burgeoning demand for food that is relentlessly adding coal to the cauldron.

There is no doubt that another Green Revolution, but of a different kind, is the only way out of the present situation. Today, the nature and number of players on the field are humongous. And, these players function in a manner that is completely detached from each other; a means of attachment is not in sight. The scientist is barely aware of the farmers' situations, while the farmer is unaware of the technologies prevailing; the banks scarcely know of the farmers' requisites, while the farmer is unmindful of the bank's provisions; the corporate interacts with the farmers oblivious to other stakeholders; policymakers and administrators rely on inaccurate farm data for framing policies or firming administrative decisions; the unpredictable market forces are infamous for harming the interests of either the farmer or the consumer; the input supplier often doubles-up as an advisor to innocent farmers, gaining through increased sales while punching holes into tiny pockets and polluting the only planet we have to live upon. The situation is precarious and the ill-effects of this chaos are quite evident. Therefore, if there has to be a solution, this chaos has to turn into order, the discord has to be replaced by harmony. It is obviously impossible to trigger a revolution when the interacting elements are all at sea with each other, with no means to bind them together. Agriculture is a team-game, which can be successful only when every player in the team is clear and confident about the rest, and plays in synchrony to achieve victory. Well, another Green Revolution is, undeniably, a necessity for the nation's food security. To make this happen, a force to bind the elements together, to clean the muddle, to cull the mayhem, has to originate.

Well, it is for certain that agriculture needs a glue, especially one that is malleable, strong and transparent, that can seamlessly integrate interactions between the diverse elements that constitute agriculture, that can be flexible to accommodate any and all dynamics of the present and the future. It is for certain that the only material to make this glue is digital. It is quite apparent that each facet of the agricultural sector is wanting to embrace the overwhelming power of the 0s and 1s. The process of digitisation has already begun. With food touted to be the next biggest thing in the world, a new app-a-day is unsurprisingly posted on one marketplace or the other, aiming to relieve agriculture from its current problems. But, are apps truly the way forward? Can apps solve the extension problems faced by today's India? What are the actual gains from apps so far? What should be the do's and do not's of the future apps? It is time that we asked these questions and evaluated the available answers. And, from the results of such evaluations, we need to shapeup digital tools so that farmers and other stakeholders of agriculture get the most out of it. While remaining focussed on

Crop Health Management (CHM), this essay shall elucidate the current situation before attempting at picturing a perspective for addressing the issues through Information and Communications Technologies (ICTs).

THE DYNAMICS OF CROP HEALTH MANAGEMENT

CHM is a complex subject. Due to the complexities, it is perhaps among the most worrisome factors to the farmer. With the exception of CHM, it has been observed that a farmer can handle most of the agricultural operations reasonably well. This inability to handle CHM has led to systematic exploitation of the farmer while jeopardising human health and environment. Therefore it may be rational to delve a little deeper into the complexities of CHM.

CHM majorly encompasses diagnosis of problems caused by, and solutions for managing, various pestiferous populations of insects, viruses, fungi, bacteria, nematodes and weeds, and nutritional deficiencies that decrease crop production and impact farmers' welfare. There are numerous species of pests that affect each crop, and not all impact in equal propensities in any given space and time. They vary with soil type, short- and long-term weather patterns, physiography, cropping history, cropping pattern, cropping practices and such external influences. They are also impacted by other competing pests, predation, emigration from and immigration into the agro-ecosystem. Further, the host plant itself responds to pests differently in different space/time situations and affects the dynamics of pests. All these add significantly to the complexity of pest dynamics. Some pests are specific to age of the host and some others are specific to parts of the host; some are specific to seasons and some others to geographical regions. Pest-pest interactions, pest-predator interactions, pest-nutrition interactions, pest resistance to pesticides and other management options, pest resurgence, etc., bear important influences on their dynamics. Spatiotemporal fluctuations in pest populations, according to which variable tags like minor, major, sporadic, persistent, endemic, epidemic, invasive, etc., are associated, are also to be noted while understanding their nature. Additionally, very little is understood about many aspects of each pest—behaviour, physiology, ecology, dispersal/migration/ spread and other such are not scientifically established for many pest species.

On the other hand, pest management options are equally complex. There are many microbial, botanical, chemical, cultural, mechanical and biological methods, and there are many techniques and tools for administering these methods to affect pest populations. Some of them are ecologically sensitive, while some others are part of the humble natural world; some are economical, while some others are expensive; some methods suit intensively managed agriculture, while some others suit extensive farming systems, and a mismatch could have dire consequences on the society, like large-scale ecosystem poisoning. Selection of pest management strategies depends on the intensity of the pest problem at a given space/time. However, assessment strategies to decide on pest intensity vary with crops, pests and physiographical features; taking it to the ground level is a difficult task. At most times, one method of management, or one strategy, would not suffice to lower a pest population; it is a combination of strategies that has been proven to be effective in most cases. This can be complicated because some methods are compatible with some others and incompatible with certain others; some methods are applicable only at a particular time of the day and some others at particular positions above/below the ground. Further, it is often observed that preferences of a farmer play a significant role in selection of management strategies. Some look for organic methods, some for inorganic, some prefer cultural and biological, while some others are open to any effective management action. All such variations should be taken into consideration while suggesting remedial actions. Moreover, there is a constant influx of new pest management tools and molecules into the market. These too need to be used appropriately so that farmers and the Nation obtain the maximum benefit. Therefore, pest management is an extremely challenging and complicated section of agriculture.

Adding to the challenges posed by the diversities of pests and their management strategies is the delivery mechanism of pest management solutions to farmers. The social fabric of our farming community, the economic position of our farmers, their level of education, the infrastructure at their disposal and their sheer numbers have erected tall hurdles along the routes of the delivery channels. Also, there are policies and regulations that vary from time-to-time and from one state to another. Of course! This is additional to the complexities of pests themselves and their management strategies. Therefore, there has been little notable success obtained in the field of CHM, unlike other subjects of agriculture, where there has been measurable amounts of achievements made since the 1970s. Success stories in the field of CHM are sporadic at most; a national revolution has been a distant dream.

CURRENT SCENARIO OF PEST MANAGEMENT IN INDIA

It is common to find farmers in India visiting pesticide-selling retail shops, with or without samples of diseased plants, and purchasing the 'remedies' sold to him. Many farmers reach the retailer only when the problem is too severe and many a time when nothing can be done. Early detection and solution is a distant cry. 'Intelligent' retailers combine an insecticide, a fungicide, a bactericide and a growth promoter, all in one pack, and sell it to the hapless farmer, so that the probability of pleasing him increases; concurrently, the retailer is benefitted through expanded sales. More often than not the farmer ends up buying excessive pesticides and unnecessary crop health enhancers, and without certainty that the concoction would resolve the problem. This is a very dangerous scenario. Just imagine the condition of the society if the medical world too had done the same— dumped a variety of medicines in the patient's body without diagnosing and quantifying the disease! It is like an unqualified pharmacist prescribing drugs without meeting the patient. A situation more chaotic than this cannot be imagined. In the agricultural world, farmers' capital expenses increases without any assurance of crop improvement (which has other serious ramifications—increased and ill-assured capital expense can result in undesirable social consequences), there is a significant increase in environment pollution and raise in health issues concerning farmers and consumers. All this is because the farming society in India does not have access to timely and accurate crop health diagnosis and information on management. On the other hand, advancements made in pest management science and technology are wasted in this situation. Researchers working towards developing novel solutions find their efforts not to be reaching the end users, which could discourage them, and simultaneously take-off responsibility from their shoulders. This situation can improve only through efficient extension; only when the farming society in India has easy and timely access to proper crop health diagnosis and remedial strategies would CHM be a revolution.

ARE THE CURRENT EXTENSION SYSTEMS AND TOOLS FAILING THEIR PURPOSE?

There are three major extension forces in the country—1) State Agricultural Departments and their affiliated functionaries, 2) Krishi Vigyan Kendras (KVKs) and other ICAR bodies and 3) NGOs/Consultants/Corporate. Of these, the State Departments, with the highest manpower dedicated for the role, occupy the most prominent position. The KVKs act as an intermediary between technology developers and the extension functionaries of the State Departments, and play a major part in training, demonstrations and provide feedbacks for technology improvement. KVKs have highly qualified personnel who serve as experts for specific subjects. The last group consists of the least coordinated extension services, which usually serves to spread certain specific products and technologies.

The State Department extension functionaries essentially act as 'one doctor for all problems'; specialisation has not made an entry so far. In addition to extension services, these functionaries are involved in implementing vast number of programs and services that are introduced by the State and Central Governments from time to time. Such involvement takes away a bulk of their time leaving very little for pest diagnosis, quantification and management. Pest dynamics is complex. It varies with crop, cropping systems and practices, soil, climate, inherent resistance to chemicals, presence/absence of other pests, etc., which makes it difficult for a non-specialist to handle the situation without any assistance. The information with the state functionaries is not sufficiently updated to handle such complex issues. They rely on assistance from KVKs for diagnosis and management. Even in a KVK, there usually is one crop protection specialist (either an entomologist or pathologist) to address queries arising in a district, which is also insufficient. Despite constant interactions between KVKs and the State machinery, it is inadequate to tackle farmers' pest problems. On one hand, the functionaries cannot reach out to each and every farmer to diagnose problems and suggest remedial strategies, and, on the other hand, every farmer cannot reach out to the State extension functionaries or KVK personnel. This situation is one of the primary reasons for the lack of timely assistance to the farmer, who invariably ends up seeking help from the retailer.

The above situation has prompted the use of mass media and other communication systems to disseminate information related to CHM. Several decades of efforts have made us realise that print, or radio, or television, or their combinations fall woefully short of meeting the requirements of the nation. They cannot provide timely solutions to farmers, and that they can be best used to convey certain general understanding of pest situations or their management. Compact disks and other sources of static information are similar to textbooks, with digitisation being the only difference. They too have failed in serving as tools for CHM. With the Internet gaining popularity, many websites were developed to carry information on pests and their management. Further, efforts were also made to take the Internet to the villages. Although this effort has led to sporadic benefits in certain areas of farming, CHM is far from being impacted by it; pest dynamics being the major bottleneck to tackle. Finally, with the rapid penetration of mobile telephony to the nooks and corners of the nation and its evolution beyond just verbal and textual communication, new opportunities have seemingly opened up in agricultural extension. Short message services are being rampantly used in dissemination of crop health information. But this too is generalised and does not assist in diagnosis and quantification of pest problems. So far, it has not been possible to provide farmer- and situation-specific information through messages. As a standalone method, messaging services may not be of great help in CHM at a national scale. Finally, the era of applications, or apps, has come to the fore. However, it appears that delivering information (like a sophisticated digitised textbook in local languages) to the hands of every farmer with a smart phone is the only objective of these apps. The apps have not evolved to enhance the way in which the end user, who, in most cases, happens to be less literate, can independently and accurately use the information. They run dangerously close to creating a potentially chaotic situation where the farmer himself/herself diagnoses the problem and takes cognisance of the suggested remedial strategies. Let us compare this with the medical world and question ourselves. Is there a medical app anywhere in the world that a patient shall log into and diagnose his or her medical condition? Further, would it be prudent to develop one such that shall also contain remedial measures? It is always mentioned that self-medication (self-diagnosis is a prerequisite for self-medication) is very dangerous. This is when it is about human health. But, ironically, we seem to be promoting self-medication when it comes to crop health. Apps and websites dealing with health, be it human or crops, can be suggestive at most. Else they can turn the crop health sector into chaos and one should tread upon this path with extreme caution. Thanks to the inefficiencies of the apps developed so far, that such a situation may not arise in this nation. A few simpler apps have attempted to directly connect experts and farmers, where sharing of images and other multimedia can yield certain cognisable and quick remedies to farmers. This is the threat of 'remote-diagnosis'. Apps through which photos can be sent across to people are increasingly popular today. But, using them as a tool to deliver crop health advisories are fraught with danger and should only be used in rare situations. For example, yellowing of leaves (which is a very common symptom) can be caused by improper/imbalanced nutrition, water stagnation, or by root diseases, or leaf diseases, or collar rot, or plant viruses, or root feeding insects, or sap-sucking insects, or by stem-boring insects. Vector-mediated problems are even more complex to diagnose. Please note that there are multiple varieties within each of these listed causal groups that further complicate the matter. Diagnosis would be risky here. Remote diagnosis using mere photographs can be erroneous and the suggested management decisions on the basis of such photographs can be dangerous. One can easily visualise the magnitude of the issue with a minimum of 50 odd problems faced by each crop and with 100 crops that are cultivated, any ICT system with an architecture that can lead to the wrong diagnosis should be discouraged. Remote diagnosis is not a scalable option and can be misused by those promoting a certain range of products or technologies (like the retailer situation as described earlier).

REAL-TIME DATA ON FARM SITUATIONS ARE CRITICAL, BUT UNAVAILABLE

Perhaps realising the inefficiencies and perils of some of the ways in which ICTs are being utilised, creditable attempts are made to create systems (unlike apps). Most of these systems fail on one point—they are designed to serve as a one-way communication, i.e., delivery to the farmer has been their objective. A true extension tool should be the one that would serve as a two-way communication system in real time. If delivery of the right information to the farmer at the right time is important, it is equally important to deliver real-time data about the farm situation to the rest of the stakeholders in the agricultural sector. The second aspect has been completely neglected by the extension systems adopted so far—from print-based to app-based, none have so far aimed at reverse flow of farm data. This has led to a divided agricultural sector where policymakers, administrators, researchers, financial institutions, corporate, traders, and others work in complete discord. Their actions can be brought together only when data regarding farm situations are shared in real time, which is necessary for the comprehensive growth of the sector and the nation.

From policymakers' perspective, data on expected agricultural production is rarely matched with crop loss estimation. This affects storage and release decisions of agricultural produce, export-import decisions, redistribution of produce, price estimation, crop-loss compensation to farmers, etc. Decisions to support new initiatives and projects are not based on real-time data, which may lead to inefficient expenditure. Real-time data on pesticide usage pattern is required to throw light on many health, export and judiciary matters concerning pesticide contamination. Real-time data on spatiotemporal dynamics of crop pests are required to conduct an authentic pest risk analysis, which is essential to promote exports in agricultural sector.

Input distribution, resource management, crop subsidies, crop insurance, damage assessment and other support schemes need to be implemented by administrators on the basis of real-time data to make them effective. Such data shall allow them to take up timely large-scale action programs before any pest reaches devastating levels. Prevailing forewarning systems can be said to be weak at most; they can be strengthened only through real-time data on pest situations. Similarly, real-time data shall allow fine-tuning of several other extension programs and eliminate the reporting hierarchy.

Real-time field data are important from the perspective of both research and education. It shall allow researchers to fine-tune research projects from proposal and evaluation angles. Feedback in real-time provides them an opportunity to evaluate the performance of their technologies/recommendations and to adopt any corrective measures with immediate effect. Detection of new pest problems, or pest developing resistance to pesticides, pest resurgence, monitoring of invasive and quarantine pests, etc. can happen only when field data are available in real-time. Data on spatiotemporal dynamics are essential to study the origin, rate of population growth and spread of any pest species in a given agro-ecosystem. Such data, if available in real-time, shall play a pivotal role in developing robust prediction maps. Today, prediction equations are developed only for a few pests, especially for those whose populations are governed by a few parameters. With real-time data associated with space and time, it would be possible to bring many more pests under these equations. Education programs can be enormously strengthened by intertwining with real-time field data that are brought to the desks of the students. Students can be provided a firsthand exposure to pest ecology and pest management actions, which is completely lacking today.

If the challenge of the changing climate has to be quelled, it is necessary to be equipped with real-time data. The change in climate is perceived in two ways — absolute change like increase in global temperature and relative change like climatic events becoming more unpredictable by the year. Although the absolute changes in climate are more predictable and are generally dealt with in advance, it is the impact of the latter that is perhaps of greater day-to-day interest, which the farmer is, at least in the present, largely concerned about. And it is here that real-time data plays a vital role.

PROPOSED SOLUTION

One will have to agree that agricultural extension systems should aim at a comprehensive growth of the sector with harmonious interactions between all its stakeholders. In order to achieve this, a two-step solution is proposed here. First, a strongly networked and dedicated extension force (not to be involved in other activities of governance) has to be established. This extension force should spread across the nation and serve as 'doctors' to the crops (Figure 1). Second,

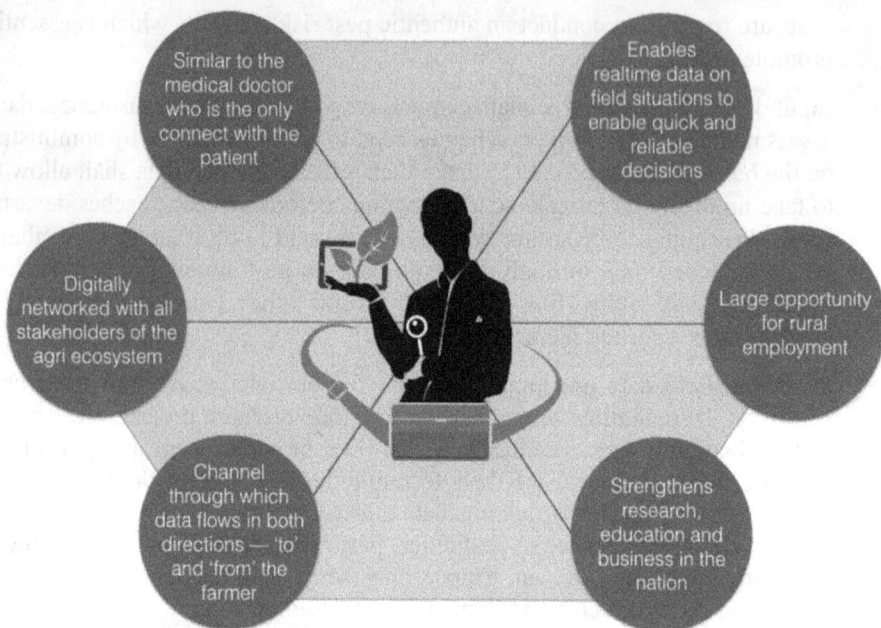

Fig. 1: Roles Envisaged for a Dedicated Extension Personnel
Serving as a Rural 'Crop Health Expert'

the extension force has to be connected in real-time with the major stakeholders of crop health management in India. They have to be digitally enabled to accurately tackle the diverse field situations and independently provide remedies that are inline with national regulations and recommendations. The extension force shall essentially drive "prescription-based" crop health management, which is akin to the medical world. In addition, every field that is visited by the extension force shall automatically turn into a valuable set of data that is available to the major stakeholders in real-time. The stakeholders (knowledge providers and decision makers) shall interact with the extension force in real-time. The knowledge providers shall continuously empower the extension force with the latest knowhow, while the decision makers shall use the real-time data generated to plan their decisions and execute actions. The entire agricultural administration network in the country and all Research Institutions and State Agricultural Universities should come under one ICT platform. The workflow has been presented as Figure 2.

Some of the main features of the ICT system that would be essential to meet the goal are mentioned here. First, the system should contain a field device with a suitable application running on it and a web portal; the two be connected through a server (with appropriate backup mechanisms in place). The field application should be completely accessible offline and updated online. It should be able to receive and send multimedia data even in 2G/GPRS network. Second, the application on the field device should be multimedia-based and designed in a

Workflow

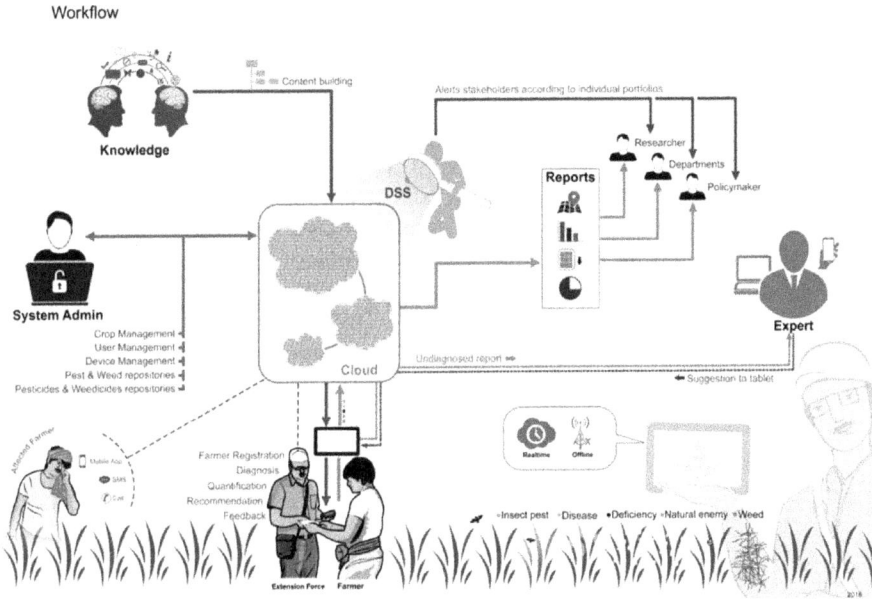

Fig. 2: A Generalised Workflow Showing the Various
Interacting Elements and their Roles

manner that would intuitively guide the extension force to accurate diagnosis and quantification of the problem, and to obtain management recommendations in the field itself; external assistance should be required in bare minimal cases. Although the extension force shall handle the device, the content should be appreciated by the farmer too so that the trust he reposes on the extension force receives a favourable boost. Third, one should be able to remotely update specific pest and management content in specific field devices, so that dissemination of information to the field user happens in real-time. Additionally, there should be scope to provide variable content that is completely localised, across devices located in different geographies for the same crop/pest situation. Fourth, the system should be horizontally and vertically scalable such that every possible field problem can be covered without any limitation including that of local language, and data shall be available across all levels of vertical hierarchy. Finally, real-time data should be available in multiple forms that can be immediately used for making decisions. These broad features shall ensure true empowerment of the extension force; dissemination of the latest pest and management content to the extension force in real-time; minimal dependency on external assistance to resolve field situations; single system that can be used throughout the nation cutting across crops, pests, management strategies and language variations; and all field data can be integrated backwards, summarised in various forms and used for making real-time decisions.

END THOUGHTS

Digitisation in agriculture is normally not about usage of digital equipment to make measurements, at least that is not what the general perception is. It is almost always about usage of ICTs to connect various stakeholders. Be it for disseminating accurate information about crop production technologies, market rates, weather, policies, etc. to farmers; for connecting the producer with the buyer while avoiding middlemen; for empowering policymakers with data on prevailing crop situations; for easing the purchase of farm inputs; for tracking of farm produce; or for anything of the like, ICTs have been steadily replacing the traditional ways since nearly a decade. Today they are clearly making an impact on many different fields of agriculture. However, ICTs should always be viewed from the perspective of integrating different stakeholders, to drive out chaos, to drive in transparency.

Capacity Building for Climate Resilient Agriculture

M.B. Chetti

Education Division, Indian Council of Agricultural Research,
New Delhi 110 001, India
E-mail: mbchetti_uas@rediffmail.com

ABSTRACT: *Adaptation to climate vulnerability Planned adaptation is essential to increase the resilience of agricultural production to climate change. Several improved agricultural practices evolved over the years for diverse agro-ecological regions in India have potential to enhance climate change adaptation, if deployed prudently. Capacity building by extensive participatory demonstrations of location specific agricultural practices helps farmers gain access to knowledge and provides confidence to cope with adverse weather conditions. Interventions needed to be built by the extension workers among farmers for climate resilient agriculture are discussed here.*

Keywords: Climate Resilient Agriculture, ICAR/CRIDA, Capacity Building.

The powerful drift of climate change already evident, the probability of further changes occurring and the increasing scale of potential climate impacts give urgency to addressing agricultural adaptation more comprehensibly. There are many potential adaptation options available for marginal change of existing agricultural systems, often variations of existing climate risk management. The implementation of these options is likely to have substantial benefits under moderate climate change for some cropping systems. However, there are limits to their effectiveness under more severe climate changes. Hence, more systemic changes in resource allocation need to be considered, such as targeted diversification of production systems and livelihoods. We argue that achieving increased adaptation action will necessitate integration of climate change related issues with other risk factors, such as climate variability and market risk, and with other policy domains, such as sustainable development. Dealing with the many barriers to effective adaptation will require a comprehensive and dynamic policy approach covering a range of scales and issues, for example, from the understanding by farmers of change in risk profiles to the establishment of efficient markets that facilitate response strategies. Multidisciplinary problems require multidisciplinary solutions, i.e., a focus on integrated rather than disciplinary science and a strengthening of the interface with decision makers. A crucial component of this approach is the implementation of adaptation assessment frameworks that are relevant, robust, and easily operated by all stakeholders, practitioners, policymakers, and scientists.

Capacity building of extension agents and farmers with respect to climate resilient agriculture involves the understanding of following dimensions.

1. Changing context of extension and rural advisory service?
2. Who are we?
3. What do we do?
4. What role do we have in capacity building for resilience in agriculture to climate change?

Climate change impacts on agriculture are being witnessed all over the world, but countries like India are more vulnerable in view of the huge population dependent on agriculture, excessive pressure on natural resources and poor coping mechanisms. The warming trend in India over the past 100 years has indicated an increase of 0.60°C. The projected impacts are likely to further aggravate field fluctuations of many crops thus impacting food security. There are already evidences of negative impacts on yield of wheat and paddy in parts of India due to increased temperature, water stress and reduction in number of rainy days. Significant negative impacts have been projected with medium term (2010–2039) climate change, e.g., yield reduction by 4.5 to 9% depending on the magnitude and distribution of warming. Since, agriculture makes up roughly 15% of India's GDP, a 4.5 to 9.0 per cent negative impact on production implies cost of climate change to be roughly at 1.5% of GDP per year. Enhancing agricultural productivity, therefore, is critical for ensuring food and nutritional security for all, particularly the resourcepoor small and marginal farmers who would be affected most. In the absence of planned adaptation, the consequences of long-term climate change could be severe on the livelihood security of the poor.

Adaptation to climate vulnerability Planned adaptation is essential to increase the resilience of agricultural production to climate change. Several improved agricultural practices evolved over the years for diverse agro-ecological regions in India have potential to enhance climate change adaptation, if deployed prudently. Management practices that increase agricultural production under adverse climatic conditions also tend to support climate change adaptation because they increase resilience and reduce yield variability under variable climate and extreme events. Some practices that help to adapt to climate change in Indian agriculture are soil organic carbon build up, in-situ moisture conservation, residue incorporation instead of burning, water harvesting and recycling for supplemental irrigation, growing drought and flood tolerant varieties, water saving technologies, location specific agronomic and nutrient management, improved livestock feed and feeding methods. Institutional interventions promote collective action and build resilience among communities. Capacity building by extensive participatory demonstrations of location specific agricultural practices helps farmers gain access to knowledge and provides confidence to cope with adverse weather conditions. Hence, capacity building effort is to be made to marshall all available farm technologies that have adaptation potential and demonstrate them in farmers' fields in most vulnerable districts of the country through a participatory approach.

Extension workers need to build the capacity of farmers through following interventions towards climate resilient agriculture.

1. Soil health is the key property that determines the resilience of crop production under changing climate. A number of interventions are made to build soil carbon, control soil loss due to erosion and enhance water holding capacity of soils, all of which build resilience in soil. Mandatory soil testing is to be done in all villages to ensure balanced use of chemical fertilizers. Improved methods of fertilizer application, matching with crop requirement to reduce nitrous oxide emission.

2. Improved, early duration drought, heat and flood tolerant varieties are to be introduced for achieving optimum yields despite climatic stresses. The varietal development to be carefully promoted by encouraging village level seed production and linking farmer's decision-making to weather based agro advisories and contingency planning.

3. Rainwater harvesting and recycling through farm ponds, restoration of old rainwater harvesting structures in dry land/rainfed areas, percolation ponds for recharging of open wells, bore wells and injection wells for recharging ground water are to be taken up for enhancing farm level water storage.

4. Water saving technologies like direct seeded rice, zero tillage and other resource conservation practices, which also reduce GHG emissions besides saving of water are to be introduced through trainings and demonstrations.

5. Community managed custom hiring centers are to be setup in each village to access farm machinery for sowing/planting. This is an important intervention to deal with variable climate like delay in monsoon, inadequate rains needing replanting of crops.

6. ICAR/CRIDA has developed district level contingency plans for more than 400 rural districts in country. Operationalization of these plans during aberrant monsoon years through the district block level extension staff helps farmers cope with climate variability.

7. Use of community lands for fodder production during droughts/floods, improved fodder/feed storage methods, feed supplements, micronutrient use to enhance adaptation to heat stress, preventive vaccination, improved shelters for reducing heat/cold stress in livestock, management of fish ponds/tanks during water scarcity and excess water are some key interventions in livestock/fishery sector.

8. Automatic weather stations at KVK experimental farms and mini-weather observatories in project villages are established to record real weather parameters such as rainfall, temperature and wind speed etc. both to issue customized agro advisories and improve weather literacy among farmers.

9. Institutional interventions either by strengthening the existing ones or initiating new ones relating to seed bank, fodder bank, commodity groups,

custom hiring centre, collective marketing, introduction of weather index based insurance and climate literacy through a village level weather station are introduced to ensure effective adoption of all other interventions and promote community ownership of the programme.

10. In each village, Climate Risk Management Committee representing all categories of farmers including women and the land less are to be formed with the approval of Gram Sabha to take all decisions regarding interventions, promote farmers participation and convergence with ongoing Government schemes relevant to climate change adaptation.

To improve networking and partnerships among key actors for climate adaptation, strengthening existing platforms and structures at all levels and exploring the role of incentives (e.g., standards) is needed. Following are a few points to be considered.

(a) To develop new, flexible financial products to support climate-resilient and inclusive agro-value chains through capacity building and innovative public-private partnerships.

(b) To increase investments in climate-resilient infrastructures such as roads, irrigation systems, storage facilities and telecommunications.

(c) To enhance the resilience of Indian agriculture covering crops, livestock and fisheries to climatic variability and climate change through development and application of improved production and risk management technologies.

(e) To demonstrate site specific technology packages on farmers' fields for adapting to current climate risks.

(e) To enhance the capacity of scientists and other stakeholders in climate resilient agricultural research and its application.

(f) Selection of promising crop genotypes and livestock breeds with greater tolerance to climatic stress.

(g) Existing best bet practices for climate resilience to be demonstrated

(h) Infrastructure at key research institutes for climatic change research to be strengthened.

(i) Adequately trained scientific manpower to take up climate change research in the country and empowered farmers to cope with climate variability.

(j) Critical assessment of different crops/zones in the country for vulnerability to climatic stresses and extreme events, in particular, intra seasonal variability of rainfall

(k) Comprehensive field evaluation of new and emerging approaches of paddy cultivation like aerobic rice and SRI for their contribution to reduce the GHG emissions and water saving.

(l) Special attention to livestock and fishery sectors including aquaculture which have not received enough attention in climate change research in the past.

Agricultural Policy and Planning

Assessment of Government Policies and Programs on Climate Change—Adaptation, Mitigation, and Resilience in South Asian Agriculture

P.K. Joshi[1] and N.K. Tyagi[2]
[1]International Food Policy Research Institute, New Delhi 110 012, India
[2]Agricultural Scientists Recruitment Board, New Delhi 110 012, India

ABSTRACT: *South Asian agriculture is highly risky and extremely climatic turbulent. By its nature, agriculture has huge cross-sector dependence, including water resources, forests, biodiversity, energy, and industry (fertilizer and chemicals, and equipment for irrigation and other operations). Climate change and agriculture are interrelated, multi-sectorial and operating on a global scale but impacting at more meso and micro level. Therefore, climate change does not affect agriculture in isolation, but its impacts are transmitted from different allied sectors. South Asia's agricultural development policies have mainly focused on improving seeds and fertilizer application, expanding irrigation, managing watersheds, and providing weather forecasts and insurance based on Information and Communications Technology (ICT). This study is an attempt to analyze the climate policies on agriculture and allied sectors, and their impacts on South Asia's capacity to mitigate, adapt and build resilience to climate change.*

IMPACTS OF IRRIGATION POLICY

South Asia is one of the most intensively irrigated agricultural regions in the world, with an estimated 40 percent of total cultivated area under irrigation. While India has large, contiguous, centrally controlled irrigation systems, Bangladesh, Nepal, and Sri Lanka have mostly small and medium-sized schemes. In all the countries of the region, groundwater resources have been developed rapidly.

Although irrigation has had significant, mostly positive, impacts, it has had some negative environmental and socio-economic effects as well. For example, during last 20 years, the average productivity of four major irrigated crops in India—rice, wheat, maize, and groundnut—has increased from 2.324 metric tons per hectare (t/ha) to 2.971 t/ha. Irrigation has generated additional food grain output of 24 million metric tons (Mt), which has reduced GHG emissions by 7.0 Mt of carbon dioxide equivalent (CO_2e). The decrease in GHG emissions due to the avoidance of converting forestland to cropland is estimated at 68.14 $MtCO_2e$. However, the energy required for pumping groundwater has produced GHG emissions estimated at 30.5$MtCO_2e$, ultimately resulting in a negative balance (mitigation benefit) of 47.8 $MtCO_2e$.

Irrigation has also been a major alleviator of poverty in South Asia. The incidence of poverty is as high as 69 percent in districts with less than 10 percent cropped area under irrigation. In contrast, it is only about 26 percent in districts where irrigation covers more than 50 percent of cropped area and just 10 percent in Punjab and Haryana, where more than 70 percent of cropped area is irrigated (World Bank 1998). Irrigation has proved to be the most effective drought-proofing mechanism and the most significant factor in adapting climate change and building resilience in agriculture sector.

IMPACTS OF FERTILIZER POLICY

Fertilizer consumption has grown rapidly in the four South Asian countries. During the last two decades, consumption increased from 1 Mt to 4 Mt in Bangladesh, 12.5 Mt to 28 Mt in India, and 3 Mt to 4 Mt in Sri Lanka, and has been fluctuating between 0.16 Mt and 1.01 Mt in Nepal. It has been clearly established that the use of fertilizers in all South Asian countries had a linear correlation with the amount of subsidy and the availability of fertilizers at given points of time at those prices.

India's fertilizer polices have had a positive effect on growth in fertilizer use, which has increased crop production and productivity. Productivity gains have strengthened India's food and nutritional security by increasing food grain availability from 455 grams/capita/day in 1981 to 463 grams/capita/day in 2011, in spite of a population increase of more than 500 million during the same period. Fertilizer use has increased GHG emissions by 58 Mt/year over the baseline year of 1990 due to increased consumption.

By contributing 13.66 Mt of additional food grain production, fertilizers avoided the conversion of 11.48 Mha of forestland under crop cultivation. This has reduced 20.13 Mt of GHG emissions which would have otherwise been generated if fertilizers consumption were restricted at 1990–1991 levels)

IMPACTS OF MICRO-IRRIGATION

Micro-irrigation did not receive much attention in South Asian countries, except some in India. In 2010, micro-irrigation (drip and sprinklers) in India covered about 4 Mha, which is only 10 percent of the projected potential area of 42 Mha (TFMI 2004). Irrigation has enjoyed high subsidy support, ranging from 50 to 75 percent under various schemes of the government.

Micro-irrigation is a triple-benefit smart technology that saves water and energy and increases yields. Worldwide, the increased production reported from micro-irrigation is on the order of 25–40 percent,

The estimated 30 percent increase in efficiency (for both water and productivity) yielded 3.483 Mt of additional production, saving 0.733 million hectare meters (Mham) of water and reducing emissions by 5.654 $MtCO_2e$. If extended to the 42 Mha of potentially suitable area, micro-irrigation would increase tenfold. The policy support in the form of a subsidy for micro-irrigation had significantly positive effects on climate mitigation and adaptation. Its extension to a larger area would stabilize the groundwater table and arrest declining water supplies.

IMPACTS OF GROUNDWATER AND ENERGY POLICIES

About 58 percent of India's 389 Mham of usable groundwater has been developed. When analyzed on regional basis, average groundwater development in the Northwest Plain is already exceeding 98 percent (highly over exploited), compared with only 43 percent in the Eastern Plain and 42 percent the Central Plain States (CGWB 2010). Groundwater development is highly unregulated, and groundwater policies, even if statutorily required, are seldom enforced. India's energy pricing policy provides for electricity that is either free or at a flat rate that is less than one-tenth of the generation cost. Inexpensive energy has exacerbated the problem of declining water tables from intensive groundwater pumping.

Total GHG emissions from groundwater pumping are estimated at 28 $MtCO_2e$. On the positive side, by contributing 16.1 Mt to agricultural production, groundwater has avoided both 13.53 Mha of deforestation and 40.29 $MtCO_2e$ of emissions. Thus, the net emission benefit due to groundwater irrigation is 10.29 $MtCO_2e$.

IMPACTS OF AGRICULTURAL POLICIES ON MITIGATION, ADAPTATION, RESILIENCE, AND SUSTAINABILITY

A set of criteria—percentage reduction or increase in baseline values of GHG emissions for mitigation and per capita increase in food grain availability for adaptation—was identified for assessing the performance of agricultural policies. Policies resulting in 0, 10, 20, and 30 percent reductions from baseline emissions and increases in food grain availability were rated fair, good, very good, and excellent respectively.

Fluctuations in food grain production represent the combined effect of all the policies. Therefore, change in variance in food grain production in the first and the last 5 years was taken as the criterion for resilience. A change of less than 5 percent received an excellent rating. Similarly, the degree of development (DD) for surface water and a groundwater abstraction ratio (GWAR) for groundwater were chosen as sustainability criteria. DDs of 0.5, 0.6, 0.75, and 0.90 were treated as normal, high, very high, and extremely high. Groundwater development was considered safe up to a GWAR of 0.65, marginal at 0.65–0.85 and unsafe beyond 0.85 (CGWRE 2009).

FUTURE POLICY DIRECTIONS

The region's future policy formulation should capitalize on best practice, by up (and out) scaling them, and removing barriers on the partly successful policies.

Water Policy

Irrigation water policy has been rated high on adaptation and mitigation, but low on sustainability, largely because of overexploitation. The overexploitation of water is linked to differences between demand and supply issues and the related issue of low-cost energy pricing, which has a considerable impact on the efficiency of water use. Therefore, achieving the sustainability goal would require structural changes in the administration of the water sector. The array of available water-smart technologies offers ample opportunity to economize on blue water (rainfall that enters lakes, rivers, and aquifers) and increase the use of green water (rainfall absorbed by vegetation and evapotranspiration back to the atmosphere) to reduce the gap between supply and demand.

Fertilizer Policy

Fertilizer policy has been rated high on adaptation and resilience, but was found wanting on mitigation and sustainability. Obviously there is a case for improving the efficiency of fertilizers to reduce their carbon footprint. The current inefficient use of fertilizers can be ascribed to several factors, including application technology, available fertilizer products, nutrient use ratio, and pricing. These issues appear to be interlinked and would require a mix of policies that addresses all of them simultaneously. Pricing has been and will continue to be a big, if, in the suggested list of reforms.

Electricity Policy

Because raising the price of energy to the point where farmers would respond by saving water would be politically difficult, the next best approach would be to create conditions that would improve the efficiency of electricity use. Promising options are separating high-voltage feeders for tubewells; promoting energy-saving devices, such as capacitors and high-efficiency motors; using ICT for groundwater withdrawal monitoring; and classifying agricultural pump sets as "appliance" under Section 14 of India's Energy Conservation Act, 2001.

ACKNOWLEDGEMENTS

The support received from the Global Research Project on 'Climate Change, Agriculture and Food Security' is acknowledged for conducting this study. Authors thank technical guidance received from Dr Pramod Aggarwal and research support from Ms Divya Pandey.

An Overview of Climate Smart Agriculture: International, National and State Policies

Gopal Naik

Indian Institute of Management, Bengaluru 560 076, India
E-mail: gopal@iimb.ernet.in

ABSTRACT: *Considering that the policy actions are essential to address climate change impact and provide food security, in this paper we analyze the genesis and the typology of Climate-Smart Agriculture Policies (CSAP), their current status and future policy directions. Specifically, we look at typology of the policies on climate smart agriculture at different geographical levels, various alternative policies that are attempted, assess their status considering the policy making processes and the trade-offs involved, and suggest directions for futures policy initiatives.*

Keywords: Climate Smart Agriculture Policies, Climate-Smart Agriculture Impact Studies.

With the inexorable nexus between agriculture and climate change, and growing evidence of climate change forecasts being actually experienced, the implications on global food supply and the resulting challenges to food and nutrition security and poverty alleviation have been paid considerable attention both at national and international levels. Also, the need to evolve concerted policy responses to address these implications has been clearly recognized. It is estimated that global food production has to increase by at least 60% by 2050 in order to meet the needs of growing population (FAO, 2013). Climate change is expected to alter fundamentally the food production pattern posing greater challenges in meeting the food security needs. Impact predictions also indicate that food production is likely to adversely impact more in regions where there is already food insecurity. Countries with low levels of economic development, weak institutions, and limited human and financial capital are likely to bear the brunt due to their limited resilience capacity. Food security in individual countries will be addressed mainly through the dynamics and interaction of agricultural markets, climate suitability, adaptive capacity and policy support (FAO, 2015). In all these factors policy has a significant role to play. Food and Agriculture Organization has identified three major components of climate-smart agriculture: sustainably increase productivity and income; strengthen adaptation and resilience to climate change and variability; and reduce agriculture's contribution to emissions leading to climate change (FAO, 2010). Climate-smart agricultural policies formulated and implemented at various governance levels incentivize and regulate actions of stakeholders at local, national and international levels that are needed to transform and reorient agricultural systems to effectively support development and ensure food security in the on-going climate change scenario. In order for these policies help achieve the objectives, there is greater need for coherence, coordination and synergy between climate change, agricultural development and food security policies (FAO, 2010). Many of the policies related to these aspects have been there for long in many countries but now will have to be viewed from the perspective of climate-smart agriculture. In this context the primary objective of this paper is to understand the policy evolution with respect to responding to climate change at international, national and local levels relevant to agriculture, prepare a typology, assess their status so far and suggest emerging policy directions.

Lipper *et al.* (2014) define climate-smart agriculture as "an approach for transforming and re-orienting agricultural development under the new realities of climate change". During the last two decades considerable knowledge has been accumulated on the nature of climate change, possible implications on food and nutritional security and on the impact of related policies that are implemented. FAO study on climate smart agriculture (FAO, 2010) concluded that climate change threatens agricultural production stability and productivity making it difficult to achieve the food production targets. Differential impact of climate change on crop productivity such as positive impact in the high altitude and negative impact on low altitude and tropical regions (IPCC, 2007; Cline, 2007) is likely to impact poverty and nutritional security. Furthermore, increased carbon dioxide concentration is likely to reduce concentration of zinc, iron and protein and increase concentration of starch and sugar in crops such as wheat, rice and soybeans (FAO, 2015). There is also a likelihood of increased water scarcity in many regions of the world posing policy challenges in terms climate adaptation and evolving coherent cross-sectoral strategies for water management at international, national and regional levels. The policies on water should not only create efficient markets but also strengthen the institutional structures to create access to underprivileged people. Climate-smart agriculture policies should able to achieve food security and development goals. Transition in agriculture production system is required to make it more productive, use inputs efficiently, enhance stability in their outputs and is resilient to shocks in climate variability. Such a transition requires major shift in the way the land, water, soil nutrient and genetic resources are managed (FAO, 2013). While some technologies that help in adaptation have mitigation co-benefits, others such as biofuels and use of nitrogenous fertilizers may pose food security trade-offs. The trade patterns are likely to change and disrupt supply chain due to more frequent extreme weather patterns. Trade may have to go with domestic adaptation strategy to reduce market and price distortions. Trade and climate change mitigation policies should include measures designed to internalize environmental cost of resources. Considering that adverse effects of climate change are greater among the poor people in developing countries, mainstreaming climate change responses within the pro-poor development strategies is important. Emphasis need to generate policy relevant evidence to examine the impact of climate change on agriculture, water and trade. Therefore achieving climate-smart agriculture requires not only appropriate international agreements but also shifts in national and local governance, legislation, policies and financing mechanism (FAO, 2015).

The method followed in this paper is to review the literature to identify various policy initiatives taken up at various geographical levels, prepare a typology of these policies, assess them in terms of policy making and adoption processes and the impact these policies have made. The paper then tries to identify emerging successful policies for future implementation at various levels.

The typology of policies is developed based on the components identified by the FAO studies. The broad components identified by the FAO are: sustainable increases in agricultural productivity and income; improving adaptation and resilience and mitigation of climate change. These policies are initiated and implemented at international, national and local levels. In this paper typology of policies is created based on these dimensions.

The framework of analysis uses the standard policy-making and adoption processes as depicted in Figure 1. The first stage of policy making is the need identification, which can be proposed by various stakeholders. In the case of climate change, scientific evidence has been an important input for stakeholders at national and international level to think about policies to address them. Once various stakeholders are convinced about the policy need, appropriate policy measure needs to be narrowed down from a set of alternatives, which often involves intensive policy debates. Effectiveness of the policy measure requires appropriate design is developed. These three measures could take place at various levels such as international, national or regional. Proper information sharing of these stages can significantly benefit policy making at all levels, local

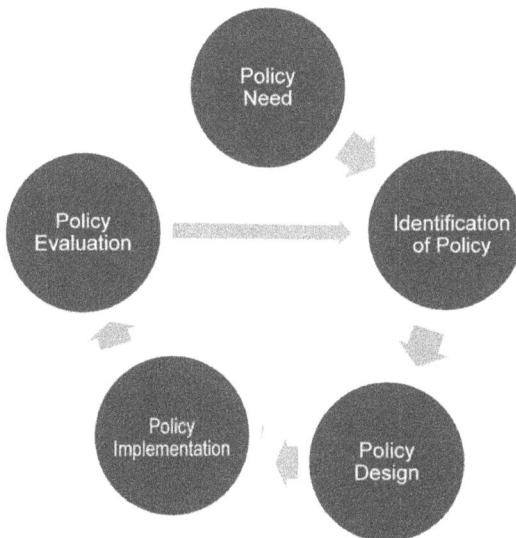

Fig. 1: Process of Policy Making and Implementation Cycle

to international. Most of policy measures need implementation by national or local governments, and the effectiveness depends on how far stakeholders are convinced, the tradeoffs involved and the effectiveness of the implementing machinery. Evaluation of implemented policies will generate useful information about the impact the policy measure is making with respect to achievement of policy objectives, and provides inputs for future course of action. In this paper we will identify the stage of policy making for various policies needs suggested in the past research and provide future directions.

We examine the climate-smart agricultural policies in terms of processes followed and their current status. The paper also tries to bring out current status of these policies and identifies areas of strengthening policy making processes and policies that have shown potential to help achieve the objectives of climate-smart agriculture.

Designing appropriate climate-smart agricultural policies require a proper under-standing of climate change impact on agriculture and related resources such as water. A typology of this understanding sets the basis for identifying and developing proper policy response. The findings of the impact studies on various dimensions of agriculture, water and trade are presented in Table 1.

Table 1: Climate-Smart Agriculture Impact Studies: Impacts on Related Dimensions

Dimension Related to Agriculture	*Impacts Observed*
Understanding Impact of Climate Change	
Water availability	Water scarcity is likely to occur in many regions
Agriculture	Negative in low latitude and tropical regions and positive in high latitude regions; higher market and price volatility; degradation of soil and water resources
Nutritional content of crops	Higher carbon dioxide concentration $[CO_2]$ is found to lower concentrations of zinc, iron and protein and raise starch and sugar content in crop plants such as wheat, rice and soybeans.
Trade	Trade will expand from mid-to high latitude regions to low latitude regions; trade disruption in supply chain logistics; trade can cushion against large production shocks arising from climate change
Inter-linkages	Climate-food-water-trade nexus; forest cover trade-off with cultivation
Understanding impact of agricultural practices on emissions with respect to	
Animal Husbandry	Increases emission particularly methane
Chemical use in agriculture	Nitrogen fertilizer use has trade off in terms of increasing productivity but increasing emission of nitrous oxide
Use of energy in agriculture	Fossil fuel use will increase emissions

The Table 1 suggests that climate change impact on agriculture could be substantial and have many dimensions. Significant impact originates from changes in temperature, rainfall pattern and carbon dioxide concentration resulting in changes in water availability. In many regions water scarcity is likely to be a serious concern. Its impact on access, agriculture production, particularly on food security, impact on trade, market and price volatility, and on vulnerable population are serious policy challenges. Challenges also arise due to strong nexus between various sectors such as climate-food-water-trade and the need for policy synergy in the light of trade-offs involved.

Mitigation of climate change has been a direct way of addressing the climate change impact. While there are many areas such as industry, commerce, consumption etc related to mitigation measures, we focus on measures that are directly relevant to agricultural practices. Table 2 provides productivity enhancing, climate change adaptation and mitigation measures suggested under the climate-smart agriculture approach.

Table 2: Climate-Smart Agriculture Measures

Objectives	*Agriculture and Land Use*	*Pre and Post Production*
Productivity Enhancing	Increased mechanization, fertilizer, irrigation	Proper storage, transportation, handling
Adaptation	Selection of varieties/breeds, crop calendar changes, efficient water use	Changes in food consumption; diversified animal feed
Mitigation	More energy efficient technology, decreased mechanization, fertilizer and livestock production, re-afforestation, biogas production	Decreased use of, and more efficient technology for refrigeration, processing, transport; switch to clean fuels
Productivity+ adaptation	Weather forecasting; matching varieties to local climates	Food safety measures
Productivity + Adaptation + Mitigation	Soil management, restoration of degraded land, improved pest-disease management	Management of food reserves, reduction food wastage by consumers
Adaptation + Mitigation	Re-afforestation with multifunction trees, efficient water storage and management	Reduced reliance of cold chain, increased energy efficiency, consumption of seasonal produce

Source: Adapted from Vermeulen *et al.* (2012).

Efforts have been made at the international level beginning with the First World Climate Conference in 1979 and setting up of Intergovernmental Panel on Climate Change (IPCC) in 1988, United Nations Framework Convention on Climate Change (UNFCCC) in 1992 and Kyoto Protocol in 1997. Cancun Agreement in 2010 emphasizes on climate change mitigation measures. Policy measures other than those related to mitigation, address multiple objectives of sectors and nations. These are indicated in Table 3 based on the past studies related to climate change impact.

Table 3: Climate-Smart Agricultural Policies at International, National and State Levels

	International	*National*	*State*
Comprehensive policies on CSA	Kyoto Protocol, Paris Agreement CAP	National Action Plan on climate change (2008) – 8 core missions NICRA–National Initiative for climate resistant agriculture (2011)	State Action Plan on Climate Change (SCAP)
Increased use of mechanization, fertilizer, irrigation, high yielding varieties	Crop Insurance Program (USA)	Nutrient based fertilizer Subsidy, creation of irrigation facility, Seed production	Support or farm mechanization, Fertilizer and seed subsidy, Watershed management, Rain water harvesting
Proper storage, transportation, handling		Warehousing development and policy	Warehousing Development and policy
Selection of varieties/breeds, crop calendar changes, Efficient water use	Water security pricing & Water use right trading (Spain), Drip irrigation, small dams (Portugal)	Drip irrigation, check dams, National Water Mission	Supply of seeds, drip irrigation, check dams
Changes in food consumption; diversified animal feed		NICRA	
Weather forecasting; matching varieties to local climates	Mixed crop and species (France)	NICRA	Weather stations, Seed bank, Insurance
Food safety measures	Codex and other standards	Food Safety Act	Food Safety Department
More energy efficient technology, decreased mechanization, fertilizer and livestock production, re-afforestation	Renewable energy use in water pumping (Portugal); Nitrate directives (CAP); Good farming practice (CAP), Energy Efficient Technology (USA), Climate Hub (USA)	Solar pumps, renewable energy, National Mission on Sustainable Agriculture	Solar pumps, Agricultural Research and Extension, State action plan on CC (SCAP)
Decreased use of and more efficient technology for refrigeration, processing, transport; switch to clean fuels	Clean energy (USA)	Use of renewable energy	

	International	*National*	*State*
Re-afforestation with multifunction trees, efficient water storage and management	Flood water storage (Hungary), Water retention landscape (Portugal), Climate resilient river (Netherlands), Water storage (China), Forest Stewardship Program (USA)	Forest development, Joint forest management, Watershed Development Programme, National Water Mission	Forest development, Joint forest management, Watershed Development Programme
Reduced reliance of cold chain, increased energy efficiency, consumption of seasonal produce	Clean energy and energy efficient technology (USA)		
Soil management, restoration of degraded land, improved pest-disease management	Obligatory crop rotation, grassland maintenance (CAP)	NICRA, National Rural Livelihood Mission	
Management of food reserves, reduction food wastage by consumers		Warehousing	
Trade	WTO		

As we can see from the table, many of measures indicated in the climate-smart agriculture are at various stages at the international, national and state levels. At the international level there is clarity about the need for policy action to mitigate, adapt and sustainably increase productivity. Common Agricultural Policy of European Union incorporated measures particularly in adaptation and mitigation areas. Several policies particularly related to productivity enhancing and adaptation have been included in the existing policies by countries and sub-regional governments. However, an integrated approach to policy is still missing at both international and national levels. Most of the policies are at the identification and design stage and lack resource commitment and effectiveness implementation. At the state level where the action really is, integrated action plan is yet to be developed, while some of the existing policies do support CSAP. Productivity enhancing and adaption measures are in consonance with the short-term needs of food security and development. In many developing countries mitigation measures are not seen with urgency considering the development stage of agriculture. While many of them have advantage of knowledge about traditional agriculture, that are very much in line with the requirement of CSA, a clear strategy to develop them into a CSAP is missing. Adapting traditional knowledge augmented by sustainable modern technology could an important area of focus for achieving CSA. An area that is not addressed sufficiently is food safety and changes in the consumption habits to reduce waste and adapt to climate resistant food production.

Additionally policy implementation and evaluation processes needs to be strengthened through proper institutional mechanism in order to identify further and design effective policy measures. At the international level, trade policies need to explicitly account for CSA from the perspective of developing countries.

Impact of these policies critically depends on the effective implementation at the grass-root level. Grass-root level implementation requires that there is sufficient information and guidance available at the local level. This points to the need for a strong extension system that needs to be developed so that the information and guidance are available at the local level.

Robust models are needed with proper representation and integration of bio-physical processes in economic models so as to convince policy makers to fill the policy vacuum. Robust trade analysis also requires integration of direct climate change impact on productivity, realistic demand changes, and constraints on resources. Also the data system needs to be developed to improve accuracy of the data.

REFERENCES

Climate change in the 2015 federal budget and Farm Bill, [Online]. Available: http://www.iatp.org/blog/201404/climate-change-in-the-2015-federal-budget-and-farm-bill.

FAO (2010). "Climate-Smart" Agriculture: policies, practices and financing for food security, adaptation and mitigation. Food and Agriculture Organization of the Unites Nations (FAO).

FAO (2013). Climate-Smart Agriculture: Source Book, Food and Agriculture Organization of the United Nations (FAO).

FAO (2015). Climate change and food systems: global assessments and implications for food security and trade. Food Agriculture Organization of the United Nations (FAO).

Lipper, L., Thornton, P., Campbell, B.M. and Torquebiau, E.F. (2014). Climate-smart Agriculture for Food Security, *Nature Climate Change,* 4:1068–1072.

Vermeulen, S.J., Campbell, B.M. and Ingram, S.J.I. (2012). Climate Change and Food Systems: *Annual Review of Environment and Resources,* 37:195–222.

Prioritizing Climate Smart Agriculture in Tribal Dominated Agrarian Society: A Case Study of Madhya Pradesh

Barun Deb Pal[1] and Parmod Kumar[2]

[1]International Food Policy Research Institute, New Delhi 110 012, India
[2]Institute for Social and Economic Change, Bengaluru 560 072, India
[2]E-mail: pkumar@isec.ac.in

ABSTRACT: *The main aim of this paper is to assess farmers' preference on feasible climate smart technologies for the state of Madhya Pradesh. This study first explores the technical and economic feasibility of various climate smart technologies across selected 4 districts of the state of Madhya Pradesh. Farmers' preferences about various climate smart technologies are assessed using stated preference method after providing detail training to the selected farmers within the selected districts. Results of this assessment reveal that the highly preferred technologies by the farmers are not necessarily highly feasible. Rather farmers prefer the technology which is input saving even if productivity does not increase much as compared to their current practices. Such behaviour of farmers is corroborated by the fact that farm land in Madhya Pradesh is by and large rain-fed and availability of agricultural machinery is concentrated among large farmers. Nevertheless, the adoption of highly feasible technologies require assured supply of inputs like water, energy and machines which are poor in supply in the state. Despite the above fact, it is to be noted that farmers are interested to replace their traditional cultivation practices with climate smart practices although less feasible with the priority of input cost saving. Summing up, this study fills the gap between scientific knowledge about the agricultural smart technologies, their technical feasibility across various regions and farmers' preference in adoption of those technologies.*

Keywords: Climate Smart Agriculture, CRIDA, NICRA.

The agriculture sector in India is heterogeneous in terms of its bio-physical characteristics, natural resource supply, and socio-economic conditions of the farmers across various states. Therefore, climate smart technologies and their feasibility will vary across regions. Hence, identifying feasible climate smart technologies for different regions and region specific prioritization will be a logical step to achieve climate smart agriculture in India.

CAUSES OF LOW PRODUCTIVITY: CLIMATE DEPENDENCY AND CLIMATE CHANGE IMPACTS

Almost half of the agriculture land in Madhya Pradesh is still rain-fed as only 39% of gross cropped area is irrigated. Therefore crop cultivation in this state is highly dependent on climatic condition and hence climate change is one of the key factors behind the low productivity of this state. Rise in minimum temperature, variability in rainfall, frost and frequent drought events are major climate related factors directly affecting productivity of crops irrespective of the season. Analysis in the literature shows that an increase in surface temperature by 3% will reduce yield of soybean and wheat by 20% to 30%. It is also observed from the fact that, due to excess rainfall the yield of soybean has been fallen by 10% in the year 2013 as compared to 2012. On the other hand, the vulnerability atlas prepared by CRIDA, GoI has identified almost 26 districts (i.e. 50%) as vulnerable (high and very high) due to climate change impact.

GOVERNMENT'S STRATEGIES TO COPE WITH CLIMATE CHANGE IMPACT

The Government of Madhya Pradesh has taken several policies to conserve necessary resources like water, soil, and fertilizer to cope with the climate change impact and achieve sustainable agricultural development for the state (Table 1). However, these initiatives are still at nascent stage and focus is given on strategic planning at the regional level. As a part of this strategic planning the state action plan on climate change has been drafted and district contingency plans have also been prepared with the objective of capacity building among the stake holders to develop feasible policies for climate change adaptation and prioritize them at the regional level to achieve climate resilient agriculture at this state.

Table 1: Important Strategies Adopted by the Government in Madhya Pradesh

Strategies	*Policies/Programmes/ Projects*	*Key Objectives*
Adoption of Improved cultivation techniques and resource conservation	District contingency plan prepared (NICRA)	Suggest farmers about change in sowing dates, crop establishment method, resource management, seed replacement and treatment and choice of crops and variety of seeds
Capacity Building of policy makers	State action plan for climate change (SAPCC)	Understand long term impact of climate change, Capacity building of the stakeholders to design policies
	Vulnerability assessment (MoEF) and Vulnerability atlas (CRIDA)	Understand climate change exposure at district level and their adaptive capacity

Strategies	Policies/Programmes/ Projects	Key Objectives
Irrigation Intensification	Irrigation Development Projects	Lining Canal to bring rain-fed area under irrigated
	Micro and Minor Irrigation	Subsidy on Sprinkler and drip irrigation to increase water use efficiency
	Installation of Tube well	Subsidy on cost of installation to increase irrigation facilities
	Balaram Tank Yojona	Rain water harvesting for better irrigation facility and ground water recharge
Soil Conservation	Integrated Nutrient and Pest Management	Balancing fertilizer use to improve productivity and soil fertility
	Soil Testing Laboratories	To fulfil nutrient deficiencies in the soil
Weather Advisories	Information and communication Technologies for weather forecasting and farm management	Prior information about rainfall and required management practices through Mobile SMS.
Extension	Field level demonstration of modern agriculture implements	Broad Bed Furrow for moisture conservation and rain water management. Ridge and furrow method for rain water management and moisture conservation. Zero tillage for moisture and nutrient management Systems of Root Intensification to increase input use efficiencies

UNDERSTANDING THE STEPS TO IDENTIFY THE GAPS IN IMPLEMENTING CSA

Steps involved in achieving climate smart agriculture as described in Figure 1 start from innovation and end at implementation. However, these steps are not unique for climate smart agriculture alone but they are general process of transforming agriculture enabling farmers' for adjustment from one situation to another. As described in the figure, innovation is the primary stage in the process which includes any innovative idea or technology or practice to reach to the final destination. However, in agriculture sector all innovative ideas or technologies may not be feasible for every location and hence, one has to identify the most feasible technology, idea or practice. The identification mentioned in this process poses two relevant questions – First, whether the innovative technologies are technically feasible? Second, even if they are feasible technically will there be economic benefit to the beneficiaries?

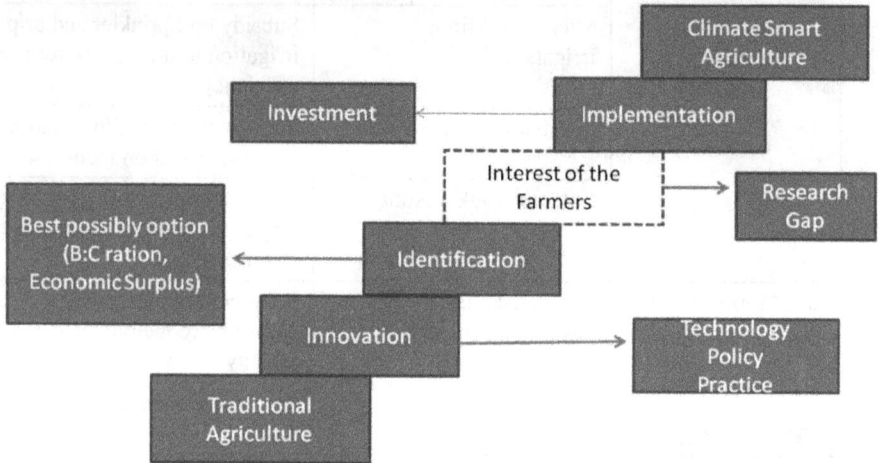

Fig. 1: Gap in Steps to Achieve Climate Smart Agriculture

To answer first question one has to understand regional heterogeneity in terms of bio-physical characteristics and map them with the biophysical characteristics necessary in adoption of any new technology. On the other hand, there are several studies available to answer the second question. These studies have described various methods like assessment of benefit cost ratio and estimate economic surplus for each innovative intervention and describe their implication to the society as a whole. Therefore, if a new technology is technically feasible and will have positive impact to the society it can be identified for implementation. So far majority of the government policies and projects were conceived and implemented by considering their feasibility and implication to the society. However, fewer attempts have been made to understand interest of the beneficiaries to adopt any new technology or policy for their development. Hence there is a major gap in the process of transformation. Hence to achieve climate smart agriculture by increasing adoption of climate smart practices, interest of farmers must be included in the process of implementation and transformation. In this context, aim of this study is not only to understand technical and economic feasibility of the climate smart agriculture, instead it focuses on assessing farmers' interest in adoption of climate smart agriculture practices and upscale them at the regional level. Therefore incorporation of farmers' interest in the steps to achieve CSA is a unique contribution of this study.

APPROACH AND METHODOLOGY

As most technologies are at demonstration level and only a few are being adopted, we have followed an ex-ante assessment approach to assess preference about the climate smart practices among the randomly selected farmers in various regions of Madhya Pradesh. Again, according to the vulnerability atlas of Madhya Pradesh, rising minimum temperature is common climate change phenomena across districts and selected districts are not exception to this. Keeping this issue in mind we have categorized the selected four districts as described in Figure 2.

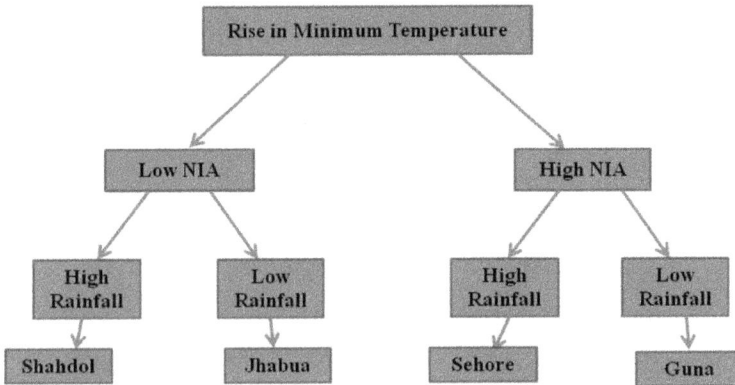

Fig. 2: Heterogeneous Profile of the Districts under Common Climate Change Exposure

SELECTED CROPS

While accessing the farmer's preference about the climate smart technology we have followed the above stated criteria. The sample size of our training was 180 farmers from four districts namely Sehore, Guna, Jhabua and Shahdol (Table 2; Figure 3). We have selected 3 developmental blocks from each districts and 1 village form each block with taking the criteria of highest number of agricultural household existing block/ village and from each village we have randomly selected 15 farmers household for the training purpose. Through the farmers training we have tried to improve farmers understanding and the awareness level about the economic and technology feasibility of climate smart technologies.

Table 2: Selected Crops from Each Districts

Districts	Cropping System	Selected Crops	% in GCA
Sehore	Kharif Soybean & Rabi Wheat Zone	Soybean Wheat Gram	52.1% 36.5% 9.3%
Jhabua	Kharif Soybean & Rabi Wheat Zone	Soybean Wheat Gram	43.7% 37.9% 5.6%
Guna	Rabi Wheat and Kharif Soybean & Rice Zone	Rice Soybean Wheat Gram	18% 33.4% 38.8% 5.5%
Shahdol	Kharif Rice & Rabi Wheat Zone	Rice Wheat	56.2% 25.3%

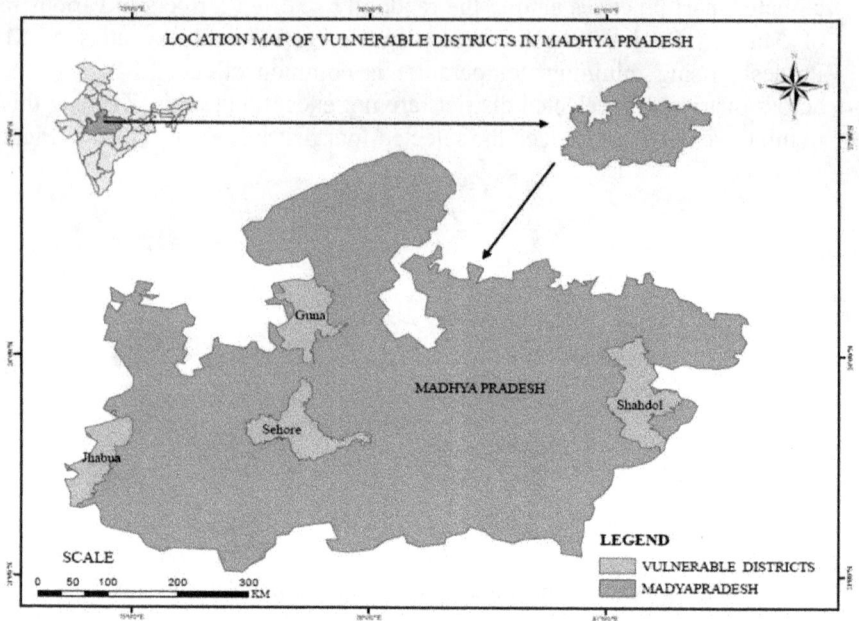

Fig. 3: Location of the Selected Vulnerable Districts in Madhya Pradesh

ASSESSMENT OF FARMERS' PREFERENCE

The farmers training was carried out in four phases (Figure 4), 1st we showed posters presentation to the farmers to give a basic idea of the feasible climate smart technologies in the region and also explained the basic characteristics of

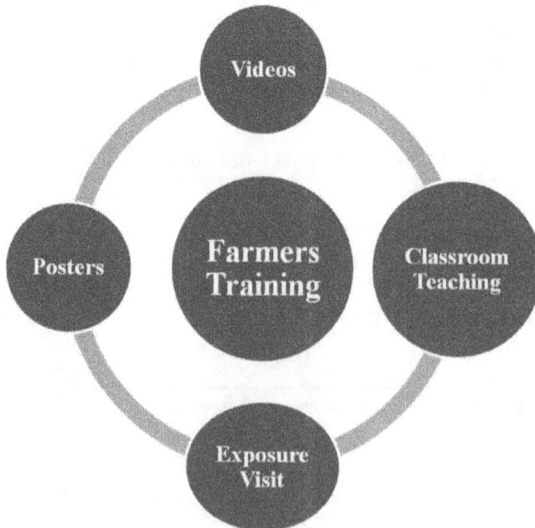

Fig. 4: The Four Phases of Farmers' Training

the technology. Secondly, we tried to apply the ICT method where we showed the videos of different technologies to the farmers where learning through video communication method has been adopted. Thirdly, the two ways learning process has been applied through classroom teaching method, where the KVK's scientist have been involved and a rigorous discussion has been made with the farmers to understand their problems. Lastly, an exposure visit was conducted to the KVK's cafeteria crop zone to rigorously understand farmers' preferences for different crop options.

Through the above stated criteria we have tried to access the farmer's preference of technologies for different crops in different regions. The observed farmers' preferences of different crops in the selected districts are discussed in the following section.

FARMERS PREFERENCE AND SCIENTIFIC FEASIBILITY OF CLIMATE SMART TECHNOLOGIES

Various climate smart technologies including input saving characteristic of each of these technologies and a comparative analysis of climate smart technologies, their scientific feasibility and level of farmers' preference for these technologies are given in Tables 3a and 3b. Using input-output data for technologies from the

Table 3a: Selected Technologies and their Relevance for Climate Smart Agriculture

Technology Codes	Descriptions of Technologies	Type of CSA
(FP) Farmer Practices	Wheat, Soybean and Gram: Line Sowing with seed cum fertilizer Drill & seed treatment Rice: Transplanted with treated seed	
INM + IPM	Nutrient & Pest management by following soil testing, deep ploughing, and balanced use of organic and chemical fertilizer and pesticides	Nitrogen Smart
BBF	Tractor drawn Broad bed and furrow which is done once in three years	Water Smart
LLL	Laser Land Leveling once in three years	Water & Energy Smart
ZT	Zero Tillage with residue mulching for wheat	Energy & Nutrient Smart
SWI	Systems of Wheat Intensification. Manual sowing of single seed at a distance of 8 cm. Seed treatment and a line showing method to be followed in this technology	Water & Energy Smart
SPNK	Sprinkler Irrigation method	Water Smart
STV	Stress Tolerant Variety seed	Weather
SDV	Short Duration Variety seed	Weather
SRI/SWI	System of Rice/Wheat Intensification	Water Smart

Table 3b: Scientific Feasibility and Farmers' Preference

CSFI: Degree of Feasible	Degree of Farmers Preference
Wheat	
Very Highly Feasible	Sehore District: (Highly Preferred Technology) SPNK + SWI + SDV
Moderately Feasible	Guna District: (Highly Preferred Technology) BBF + SDV
Moderately Feasible	Jhabua District: (Highly Preferred Technology) BBF + INM + IPM + SDV
Moderately Feasible	Shahdol District: (Highly Preferred Technology) ZT + SDV + ST
Soybean	
Moderately Feasible	Sehore District: (Highly Preferred Technology) BBF + INM + IPM + SDV
Highly Feasible	Guna District: (Highly Preferred Technology) BBF + INM + IPM + SDV
Highly Feasible	Jhbua District: (Highly Preferred Technology) BBF + INM + IPM + SDV
Gram	
Moderately Feasible	Sehore District: (Highly Preferred Technology) BBF + INM + IPM + SPNK
Least Feasible	Guna District: (Highly Preferred Technology) SPNK + SDV + FP
Least Feasible	Jhabua District: (Highly Preferred Technology) BBF + INM + IPM + SDV + SPNK
Rice	
Unfeasible	Guna District (Highly Preferred Technology) DSR + LLL
Highly Feasible	Guna District (Moderately preferred Technology) SRI + ST
Least Feasible	Shahdol District (Highly Preferred Technology) LLL + INM + IPM + ST + FP

demonstration field, research stations and KVKs we prepared a climate smart feasibility index for each technology corresponding to each crop. The climate smart feasibility index for technologies has been estimated in terms of their impact on water use efficiency, energy use efficiency, nitrogen use efficiency and productivity. Higher feasibility implies higher level of climate smart agricultural transformation for the selected crops and selected regions. On the other hand, farmers' preference has been assessed among the technologies irrespective of their level of feasibilities. Therefore, the above mentioned two methods are independent of each other in order to avoid the preference bias. Hence, it is clearly observed from details presented in the table that highly feasible technologies are not highly preferred by the farmers, rather farmers' prefer technologies based on their farming experience and constraints associated with the adoption of these technologies. However, it was observed during our field trip and training programmes with farmers that they were willing to transform their farming practices to move towards climate smart agriculture albeit in the beginning they were ready for going into low level climate smart agriculture that was feasible at a lower cost.

CONCLUSIONS AND POLICY RECOMMENDATIONS

It was observed during the course of this study that farmers are interested to adopt modern technologies to increase their profitability. However, identifying technologies to link their profitability with sustainability is crucial. Due to regional heterogeneity identifying feasible technology and understanding farmers' willingness is required to achieve the appropriate adoption of such climate smart agriculture technologies. In this context this study makes a unique contribution by prioritizing climate smart practices across various regions in Madhya Pradesh to achieve the goal of climate smart agriculture in the state. Key policy recommendations derived from this study are as follows:

1. Link farm mechanization with regional priorities and needs. Hence, zero-tillage and laser land leveling (LLL) can be focused in rice wheat zone for better productivity and resource conservation. Broad bed and furrow (BBF) can be promoted in soybean-wheat zone for better water management and productivity.

2. Irrigation intensification through tank irrigation in the low irrigated but high rainfall areas will be helpful to achieve irrigation intensification without much threat into the ground water availability.

3. It is observed that farmers apply more fertilizer and seed in the rain-fed region with the perception of higher yield. On the other hand, most of the farmers (around 75% out of 180 selected farmer households in the selected regions) were not aware and not trained in INM and IPM practices. Therefore training on nutrient and pest management to the rain-fed region and distribution of information about the existing policies and support by the government would help to increase adoption of INM and IPM practices at the farmers' field.

ACKNOWLEDGEMENT

This study is a part of CGIAR research program on Climate Change, Agriculture and Food Security (CCAFS) and has been conducted by the International Food policy Research Institute (IFPRI) in collaboration with Institute for Social and economic Change (ISEC), Bangalore, Rajmata Vijayaraje Scindia Krishi Vishwavidyalaya (RVSKVV), Gwalior. We acknowledge the CGIAR Fund Council, Australia (ACIAR), Irish Aid, European Union, International Fund for Agricultural Development (IFAD), Netherlands, New Zealand, Switzerland, UK, USAID and Thailand for funding to the CGIAR Research Program on Climate Change, Agriculture and Food Security (CCAFS)".

REFERENCES

GoI (2015). Directorate of Economics and Statistics, Department of Agriculture. http://eands.dacnet.nic.in/ last accessed on September 2015.

Taneja G., Pal, B.D., Joshi, P.K., Aggarwal P.K. and Tyagi, N.K. (2014). Farmers Preference on Climate Smart Agriculture, an Assessment in Indo-Gangetic Plain in India. IFPRI Discussion paper, 01337, Washington DC, USA

FAO (2010). Climate-Smart Agriculture: Policies, Practices and Financing for Food Security, Adaptation and Mitigation. Rome: FAO. www.fao.org/docrep/013/ i1881e/i1881e00 .pdf.

GoMP (2015). Department of Agriculture, Government of Madhya Pradesh. http://www.mpkrishi.org/krishinet/Compendium/Basic_agroclimatic.asp last accessed on April 2015.

Go, (2014). Economic Survey of Madhya Pradesh, https://data.gov.in/dataset-group-name/economic-survey-madhya-pradesh-2013-and-2014. last accessed August 2015.

Challenges and Appropriate Agriculture Policy in India

Narendra S. Rathore

Education Division, Indian Council of Agricultural Research,
New Delhi 110 001, India
E-mail: ddgedn@gmail.com; nsrdsr@gmail.com

ABSTRACT: *Agriculture is the backbone of Indian economy and is the principal source of livelihood for more than 58 per cent of the country's population. It contributes around 14 percent to the GDP and ten percent of total exports of the nation. Over 60 per cent of India's land area is arable making it the second largest country in terms of total arable land and the GDP of Indian agriculture industry ranks twelfth in the world. India also ranks second in terms of farm output globally. But at the same time, Indian farming is facing numerous social, economical, cultural and environmental problems. Agriculture in our country is still labor-intensive. Human labor accounts for 2/3 of the power input in agriculture, draft animals ¼, and machine power less than 1/10. Agricultural technologies are critical components to meet the challenges facing increased crop production. In the present context, the chief challenges faced by agricultural disciplines are ensuring an adequate and safe food supply for an expanding world population, protecting or remediating the world's natural resources including water, soil, air and energy and tackle and mitigate the impacts of climate change. The R&D for agricultural sectors should be restructured to keep in line with the progress of Science & Technology and be ready to provide solutions for the new needs of the world. Agriculture development programs can be justified only if they have a holistic approach and offer a better promise for the future. We must consider carefully the social and economic impact of engineering decisions that influence the way people live.*

Keywords: Agriculture Policy, Agriculture Development Programmes, Natural Resources.

At present, India's population is about 128 crores and increasing every year by about 2.50 to 2.75 crores. This constitutes around 19.5 per cent of the total world's population and stands second globally for population size. India is the only country in the world, where more than 80,000 children are taking birth per day and because of normal death rate, death at neo-natal stage, death because of other health problems and natural calamities and accident etc, more than fifty per cent of lives are lost. Despite this a population of 55000–60000 every day is added to the national population. In comparison, China with a current population of 138 crores, is adding only 0.5 crores to its population every year due to increased public awareness. Therefore, it is expected that by the year 2022, our country will stand on first position across the globe as far as population is concerned. Further, this country is having only 4 per cent of water, 4 per cent of crude oil and 2 per cent of natural gas of the world. Because of unavailability of quality food in our country, more than 43 per cent of children of below 5 yrs are malnourished and about 58.7 per cent of pregnant women, 63.2 per cent lactating mothers and 69.5 per cent of pre-school children are anaemic. There is direct relationship between agriculture and population. Considering the present condition, there may be a number of challenges in front of Government in the field of Agriculture, Education, Health and Energy, etc. In the year 2013-14, our country touched the figure of 263 MT in grain production, which includes 104 MT rice and about 95.4 MT wheat. However, based on the performance of National Food Security Mission run by the Government of India, we will have to add 3 per cent production every year. However, an examination of the trends over the last 10 years' production data reveals that 3 per cent annual increment in production was never touched. In 2014-15, unfortunately our production reduced to 252 MT because of low rainfall, which was about 14 per cent less than average rainfall and in this year (–) 0.25 per cent growth rate was observed, similarly in year 2015-16, there was about 12 per cent less than average rainfall, which produced only 253 MT grain production leading to 1.2 per cent growth rate.

There is one more alarming situation faced by agriculture which is the increasing demand for land due to urbanization. It is estimated that about 5 per cent agriculture land is converted for infrastructure development every year. It is estimated that in 2025 when the population crosses 150 crores we will need about 325 MT of food production. *Will the existing infrastructure, technology inputs and farmers' commitment lead to the targeted levels of agricultural production?*

We are proudly claiming that minimum support price for the farmer in the last 10 years has doubled and which is at present, around ₹ 900 per quintal on the average grain basis. Further, in last 5 years, more than 11 crore credit cards have been distributed to the farmers. *Are these data are sufficient for our future Agriculture growth, where more than 67 crores of population is dependent on Agriculture and agro-based industries?* On an average, per capita holding in our country is less than one hectare and economic condition of farmers is also not in a position to adopt mechanization and other technologies for accelerating growth. In this case, Kisan Credit Cards is the only option. There is also a need to increase the export potential of our country. At present, it is only 1 per cent business from agriculture sector as far as total international trade of our country is concerned. *Is our country in a position to support farmers to attain quality para-meters for enhancing international trade in Agriculture?*

Agricultural education is also in a very poor shape in our country. Recently UN reported that more than 37 per cent of world's illiteracy is in India, which constitutes more than 28 crores of population of India. In the present context, around 52 per cent of children are not getting opportunity for primary education and only 10 per cent of total child population is getting an education at secondary level. One more alarming figure is that only 21 per cent of children with secondary education are getting opportunity for college education. As a whole, only 2 per cent of young population in our country is getting under-graduate education. Only 5.62 per cent Indians are graduates where as only 0.36 per cent are graduates in agriculture. Today we need skills, not just degrees, with the ability to assimilate, adapt, apply and develop new technologies. Today we need high quality graduates equipped with problem solving and creative skills and ability to think and improve productivity of different sector. Apart from the technical and generic skills, our graduates need leadership and entrepreneurial skills to build leading teams, and put innovations into practice and respond to competitive environments. There is need to articulate four **T's** *i.e.* **Tradition, Technology, Talent and Trading** to make better alternative field for livelihood and sustainable development. India is believed to be a young country, where more than 57 per cent of population is below 30 years and only 0.1 per cent of this population has opportunity for agriculture education. There are a number of new technologies in the agriculture field, such as Conservation Agriculture, Precision Agriculture, Specialty Agriculture and Secondary Agriculture, which need a different type of education. In a huge country like India, only 73 Agriculture Universities are currently meeting the educational needs of the sector. It needs to be increased to ten times in order to meet the requirement of skilled agriculture personnel by 2020 to the tune of 1,50,000 every year.

We are the largest producers of milk touching 160 MT in 2016. Similarly, we are the second largest as far as cereals, fruits and vegetable production is. India is producing around 10 and 15 per cent of fruits and vegetables of the world, respectively. However, there is only 2 per cent of fruit and vegetable processing in this sector, which need to be enhanced to 20 per cent. It is also estimated that more than 10–15 per cent of fruits and vegetables are lost due to lack of the availability of right type of post-harvest management at village level. Therefore, there is a need to have a fresh look on Secondary Agriculture in our country including implementation of adequate post-harvest management, processing and value addition to make agriculture profitable to village youth. There is a need to have a smart village concept in the same line as in Israel. Looking at the water crisis, soil condition and area available for farming, it is high time to articulate component of high-tech cultivation, modern irrigation techniques, hydroponic cultivation and vertical farming for increasing total production. There is a good scope for implementation of farm mechanization in our country. It has been observed that cost of production to a level of 15 per cent can be reduced through mechanization and increase production by 10 per cent in our country. Therefore, there is a need to have small farm implements suitable for small and marginal farmers at the block level through custom hiring mechanism, so that more and more farmers can get the benefit.

Climate change and enhancement of green house gas production is a bad sign for Agriculture production. India is at the third position in the world as far as green house gas emission is concerned. The problem is further accelerated because of natural calamities and global warming. The rainfall is also becoming erratic and non-uniformly distributed, therefore, there is a need to have some strategic planning to mitigate the impacts of climate change. There is an urgent need to articulate technology of organic farming in our country because we are consuming high levels of chemical fertilizers and insecticide. We have crossed the use of pesticides to a level of 90,000 tonnesin 2015 as against 425 tonnes in 1950.

Presently, our country is producing around 2,87,000 MW of electricity, out of which 7 per cent is available for Agriculture and rural areas. One more alarming point is that in our country, per capita electricity consumption per annum at urban level is only 388 units, whereas at village level, it is hardly 128 units compared to world average of 2,400 units. In our country, more than 32 per cent transmission and distribution losses have been identified as against the world figure of 15 per cent. Still, there is more than 15 per cent gap between electricity demand and supply at village level. The Government will have to provide right type of energy sources for agricultural operations in rural areas.

It is time India emulated technology achievements of developed countries. It is noteworthy that Russia has already grown vegetables and cereals in the space and NASA has claimed that by 2018 and 2020 respectively, they will produce water at moon and oxygen on Mars. There is a big challenge for Government to have fresh look and leverage modern science and technology inputs for innovation in agriculture field, so that we will be self-sufficient as far as food grain production is concerned.

AGRICULTURE POLICY IN INDIA

There are a number of Agricultural policies which are in vogue and need further attention. Most Agriculture policies are diverted towards PQRST formula, which includes increasing production, productivity, profitability to farmers, quality production, increasing remuneration, sustainable development and federating farmers into business group. These are included as follows:

1. *Public investment for increasing productivity:* This includes land reforms (Land to the tiller, consolidation of land holdings), public investment for expansion of surface irrigation, strengthening and coordination of agricultural R&D (Extension part of rural and community development programs) and focus on high potential regions. However, there is need for a coordinated approach to draft a policy, leading to setting up of the National Commission on Agriculture.

2. *Food and price policy:* It includes technological and institutional interventions leading to the green revolution, Government interventions for food security (Procurement and public distribution of food grains), Establishment

of Agricultural Price Commission (now CACP), Food Corporation of India and National and State Seeds corporations and enactment of Seed Act (1966) and finally Regulation of agricultural markets. But it is low input-low output price policy in present context, which needs a critical re-look.

3. *Interventions for farm credit:* As far as interventions for farm credits are concerned it includes priority sector lending, strengthening of cooperative structure and expansion of cooperative network (multi-purpose primary coop. societies), establishment of Regional Rural Banks in 1976 and establishment of the National Bank for Agriculture and Rural Development in 1982. However there is need to have Multi-agency approach with focus on agriculture. Altogether a new road map is required for farmer-credit mechanism.

4. *Agricultural R&D policy:* It covers a number of reforms i.e. expansion of agricultural R&D capacity, establishment of state universities since 1960s, expansion and coordination of ICAR system, doubling of research capacity, creation of the national extension system with T&V and KVKs, technology management agency (ATMA) at district level and attracting private R&D. Further there is need to have Open-access to public material and technology

5. *Business sector development:* It encompasses rising participation of private sector in agriculture in general and agribusiness in particular, incentives, demand for products, infrastructure and business environment (Regulatory environment and Manpower to manage the business).

6. *Intellectual property rights:* It includes policy of 'open access' to the public material that paved the way for private seed production, plant variety protection since 2005, with researchers and farmers rights, allowing process and product patents, technology specially in the area of agro-chemicals and demand for PVP from both public and private sectors.

7. *Focus on farmer's welfare:* It comprises shifting focus on productivity to income of the farmers (Doubling farm income in next five years), human capital development and social safety net programs (skill, information, health etc), risk management in agriculture {PM Fasal Bima Yojana (crop insurance)}, equity concerns (Regional equity, rural youth and farm women).

8. *Agrarian structure and challenge:* It includes a number of reforms such as production to distribution, value addition and quality (nutrition), infrastructure and incentives, investment in sustainability of agriculture (land and water binding constraints and Productivity of dryland regions).

9. *Small holder agriculture:* It includes agricultural growth essential for food and nutrition security and poverty reduction (also necessary for sustaining overall economic growth), small holder agriculture not necessarily impediment for growth, public investment in farm support service and infrastructure, need support for agriculture as incomes are low and role of rural non-farm sector.

Doubling Farmers' Income under Climate Change

Uma Lele[1]

The World Bank

ABSTRACT: *The paper outlines determinants of farmer incomes under changing climate, i.e., growing risk and uncertainty from volatility in agricultural production, as well as in domestic and international commodity markets. It illustrates the challenge of determining the appropriate blend of the public and the private sector actions going forward and the challenges in scaling up and replicating some of the successful pilot innovations in India's vast and diverse agriculture under a decentralized democracy. The incremental changes adopted by the Government of India and state governments are insufficient in a country with a wide range in agricultural productivity, growth performance and dependence on agriculture. Fragmented policy initiatives need to be replaced by a comprehensive, integrated, climate-smart strategy at the central level and in each of the states.*

Keywords: Structural Transformation, Productivity, Crop Diversification.

INTRODUCTION

Prime Minister Modi's dream of doubling Indian farmer incomes declared in February 2016[2] is now India's national policy. Finance Minister Jaitley announced a variety of programs in support of the policy in the budget presentation (2016–17), including increased allocations to irrigation and crop insurance. The central government has also launched an electronic National Agriculture Market, or e-NAM, connecting 21 markets from eight states in the first phase, and by March 2018, e-NAM will be extended to 585 mandis across India. The policy has stimulated considerable debate, including on "what income"—whether farm income or all income, nominal or real. Following Chand (2016) we assume the focus is on farm income, i.e., income from cropping and livestock. Chand (2016) and Swaminathan (2016), among others, argue that the goal is realizable, provided policies change from patchy and sporadic to more firm reforms, e.g. nearly doubling government expenditures, vigorous

---

[1] I am grateful to Sambuddha Goswami for background research.

[2] Including (i) large investments in irrigation (ii) quality seeds (iii) soil health (iv) cold chain and warehousing to prevent losses (v) value addition through food processing (vi) creation of a pan-India national market for farm produce, and (vii) risk mitigation through crop insurance and agricultural diversification into areas like poultry, beekeeping and fisheries.

implementation of the 2003 Agricultural Produce Market Committee (APMC) marketing act, secure land leases, among others (Chand 2016); and soil health care, water harvesting and management, appropriate technology and inputs, credit and insurance, opportunities for remunerative and assured marketing, knowledge, skill, credit and land ownership and empowerment of women farmers (Swaminathan 2016). Most of these require states to act. Others are more skeptical (Gulati and Saini 2016a,b; Desai 2016; Sharma 2016). Gulati and Saini (2016b) argued that it would be a "miracle of miracles" without major reforms, calling for massive investments in research and development (R&D), and stressing the role R&D and Minimum Support Prices (MSPs) have played in China. Waghmare (2016) argued that MSPs are irrelevant in the case of most crops and regions in India. In the case of pulses, Joshi (2017) noted that MSPs help traders more than farmers, leaving open the role of MSPs, India's favorite tool, in pulse production. A conference of the National Bank for Agriculture and Rural Development (NABARD 2016) summarized these arguments and presented recommendations, strongly emphasizing institutional credit, farmers' organizations in value chains and processing, and the role of investments in infrastructure such as cold storages. Which of these recommendations should be prioritized to double farmer income on a sustained basis and where should the actions come from?

CHALLENGES AND POLICY IMPLICATIONS

This paper outlines determinants of farmer incomes under changing climate, i.e. growing risk and uncertainty from volatility in agricultural production, as well as in domestic and international commodity markets. It illustrates the challenge of determining the appropriate blend of the public and the private sector actions going forward and the challenges in scaling up and replicating some of the successful pilot innovations in India's vast and diverse agriculture under a decentralized democracy. There are well documented differences among Indian states in agricultural productivity and incomes, and several paradoxes: e.g. mountains of food and rising food prices with a large presence of undernourished people; huge unexploited potential in eastern India, with the presence of abject poverty, weak institutions and human capacity. The new policy is moving towards correcting some of these anomalies, e.g. rightly moving towards more cash transfers and less public sector physical distribution of inputs and outputs. There are multiple reasons for these paradoxes, however, and no silver bullets. Historically, India has had a mixed record of political commitment to agriculture, and its implementation record is also mixed. India's challenge is to generate the necessary political commitment and develop administrative capacity over its vast agricultural lands, at multiple levels ranging from farm households, panchayats, districts, and states, in order to establish strong, pluralistic institutions and a functioning physical infrastructure. Those institutions need the necessary skill mix and incentives to build a transformational, scientific approach to agriculture, make governance of agriculture accountable to farming communities, and deliver more and better results more uniformly across states. This will call for a broadly consultative, participatory "adaptive management" style to the policymaking process, with an emphasis on implementation, routine monitoring, and evaluation of the results to learn by doing and make results-based improvements. Environmental transitions, such as rising temperatures, more frequent droughts and floods, growing water scarcities, increased incidence of pests and diseases, and land degradation require an adaptive approach with skill mixes in the public, private, and civil society to deal with them.

The paper does not include the impact of demonetization. In the short run, the impacts in remote rural areas seem to be negative. In the long run, it seems to be accelerating digitization, beneficial for the economy (New Indian Express, 2017). More systematic research will be needed to know the impacts. Suggesting that India is moving in the right direction is the Global Competitiveness Index of the World Economic Forum (2016–17) report of sharp improvement in India's ranking, which stood at 39 in 2016 among 140+ countries (WEF 2016), down from a rank of 79 in 2012 (World Bank, 2014). This progress on competitiveness needs to be accelerated.

Past Indian agricultural policy focused on farm productivity, food security (mainly viewed as cereal self-sufficiency) and poverty reduction, with the heavy hand of government doing rather than guiding. Sixty percent of Indian agriculture is rainfed, with consecutive drought years in 2013–14 and 2014–15. As many government commissions and experts have noted, however, even though productivity in well irrigated areas is stagnating and resource degradation is all too evident, barring "Har Khet ko Pani" (i.e., proposed expansion of irrigation to all lands), rain-fed agriculture or weather shocks have not received the attention deserved in the discussions on India's long-term agricultural policy or performance.

DETERMINANTS OF FARMER INCOMES

Coherent, climate-smart, integrated, national- and state-level agricultural policy and strategy frameworks are needed, not just "bringing different mission mode initiatives together," as NABARD (2016) suggests. All too evident, despite much progress, is that India lags behind neighboring countries, including the early Asian Tigers, (South Korea, Taiwan, Malaysia, Singapore) as well as China, Indonesia, and lately, Vietnam. To understand these cross-country differences and the challenges of increasing Indian household incomes, we first need to define household income and its strategic implications for an appropriate balance between national, state, and local actions. Again, following Chand (2016), increasing farmer incomes can be thought of as consisting of farm and non-farm income components, each with the following determinants:

- *Farm Income* = f [Output × prices – cost of production]
- *Annual Output per Farm Household* determined by: (1) yields per unit of land per crop, (2) double or triple cropping per unit of land (3) shift from low to higher value crops, and (4) income from livestock and fisheries.

- **Farm Size** = f [Share of population in agriculture, land distribution, land access to hire in or out more land and other factors of production].
- **Prices** determined either by Government Minimum Announced Procurement prices (MSPs) for scheduled crops or Market Prices for crops, livestock, and fisheries, for which there are no declared scheduled prices and market access.
- **Cost of Production** determined by Input Costs (including Input Subsidies).
- **Non-Farm Income** = f [(Non-Farm Employment and wages) + (Entrepreneurial income from trade) + (Transfer Payments/Social Safety nets such as Mahatma Gandhi National Rural Employment Guarantee Act (MGNREGA) – Debt)]

CURRENT LANDSCAPE AND ISSUES GOING FORWARD

Of the 156 million rural households, GOI surveys classified 58 percent, about 90 million, as "agricultural" households,[3] and 40 percent from farming (cultivation + farming of animals) as principal income source for the agricultural year, July 2012–June 2013. At the all-India level, average monthly income (cultivation + farming of animals + salary/wages + non-firm business) per agricultural household was ₹ 6,426, with farming (cultivation + farming of animals) income accounting for about 60 percent of the average monthly income (Figure 1) (GOI 2014).

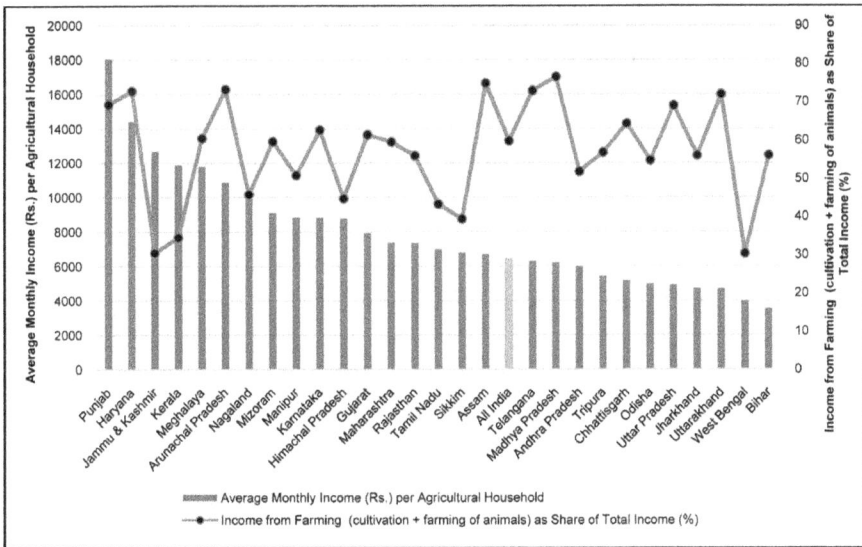

Fig. 1: Average Monthly Income (₹) and Income from Farming (cultivation + farming of animals) as Share of Total Income (%) by Agricultural Household by States (July 2012–June 2013).

Source: Based on data from GOI (2014).

---

[3] The Situation Assessment Survey 2013 defines an agricultural household as a household receiving some value of produce more than ₹ 3000 from agricultural activities and having at least one member self-employed in agriculture either in the principal status or in subsidiary status during last 365 days.

The "doubling farmer income" policy must address both farm and non-farm income, and this is why comparisons with neighboring countries are of interest, because in those countries, both agricultural and non-agricultural income have increased, on balance, because of more and greater productive investments relative to subsidies and safety nets.

Second, India's agricultural household (agricultural and non-agricultural) income ranges from ₹ 3,558 in Bihar to ₹ 18,059 in Punjab (Figure 1). Poverty rates are more than double the national average of 22.5 percent in Jharkhand in 2011–12, compared to only 0.5 percent in the state of Punjab.

South Korea, Taiwan, Malaysia, Singapore, China, Indonesia and Vietnam each started with similar or less favorable human, institutional, and physical capital and resources in the 1960s, but now outstrip India using several performance measures (Lele *et al.,* 2017, forthcoming). Agricultural Total Factor Productivity (TFP) in Figure 2 captures some of the key indicators including land intensification, diversification to higher value crops or animals, and value added beyond the farm through value chains (Fuglie, 2012). Further details on these indicators are available (Lele *et al.,* 2013; World Bank, 2014; Lele *et al.,* 2015). As a result of their better overall agricultural performance, barring Vietnam India's neighbors in East and South East Asia are ahead in the growth of agricultural value added per worker and in the process of structural transformation, i.e. shift of labor from agriculture to non-agriculture. Yet, except for Brazil and Indonesia after the 1997 Asian crisis, the ratio of non-agricultural per worker

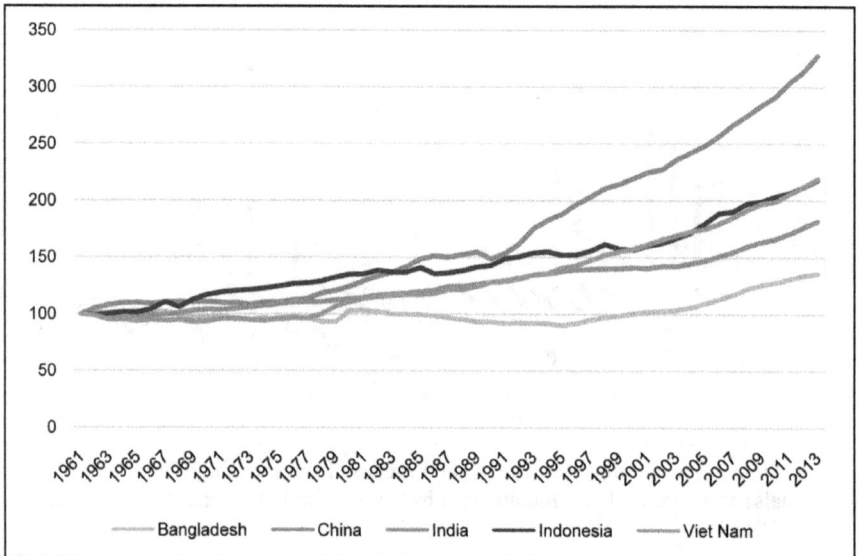

Fig. 2: Agricultural Total Factor Productivity (TFP) Index Growth (1961–2013) (Base Year 1961 = 100) for Bangladesh, China, India, Indonesia, and Vietnam

Source: Based on data available at https://www.ers.usda.gov/data-products/international-agricultural-productivity.

productivity to agricultural productivity has increased much more in China and Vietnam than in India (Lele *et al.,* 2013; World Bank, 2014). Asian countries have taken longer to achieve structural transformation (i.e., decline in the share of labor in agriculture relative to the decline in the share of agriculture in aggregate GDP) than their industrial counterparts, but India is behind still a larger share of India's population depends on agriculture relative to agriculture's contribution to GDP. Labor productivity in agriculture is lower than in China, even taking into account the debate about China's labor in agriculture. FAO data suggests much slower labor transfers out of agriculture in China than for example do International Labor Organization (ILO) data.

In a follow-up study of structural transformation using ILO data, which also provides sectorial breakdown of employment for 139 countries, Lele *et al.* (2017, forthcoming) show that agricultural labor productivity is directly related to the growth of employment in the service and the industrial sector. Again, East Asia has done better on labor productivity, particularly in the industrial sector, than South Asia. India's ratio of labor productivity growth in the service sector relative to agriculture is by far the best, compared to the East Asian countries, but India's industrial labor productivity is by far the lowest against agriculture.

As elsewhere, agricultural household incomes in India are closely related to the share of irrigated area in the state. Yet there are significant differences in irrigation efficiency across among countries and Indian states. Using Fuglie (2012) data, Indian agriculture has been shown to be less efficient than agriculture in China and Indonesia (World Bank 2014). Using state and district level data, Kshirsagar and Gautam (2013) show that whereas a few lagging Indian districts "caughtup" with better-performing districts, a substantial number fell further behind, despite registering positive but low levels of growth. These differences across districts are rooted in strategy and the enabling environment. Improved seeds, the other key element of the Green Revolution technology, have spread faster and wider, but alone, have not narrowed the productivity differentials across districts. Market density has fallen more in the low-yield/stable districts, constraining producer incentives. In the "growth" districts, irrigation and fertilizers have contributed to significant changes in productivity. In others, cropping patterns need to be more attuned to the environment. Crop management practices need to improve. This means effective improved extension, input delivery, and market access are critical.

The Government of India (GOI) has proposed rapid expansion in irrigation and promotion of solar energy as policy priorities, but experience to date suggests that ground water extraction has been adouble-edged sword. Growth of rural electrification, combined with subsidies, has led to unsustainable use of water. Rapid decentralized growth of solar energy making water exploitation easier, but it can further contribute to unsustainable use of water, if not accompanied by other policies. The GOI's investment in mapping of all aquifers, an important initiative, will make it possible to quantify the relationship between rainfall and groundwater levels under alternative modes of irrigation and farming, which should enable prioritization of prospective water and irrigation investments. Never the less a combination of rewards for conservation and effective regulatory institutions at different administrative levels are essential to resolve growing water conflicts, and formulate and enforce rules.

Beyond such "traditional" development challenges, all countries face new challenges of climate change with agriculture contributing to climate change and being affected by it (IPCC 2014). This increases the importance of managing a range of environmental transitions. Some transitions are a result of natural, external forces, emanating from climate change, such as the increased frequency of floods and droughts and soil degradation. Others are a result of policy or policy failures, from excessive use of water, pesticides, and chemicals, leading to adverse impacts on productivity, biodiversity, and human health, well documented Asian countries, (e.g. increase in the incidence of cancer; see Rola and Pingali, 1993). These transitions call for a fundamentally different process of policy formulation and implementation, away from rigid, uni-directional, top-down segmented targets to an iterative process of "learning by doing" and "adaptive management." It calls for considerable investment in human capital, embracing a culture of working with multiples stakeholders in different sectors at multiple levels, posing a coordination challenge (Loorbach and Rotmans, 2006; Shove and Walker, 2007). India's area under conservation agriculture, using three principles, minimal soil disturbance, permanent soil cover and crop rotations is for example, of 1.5 million hectares, out of about 74 million hectares under rice and wheat, illustrating the challenges in scaling up.

Aggarwal (2016) demonstrated the increasing trend in extremes of rainfall, drought, short-term floods, heat events, and greater coefficients of variation and multiple weather-related risks in the same season, as well as declining rainy periods and more variable rainfall in India's breadbasket in Northwest India since 1960. Kshirsagar and Gautam (2013) showed the strong relationship between trend deviation in rainfall and trend deviation in agricultural productivity. Covering a long period of 1961–2009, through simulations, they also demonstrated a sharp deviation between actual and counterfactual trends had rainfall been normal. Impacts of the post-1990s reforms were eroded by poor rains in consecutive years, starting from the mid-1990s. They further explore the current levels of state wise actual yields of wheat and rice during 2004–2010, relative to potential yields and attainable yields. A counterfactual shows how much higher rice and wheat yields would have been if all the required inputs were available at the right time in the right place. This not only calls for a better delivery system, but substantially greater investment in research to develop multi-resistant crops. Figure 3 shows the vastly greater investment in agricultural research in China, compared to India; agricultural research expenditure as share of agricultural GDP is declining in India, but increasing in China. Furthermore, in a bold move China recently acquired Syngenta, one of big five private research companies for $40 billion. While differing on specifics, particularly on pricing policy issues, Swaminathan (2016), Chand (2016), and Gulati and Saini (2016 a, b), among others, are all arguing in favor of more investment in India's agricultural and rural sector, ranging from research, education and extension to public–private–civil society partnerships.

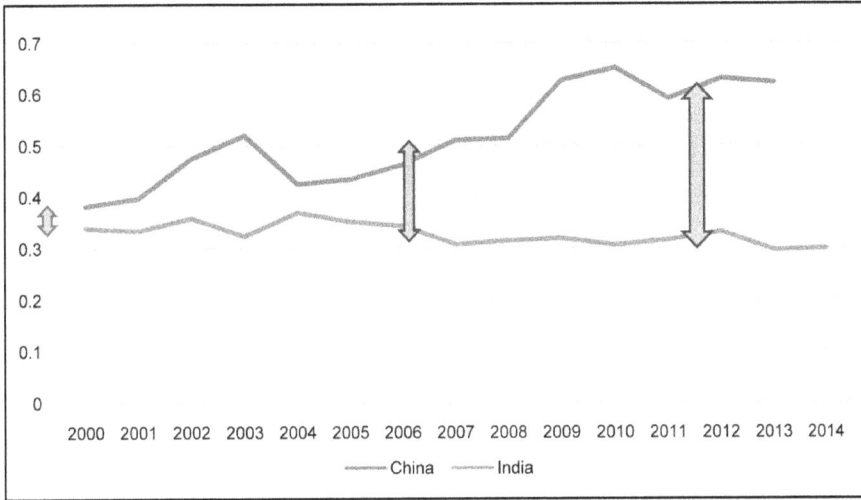

Fig. 3: Agricultural Research Expenditure as Share of Agricultural GDP (%) (2000–2014)

Source: Based on data from https://www.asti.cgiar.org
Note: China's data available until 2013.

Public policy also continues to under-provide for livestock development. The sector has been growing rapidly and offers multiple wins: increased resilience to climate change, increased income, greater employment particularly of women and socially marginal households, and better nutrition. Although China is the largest emitter of green house gases (GHGs), relative to the size of its agriculture, China's emissions from agriculture, are small: 0.712 billion tonnes CO_2 eq. in 2014 compared to India's 0.627 billion tonnes CO_2 eq. In India, enteric fermentation accounted for 45 percent of the sector's total GHG outputs in 2014; manure left in the field was 10 percent. Emissions from the application of synthetic fertilizers accounted for 18 percent of agricultural emissions in 2014. GHG resulting from biological processes in rice paddies make up 15 percent of total agricultural methane emissions. With the spread of Ration Balanced Programme, huge gains in GHG emission reductions are possible. The National Dairy Development Board (NDDB) (2016) reports that 1.5 million herd of cattle were brought under the Ration Balancing Programme, with about 12 percent reduction in cost of feeding per kilogram of milk and 12 percent reduction in methane emission. With 120 million breedable cattle, including cows and buffalo, this suggests only 1.25 percent of the cattle are covered by the Ration Balanced Programme. Livestock's potential for doubling farmer incomes is immense. It will also help to scale up the 2000 climate-smart villages India has established with the help of the CGIAR (Aggarwal, 2016).

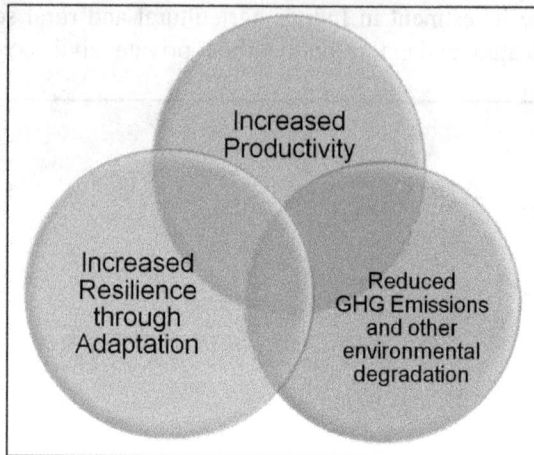

Fig. 4: Doubling Farmer Income Calls for Climate-Smart Agriculture

Authors' Depiction of FAO's definition of climate smart agriculture (CSA), "agriculture that sustainably increases productivity, enhances resilience (adaptation), reduces/removes GHGs (mitigation) where possible, and enhances achievement of national food security and development goals". FAO 2010.

CONCLUSION

The incremental changes adopted thus far by the GOI and state governments are insufficient in a country with such a wide range in agricultural productivity growth performance and dependence on agriculture. The myriad fragmented policy initiatives need to be replaced by a comprehensive, integrated, climate-smart strategy at the central level and in each of the states. The current approach should be followed by a transformational process of agricultural policymaking and implementation, building ownership among all key stakeholders—from the scientific community, private sector, civil society and farming communities, and taking advantage of India's powerful IT industry—who must play their part. The central and state strategies need a clear, state-specific set of priorities, strong focus on implementation, and routine monitoring and dissemination of broadly agreed upon performance indicators and their determinants. Adaptive learning within and across states, with accountability for results, should become the hallmark of development culture, replacing rigid, top-down targets. As former Chief Minister (CM), Prime Minister Modi, like some other CMs, had a strong record of implementation and demonstrated results in agriculture. He may need to lead this paradigm shift of change in the culture of accountability for sustainable results.

REFERENCES

Aggarwal, P. (2016). "Climate-Smart Agriculture: Addressing Weather-Risks Related Agrarian Distress." CGIAR Research Program on Climate Change, Agriculture and Food Security, Borlaug Institute for South Asia, CIMMYT, New Delhi.

Chand, R. (2016). "Doubling Farmer's Income: Strategy and Prospects." Presidential Address. The Indian Society of Agricultural Economics, *76th Annual Conference, Assam Agricultural University,* Jorhat, Assam, November 21–23.

Desai, A.V. (2016). "Budget 2016: Jaitley's Promise to Double Farmers' Income in 5 Years Is Next to Impossible." First Post, March 2. http://www.firstpost.com/business/budget-2016-jaitleys-promise-of-double-income-for-farmers-in-five-years-is-next-to-impossible-2651358.html

FAO (2010). "Climate-Smart" Agriculture: Policies, Practices and Financing for Food Security, Adaptation and Mitigation. Rome: Food and Agriculture Organization (FAO).

Fuglie, K.O. (2012). "Productivity Growth in the Global Agricultural Economy and the Role of Technology Capital." In Productivity Growth in Agriculture: An International Perspective, edited by K.O. Fuglie, S.L. Wang, and V.E. Ball, p. 335. Wallingford, UK: CAB International.

GOI (Government of India), (2014). Key Indicators of Situation of Agricultural Households in India. NSS 70th Round, Ministry of Statistics and Programme Implementation, *National Survey Sample Office (NSSO),* New Delhi.

Gulati, A. and Saini, S. (2016a). "Farm Incomes: Dreaming to Double." Indian Express, July 28. http://indianexpress.com/article/india/india-news-india/farm-incomes-dreaming-to-double-2939405

Gulati, A. and Saini, S. (2016b). "From Plate to Plough: Raising Farmers' Income by 2022." *Indian Express*, March 28. http://indianexpress.com/article/opinion/columns/from-plate-to-plough-raising-farmers-income-by-2022-agriculture-narendra-modi-pradhan-mantri-fasal-bima-yojana

IFPRI/CGIAR n.d. Agricultural Science and Technology Indicators Data. https://www.asti.cgiar.org

IFPRI (International Food Policy Research Institute)/CGIAR n.d. Agricultural Science and Technology Indicators Data. https://www.asti.cgiar.org

IPCC (Intergovernmental Panel on Climate Change), (2014). Climate Change 2014: Impacts, Adaptation, and Vulnerability. Part A: Global and Sectoral Aspects. Contribution of Working Group II to the Fifth Assessment Report of the Intergovernmental Panel on Climate Change. New York: Cambridge University Press.

Joshi, P.K., Kishore, A. and Roy, D. (2016). "Making Pulses Affordable Again: Policy Options from the Farm to Retail in India." *Economic and Political Weekly,* 52(1): 37–44.

Kshirsagar, V. and Gautam, M. (2013). "Agriculture Productivity Trends in India: A District Level Analysis." Background paper of the World Bank Report *Republic of India: Accelerating Agricultural Productivity Growth,* 2014.

Lele, U., Agarwal, M. and Goswami, S. (2015). Growth of Land, Labour and Total Factor Productivity in Agriculture in Brazil, India, Indonesia and China. In *World Scientific Reference on Asia and the World Economy*, Vol. 1, ed. Whalley, J. and Agarwal, M., London: World Scientific Publishing Company.

Lele, U., Agarwal, M., Timmer, P. and Goswami, S. (2013). "Patterns of Structural Transformation and Agricultural Productivity Growth: With Special Focus on Brazil, China, Indonesia, and India." Background paper of the World Bank Report *Republic of India: Accelerating Agricultural Productivity Growth,* 2014.

Lele, Uma, Sambuddha Goswami and Gianluigi Nico. (2017) (forthcoming). "Structural Transformation and the Transition from Concessional Assistance to Commercial Flows: The Past and Possible Future Contributions of the World Bank." In *Agriculture and Rural Development in a Globalizing World: Challenges and Opportunities*, edited by Prabhu Pingali and Gershon Feder, Chapter 15. London: Routledge.

Loorbach, Derk and Jan Rotmans (2006). "Managing Transitions for Sustainable Development." In Understanding Industrial Transformation: Views from Various Disciplines, Environment & Policy, Vol. 44, edited by Xander Olsthoorn and Anna J. Wieczorek, pp. 187–206. Dordrecht, The Netherlands: Springer.

NABARD (National Bank for Agriculture and Rural Development), 2016. "Doubling Farmers' Income by 2022." Monograph. https://www.nabard.org/Publication/NABARD%20Monograph%20%20Doubling%20Farmers.pdf

NDDB (National Dairy Development Board), (2016). Annual Report 2015–16. Anand. http://www.nddb.org/sites/default/files/NDDB_AR_2015-16Eng.pdf.

New Indian Express (2017). "Cash Crunch in Rural Areas to Normalize Soon, RBI Governor Tells PAC." *New Indian Express,* January 20.

http://www.newindianexpress.com/business/demonetisation/2017/jan/20/cash-crunch-in-rural-areas-to-normalise-soon-rbi-governor-tells-pac-1561802.html

Rola, A.C. and Pingali, P.L. (1993). *Pesticides, Rice Productivity and Farmers' Health: An Economic* Assessment. Los Baños, Philippines: International Rice Research Institute and Washington, DC: World Resourcs Institute.

Sharma, D. (2016). "Hoping Against Hope, No Signs of Doubling Farmers' Income in the Next Five Years." Ground Reality blog, March 2.

http://devinder-sharma.blogspot.in/2016/03/hoping-against-hope-no-signs-of.html

Shove, E. and Walker, G. (2007). "Commentary: CAUTION! Transitions Ahead: Politics, Practice, and Sustainable Transition Management." Environment and Planning A 39: 763–770. Swaminathan, M.S. (2016). How to double farmers' income, March 23. http://www.mydigitalfc.com/op-ed/how-double-farmers-income-043

Swaminathan, M.S. (2016). "How to Double Farmers' Income." Financial Chronicle, March 23. http://www.mydigitalfc.com/op-ed/how-double-farmers-income-043

USDA (United States Department of Agriculture), (2017). USDA Economic Research Service. International Agricultural Productivity. Data Set: Agricultural Total Factor Productivity Growth Indices for Individual Countries, 1961–2013.

https://www.ers.usda.gov/data-products/international-agricultural-productivity/

Waghmare, A. (2016). "Why It Is Hard to Double Farmers' Income by 2022." *India Spend,* March 30. http://www.indiaspend.com/cover-story/why-it-is-hard-to-double-farmers-income-by-2022-2022.

WEF (World Economic Forum), (2016). *The Global Competitiveness Report 2016–2017.* Geneva: World Economic Forum.

World Bank (2014). Republic of India: Accelerating Agricultural Productivity Growth. Washington, DC: World Bank.

Subject Index

Author Index